D1612123

POWER MOSFETS

POWER MOSFETS

Theory and Applications

DUNCAN A. GRANT

JOHN GOWAR

A Wiley-Interscience Publication

John Wiley & Sons

New York ▪ Chichester ▪ Brisbane ▪ Toronto ▪ Singapore

Copyright © 1989 by John Wiley & Sons, Inc.

All rights reserved. Published simultaneously in Canada.

Reproduction or translation of any part of this work beyond that permitted by Section 107 or 108 of the 1976 United States Copyright Act without the permission of the copyright owner is unlawful. Requests for permission or futher information should be addressed to the Permissions Department, John Wiley & Sons, Inc.

Library of Congress Cataloging in Publication Data:

Grant, Duncan Andrew.

Power MOSFETS: theory and applications/Duncan Andrew Grant, John Gowar.

p. cm.

"A Wiley-Interscience publication."

Includes bibliographies and index.

ISBN 0-471-82867-X

1. Metal oxide semiconductor field-effect transistors. I. Gowar, John, II. Title.

TK7871.95.G73 1989

621.3815′28—dc 19

88-31169

CIP

Printed in the United States of America

10 9 8 7 6 5 4 3 2 1

Preface

The advent of the power MOSFET ranks as one of the most significant developments in power electronics in recent years. While the *V*-groove devices which appeared in the late seventies looked set to find an important place in the market, particularly in the area of high-frequency power conversion, the overall dominance of the power bipolar transistor did not seem seriously threatened. However, when the more easily manufacturable vertical DMOS devices appeared in volume in 1978, the scene was set for a revolution. The power MOSFET rapidly achieved a reputation for being forgiving and easy to design with, but universal acceptance was delayed by its relatively high cost. Now that manufacturers have been able to achieve economies of scale and to reduce prices, in some cases to below that of the equivalent bipolar transistor, the last barriers have been removed to the MOSFET becoming the dominant power device.

This book details the theory of power MOSFETS and their applications. In particular it seeks to give the power MOSFET user a good understanding of the origin of MOSFET characteristics and the features which determine its behavior. Then it looks at the interaction of the device with other elements in the circuit and how device characteristics influence circuit design. A broad look at power-MOSFET applications completes the picture.

Several circuits are described at some length to highlight the practical details of power-MOSFET use. These circuits should not be taken as guaranteed designs or infallible recipes for success, but rather they are intended as discussion exercises for illustrating aspects of power-MOSFET application.

DUNCAN A. GRANT
JOHN GOWAR

Bristol, United Kingdom
February 1989

Acknowledgments

As well as to those sources identified in the text, the authors are grateful to International Rectifier Corporation of El Segundo, California, for permission to use many of the illustrations and some sections of text which appear in this book. Furthermore, we are grateful to a number of people on whose work sections of this book have been based. These are:

Ted Colley (International Rectifier (GB) Ltd.)—Chapter 12, Motor Drives.

Phil Thibodeau (Hughes Aircraft Co. MCD.)—Chapter 19, Radiation-Hardened MOSFETS.

Sean Young (International Rectifier Corp.)—Chapter 9, The Power MOSFET as a Switch, and sections 19.14, 19.15, 19.17 and 19.18.

Bob Ghent (International Rectifier Corp.)—Appendix 8, Electrostatic Discharge Protection.

We are grateful to the following for their contributions and advice:

Brian Taylor (International Rectifier (GB) Ltd.)
Steve Clemente (International Rectifier Corp.)
Ken Wagers (International Rectifier Corp.)
Perry Merril (International Rectifier Corp.)
Peter Wood (International Rectifier Corp.)
Ajit Dhubhashi (International Rectifier Corp.)
Steve Brown (International Rectifier Corp.)
Vito Rinaldi (International Rectifier Canada)
Giuseppe Castino (International Rectifier Compania Italiana, Italy)
Peter Wilson (Formerly of International Rectifier (GB) Ltd., now of SGS–Thomson Microelectronics Ltd.)
Robert Pearce (University of Bristol, U.K.)
Jon Clare (University of Bristol, U.K.)
Dick Clements (University of Bristol, U.K.)
Ken Sander (Formerly of University of Bristol, U.K.)

Contents

Preface v

Acknowledgments vii

1 GENERAL INTRODUCTION 1

1.1 The Development of Low-Power MOSFETS, 1
1.2 The Development of Power MOS Devices, 5
1.3 The Merits of the Power MOS Transistor, 17
Summary, 21
References, 21

2 PRINCIPLES OF OPERATION 1: THRESHOLD VOLTAGE 25

2.1 Levels of Device Modeling, 25
2.2 The Polysilicon–Silicon Dioxide–Silicon Sandwich, 28
2.3 The Threshold Potential, 34
2.4 Inversion and Accumulation Layers, 40
2.5 Electron Energy Levels along the Channel, 44
Summary, 51
References, 51

3 PRINCIPLES OF OPERATION 2: STATIC CHARACTERISTICS 53

3.1 Introduction, 53
3.2 Control of the Drain Current, 55
3.3 The Subthreshold Region, 65
3.4 Device Transconductance, 68
3.5 The Drain Drift Region, 74
3.6 Parasitic Components, 79
3.7 Temperature Effects, 91
3.8 p-Channel and Other Devices, 92
Summary, 94
References, 95

4 PRINCIPLES OF OPERATION 3: TRANSIENT AND HIGH-FREQUENCY BEHAVIOR 97

4.1 Introduction, 97
4.2 Turn-on, 104
4.3 Turnoff, 123
4.4 Gate Charge, 135
4.5 Limitations of the Polycrystalline Silicon Gate at High Frequency, 139
Summary, 142
References, 143

5 FABRICATION AND RELIABILITY 145

5.1 Fabrication, 145
5.2 Quality and Reliability, 151
Summary, 161
References, 162

6 DISSIPATION AND HEATSINK DESIGN 163

6.1 Introduction, 163
6.2 Switching Losses, 163
6.3 Leakage-Current Losses, 164
6.4 Gate Dissipation, 165
6.5 Conduction Power Loss, 165
6.6 Steady-State Thermal Resistance, 167
6.7 Transient Thermal Impedance, 168
6.8 Forced Cooling, 170
References, 171

7 PARALLEL AND SERIES OPERATION 173

7.1 Parallel Operation, 173
7.2 Instability during Switching of a Single MOSFET, 186
7.3 Series Operation, 188
References, 190

8 GATE DRIVE CIRCUITS 191

8.1 Basic Principles, 191
8.2 Choosing the Gate-Driver Voltage, 191
8.3 OFF-state Gate Voltage, 192
8.4 ON-state Gate Voltage, 193

8.5 Gate-Driver Impedance, 194
8.6 Asymmetric Gate Drive Circuits, 195
8.7 Complementary Emitter-Follower Drive Circuit, 196
8.8 Integrated-Circuit Gate Drivers, 197
8.9 Driving MOSFETS from Operational Amplifiers and Comparators, 197
8.10 Stored-Energy Gate Drive Circuits, 199
8.11 Driving MOSFETS from TTL, 200
8.12 Driving MOSFETS from CMOS, 203
8.13 Transformer-Isolated Gate Drivers, 204
8.14 Optically Isolated Gate Drivers, 210
8.15 The High-Side Switch, 212
8.16 H-Bridge Driver, 213
References, 214

9 THE POWER MOSFET AS A SWITCH 215

9.1 Introduction, 215
9.2 Basic Switching Configurations, 215
9.3 Inductive Load Switching, 216
9.4 High Inrush-Current Loads, 218
9.5 Logic Control of Loads, 219
9.6 Relay Driver, 220
9.7 Printer Hammer Drive using MOSFETS, 221
9.8 Lamp Applications, 223
9.9 Low-Power Lamp Dimmer, 225
9.10 AC Load Control, 225
References, 227

10 AUTOMOTIVE APPLICATIONS 229

10.1 Introduction, 229
10.2 High-Side Switching, 229
10.3 Voltage-Rating Requirements, 231
10.4 Threshold Voltage, 232
10.5 Reverse Battery Voltage, 232
10.6 Operating Temperature, 233
10.7 Applications, 235
10.8 Automobile Ignition Systems, 236
References, 237

11 POWER SUPPLIES 239

11.1 Introduction, 239
11.2 Operating Frequency, 239

11.3 SMPS Circuits, 241
11.4 Current-Mode Control with Current-Sensing MOSFETS, 243
11.5 Resonant Power Supplies, 246
11.6 Range of Resonant Frequencies, 246
11.7 Resonant Schemes, 248
11.8 Synchronous Rectification, 250
References, 255

12 MOTOR DRIVES **257**

12.1 Introduction, 257
12.2 DC Motor Drives, 257
12.3 DC Motor Drives—Chopper Control, 258
12.4 Braking, 262
12.5 Reversible Full-Bridge Chopper Drive, 264
12.6 Brushless Motor Drive, 267
12.7 Stepping-Motor Drives, 272
12.8 Battery-Driven Motors, 275
12.9 Six-Step Three-Phase Drive, 277
12.10 Six-Step Three-Phase Drives with Harmonic Elimination, 278
12.11 Silent Three-Phase PWM Drives, 280
References, 280

13 PWM INVERTERS **283**

13.1 Introduction, 283
13.2 Current Waveforms in PWM Inverters, 283
13.3 Using the Body–Drain Diode, 284
13.4 Diode Recovery Current, 288
13.5 MOSFET Turn-on, 289
13.6 Controlling the Rate of Turn-on, 290
13.7 Using Inductance to Control Diode Recovery Current, 293
13.8 Diode Recovery Losses and Switching Losses, 294
13.9 Effect of di/dt and Temperature on Q_{RR}, 295
13.10 Calculation of Switching Losses, 296
13.11 Conduction Losses, 301
13.12 Fast-Recovery Diode MOSFETS, 305
13.13 Avalanche Requirements, 306
13.14 Gate-Driver Design, 307
13.15 PWM Waveform Generation, 309
13.16 Inverter Ratings versus MOSFET Type, 310
13.17 Comparisons of Bipolar Transistors and MOSFETS, 310
References, 311

14 HIGH-FREQUENCY APPLICATIONS **313**

14.1 Solid-State Ballasts, 313
14.2 Class-D Amplification, 313
14.3 Induction Heating, 319
References, 320

15 LINEAR APPLICATIONS **323**

15.1 Linear Postregulators, 323
15.2 Audio Amplifiers, 324
References, 330

16 MOSFETS AND BIPOLAR TRANSISTORS **333**

16.1 Comparison of MOSFETS and Bipolar Transistors, 333
16.2 MOSFET-and-Bipolar Combinations, 344
References, 353

17 PACKAGING, TESTING, RELIABILITY, AND HANDLING **355**

17.1 Packaging, 355
17.2 Testing, 365
17.3 Reliability, 368
17.4 Handling Power MOSFETS, 374
References, 374

18 MODELING POWER MOSFETS **377**

18.1 Introduction, 377
18.2 SPICE Modeling of Power MOSFETS, 377
References, 381

19 SPECIAL-PURPOSE MOSFETS **383**

19.1 Current-sensing MOSFETS, 383
19.2 Radiation-Hardened Power MOSFETS, 393
19.3 Low-Threshold-Voltage MOSFETS, 401
References, 402

20 OTHER MOS POWER DEVICES 405

20.1 Introduction, 405
20.2 The IGBT, 405
20.3 MOS-Controlled Thyristor (MCT), 412
20.4 Power Integrated Circuits (PICS), 415
References, 420

APPENDIX 1. BASIC ELECTROSTATIC THEORY OF THE DEPLETION LAYER 425

APPENDIX 2. THE FORMATION OF DEPLETION, INVERSION, AND ACCUMULATION LAYERS AT THE SILICON SURFACE 431

APPENDIX 3. MORE RIGOROUS THEORY OF THE FORMATION OF INVERSION AND ACCUMULATION LAYERS 437

APPENDIX 4. CHANNEL TRANSCONDUCTANCE 445

A4.1 The Turned-on Condition, 445
A4.2 After Pinchoff, 448
References, 453

APPENDIX 5. CELL GEOMETRY 455
Reference, 457

APPENDIX 6. ON-RESISTANCE AND BREAKDOWN VOLTAGE 459

A6.1 Basic Equations, 459
A6.2 Functional Relationship, 460
A6.3 Optimization with Uniform Doping, 463
A6.4 Optimum Doping Profile, 465
References, 465

APPENDIX 7. A POWER-MOSFET DATA SHEET 467

A7.1 Introduction, 467
A7.2 Notes on Data-Sheet Features, 467
References, 482

APPENDIX 8. ELECTROSTATIC-DISCHARGE PROTECTION **483**

A8.1 Introduction, 483
A8.2 The Nature of ESD, 483
A8.3 Generation of Static Electricity, 484
A8.4 MOSFET Failure Mode, 484
A8.5 Complete Static Protection, 486
A8.6 Insulating Material, 487
A8.7 Antistatic Material, 487
A8.8 Static-Dissipative Material, 487
A8.9 Conductive Material, 487
A8.10 Ionizers, 488
A8.11 Floors, 488
A8.12 Table Tops, 489
A8.13 Containers, 490
A8.14 Personnel, 490
A8.15 Grounding, 491
A8.16 Test Equipment, 491
A8.17 Complete ESD Protection, 491
References, 492

APPENDIX 9. POWER MOSFET TEST CIRCUITS **493**

A9.1 Introduction, 493
A9.2 BV_{DSS}: Drain–Source Breakdown Voltage, 493
A9.3 $V_{GS(th)}$: Threshold Voltage, 493
A9.4 $R_{DS(on)}$: Saturated On-Resistance, 494
A9.5 C_{iss}, C_{oss}, and C_{rss}: Input, Output, and Reverse Transfer Capacitances, 494
A9.6 $t_{d(on)}$, t_r, $t_{d(off)}$, and t_f: Turn-on Delay Time, Rise Time, Turnoff Delay Time, and Fall Time, 495
A9.7 Q_g, Q_{gs}, and Q_{gd}: Total Gate Charge, Gate-Source Charge, and Gate–Drain Charge, 495
A9.8 V_{SD}: Body–Drain Diode Conduction Voltage, 496
Reference, 496

Index **497**

POWER MOSFETS

CHAPTER 1

General Introduction

1.1. THE DEVELOPMENT OF LOW-POWER MOSFETS

The metal–oxide–semiconductor field-effect transistor (MOSFET) is the most commonly used active device in the very large-scale integration of digital integrated circuits (VLSI). During the 1970s these components revolutionized electronic signal processing, control systems and computers. Since 1978, devices based on the same fundamental principles and fabricated using similar manufacturing techniques, but having a significantly modified geometry, have made an equivalent impact on power electronics applications. These are the vertical, double-diffused MOSFETS, whose characteristics and applications are the principal concern of this book.

The basic principle of the MOSFET has a long history. The conductance of a piece of semiconductor depends on the number of free carriers it contains and on their mobility. The effective number of free carriers can be modified by establishing a static electric field in the semiconductor in the direction transverse to the current flow. Such a field can in principle be capacitively coupled through a thin insulating dielectric layer. This concept was the basis of devices proposed and patented in the 1920s and 1930s by Lilienfeld [1] and Heil [2].

Early attempts to exploit this *field effect* failed because of the influence of impurities at the surface, which may become charged, and because of surface electronic states, which can trap and thus immobilize the normally mobile charge in the semiconductor. Between them, these two effects block the influence of the applied voltage. It was while they were investigating surface states in 1947 that Bardeen and Brattain [3] formed two adjacent pn junction diodes in a single crystal of germanium and discovered the point-contact, bipolar transistor effect. The invention and realization of the bipolar junction transistor (BJT) followed directly [4–6].

Careful control of the concentration of surface states and the minimization of charged impurities at the semiconductor surface is crucial to the operation of most semiconductor devices. This includes the bipolar junction transistor itself. Especially vulnerable is the region where the pn junctions intersect the surface. Improvements in semiconductor technology intended primarily to enhance the performance of BJTS eventually enabled the transverse field required for the field effect to be established in a different way, by means of a reverse-biased pn junction.

The principle of this device, the *junction field-effect transistor* (JFET) had been proposed by Shockley in 1952 [7]. A modern realization is illustrated in Figure

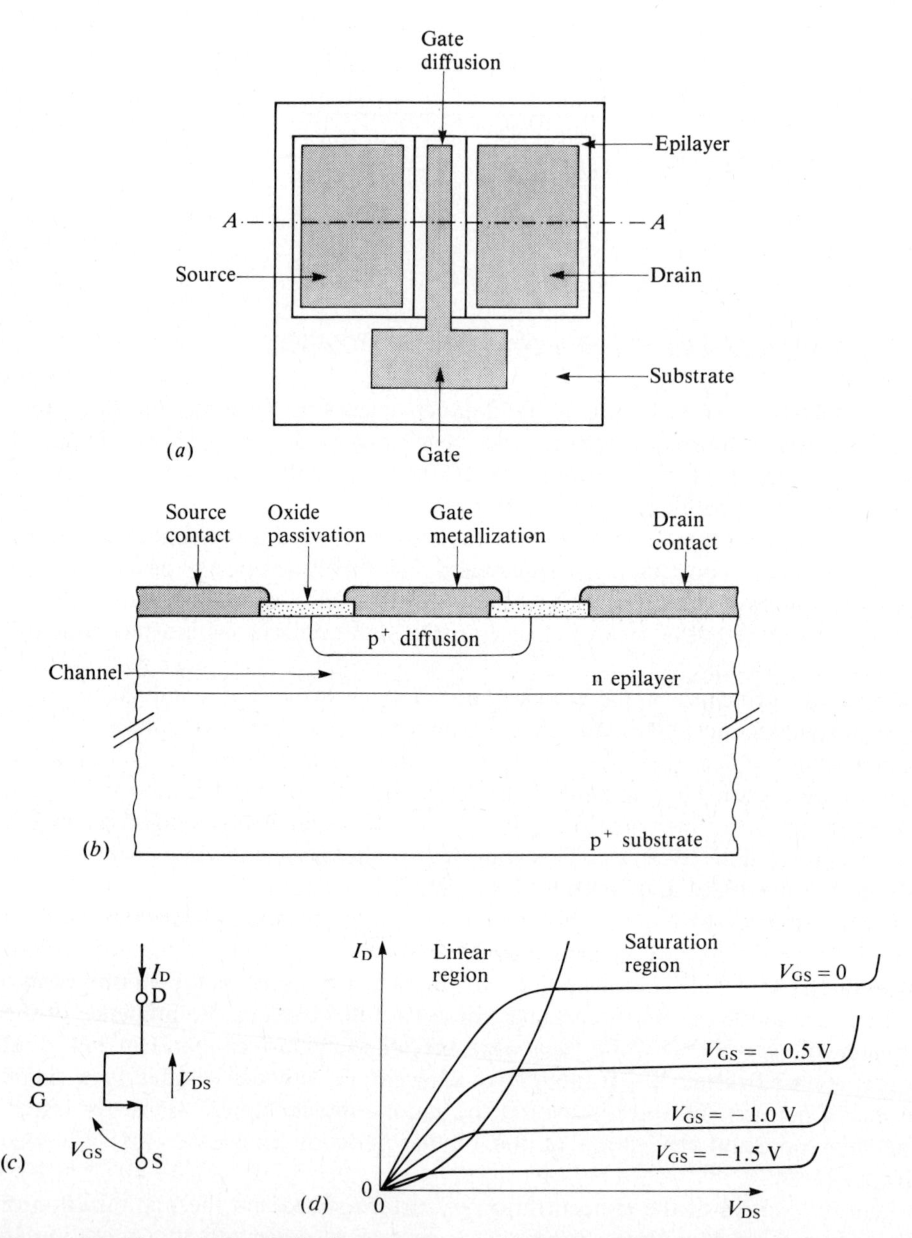

Figure 1.1. The junction field-effect transistor. (*a*) Plan view; (*b*) section *A–A′* through active region of device; (*c*) circuit symbol; (*d*) current–voltage characteristic. Increasing the negative gate voltage expands the depletion layer between gate and channel, and increases the channel resistance. If V_{DS} is increased, the depletion layer at the drain end expands to pinch off the channel. I_D then saturates. With p-channel devices the polarities are reversed, and in the circuit symbol the arrow on the source lead is reversed.

1.1. This is a *depletion-mode* device. That is, it is normally ON or in the freely conducting state, and it can be turned OFF by a voltage of suitable polarity applied to the gate. This reverse-biases the gate-channel pn junction and so expands the depletion layer. As a result the effective cross-sectional area of the conducting channel is reduced and its resistance is increased. Note that the channel material may be either p-type or n-type. In the case of the n-channel device illustrated, a negative gate voltage will tend to pinch off the conduction channel between source and drain. With a p-type channel a positive gate voltage would be needed for pinchoff. In Figure 1.1 the gate and substrate regions are shown as "p^+" to indicate that they are normally heavily doped regions.

The successful manufacture of the JFET depended on the introduction of *planar* silicon technology in 1960 [8]. This technique, which led directly to the development of integrated circuits, involves a combination of processes. These take place on a plane surface of single crystal silicon, and depend on a number of unique properties of this material. The processes include:

1. Epitaxial crystal growth from the vapor phase.
2. The formation of thermally grown silicon dioxide layers for surface passivation, for masking, and as an insulating layer over which metal interconnections may be run.
3. The photolithographic etching of "windows" in the oxide layer.
4. The introduction of controlled levels of impurity into selected regions of the silicon through these windows in the oxide, either by diffusion from the vapor phase or by ion implantation.
5. The deposition by evaporation or sputtering of a thin conducting film (usually Al) and the etching of this film to form a pattern of interconnections on the surface.

These processes are illustrated schematically in Figure 1.2.

Further technological refinements in the early 1960s enabled a different type of field-effect transistor to be made. This was the *metal–oxide–semiconductor field-effect transistor* (MOSFET) [9–11]. The principle of this device is much closer to the original ideas of Lilienfeld and Heil. The controlling gate electrode is now separated from the semiconductor by a thin insulating layer of gate oxide, as shown in Figure 1.3. The gate potential determines the number of carriers in the channel between source and drain, and hence its conductivity. With this device it is particularly important that the gate oxide (silicon dioxide) thickness should be controlled most carefully and that the concentrations of impurities and surface-state energy levels at the silicon–silicon dioxide interface should be kept very low indeed. Then, either p-channel or n-channel devices may be made, and either may be used as normally-ON (*depletion-mode*) devices like the JFET, or as normally-OFF (*enhancement-mode*) devices.

The four possible device types, together with their circuit symbols and current–voltage characteristics, are illustrated in Figure 1.4. A detailed physical explanation of these characteristics is given in Chapter 3. However, the basic device operation can be understood very easily. Consider, for example, n-channel devices. The argument is the same for p-channel devices with the dopant types interchanged and the polarities reversed.

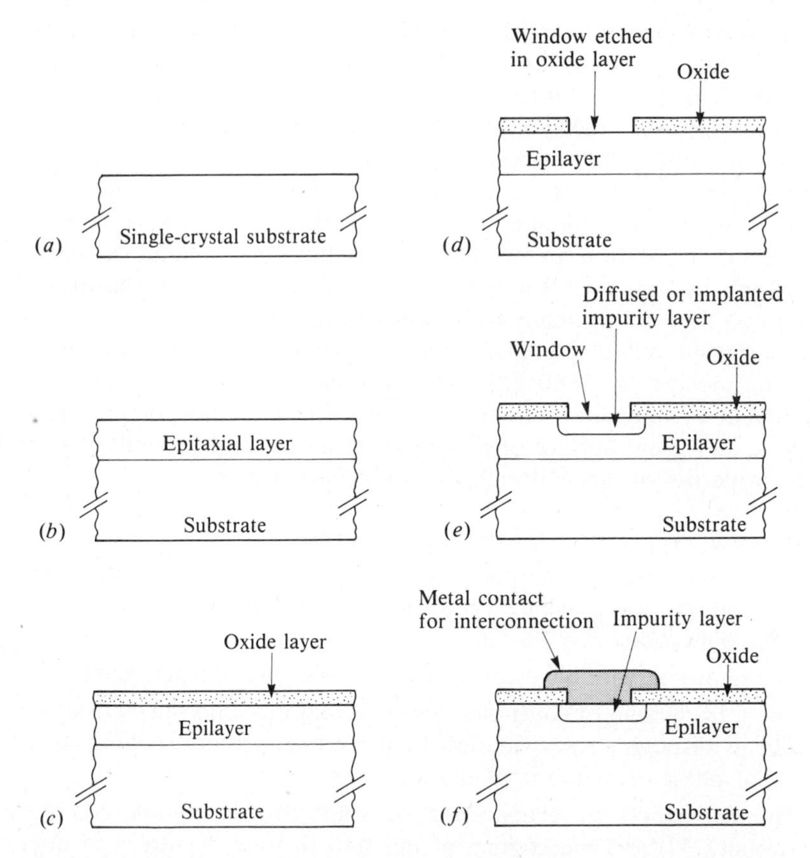

Figure 1.2. Planar silicon technology processes. (*a*) Polished single crystal slice; (*b*) growth of epitaxial layer; (*c*) formation of oxide; (*d*) photolithographic etching of oxide windows; (*e*) diffusion of impurities; (*f*) evaporation and etching of metal contacts.

The depletion-mode device has a conducting n-type channel normally connecting source and drain. Application of a negative voltage to the gate electrode causes a depletion layer to be established in the semiconductor near the oxide–semiconductor interface. This is simply because the majority carriers, the electrons, are repelled from this region by the negative gate voltage. As the gate is made more negative, the depletion region extends further across the channel until it is finally pinched off. With a positive drain–source voltage V_{DS}, this occurs first at the drain end of the channel. The drain current I_D then becomes largely independent of further increase of V_{DS}.

In the enhancement-mode device the path between source and drain takes the form of two back-to-back series diodes. Thus, negligible current normally flows. Application of a positive gate voltage exceeding a certain threshold value, V_T, attracts negative carriers to the surface of the semiconductor in sufficient numbers to form an *inversion layer*. The surface layer then behaves as a piece of n-type material and forms a conducting bridge or channel between source and drain. Further increase in gate voltage increases the conductance of the channel.

There are applications for all types of MOSFET, but the n-channel,

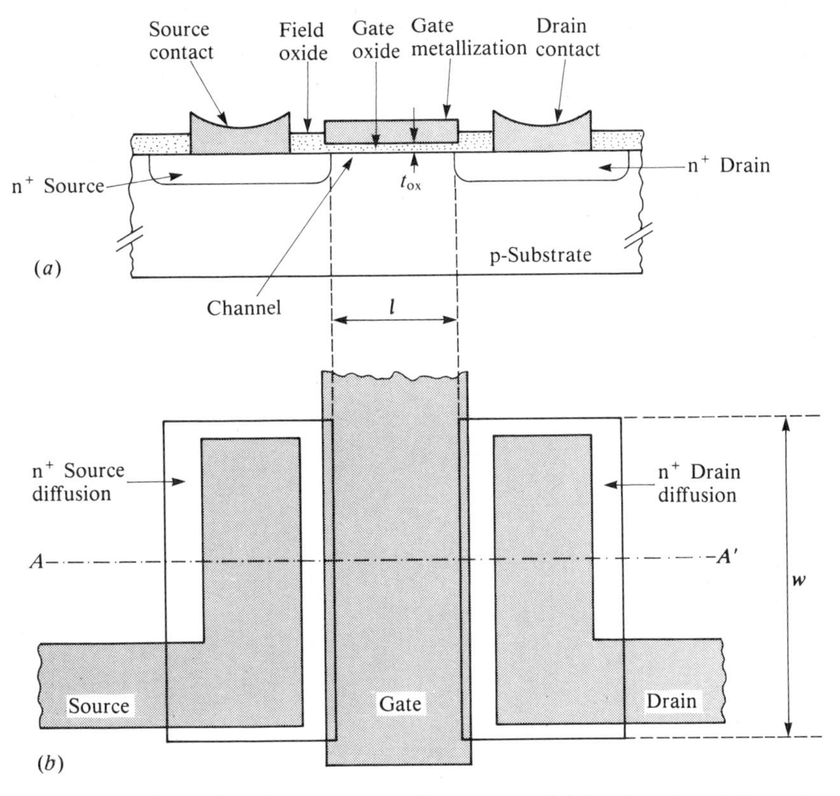

Figure 1.3. An *n*-channel, enhancement-mode MOSFET. (*a*) Section *A–A′* through active region; (*b*) plan view. Important device parameters are the gate length *l*, the gate width *w*, and the gate oxide thickness t_{ox}.

enhancement-mode device is especially important for three reasons. First, the conductivity is by means of electrons, which have a higher mobility than holes. Thus, for given dimensions, the channel resistance is lower and the carrier transit time is shorter in an n-channel device. Secondly, as an enhancement-mode device, it draws negligible current and hence dissipates no power in its normal, OFF state. Thirdly, MOSFETS are self-isolating, because all the junctions with the substrate are reverse biased. However, the surface conditions demanded for the successful operation of an n-channel, enhancement-mode MOSFET are the most onerous, and this was the last type of MOSFET to become commercially available. It is now, of course, an essential component in any NMOS or CMOS integrated circuit.

1.2. THE DEVELOPMENT OF POWER MOS DEVICES

Not surprisingly, many of the merits of MOSFETS are as important in devices designed for power applications as they are in those designed for large-scale integration. Of course, the decision as to what constitutes a power device is quite

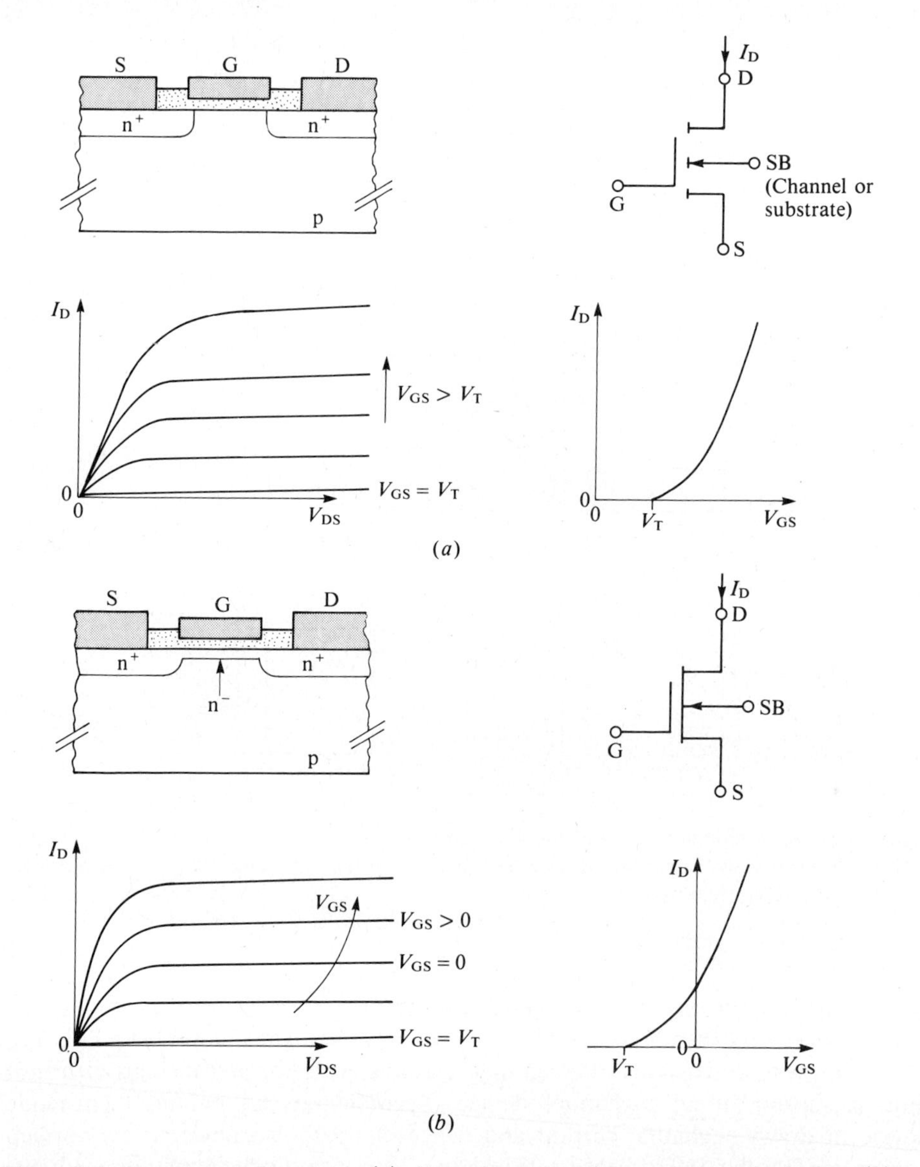

Figure 1.4. Four types of MOSFET. (*a*) An n-channel, enhancement-mode device; (*b*) an n-channel, depletion-mode device; (*c*) a p-channel, enhancement-mode device; (*d*) a p-channel, depletion-mode device. The cross-section through the active region of the device, the circuit symbol, the output characteristics and the transfer characteristic are shown in each case.

arbitrary. For purposes of definition we shall apply this term to any device capable of switching at least 1 A. There are specific requirements for p-channel and for depletion-mode power devices. However, the n-channel, enhancement-mode device is much the most important. The question is, how can the MOSFET best be adapted for such use? In this section we follow the main steps in the development of power MOSFETS.

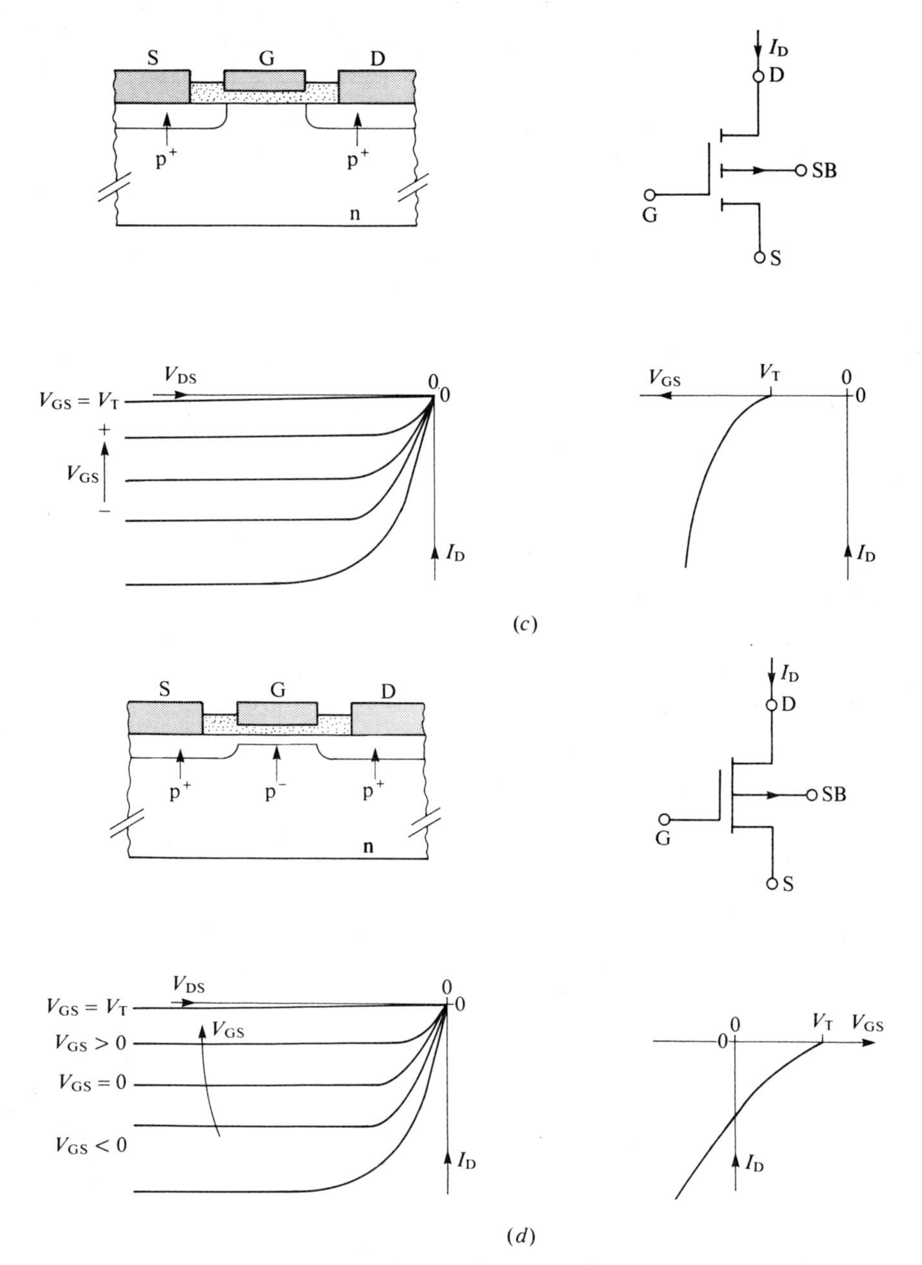

Figure 1.4. (*Continued*)

The operating principles and characteristics of MOSFETS are discussed in Chapters 2 and 3. There, it is shown that two key parameters governing their performance are the gate length l and the thickness of the gate oxide, t_{ox}. In planar MOSFETS of the type shown in Figure 1.3, the minimum gate length is determined by the minimum photolithographic feature size that can be obtained in processing. Successive technological improvements have enabled l to be reduced from 20 μm to 1 μm or less. Corresponding reductions in gate oxide thickness have been possible whilst still preserving the necessary uniformity and integrity of the insulating layer. These are the technical achievements that have permitted very high component counts (1,000,000 active devices on a single chip) to be obtained in VLSI circuits. Future reductions to submicrometer dimensions are likely [12].

For power applications many individual devices may be connected together in parallel during the final metallization process. However, there are two major reasons why the planar structure of Figure 1.3 is unsatisfactory if it is simply scaled up for higher powers. First, the drain–source spacing has to be increased in order to obtain a high voltage blocking capability. The reason is this. Once the channel is pinched off, a depletion layer forms between the channel and the drain. The width of the depletion layer varies with the drain–source voltage V_{DS}. If l were kept small, the drain depletion layer would eventually punch through to the source at high values of V_{DS}. Simply increasing the channel length of a normal MOSFET, as in Figure 1.5*a*, is not satisfactory either. The channel length still varies as the depletion layer expands and contracts with changes in V_{DS}. The arrangement shown in Figure 1.5*b* overcomes this problem. By having the channel more heavily doped than the adjacent drain region, the depletion layer occupies the drain region preferentially. The channel length can be kept short, and it is not modulated significantly by V_{DS}. An analysis of this effect is given in Appendix 1.

The principle of the lightly doped drain (LDD) structure is widely used in modern VLSI chips. However, for power devices, it is still the case that a great deal of the valuable surface area of the silicon is taken up with supporting the drain voltage. The second disadvantage of the lateral power MOS transistor arises from the need to make all three connections (to source, gate, and drain) on the same, upper surface. While this facilitates the monolithic integration of components, it complicates the metallization required for a single power device. Both effects reduce the area of silicon usefully used to form the active transistor region. There is thus a low silicon utilization factor. As a result, lateral power MOSFETS are rarely used as discrete devices, except in some linear applications. They are being used increasingly in power integrated circuits.

During the 1970s some radically different MOSFET configurations were evolved. These eventually enabled the two major disadvantages of the lateral power MOS transistors to be avoided. The essential step was to use the substrate material to form the drain contact [13]. As a result the current flows "vertically" through the silicon from drain to source. This idea had been anticipated in the multichannel junction FET [14–16], the forerunner of the static induction transistor (SIT), whose basic structure is shown schematically in Figure 1.6.

One suggested way of achieving the vertical structure with a MOSFET was to put the gate contacts on the sides of mesas produced using an isotropic etch [17, 18]. Such a device is illustrated in Figure 1.7.

A technique that was more successful and that led to the production of the first commercial power devices [19–22] was to use an anisotropic etch to produce a V-shaped groove in the silicon surface. This technique [23] had already been used as a means of isolating BJTS in an integrated circuit [24], and in some novel designs of high-frequency lateral MOS transistors [25, 26]. In order to make a

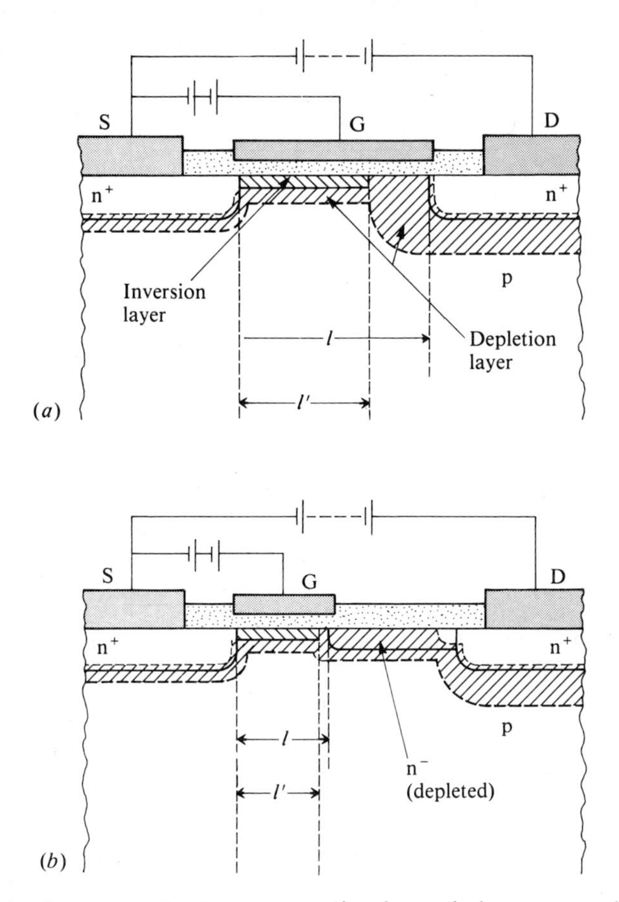

Figure 1.5. Lateral MOSFET structures supporting large drain–source voltages. (*a*) Conventional structure showing the effect of the drain–substrate depletion layer in reducing the effective channel length from l to l'. The formation of the inversion and depletion layers is discussed in Chapter 2. (*b*) Use of a lightly doped drain diffusion to minimize channel-length modulation.

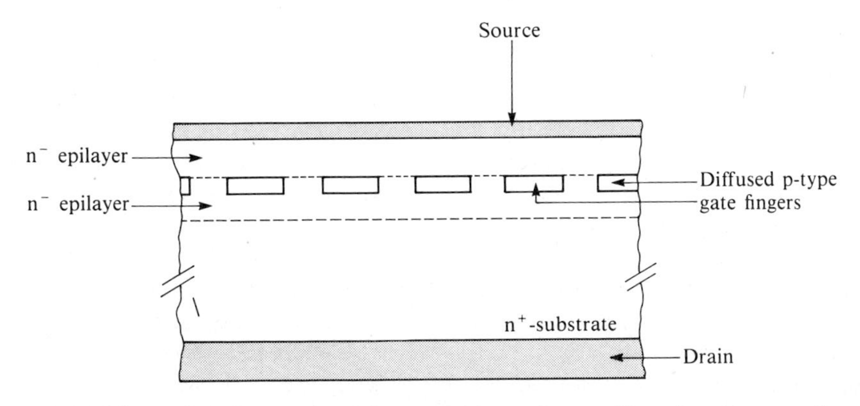

Figure 1.6. Schematic cross-section of a multichannel FET. Negative bias on the gate fingers leads to pinchoff.

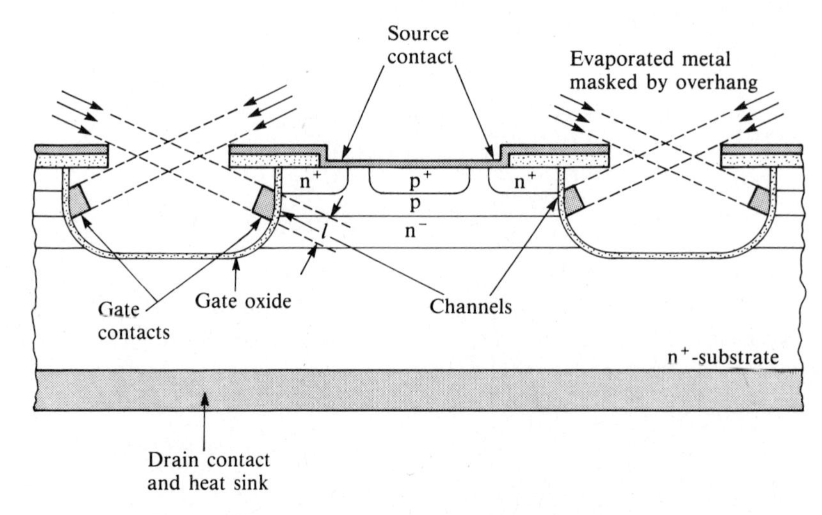

Figure 1.7. A vertical-geometry MOSFET. The diagram shows a section through one of several parallel mesa "fingers" each supporting a source contact. There are two channels from each source contact to the drain, and each is controlled by one of the evaporated gate electrodes. The gates are connected together at one end, and the sources at the other end. In the device described in Refs. 17 and 18, the channel length was 1 μm and the distance between the centers of the mesa ridges was 33 μm. The oxide overhangs were used to define the gate electrodes during the metallization of the source contacts.

power FET, the V-shaped groove is etched into the surface after successive p and n^+ diffusions have been completed. This produces the structure shown in Figure 1.8*a*. Such devices have become known as VMOS transistors, though whether the "V" should be thought to stand for "vertical" or to represent the groove shape is open to debate. In order to avoid ambiguity we shall refer to them as vertical V-groove MOS (VVMOS) devices.

The angle of the groove is determined by the crystal structure of silicon. Most power semiconductor devices are fabricated on slices cut from boules whose axes lie in the $\langle 111 \rangle$ crystal direction. It is easier to obtain a low concentration of defects in crystals oriented in this way with the result that high-quality material is cheaper. This is an important consideration in power devices, which use a large amount of silicon for each device. The $\{111\}$ faces have the highest atomic surface density and are less susceptible to surface contamination. They are used for most bipolar integrated circuits and discrete devices. For MOS ICS and devices, other considerations apply, and they are invariably fabricated on $\{100\}$ faces. This allows the channel to be oriented along a $\langle 100 \rangle$ direction, in which the electron mobility is higher. More importantly, the lower surface atomic density means that the concentrations of fixed interface charge and interface trapping levels are more easily reduced to acceptable levels. This is essential, since they both adversely affect MOS transistor action, as discussed in Section 2.3. Finally, silicon cleaves naturally along $\{110\}$ planes. Thus $\{100\}$ slices can be readily cleaved at right angles to the surface into square dice.

For VVMOS devices, it is necessary to use $\{100\}$ slices. On the heavily doped n^+ substrate a lightly doped n^- epitaxial layer is grown, and into this, successive p and n^+ layers are diffused, just as though an npn bipolar junction transistor were being made. The oxide is removed, and an etchant such as potassium hydroxide is applied. This etches much more rapidly in directions parallel to the {111} planes than into them. Thus, when another {111} plane is reached, etching effectively ceases. It is the natural angle between these {111} and {100} crystal planes that causes the sides of the groove to be at an angle of 54.7° to the surface of the slice.

After etching, the gate oxide is grown on the exposed face of the groove, and the gate and source metallizations are deposited. The parasitic, parallel, npn bipolar transistor that is characteristic of the MOS structure is shown very clearly in this device. To minimize its effect on the proper behavior of the n-channel MOS transistor, the source contact is normally shorted to the channel, as shown in Figure 1.8*a*. This still leaves a parasitic pn junction diode connecting source and drain.

The channel length l is determined by the relative depths of the successive diffusions. The base width of a BJT is determined in the same way. A typical value would be one or two micrometers. A penalty that has to be paid for the VVMOS structure is the reduced electron mobility in the channels under the {111} faces of the V-grooves, in comparison with that of normal {100} MOS devices.

Because the epitaxial n-region is more lightly doped than the p-channel, the depletion layer extends only a little way into the channel, and most of the voltage is dropped in the epilayer. The required current rating is achieved by connecting many grooves in parallel, as shown in Figure 1.8*b*.

Each gate of the VVMOS transistor controls the current from the two sources, one on either side of the groove, as it flows to the common drain. Current

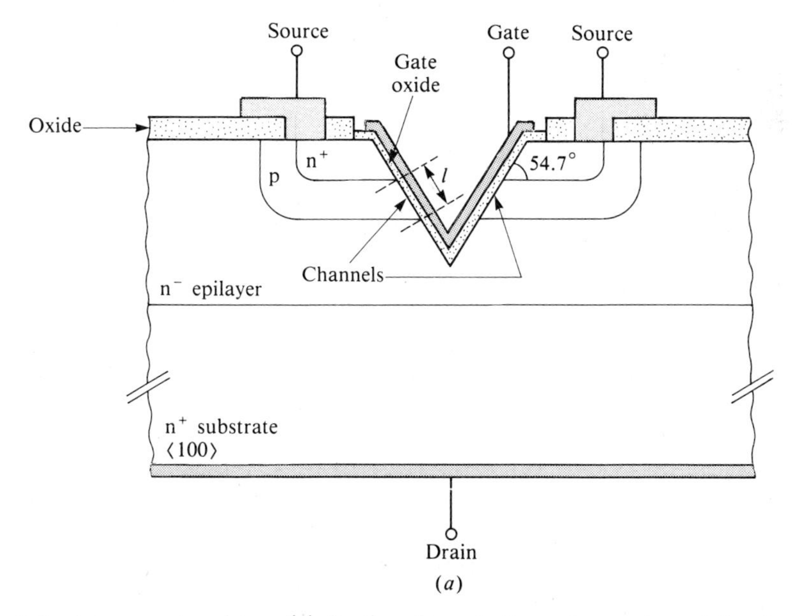

Figure 1.8. A VVMOS transistor. (*a*) Section through the active region of the device; (*b*) section and plan views showing the parallel connection of sources and gates required for high-current operation.

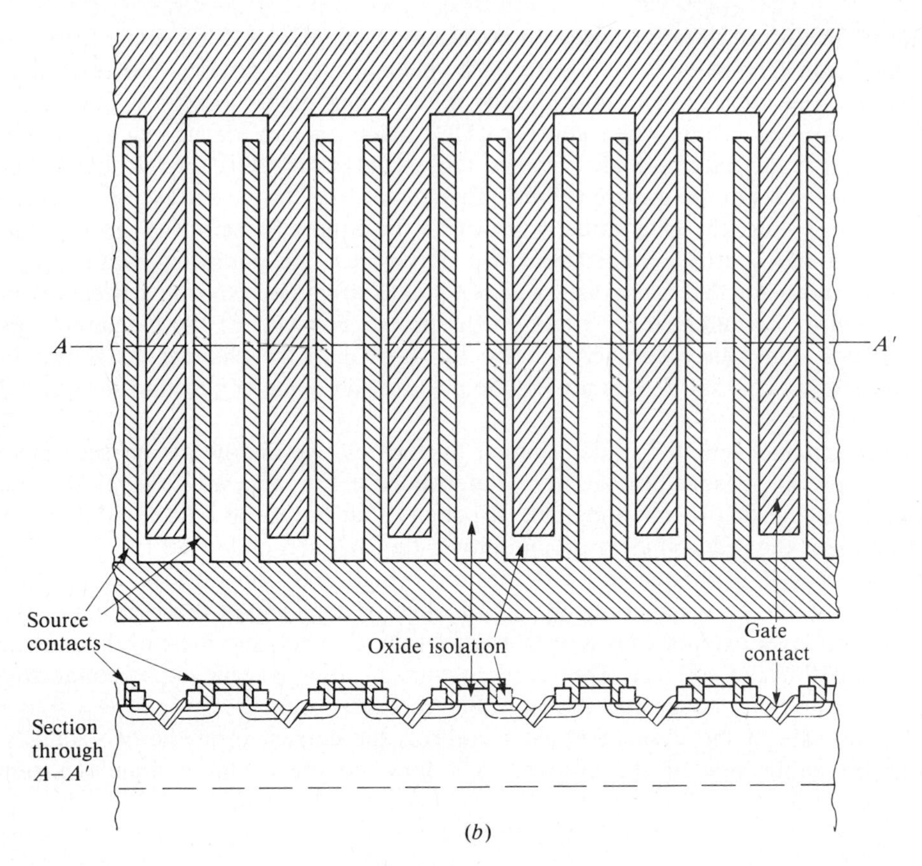

Figure 1.8. (*Continued*)

crowding at the apex of the groove, as shown in Figure 1.9*a*, can limit the useful current rating of the device. In the OFF state the sharp apex causes a local region of high electric field to develop, and this may limit the voltage rating. By arranging for the groove etch to stop before an apex is formed, the structure shown in Figure 1.9*b* can be obtained and both of these problems are reduced.

Because of problems in controlling the critical etching processes, VVMOS FETS are difficult to produce. To a large extent they have been superseded by a different type of vertical MOS transistor, upon which we shall concentrate.

We have seen how it is possible to use two successive diffusions, first a p diffusion using boron, then an n diffusion using phosphorus, to produce two closely spaced pn junctions at different depths below the silicon surface. This is essential to the manufacture of BJTS and can be used to define the channel length in VVMOS transistors. The different *lateral* spreads of the two diffusions can be used in exactly the same way. This technique was applied first to the lateral MOS device described earlier and shown in Figure 1.5*b* [27]. The device structure then takes the form shown in Figure 1.10. We shall refer to it as a "lateral DMOS transistor" (LDMOS), where the "D" indicates the double diffusion process. The channel length is no longer dependent on the resolution of the photomicrolithog-

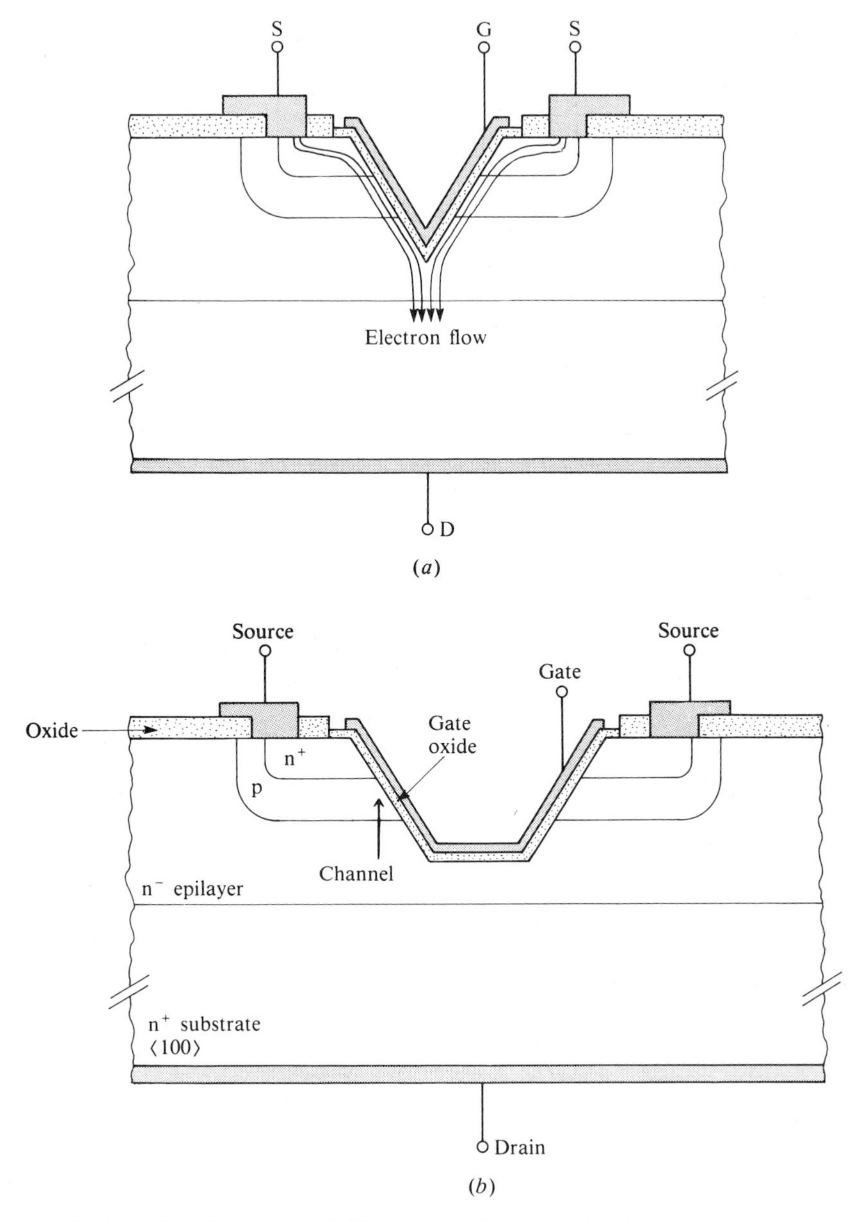

Figure 1.9. Benefits of a truncated V-groove. (*a*) Conventional V-groove showing current crowding at the apex; (*b*) truncated V-groove.

raphy, but on the control of the lateral spreads of successive phosphorus and boron diffusions through the same oxide window.

The vertical, double-diffused MOS structure [28–30], which has become so important and which we shall refer to as VDMOS, fuses together these two concepts. As Figure 1.11 shows, it uses the double-diffusion technique to determine the channel length l, and it supports the drain voltage vertically in the

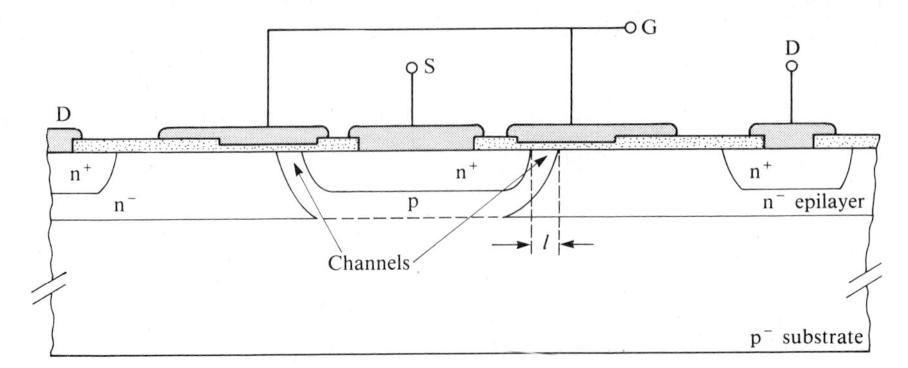

Figure 1.10. The lateral DMOS transistor.

n^--epilayer. The current flows laterally from the source through the channel, parallel to the silicon surface, and then turns through a right angle to flow vertically down through the drain epilayer to the substrate and the drain contact. The p-type "body" region, in which the channel is formed when a sufficiently positive gate voltage is applied, and the n^+ source contact regions are diffused successively through the same window etched in the oxide layer. The channel length can be controlled to submicrometer dimensions if required. Because of the relative doping concentrations in the diffused p-channel region and the n^- epilayer, the depletion layer which supports V_{DS} extends down into the epilayer rather than laterally into the channel. The details of the depletion, accumulation, and inversion layers that are formed in the different regions are discussed in Chapter 2.

A development that was of great importance in integrated-circuit MOS technology in the 1970s was the use of heavily doped polycrystalline silicon, rather than aluminum, to form the gate electrode [31]. It has a number of

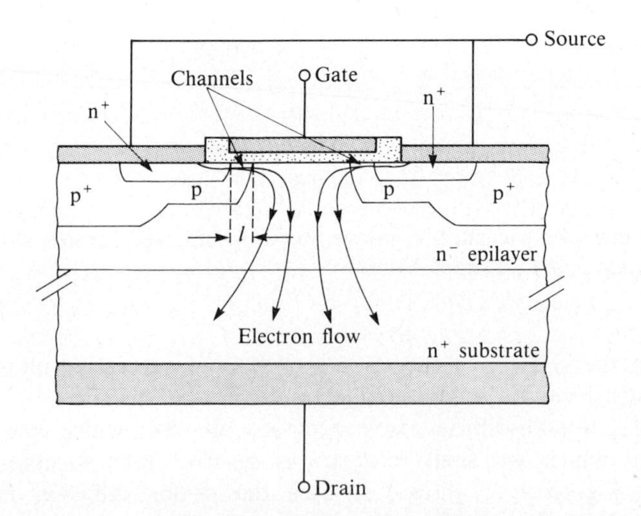

Figure 1.11. Schematic illustration of a vertical DMOS transistor.

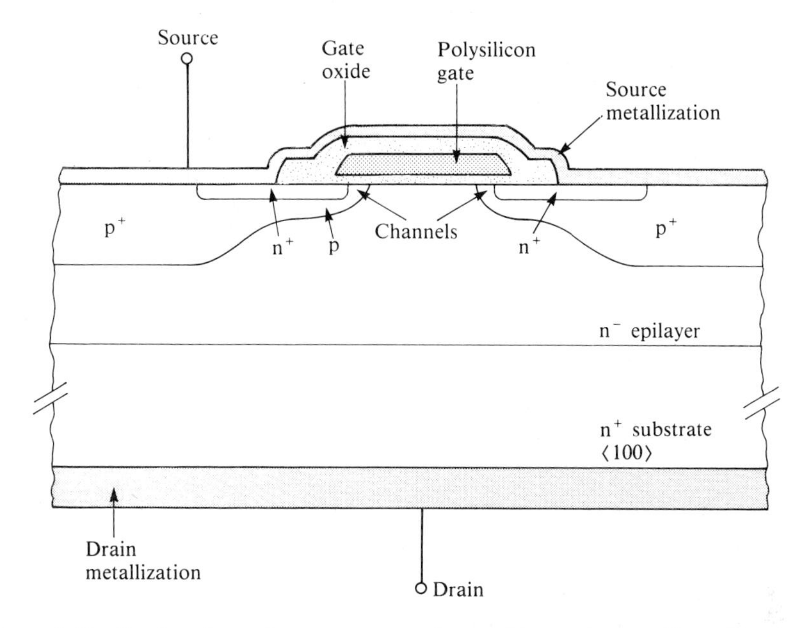

Figure 1.12. Section through a VDMOS transistor with a polycrystalline silicon gate.

advantages for the power FET also:

1. It simplifies the connection metallization: an oxide layer can be formed over the poly-Si, and the source metallization may then be extended over the whole of the upper surface. This is shown in Figure 1.12.
2. The poly-Si layer can be deposited with great accuracy, and the gate oxide is more stable and less prone to contamination than when an aluminum gate contact is used. As a result there is better control of the threshold voltage.
3. The source is self-aligned automatically with the gate edge.

The one disadvantage of polycrystalline silicon is that it has a higher resistance than aluminum. A typical value is 30 Ω per square. This can give rise to a gate resistance of 1–2 Ω, enough to limit performance at high frequencies because of the need to charge the gate capacitance. This is one aspect in which the performance of devices with aluminum gates may be superior.

Best utilization of the silicon is obtained by fabricating the FETS in a cellular structure, as shown in Figure 1.13. Several different schemes are employed, and their relative merits are discussed in Section 3.4. High-current capability is obtained by connecting many cells together in parallel.

It is important to appreciate that the success of the VDMOS FET has not depended on one major inventive step, but rather is the result of combining together a number of significant features. These include the vertical geometry, the double diffusion process, the polycrystalline silicon gate, and the cellular structure. Following these, several important but more subtle technological

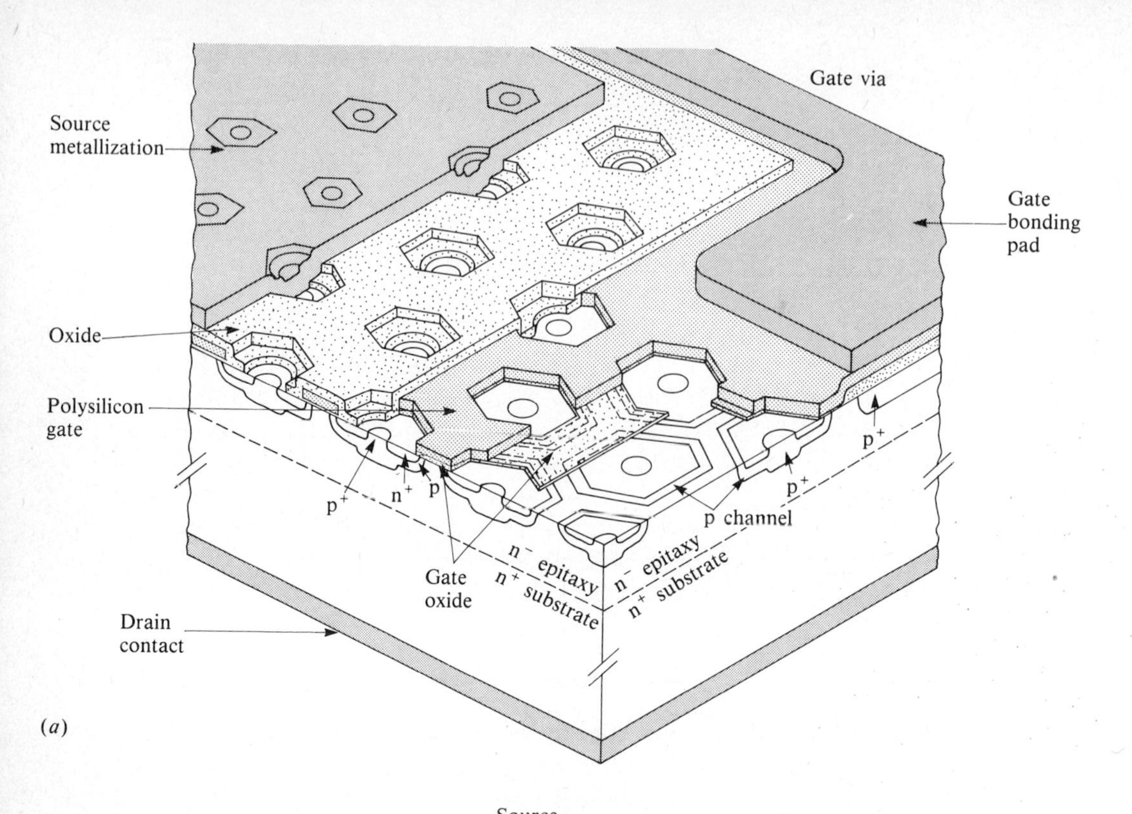

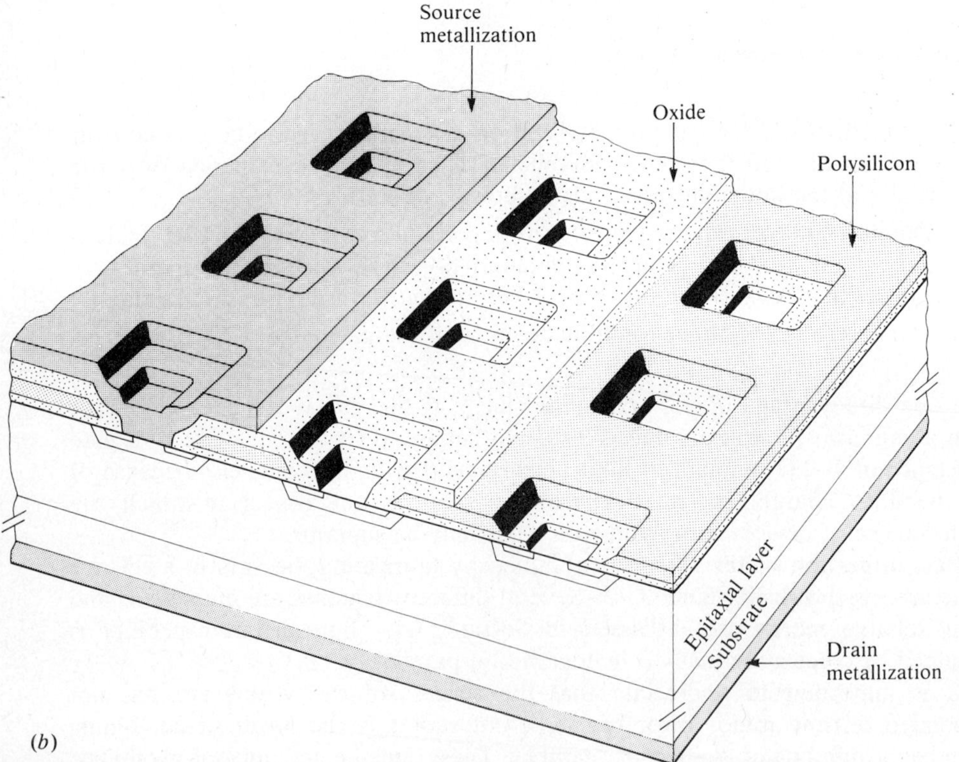

Figure 1.13. Two examples of cellular structures used in VDMOS FETS. (*a*) Hexagonal; (*b*) square. The cell pitch is typically 35 to 40 μm, reducing to less than 20 μm in some more recent low-voltage devices.

refinements have been introduced in different ways by different manufacturers and so have steadily improved device performance.

1.3. THE MERITS OF THE POWER MOS TRANSISTOR

Having catalogued at some length the history of the development of power MOS transistors, it is pertinent to ask what the driving force was. What benefits can power FETS hope to bring to power-electronics applications? In an attempt to answer this for both analogue and digital requirements we first list the characteristics that would be expected in an ideal power switch and an ideal linear voltage amplifier. Against these ideals we can compare the VDMOS FET with bipolar devices such as BJTS or the several different types of thyristor that are now available, particularly the gate-turnoff (GTO) thyristor.

For switching we should like:

1. Zero admittance when OFF.
2. Zero impedance when ON.
3. Instantaneous transitions from ON to OFF and OFF to ON with zero switching losses.
4. Low triggering power, but good noise immunity.

For linear voltage and power amplification we expect:

1. Infinite input impedance.
2. Zero output impedance.
3. High and uniform voltage gain over a wide range of frequencies.
4. Infinite power gain.

In both cases we should look for other attributes that are less easy to quantify but are of the greatest importance, such as:

1. Ruggedness to electrical, thermal and mechanical maltreatment.
2. Long life with stable characteristics.
3. Low cost.
4. Ease of handling and use in circuit design.

The ranges of voltage, current, and switching frequency over which the different types of semiconductor device can operate are illustrated in Figures 1.14 and 1.15. The parameter in which the VDMOS FET has a marked superiority over other device types is the important one of frequency or switching speed. Not only does it maintain gain to much higher frequency, it also has a linear transfer characteristic. In common with other types of MOSFET, its input resistance is very high. Negligible power is needed to maintain it in the ON state, and it can often be driven directly from a CMOS or TTL logic output. Like other MOSFETS, it is liable to catastrophic failure through electrostatically induced gate overvoltage, but the

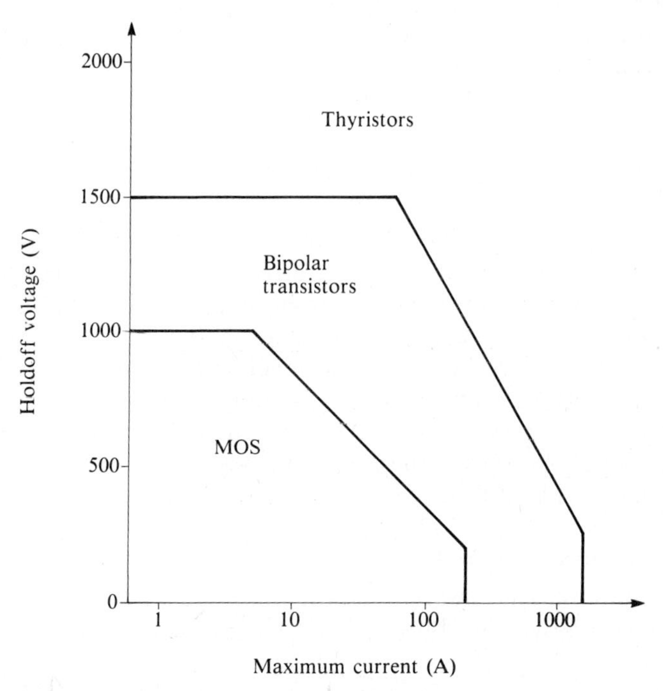

Figure 1.14. Voltage–current limitations of MOSFET, BJT, and thyristor technologies for single-chip devices.

thicker gate oxide and greater gate capacitance of power devices means that they are less vulnerable than integrated circuits. Nevertheless, they should be handled with care prior to installation, and throughout operation the gate must be protected from excess voltage spikes. Although VDMOS FETS can be mass-produced using many of the techniques developed for the fabrication of VLSI circuits, these processes are inherently more complex than the simple diffusion processes used to make equivalent bipolar devices. As a result, transistors and thyristors have retained a general cost advantage over FETS of comparable rating, although for some ratings, notably low voltages, FETS may now have the advantage.

Vertical MOSFETS, while they are much improved in their voltage and current capabilities over lateral MOS devices, do not yet match the combined high voltage and high current ratings possible with bipolar devices. One reason for this is the steep rise in the ON-state drain–source resistance, $R_{DS(on)}$, with the FET voltage rating. This is discussed in Section 3.5. The minimization of the ON-state voltage drop is an important consideration in power devices. For voltage ratings less than about 100 V, VDMOS FETS and bipolar devices occupying the same silicon area have similar ON-state voltage drops. This is not true for higher voltage ratings, particularly above about 200 V. However, the lower voltage drop of the bipolar devices is obtained at a considerable cost in operating speed. In BJTS the low ON-state voltage is obtained by driving them hard into saturation. Thyristors automatically achieve this condition. However, these are minority-carrier controlled devices. They depend on the current being carried through certain regions by minority-carrier diffusion. To support the flow of current, large excess carrier

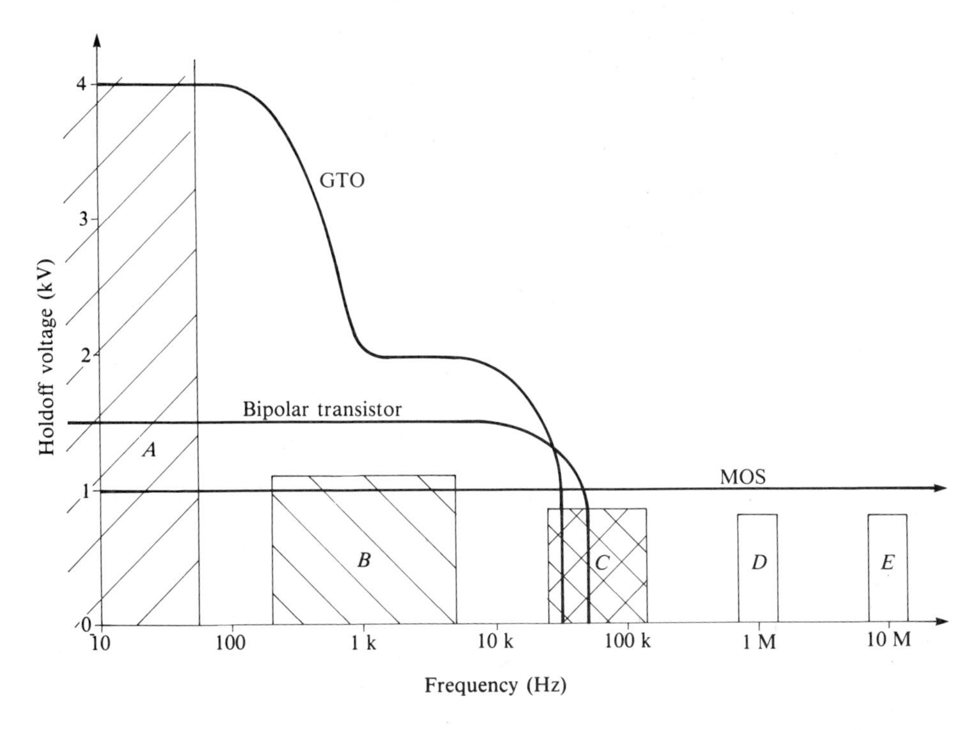

Figure 1.15. Available voltage–frequency ratings of MOS, BJT, and GTO technologies. Shown on the diagram are the ranges of values required in different applications: *A*, phase control and dc switching; *B*, sinusoidal inverters, ac motor control and uninterruptible power supplies (UPS); C, switched-mode power supplies; D, resonant and switched-mode power supplies; E, resonant dc-to-dc power supplies.

concentrations have to be established. These take time to build up during the turn-on process and to decay away during turnoff, typically about 1 μs for 1-A devices and tens of microseconds for higher current levels. The effect of this is normally represented by a large diffusion capacitance, which has to be charged and discharged. With FETS, only the majority carriers are involved in the conduction process; with n-channel devices, only the electrons. In consequence their speed of operation is some tens of nanoseconds or less and is limited only by transit-time effects and parasitic reactance.

The FET ON resistance has the very desirable feature that it increases with increasing temperature. This aids the establishment of a uniform current density throughout the device and means that FETS operated in parallel share the current. Current hogging is avoided. Equally important is the fact that *second breakdown,* which so limits the power handling capacity of BJTS, is normally avoided. This effect arises from thermal runaway when devices are subjected simultaneously to high voltage and high current. A local hot spot develops and current is concentrated into that region as a result of the negative temperature coefficient of resistance characteristic of bipolar devices. The result is a reduction in the *safe operating area* (SOA), as shown in Figure 1.16. With FETS the SOA is limited only by the total dissipation in the device. Well-designed VDMOS FETS are also able to sustain large transient pulse power levels resulting from currents flowing under conditions of controlled avalanche breakdown. This can be a most useful attribute in inductive switching circuits, such as those used for power supplies and motor power control. All of these topics are dealt with in more detail in later chapters.

The power MOSFET, then, offers the engineer unrivaled opportunities for the control of power at high frequencies, within the voltage and current limits shown in Figure 1.14. Detailed developments in device design are continually extending the boundaries shown in the figures to higher levels. The only real disadvantage of the power MOSFET is the relatively high ON-state resistance of high-voltage devices. This can be avoided only by using bipolar conduction processes and hence sacrificing some of the high-frequency performance.

The device shown in Figure 1.17 attempts to strike a different balance between these parameters [32–34]. It is known as the *insulated-gate bipolar transistor* (IGBT) and also under several trade names: Conductivity Modulated FET (COMFET™); Insulated Gate Transistor (IGT™); (GEMFET™). A detailed description of its operation is given in Chapter 19. It is fabricated on a p^+, rather than an n^+, substrate. As a result, holes are injected into the n^- region during forward conduction. A lower $V_{DS(on)}$ may result for a chip of given area, especially for higher-voltage devices. But the turn-off time is increased.

The IGBT is just one example of the many possible variations of the basic n-channel MOSFET. Each one may have merits in certain applications, but it is the device that leads to a wide range of novel applications that is important, which

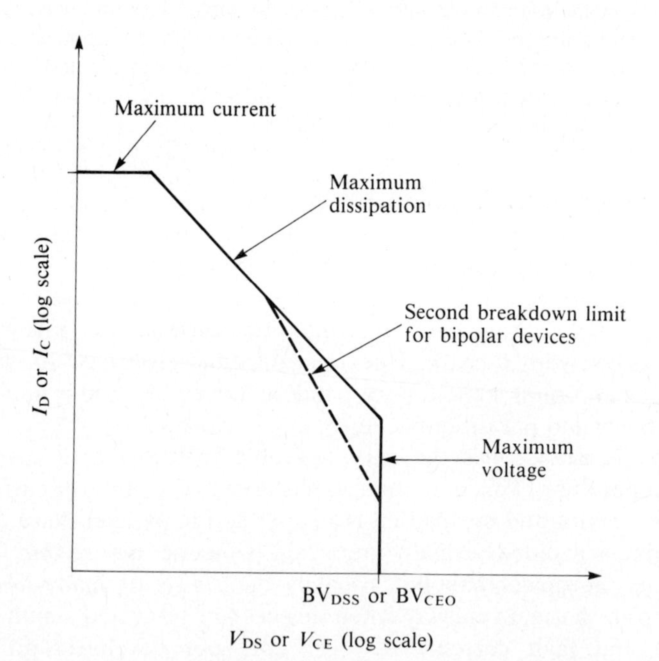

Figure 1.16. Safe operating area, showing reduced performance of bipolar devices caused by second breakdown. This schematic diagram indicates the maximum simultaneous voltage and current levels permitted. Note that logarithmic scales have been used. The current limit is set by the chip size, and by the bond wires and contacts. The voltage limit is set by avalanche breakdown.

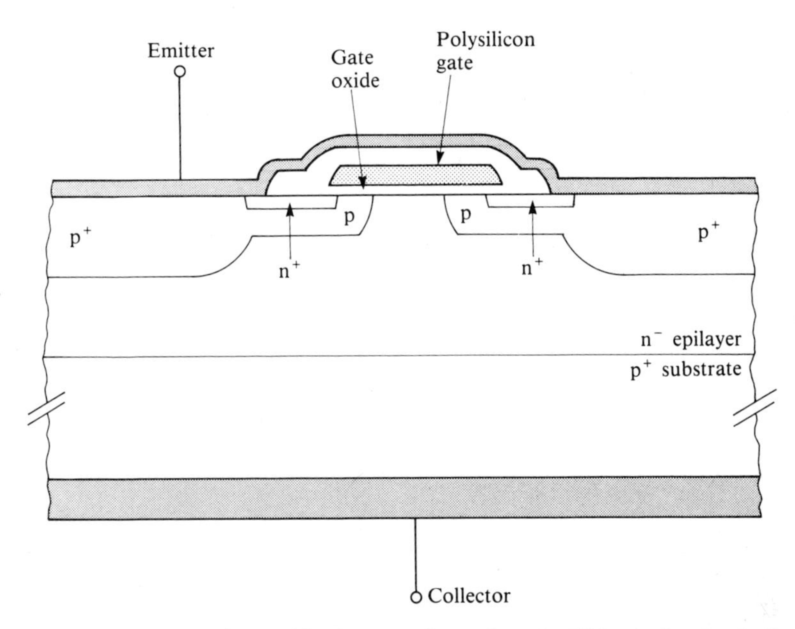

Figure 1.17. The insulated-gate bipolar transistor (IGBT). This device is similar to the vertical DMOS FET of Figure 1.12 except that it is made on a p^+ rather than an n^+ substrate. It thus has the p^+npn^+ structure of a thyristor, but with an insulated gate. Its ON-state voltage drop is low, while it preserves the high input resistance of an MOS device.

brings us back to VDMOS FETS. Many of the applications for which these are best, and in some cases uniquely, suited are set out in the second part of the book, in Chapters 8 to 15.

SUMMARY

Adaptation of MOS technology for power applications requires that a vertical geometry be used. The first successful vertical power MOSFETS used an etched V-groove to define the gate region. These have been superseded by the vertical, double-diffused devices (VDMOS). A cellular geometry enables a very wide gate to be obtained, so that the transconductance is high and the ON-state resistance is low. The unique characteristics of these devices include very fast switching speeds and inherently high reliability. Their fabrication brings together several semiconductor technologies. Some, such as ion implantation and the vapor deposition of epitaxial and polycrystalline silicon, are normally associated more with integrated circuits than with power devices.

REFERENCES

1. J. E. Lilienfeld, U.S. Pat. 1,745,175 (1930); U.S. Pat. 1,877,140 (1932): U.S. Pat. 1,900,018 (1933).

2. O. Heil, Br. Pat. 439,457 (1935).
3. J. Bardeen and W. H. Brattain, "The transistor—a semiconductor triode." *Phys. Rev.* **74,** 230–231 (1948).
4. W. Shockley, U.S. Pat. 2,569,347 (1951).
5. W. Shockley, "The theory of p–n junctions in semiconductors and p–n junction transistors." *Bell Syst. Tech. J.* **28,** 435–489 (1949).
6. W. Shockley, M. Sparks, and G. K. Teal, "p–n junction transistors." *Phys. Rev.* **83,** 151–162 (1951).
7. W. Shockley, "A unipolar field-effect transistor." *Proc. IRE* **40,** 1365–1376 (1952).
8. J. A. Hoerni, "Method of manufacturing semiconductor devices". U.S. Pat. 3,025,589 (1962).
9. D. Kahng, U.S. Pat. 3,102,230 (1963).
10. M. M. Atalla, U.S. Pat. 3,056,888 (1962).
11. S. R. Hofstein and F. P. Heiman, "The silicon insulated-gate field-effect transistor." *Proc. IEEE* **51,** 1190–1202 (1963).
12. R. H. Dennard, F. H. Gaensslen, H. Yu, V. L. Rideout, E. Bassons, and A. R. LeBlanc, "Design of ion-implanted MOSFETS with very small physical dimensions." *IEEE J. Solid-State Circuits* **SC-9,** 256–268 (1974).
13. Y. Tarui, Y. Hayashi, and T. Sekigawa, "Diffusion self-aligned MOST: A new approach for high speed devices." *Proc. Conf. Solid State Devices* **1,** 105–110 (1969).
14. R. Zuleeg and V. O. Hinkle, "A multichannel field effect transistor." *Proc. IEEE* **52,** 1245–1246 (1964).
15. R. Zuleeg, "Multichannel field effect transistor theory and experiment." *Solid-State Electron.* **10,** 559–576 (1967).
16. J. I. Nishizawa, T. Terasaki, and J. Shibata, "Field effect transistor versus analogue transistor (static induction transistor)." *IEEE Trans. Electron Devices* **ED-22,** 185–197 (1975).
17. T. M. S. Heng and H. C. Nathanson, "Vertical MOS transistor geometry for power application at microwave frequencies. *Electron Lett.* **10,** 490–492 (1974).
18. J. G. Oakes, R. A. Wickstrom, D. A. Tremere, and T. M. S. Heng, "A power silicon microwave MOS transistor." *IEEE Trans. Microwave Theory Tech.* **MTT-24,** 305–311 (1976).
19. Siliconix, "MOSFET power soars to 60 W with currents up to 2 A." *Electronic. Des.* **23,** 103–104 (1974).
20. M. Vander Kooi and L. Ragle, "MOS moves into higher power applications." *Electronics* **49,** 98–103 (1976).
21. A. D. Evans, D. Hoffman, E. S. Oxner, W. Heinzer, and L. Shaeffer, "High power ratings extend VDMOS FETS' domination." *Electronics* **51,** 105–112 (1978).
22. K. P. Lisiak and J. Berger, "Optimization of nonplanar MOS devices." *IEEE Trans. Electron Devices* **ED-25,** 1229–1234 (1978).
23. D. B. Lee, "Anisotropic etching of silicon." *J. Appl. Phys.* **40,** 4569–4574 (1969).
24. T. J. Rogers and J. D. Meindel, "Epitaxial V-groove bipolar integrated circuit processes." *IEEE Trans. Electron Devices* **ED-20,** 226–232 (1973).
25. F. E. Holmes and C. A. T. Salama, "V-groove MOS transistor technology." *Electron Lett.* **9,** 457–458 (1973); "VMOS—a new MOS integrated circuit technology." *Solid-State Electron.* **17,** 791–797 (1974).
26. C. A. T. Salama, "A new short channel MOSFET structure (UMOS)." *Solid-State Electron.* **20,** 1003–1009 (1977).

27. H. J. Sigg, G. D. Vendelin, T. P. Cauge, and J. Kocsis, "D-MOS transistor for microwave applications." *IEEE Trans. Electron Devices* **ED-19,** 45–53 (1972).

28. H. W. Collins and B. Pelly, "HEXFET, a new power technology, cuts on-resistance, boosts ratings." *Electron. Des.* **27,** 36–40 (1979).

29. A. Lidow, T. Herman, and H. W. Collins, "Power MOSFET technology." *Tech. Dig.—Int. Electron Devices Meet.* pp. 79–83. (1979).

30. R. W. Coen, D. W. Tsang, and K. P. Lisiak, "A high performance planar power MOSFET." *IEEE Trans. Electron Devices* **ED-27,** 340–343 (1980).

31. F. Fagin and T. K. Klein, "Silicon gate technology." *Solid-State Electron.* **13,** 1125–1144 (1970).

32. J. Tihanyi, "Functional integration of power MOS and bipolar devices." *Tech. Dig.—Electron Devices Meet.* pp. 75–78 (1980).

33. B. J. Baliga, M. S. Adler, P. P. Gray, R. P. Love, and N. Zommer, "The insulated gate rectifier." *Tech. Dig.—Int. Electron Devices Meet.* Abstr. 10.6, pp. 264–267 (1982).

34. J. P. Russell, A. M. Goodman, L. A. Goodman, and J. M. Neilson, "The COMFET—a new high conductance MOS-gated device." *IEEE Electron Devices Lett.* **EDL-4,** 63–65 (1983).

CHAPTER 2

Principles of Operation 1: Threshold Voltage

2.1. LEVELS OF DEVICE MODELING

Power MOSFETS, like other electronic devices, may be modeled at different levels of sophistication. First, their terminal characteristics may be represented by an equivalent circuit made up of linear and nonlinear ideal components such as resistors, capacitors, current and voltage sources, and switches. This level of representation is used for the most part in the later chapters in analyzing the suitability of devices in different circuit configurations for different applications. It is particularly valuable when considering small signal variations about some steady bias point on the device characteristics, because then a linear equivalent circuit may often be used. Such linear and nonlinear equivalent circuits are readily analyzed using the various computer circuit analysis and computer aided design (CAD) packages that are widely available. The applicability to power MOSFETS of one of the most common of these, SPICE, is discussed in later chapters.

More sophisticated device modeling really needs to be tackled in two stages. The first is to model the process so that an accurate representation of the actual physical and chemical structure of the device is obtained, based on the known or intended fabrication schedule. The second is to determine the distribution of static charge and mobile charge (free electrons and holes) within the device, in terms of the known device structure and the known boundary conditions (the electrode potentials and currents and the ambient temperature). These have to be consistent with the current and potential distributions within the device, as determined by the laws of electrical and thermal conduction and of electrostatics. A number of computer models have been developed to represent both of these stages with various degrees of sophistication [1–3]. The aim of such work must be the accurate prediction of device terminal characteristics, and their representation by means of suitable equivalent circuits. To some extent we have applied this representation in Chapter 1, in a very simple and qualitative way, to gain a basic understanding of MOSFET behavior.

In this chapter and Chapter 3, we take a somewhat different approach, modeling the electron energy-level structure or band diagram for the different conducting, semiconducting, and insulating regions. This enables the causes of the various charge and carrier distributions, and their effects, to be pictured more clearly. The hope is that this will give a deeper insight into device behavior and aid the understanding of some of the limitations that arise. Whilst this is desirable, we have to beware in a book of this kind, which is mainly directed towards understanding the suitability of devices for different applications, of delving too deeply into the fundamental physics of device behavior. The reader who wishes to skip this material will find the principal points listed in the end-of-chapter summaries. Readers seeking more background on semiconductor materials and devices are referred to the many, more general books covering these topics (e.g., [4–6]).

A close look at the structure of the active region of a power VDMOS transistor shows it to comprise four distinct regions between the n^+ source diffusion and the n^+ substrate that forms the drain contact. These are represented in Figure 2.1 as the regions Ⓐ, Ⓑ, Ⓒ, and Ⓓ. They are:

- Ⓐ: the surface layer of the p-type "body" regions where conducting channels can be formed by inversion.
- Ⓑ: the surface layer of the n-type regions between the p diffusions.
- Ⓒ: the regions between the p diffusions.
- Ⓓ: the main "drain drift" region in the n^- epilayer.

Note that we have been designating regions that are heavily doped with donors or

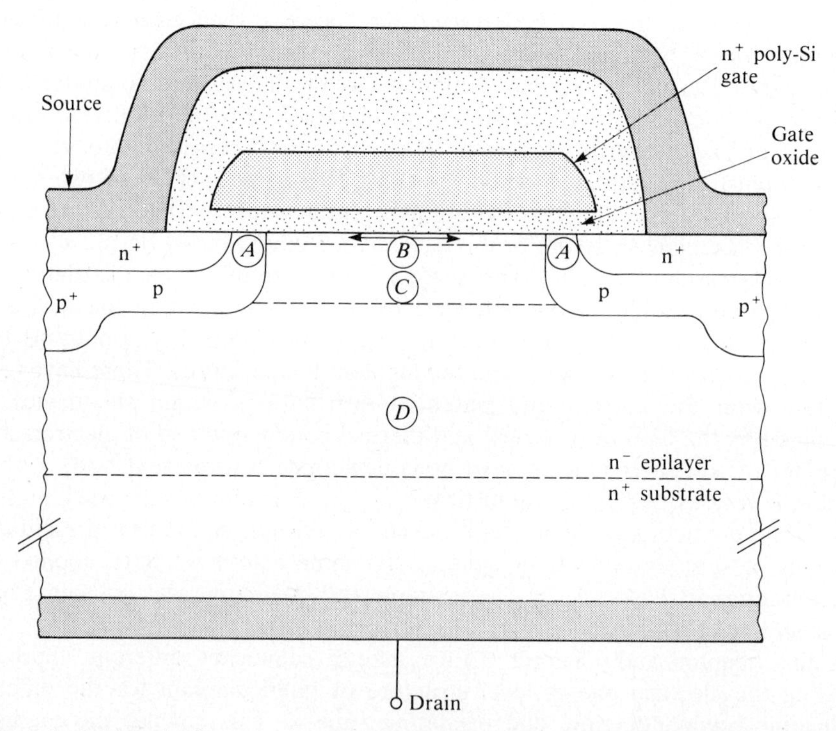

Figure 2.1. Schematic diagram showing the active regions of a VDMOS transistor. Ⓐ, the surface layer of the p-type body region, where the inversion channel is formed; Ⓑ, the surface layer of the n-type epilayer; Ⓒ, the epilayer between the diffused p-type regions; Ⓓ, the main drain drift region in the epilayer below the p-type diffusions.

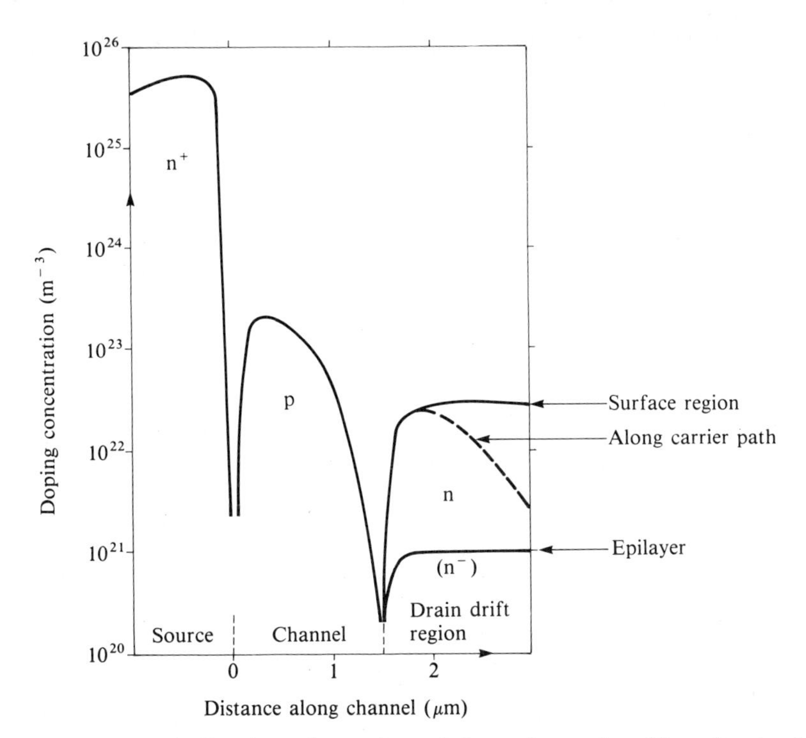

Figure 2.2. Doping profile along the surface of the active region. Note that in the drain drift region the doping concentration may also vary in the direction normal to the surface. The minimum donor concentration is 10^{21} to $10^{22}\,m^{-3}$, depending on the voltage rating required. It is reached in the main part of the epilayer, Ⓓ in Figure 2.1. The doping level then rises sharply to about $10^{25}\,m^{-3}$ in the substrate. This is further illustrated in Figure 3.13.

acceptors as n^+ or p^+, respectively. In general we shall use this to indicate low-resistivity regions where the impurity concentration (n_D or n_A) exceeds about $10^{24}\,m^{-3}$ ($10^{18}\,cm^{-3}$), so that the resistivity is less than about $2 \times 10^{-4}\,\Omega\,m$ ($0.02\,\Omega\,cm$). Conversely, lightly doped, high-resistivity material [with n_D or n_A less than about $10^{21}\,m^{-3}$ ($10^{15}\,cm^{-3}$) and hence a resistivity greater than about $0.1\,\Omega\,m$ ($10\,\Omega\,cm$)] is shown as n^- or p^-. An illustration of the doping profile of a typical device is given in Figure 2.2.

The conditions that arise in the two surface regions, Ⓐ and Ⓑ, are discussed in detail in this chapter. The conditions in the drain drift regions, Ⓒ and Ⓓ, are discussed in Chapter 3 for both the ON and the OFF states of the device. Note that it is the channel region that controls the flow of current in the device, while any blocking voltages are developed across the drain drift region. The principal purpose of Chapters 2 and 3 is to determine the way that the static characteristics of the MOSFET depend on its detailed structure and hence on the device design. It is these relationships that determine the required fabrication schedule such as the one outlined in Chapter 5.

2.2. THE POLYSILICON–SILICON DIOXIDE–SILICON SANDWICH

It can be seen from Figure 2.1 that the sandwiches formed by the polycrystalline silicon gate, the gate oxide, and the variously doped regions in the single-crystal silicon epilayer are most important regions which determine many of the critical properties of the VDMOS device. The poly-Si gate is normally strongly n-type, being very heavily doped with phosphorus. The epitaxial layer is deposited as a lightly doped n-type region Ⓓ some 10 to 50 μm thick. The doping level may sometimes be increased towards the surface, in regions Ⓑ and Ⓒ. The channel or body region Ⓐ is more heavily doped p-type. In the remainder of this chapter we give an account of the theory underlying the behavior of the surface regions Ⓐ and Ⓑ under different conditions of gate bias.

Figure 2.3 illustrates a poly-Si–SiO_2–Si sandwich for the special case in which the two semiconductor regions are identical and equally doped, and there is no interface charge present. This situation does not arise in any practical device, but it is important theoretically because it gives rise to the so-called *flat-band* condition. This is illustrated in the electron energy level diagram of Figure 2.3*b*. On this diagram are marked the lowest allowed energy level in the conduction band, E_C, the highest allowed energy level in the valence band, E_V, the Fermi energy E_F, and the mid-band-gap energy E_i. Inverted, the diagram may also be thought of as representing the local electrostatic potential inside the semiconductor. This is indicated on the right-hand side of the diagram. Because the system is in equilibrium (that is, no bias voltages are applied and no current is flowing), the Fermi energy E_F is uniform throughout the material. If an electron energy level were to exist at the Fermi energy, it would have a 50% probability of being occupied. At energies more than a few times kT above E_F, electron energy levels have a very small probability of being filled, whereas at energies more than a few times kT below E_F they are very unlikely to be empty.

The actual probability distribution function, representing the probability that a level at energy E is filled, is the Fermi function:

$$P(E) = \left[1 + \exp\left(\frac{E - E_F}{kT}\right)\right]^{-1} \tag{2.1}$$

The probability distribution function representing the probability of a level being empty is

$$1 - P(E) = \left[1 + \exp\left(\frac{E_F - E}{kT}\right)\right]^{-1} \tag{2.2}$$

Here, k is Boltzmann's constant (1.38×10^{-23} J/K) and T is the absolute temperature. Thus at room temperature (25°C, 298 K), $kT = 0.0257$ eV. In comparison, the band gap energy $E_G = E_C - E_V$ is 1.1 eV for silicon at room temperature.

The midpoint energy of the band gap, E_i, is a convenient reference energy. The difference between E_i and E_F is logarithmically dependent on the dopant

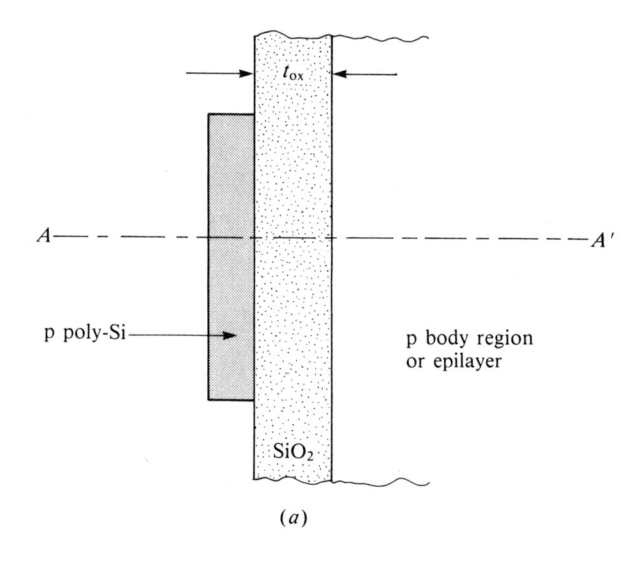

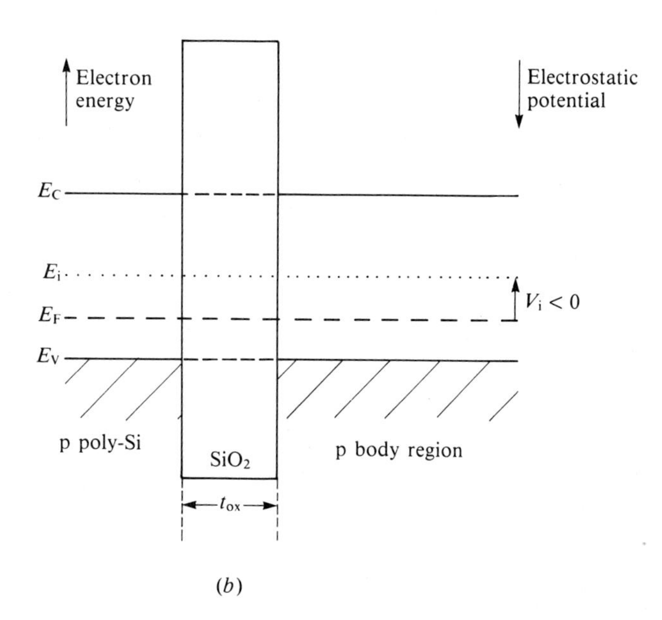

Figure 2.3. The polycrystalline-silicon–silicon dioxide–silicon system: flat-band condition. (*a*) Section showing semiconductor surface, oxide layer, and polycrystalline silicon layer. (*b*) Energy-band diagram along section AA' normal to the surface in the particular case when both the poly-Si and the bulk Si are doped with the same impurity concentration, in this case p-type. This is the *flat-band* condition.

concentration in the semiconductor. In p-type material,

$$E_F - E_i = eV_i = -kT \ln(p/n_i) \approx -kT \ln(n_A/n_i) \qquad (2.3)$$

where the hole concentration p is assumed to be equal to the concentration of acceptors, n_A, and n_i is the intrinsic carrier concentration. With n_i in m^{-3} and T in kelvin,

$$n_i = 4.83 \times 10^{21} T^{3/2} \exp(-E_G/2kT) \qquad (2.4)$$

For silicon at room temperature, $n_i \approx 1.5 \times 10^{16}\, m^{-3}$ ($1.5 \times 10^{10}\, cm^{-3}$). Note that as defined in Equation (2.3), V_i represents the potential of the midband energy level measured with respect to the Fermi energy. In p-type material, it has a negative value, as indicated on Figure 2.3*b*. In n-type material,

$$E_F - E_i = eV_i = kT \ln(n/n_i) \approx kT \ln(n_D/n_i) \qquad (2.5)$$

where now the electron concentration is assumed to be equal to n_D, the donor concentration. Note that the value of V_i is positive in n-type material.

Figure 2.4 illustrates the same poly-Si–SiO_2–Si interface for the more relevant case of region Ⓐ of Figure 2.1. This is the p-type body region where the FET

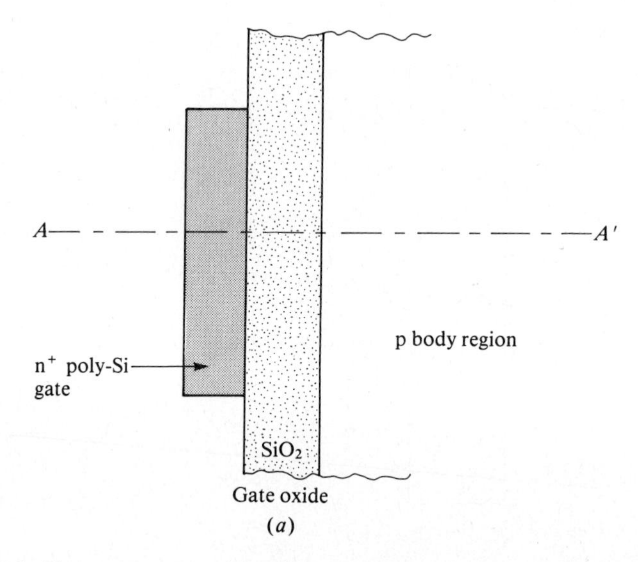

Figure 2.4. The gate–gate-oxide–silicon region under normal conditions. (*a*) Section showing p-type bulk semiconductor, gate oxide, and heavily doped n^+-polycrystalline silicon gate. (*b*) Energy-band diagram along section AA' normal to the surface showing the effect of the contact potential between the n^+-polycrystalline silicon gate and the p-type body region. Note that depletion layers are formed in both semiconductor regions. (*c*) Energy-band diagram along AA' when the gate is so heavily doped that the semiconductor has become degenerate. The Fermi energy is then close to the band edge over a wide a range of doping levels. In the following discussion, we shall assume that the energy levels, E_C and E_F, are coincident. In the band diagrams mobile charge is represented as ⊖, ⊕, indicating free electrons and holes respectively. Fixed charge is represented as −, +, indicating, for example, filled acceptor and ionized donor levels, respectively.

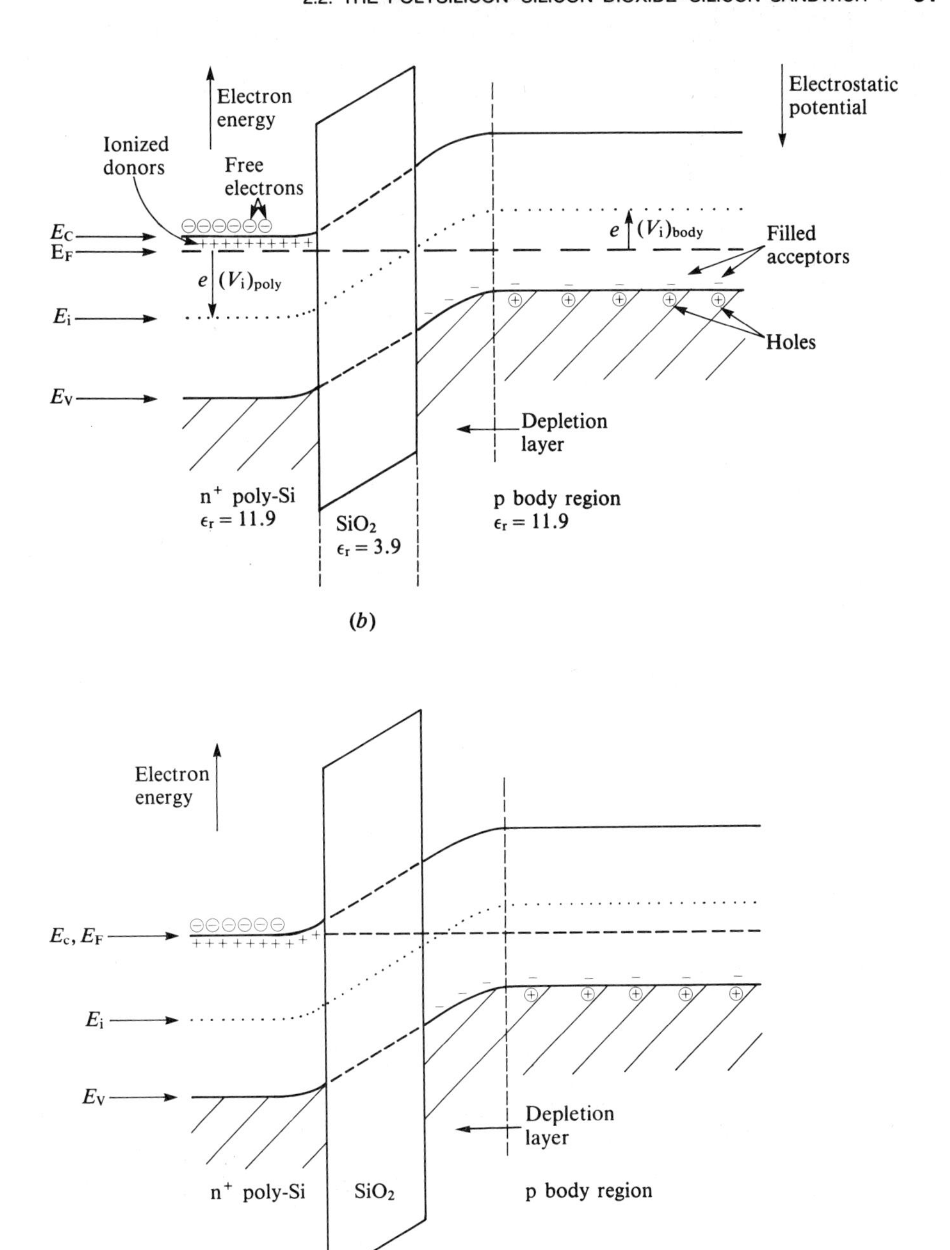

Figure 2.4. (*Continued*)

conduction channel is formed. Again, the equilibrium situation is shown on the diagram. As a result, the Fermi level is uniform everywhere. Now, because of the different doping of the n-type poly-Si gate and the p-type bulk semiconductor, a *contact potential* develops rather as it would across an ordinary pn junction. For the doping levels normally encountered in semiconductor devices, the magnitude of the contact potential, V_{CP}, is given by

$$V_{CP} = (V_i)_{Poly} - (V_i)_{Body} = \frac{kT}{e} \ln \frac{n_A n_D}{n_i^2} \tag{2.6}$$

This is, of course, the same as the diffusion potential that would be obtained at a pn junction formed from these same materials. Remembering that electrostatic potential has the opposite polarity to electron energy, we see that the n^+ poly-Si is positive with respect to the p-body region. As defined here, $(V_i)_{Poly} > 0$, $(V_i)_{Body} < 0$, and $V_{CP} > 0$. Note that, as in a normal pn junction, the contact potential has the effect of depleting the adjacent regions of their majority carriers. In this case, depletion layers form in the semiconductor regions bordering the oxide layer.

Equations (2.3), (2.5), and (2.6) cease to be valid when the doping concentration exceeds about $10^{25}\,m^{-3}$ ($10^{19}\,cm^{-3}$), at room temperature. Then, the Fermi level approaches the band edge, and the material is said to be *degenerate*. This is usually the case for the heavily doped source and drain regions, and for the polysilicon gate of a typical power MOSFET. Several effects complicate detailed theoretical analysis in such material, but the Fermi energy does not in practice move far from the normal band edge. In Figure 2.4*c* and in all subsequent band diagrams, the Fermi level in the source, drain, and gate material is assumed to lie at the band edge, E_C in n-channel devices. Equation (2.6) then becomes,

$$V_{CP} = \frac{1}{2}\frac{E_G}{e} + \frac{kT}{e} \ln \frac{n_A}{n_i} \tag{2.6a}$$

It is commonly found that a layer of positively charged ions gathers at the Si–SiO_2 interface. There are several possible reasons for this. One that was particularly troublesome for many years was the presence of alkali-ion contamination in the oxide. The ions have a significant mobility within the oxide, and tend to migrate towards the semiconductor surface. For MOS devices, the result of this is that several important parameters vary during the life of the device. Great care is therefore taken during processing to eliminate possible sources of sodium and potassium contamination. There may also be some fixed charge within the oxide layer, but this is normally negligible. Also in the oxide, within a few atomic layers of the interface, there is a significant surface density of fixed interface charge, which results from incomplete Si–O bonds, as the crystal structure becomes nonstoichiometric towards the surface. Finally, at the interface itself, there is a certain surface density of trapping levels. These are electron energy levels within the semiconductor band gap. The amount of charge trapped in these levels is a function of the surface potential, and it must be reduced to a negligible level for MOS transistor action to be satisfactory. This is achieved during processing by annealing the oxidized surface in hydrogen or forming gas. A high-temperature anneal at the end of processing may also help.

With a p-type semiconductor, the electric fields in the surface region cause the interface charge to be predominantly positive. We shall assume that a layer of fixed positive charge having a surface density Q_{SS} is normally present at the interface. Its effects are illustrated in Figure 2.5. Figure 2.5*a* shows what would

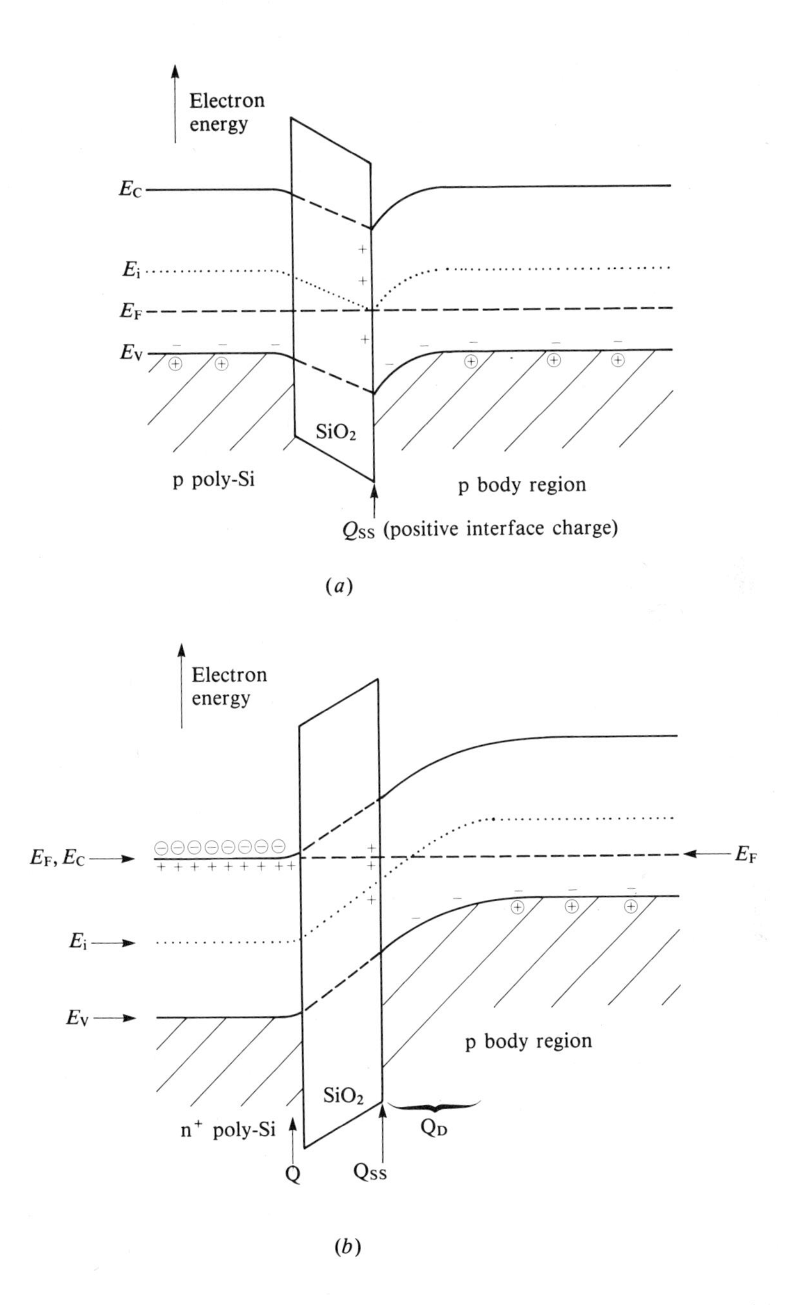

Figure 2.5. The effect of positive surface charge at the Si–SiO_2 interface. (*a*) On the "flat-band" system shown in Figure 2.3. (*b*) On the system shown in Figure 2.4*c*. The potential change in the silicon ΔV, is increased, and the field in the oxide is reduced by the effect of Q_{SS}.

otherwise be the flat-band case of Figure 2.3. The positive interface charge creates depletion layers in the p-type semiconductor as majority holes are repelled away. Figure 2.5*b* shows the effect of positive interface charges on the normal n^+ poly-Si gate structure. It increases the size of the depletion layer formed in the bulk semiconductor by the contact potential, and reduces the electric field in the gate oxide.

2.3. THE THRESHOLD POTENTIAL

We next consider the situation when the gate is biased positively with respect to the bulk semiconductor. This results in a further increase in the potential difference ΔV between the potential at the semiconductor–oxide interface and that in the bulk semiconductor. This is shown in Figure 2.6*a*. When the increase in ΔV is sufficient to take the mid-band-gap energy E_i below $(E_F)_{Si}$ at the silicon surface, then the concentration of free electrons at the surface exceeds the

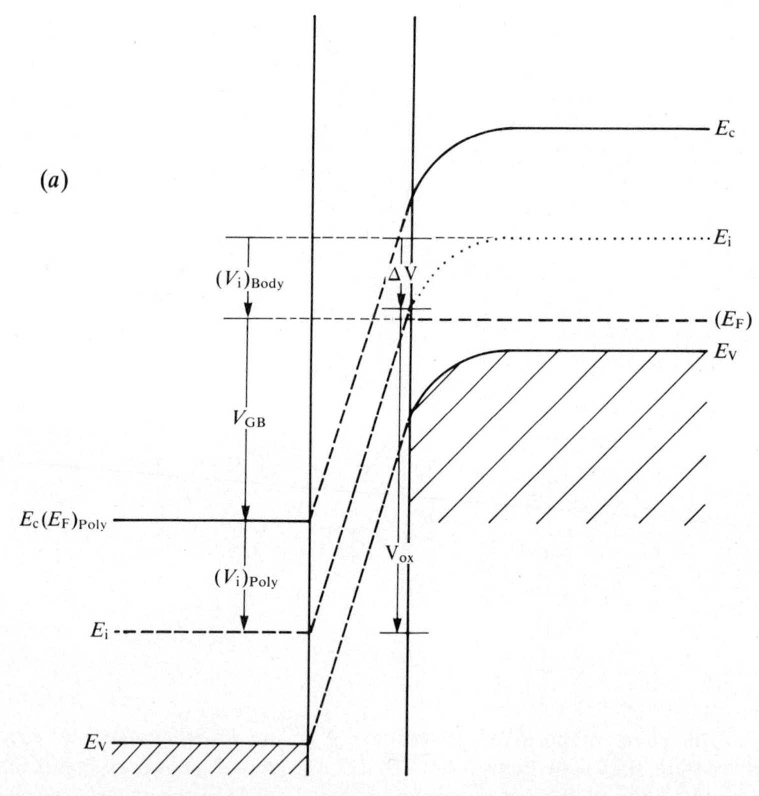

Figure 2.6. The effect of positive gate voltage. (*a*) A small voltage, less than the threshold voltage ($V_{GB} < V_T$); (*b*) the threshold voltage, V_T, which just brings the silicon surface layer to the point of strong inversion; (*c*) a voltage greater than V_T, causing additional charge to build up on either side of the gate oxide. For clarity, the only charges shown are those that are not locally neutralized.

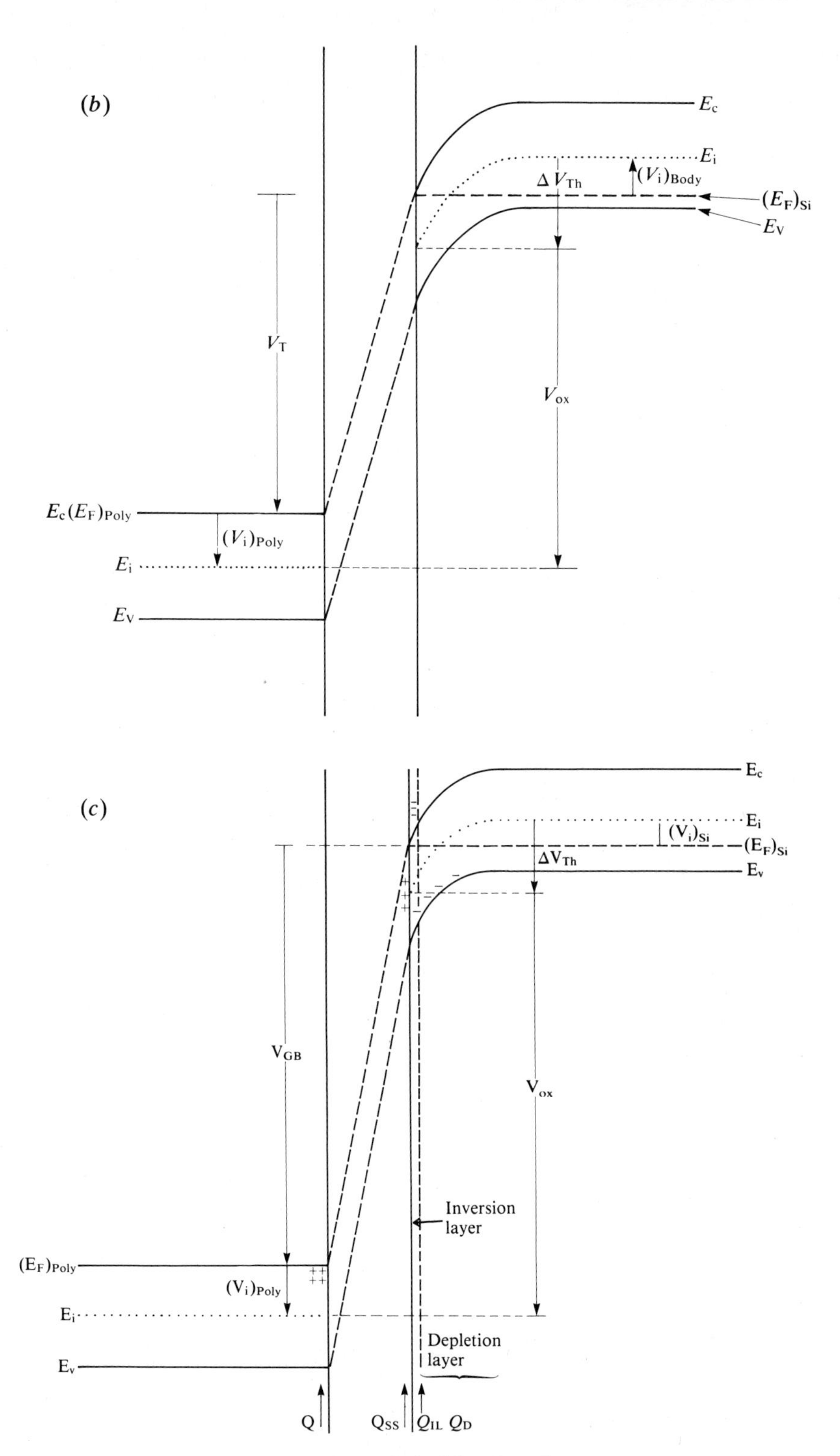

Figure 2.6. (*Continued*)

concentration of holes. The surface layer starts to take on the characteristics of n-type material. It is said to have *inverted.*

Further increase in the positive gate voltage V_{GB} may take E_i at the surface to a level as far below $(E_F)_{Si}$ as it is above $(E_F)_{Si}$ in the bulk material. This condition is illustrated in Figure 2.6*b*. Note that now $\Delta V = \Delta V_{Th} = 2(-V_i)_{Body}$. What this means is that the concentration of free electrons in the surface layer now equals the concentration of holes in the bulk material. It represents the onset of what is known as *strong inversion.* At this point the applied gate voltage V_{GB}, which is the potential difference between the Fermi levels in the poly-Si gate and in the bulk silicon, is $V_{GB} = V_T$, and is known as the *threshold voltage.* Because the source and the bulk semiconductor are shorted by the source contact in power MOS devices, $V_{GB} = V_{GS}$, under static conditions. It should be understood that we are discussing a theoretical definition of threshold. The relationship between V_T and the threshold voltage defined from measurements on practical devices is something we shall consider further in Chapter 3.

Any increase of V_{GB} above V_T is dropped almost entirely across the oxide layer, as shown in Figure 2.6*c*. This is because, once threshold is reached, there is no significant further increase in the voltage dropped across the depletion layer, and ΔV remains equal to $\Delta V_{Th} = 2(-V_i)_{Body}$. Likewise, the width of the depletion layer, l_p, and the quantity of fixed negative charge per unit area, Q_D, contained within it, also remain sensibly constant. What happens is that negative charge in the form of free electrons builds up in the inversion layer, and an equal and opposite positive charge builds up on the gate side of the gate oxide. Thus, the oxide layer acts just like the dielectric in a parallel-plate capacitor. The charge distributions are shown in Figure 2.7. It should be noted that in Figures 2.4 to 2.7 no attempt has been made to represent the polarization effects caused by the different relative permittivities of the oxide ($\epsilon_r = \epsilon_{ox} = 3.9$) and the semiconductor ($\epsilon_r = \epsilon_{Si} = 11.9$).

It can be seen in Figure 2.6 that:

$$V_{GB} = V_{ox} + \Delta V - V_{CP} \tag{2.7}$$

where V_{ox} is the potential difference across the gate oxide, ΔV is the potential difference across the depletion layer in the bulk semiconductor, and V_{CP} is the contact potential given by Equation (2.6). In Appendix 2 it is shown that the fixed charge stored per unit area in the depletion layer, Q_D, is related to ΔV by

$$Q_D = (2e\epsilon_0\epsilon_{Si}n_A\,\Delta V)^{1/2} \tag{A2.6}$$

and that the voltage dropped across the oxide is:

$$V_{ox} = \frac{(Q_D - Q_{SS})}{C_{ox}} \tag{A2.7}$$

where C_{ox} is the capacitance per unit area of the gate oxide, $C_{ox} = \epsilon_0\epsilon_{ox}/t_{ox}$, where t_{ox} is the gate oxide thickness.

At threshold and above,

$$Q_D = 2e\epsilon_0\epsilon_{Si}n_A\Delta V_{Th})^{1/2} \tag{2.8}$$

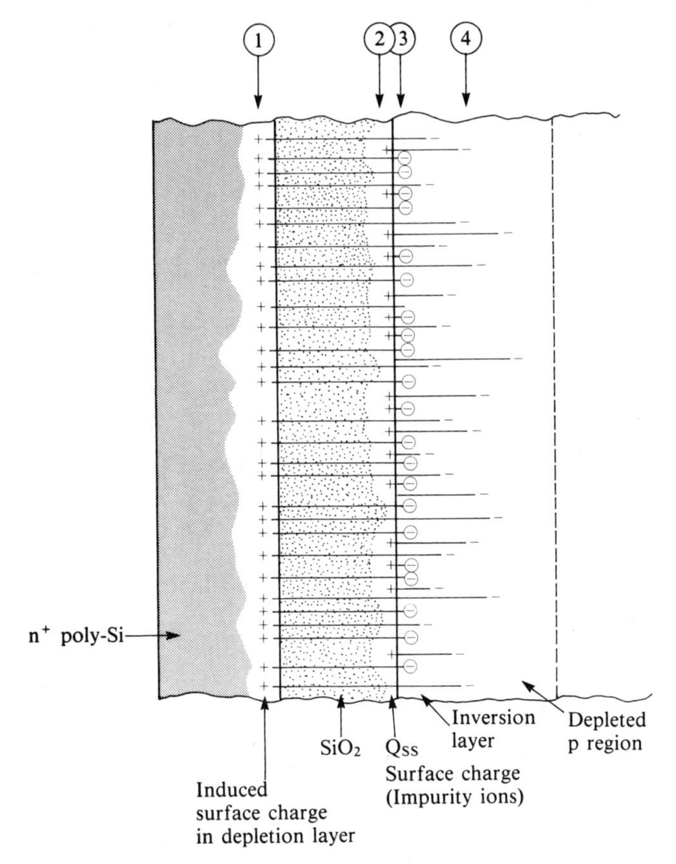

Figure 2.7. Charge distributions with inversion layer formed. This figure is like Figure 2.4*a*, but with the gate biased beyond threshold. An inversion layer forms, and the charge distributions from left to right in the figure are ① the positive charge Q of ionized donors (not compensated by free electrons) at the surface of the n^+-poly-Si; ② the positive surface charge Q_{SS} caused by ionized impurities at the Si–SiO_2 interface; ③ the negative charge Q_{IL} of the free electrons in the inversion layer; ④ the fixed negative charge Q_D of the filled acceptor levels (not compensated by holes) in the depletion layer in the p-type body region. In the region as a whole there must be overall charge neutrality. Thus, $Q + Q_{SS} = Q_{IL} + Q_D$.

where

$$\Delta V_{Th} = -2(V_i)_{Body} = \frac{2kT}{e} \ln(n_A/n_i) \tag{2.9}$$

Substituting Equations (A2.6) and (A2.7) into Equation (2.7), we see that the threshold voltage occurs when

$$V_{GS} = V_{GB} = V_T = \frac{Q_D - Q_{SS}}{C_{ox}} + \Delta V_{Th} - V_{CP} \tag{2.10}$$

with Q_D determined by Equation (2.8), ΔV_{Th} by Equation (2.9), and V_{CP} by Equation (2.6) or (2.6a).

Typical values of the relevant parameters for a power VDMOS FET at room temperature are as follows:

$$(n_A)_{Body} = 1 \times 10^{23}\ m^{-3}\ (1 \times 10^{17}\ cm^{-3})$$

(this is the maximum value in the channel region, near to the source end),

$$(n_D)_{Poly} = 1 \times 10^{26}\ m^{-3}\ (1 \times 10^{20}\ cm^{-3})$$

(this is the same as the source doping concentration and is near to the solid solubility limit for phosphorus in silicon),

$$t_{ox} = 100\ nm\ (1000\ \text{Å})$$

$$Q_{SS} = 2.5 \times 10^{15}\ electrons/m^2\ (2.5 \times 10^{11}\ electrons/cm^2)$$

$$= 4.0 \times 10^{-4}\ C/m^2\ (40\ nC/cm^2)$$

These lead to:

$$\Delta V_{Th} = 0.81\ V$$

$$Q_D = 16.5 \times 10^{-4}\ C/m^2\ (165\ nC/cm^2)$$

$$V_{CP} = 0.96\ V$$

$$C_{ox} = 3.45 \times 10^{-4}\ F/m^2\ (34.5\ nF/cm^2)$$

$$V_{ox} = 3.62\ V$$

$$V_T = 3.47\ V$$

The threshold voltage may be increased by increasing $(n_A)_{Body}$, by increasing t_{ox}, and by obtaining a lower value of Q_{SS}. For some applications, for example when the MOSFET is to be driven directly from a logic output (say TTL), there is a need to reduce V_T. This is not easy to achieve because the three principal factors that control V_T are each determined by other, usually more important, considerations. Reducing t_{ox} makes the transistor more vulnerable to voltage breakdown. To try to increase Q_{SS} artificially would be foolish. The doping concentration in the channel could be lowered, but this is constrained by the need to avoid the channel modulation effects caused by the drain–source voltage. In this respect the double-diffusion process does not give the best doping profile along the channel. As shown in Figure 2.2, the acceptor concentration is highest at the source end where the threshold voltage is determined, and is lower at the drain end where the channel length modulation occurs [7].

The room temperature variations of V_T with n_A and Q_{SS} are shown in Figure 2.8. As the temperature rises, the threshold voltage is found to decrease at the rate of approximately 6 mV/K. Differentiation of Equation (2.10) confirms this value, when account is taken of all the temperature-varying terms, including the

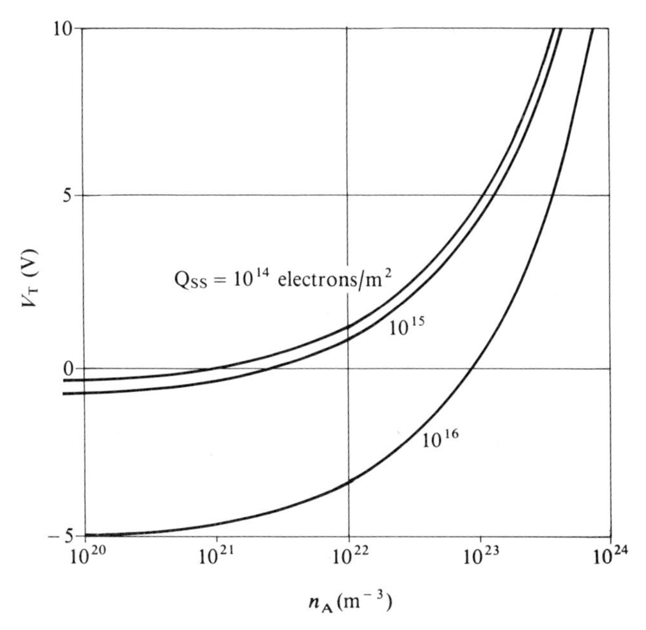

Figure 2.8. Threshold-voltage variation with channel doping concentration and interface charge ($T = 300$ K).

band-gap energy [see Ref. 4, p. 15]. The example just given may be worked through for temperatures of $-55°C$ (218 K, when $kT/e = 0.0188$ V and $n_i = 5 \times 10^{12}$ m^{-3}, giving $V_T = 3.64$ V), and $+175°C$ (448 K, when $kT/e = 0.0386$ V, $n_i = 5 \times 10^{19}$ m^{-3}, and $V_T = 2.56$ V). For reasons that are discussed in Section 3.3, the practical threshold voltages usually quoted for power MOSFETS are rather smaller than the theoretical values obtained here.

In p-channel devices the dopant types and the polarities of the applied potentials are reversed. In particular, the channel has to be formed in n-type material. The contact potential and any positive interface charge both then tend to increase the threshold voltage required for strong inversion, as discussed in the next section. As a result V_T is expected to be higher in a p-channel device than in an otherwise similar n-channel device.

Finally we should examine the effect on the threshold voltage of the use of a metal gate. The change of gate material affects V_{CP}. The contact potential between two materials is normally expressed as the difference between their work functions. (The work function ϕ is the energy difference between the Fermi level and the vacuum just outside the material.) Because aluminum and heavily doped silicon both have work functions in the region of 4.1 eV, this change alone is not expected to cause a significant change to the threshold voltage. Deliberate changes to the doping level of the substrate, and inadvertent variations of Q_{SS}, each alter V_T in the usual way predicted by Equation (2.10). Energy-level diagrams for a system with an aluminum gate, under equilibrium conditions and at threshold, are shown in Figure 2.9.

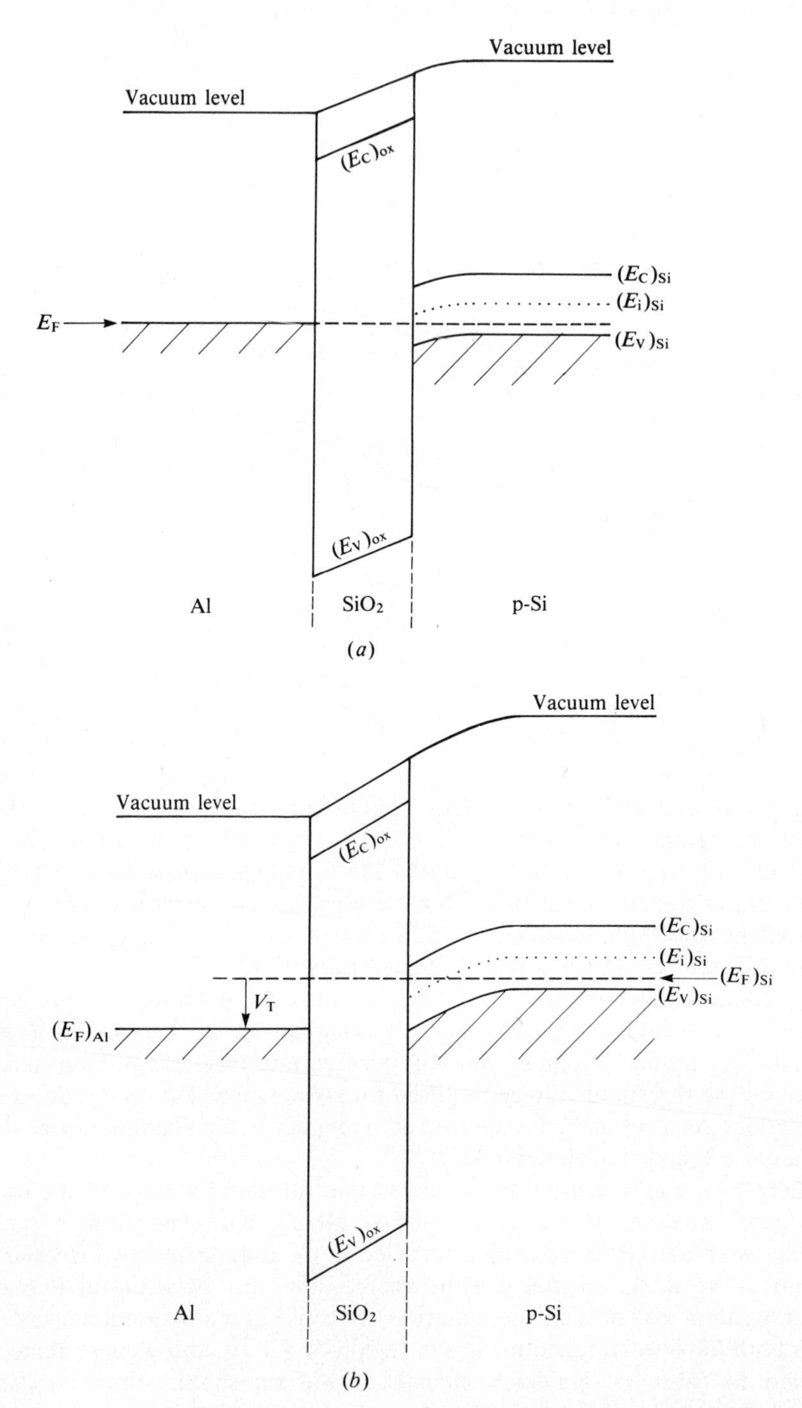

Figure 2.9. Energy-level diagrams for an *n*-channel MOSFET with an aluminum gate. (*a*) Equilibrium; (*b*) At threshold.

2.4. INVERSION AND ACCUMULATION LAYERS

In Appendix 2, the mobile electronic charge per unit area in the inversion layer, Q_{IL}, is shown to increase linearly with V_{GB}, above the threshold voltage. Thus:

$$Q_{IL} = C_{ox}(V_{GB} - V_T) \tag{A2.13}$$

Because, in a power MOSFET, the source and bulk semiconductor are always shorted together, we may write this as

$$Q_{IL} = C_{ox}(V_{GS} - V_T) \tag{2.11}$$

In Equation (2.10), the final two terms almost cancel. It is interesting to consider the implication of neglecting them. When $V_{GB} = V_T' = (Q_D - Q_{SS})/C_{ox}$, the free-electron concentration at the silicon surface equals that in the n^+-gate. Under these conditions, therefore, there is a well-established conducting channel at the surface. However, V_T' is only a few tenths of a volt more than V_T. This sensitivity of the channel conductance to small changes in surface potential justifies our earlier assertion that almost all of any applied gate voltage in excess of V_T is dropped across the gate oxide. A more rigorous one-dimensional analysis of the interface region is presented in Appendix 3.

Now consider region (*B*) of Figure 2.1. Here, the silicon epilayer is lightly doped n-type. As a result, the effects of the contact potential, V_{CP}, the positive interface charge, Q_{SS}, and any positive gate bias voltage, V_{GB}, all help to create an *accumulation layer* at the surface of the n^- epilayer. This is illustrated in Figure 2.10*a* to *c*. This means that the surface layer now has a higher concentration of free electrons than does the bulk material. In consequence the resistivity at the surface is lower. The effects of negative gate bias are shown in Figure 2.10*d* and *e*. It tends to offset the combined effect of the contact potential and the positive interface charge. If sufficiently large, it causes first a depletion layer to form at the surface and then, given a supply of holes adequate to maintain equilibrium, an inversion layer.

With the vertical DMOS transistor, we shall find that the surface layers in the p-type body region and the n-type drain drift region form accumulation, depletion, and inversion layers depending on the state of bias of the device. It must be remembered that the relevant bias voltage is that between the gate and the *local* value of potential in the bulk material immediately under the gate oxide. When there is a lateral voltage drop along the surface layer, this bias voltage changes from point to point.

It will be clear from what has been said that the ac capacitance between the gate and the substrate is much greater when an inversion or accumulation layer is present at the semiconductor surface than when there is a depletion layer. With the inversion or accumulation layer the capacitance per unit area approximates to C_{ox}. With the depletion layer alone it is reduced because the capacitance per unit area of the depletion layer is effectively in series with that of the oxide. In VDMOS devices an accumulation layer forms at the surface of the n^- drain drift region whenever V_{GS} is greater than V_{DS}. At that point a significant increase in the differential gate–drain capacitance can be observed. This is illustrated in Figure

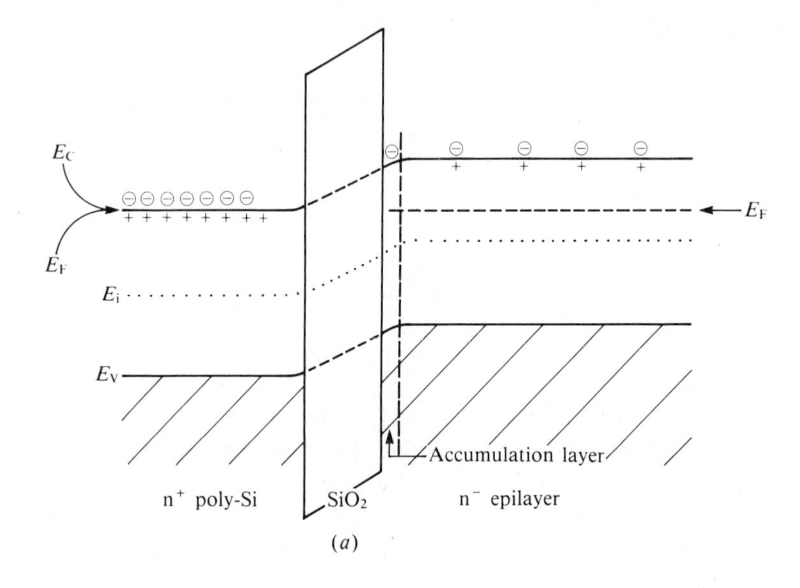

(*a*)

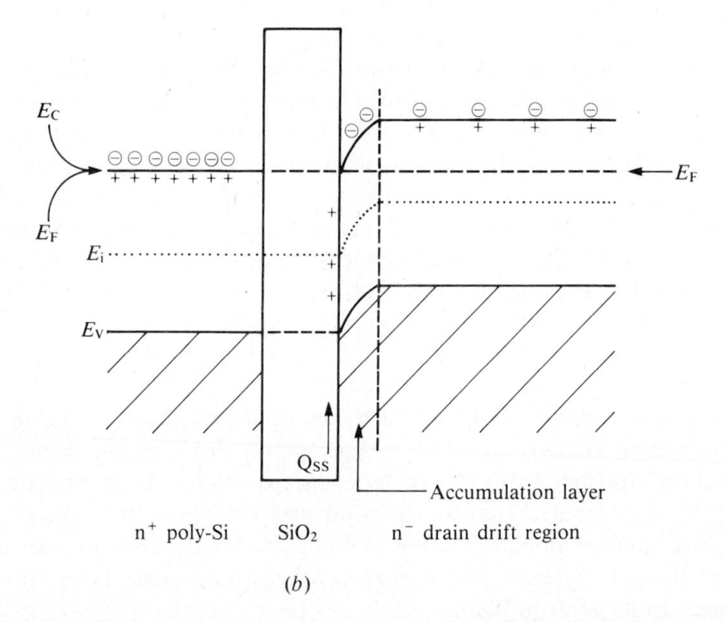

(*b*)

Figure 2.10. Energy-level diagrams across the n^+-Gate–Gate-Oxide–n^--epilayer region. (*a*) Effect of contact potential with no interface charge or applied bias; (*b*) change caused by positive interface charge; (*c*) further change caused by positive gate bias; (*d*) effect of negative gate bias; (*e*) formation of inversion layer with larger negative gate bias voltage applied, assuming a local source of holes. Note that in *a, b,* and *c* an accumulation layer forms at the surface of the n-type bulk silicon. The contact potential, the positive interface charge, and any positive applied gate voltage all assist this. Application of a negative gate potential offsets the effect of both contact potential and interface charge. It leads first to the formation of a depletion layer. Eventually an inversion layer would form, if holes could be supplied at the rate needed to maintain carrier equilibrium at the surface.

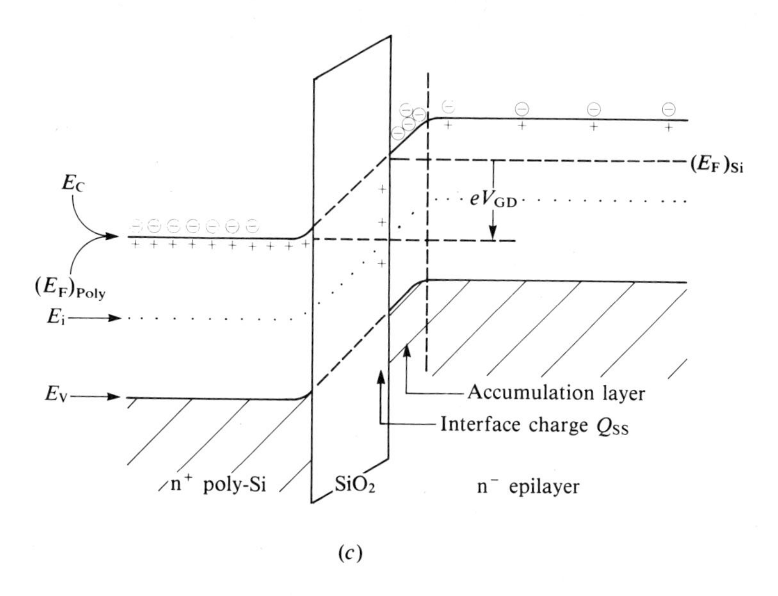

(c)

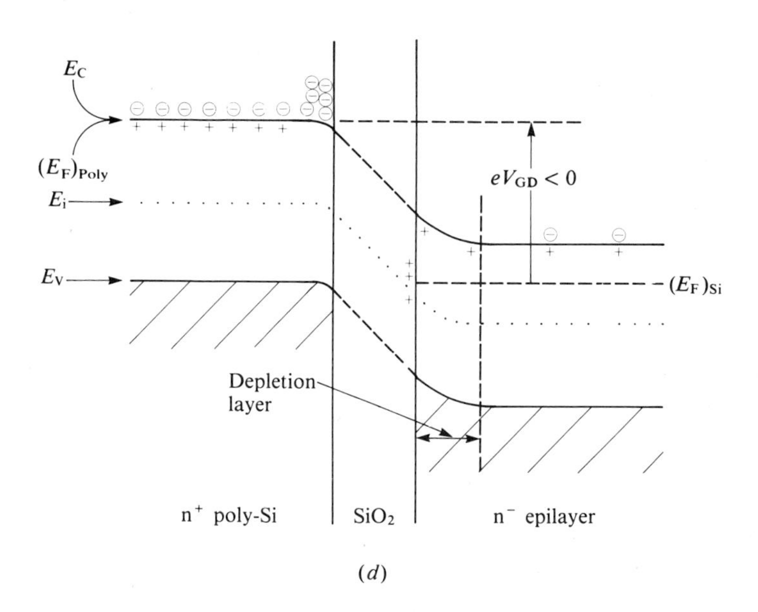

(d)

Figure 2.10. (*Continued*)

2.11. It is theoretically possible for an inversion layer to form under the gate oxide when $V_{DG} = V_{DS} - V_{GS}$ exceeds the threshold voltage of the drain material. In practice this rarely happens, because there is no ready source of holes to supply the inversion layer. This would require the body–drain diode to be forward biased. So we should need $V_{DS} < 0$ at the same time as V_{DG} was greater than the inversion threshold voltage.

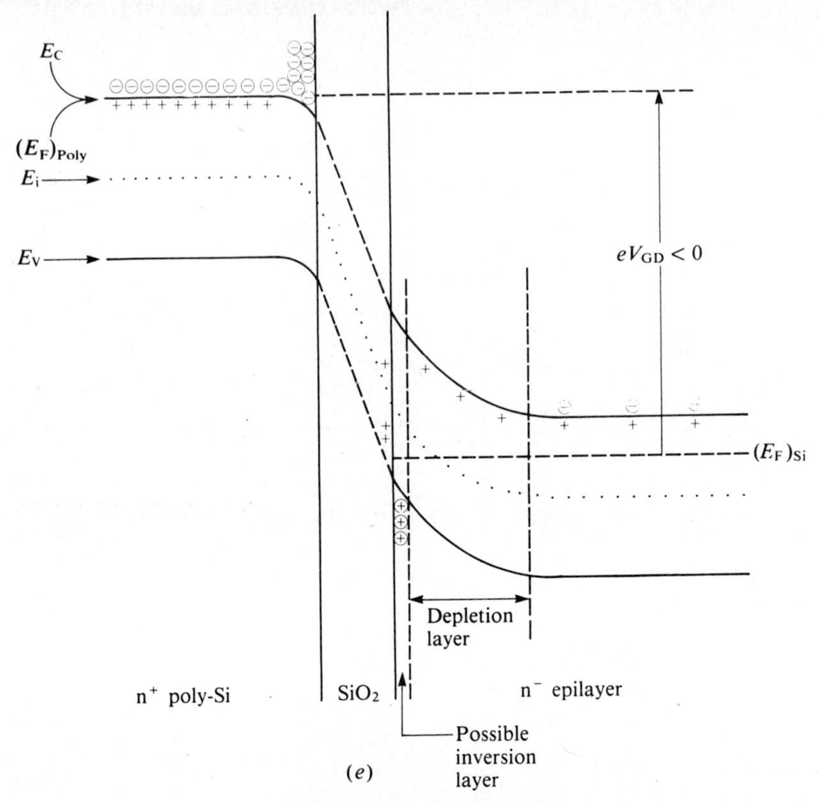

Figure 2.10. (*Continued*)

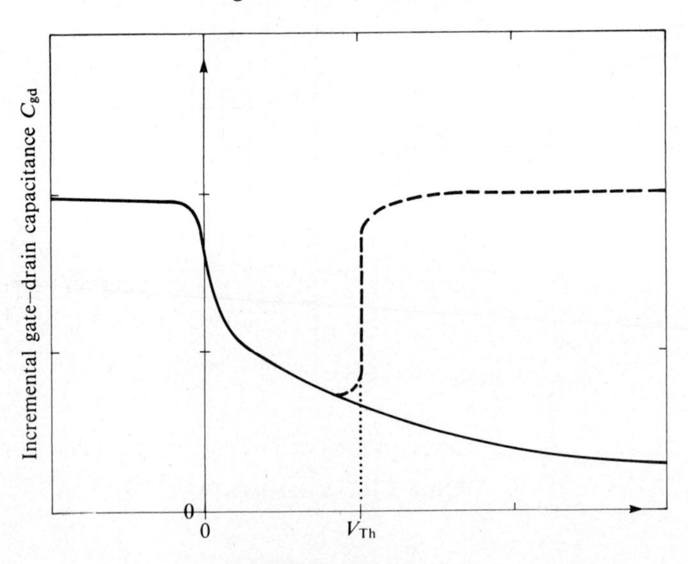

Figure 2.11. Variation of the gate–drain capacitance ($C_{gd} = C_{rss}$) with the drain–gate voltage (V_{DG}). The graph shows the abrupt increase in C_{gd} for $V_{DG} < 0$, as the accumulation layer forms, and the steady decrease as the depletion layer under the gate oxide expands with increasingly positive values of V_{DG}. When V_{DG} exceeds the threshold voltage V_{Th}, an inversion layer can in theory be formed, causing C_{gd} to increase again. In practice, the body–drain diode would have to be forward biased to supply the holes required, so this condition does not normally arise.

2.5. ELECTRON ENERGY LEVELS ALONG THE CHANNEL

So far, in Section 2.2, we have examined the electron energy levels in the semiconductor as a function of the vertical depth below the semiconductor–oxide interface. Now, we shall briefly consider the potential variation *along* the channel, from the source to the drain drift region, just beneath the semiconductor surface. We shall do this for the three possible operating conditions illustrated in Figure 2.12. Figure 2.12*a* shows a simple circuit that can be used to obtain the typical output characteristics shown in Figure 2.12*b*. These are similar to those shown in Figure 1.4*a* and refer to the normal type of enhancement-mode device. For each curve, V_{GS} = constant. Point *A* on the characteristics represents the fully turned-ON state, with V_{GS} biased well above V_T and a good conduction channel linking drain and source. At point *B*, $V_{GS} < V_T$ and no channel is formed. At point *C*, the channel is pinched off. This is a bias condition that may be used for linear amplification.

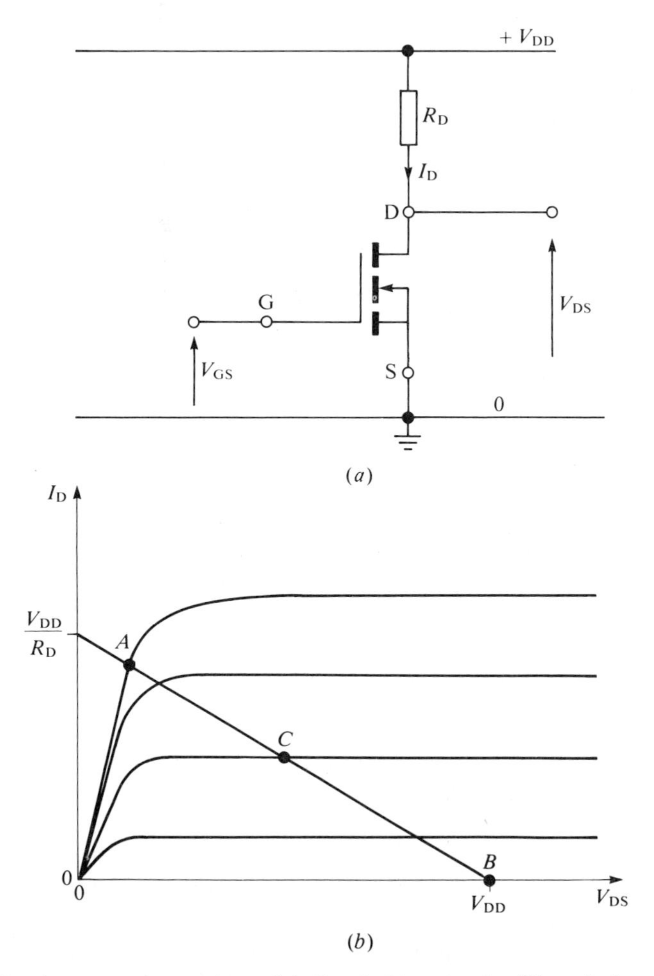

Figure 2.12. Device operating regions. (*a*) Simple bias circuit; (*b*) typical operating points on the load line that represents R_D. At *A* the device is fully ON. At *B* it is OFF. The bias condition *C* is one suitable for linear amplification.

There has been much confusion over the names given to the three MOS operating points identified on Figure 2.12*b*. We would prefer to describe the transistor at point *A* as *saturated*. This is the name given to the corresponding condition in a bipolar junction transistor. However, in order to avoid confusion, we shall use the terms *turned* ON or *fully turned* ON to describe this state. We shall refer to point *B* as the *forward blocking* or *turned* OFF state, and to point *C* as the *constant-current* or *linear* condition. The reader should be aware that the region to the left of the locus of points linking the start of the constant-current condition has in the past been called the "linear" region. A less appropriate name could scarcely be imagined, but it derives from the fact that here the drain current I_D changes linearly with V_{GS}. Likewise, the constant-current region has been called the "triode" region, and also the "saturation" region, because the carriers often reach their saturation drift velocity. To reiterate, we prefer:

point A = turned-ON,
point B = blocking or turned-OFF,
point C = constant current.

Electron-energy diagrams along the channel are shown in Figure 2.13 for each of the three cases. Also shown is the diagram for the equilibrium situation when no bias voltages are applied and no current flows. Then, E_F is uniform throughout. It can be seen that electron flow along the channel is essentially regulated by the height of the potential barrier seen by the electrons in the source as they enter the channel. The height of this barrier is in turn controlled by the gate potential. The flow of channel current requires an ohmic voltage gradient in the channel. This is represented on the diagram as a slope of the whole band structure. The model for device behavior that we are using here might be termed a "hydraulic" model, in which the electrons are thought of as a fluid whose flow can be controlled by the raising and lowering of a weir, as shown in Figure 2.14. Similar models may be constructed [8] to represent the behavior of bipolar junction transistors (the potential barrier at the emitter–base junction regulates the flow of carriers from the emitter through the base to the collector) and the behavior of thermionic vacuum tubes (the potential barrier between the grid wires regulates the flow of electron current from cathode to anode). The model does not clearly represent the thermal energy distribution of the carriers, unless, perhaps, we think of the source as a cauldron of boiling liquid. In bipolar devices the potential barrier regulates the flow of majority carriers, which maintain a quasi-equilibrium. Further conditions have to be satisfied by the minority carriers.

In Figure 2.13*a* the equilibrium case for a MOSFET with a uniformly doped channel is shown. This is repeated in the other figures for reference. Figure 2.13*b* corresponds to point *A* on the characteristics and shows a fully developed inversion layer connecting the n^+ source to the accumulation layer at the surface of the n^- epilayer. The ohmic voltage drop in the channel and epilayer regions is shown by the gradient of the Fermi level. Figure 2.13*c* corresponds to point *B* and shows the channel and epilayer regions to be fully depleted. The gate voltage is negative with respect to the surface potential of the epilayer. Note that in this situation, an inversion layer at the n^- surface does not form because there is no source of holes to supply the necessary charge and so maintain thermodynamic equilibrium. Any holes generated leak rapidly away to the source. In cases *A* and *C* the free electrons in the inversion layer that forms in the channel are supplied from the source across the intervening potential barrier. At low values of the source current, approximate thermodynamic equilibrium is maintained with the source electrons. But as soon as a drain voltage of any magnitude is applied and significant drain current flows, this equilibrium is lost at points beyond the potential minimum. Then, the charge in the channel has to satisfy two physical laws. The first is charge continuity. The second is Poisson's equation. It is these laws that determine the device characteristics as described in Chapter 3.

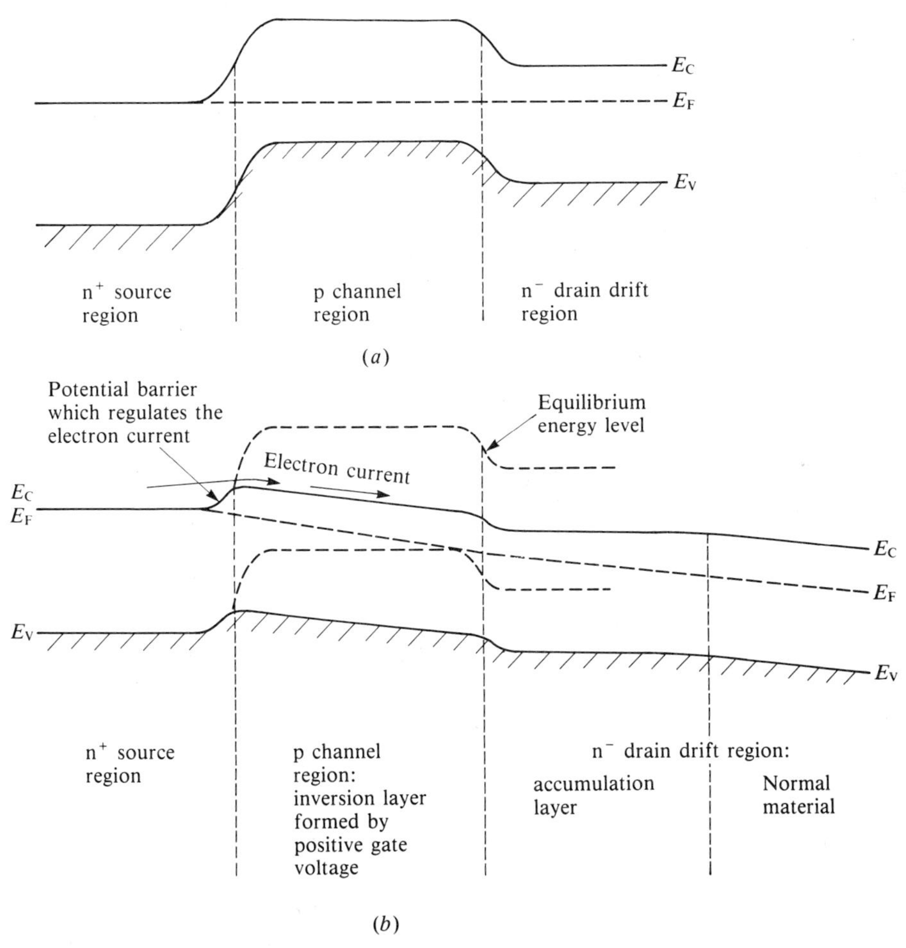

Figure 2.13. Energy-band diagrams along the channel. (*a*) Equilibrium. (*b*) Device turned ON, point *A*. (The slope of the "quasi Fermi level" indicates the ohmic voltage gradient. Electrons in the channel are unable to maintain thermodynamic equilibrium with those in bulk semiconductor regions. Note the formation of the accumulation layer at the surface of the drain drift region, shown by the raising of the Fermi level within the band gap.) (*c*) Device OFF, point *B*. (*d*) Device in the constant-current condition, *C*.

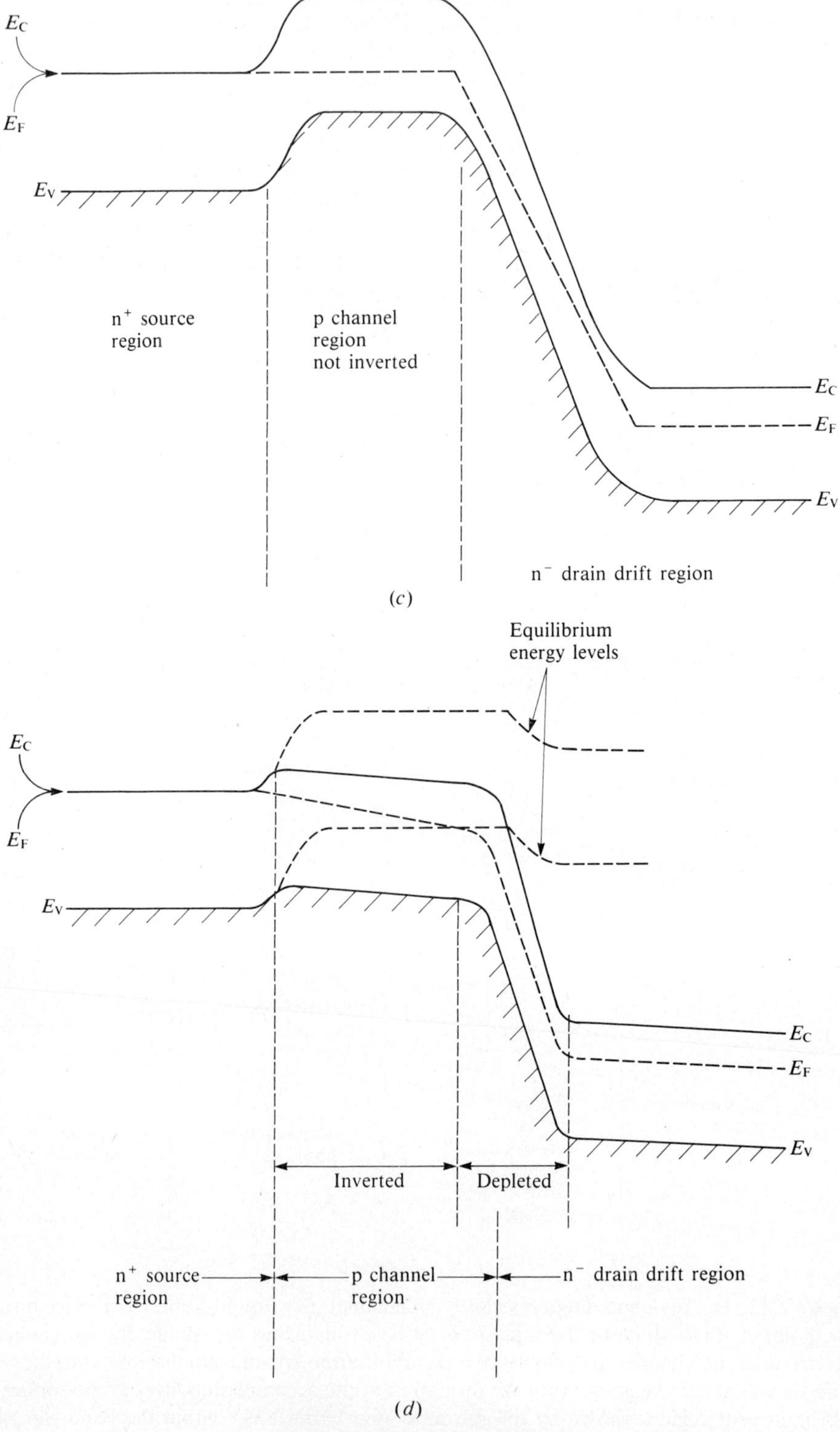

Figure 2.13. (*Continued*)

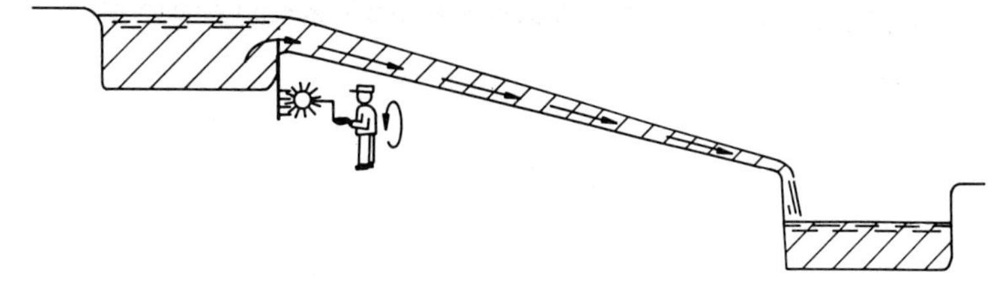

Figure 2.14. A hydraulic analogy to the operation of electronic amplifying devices.

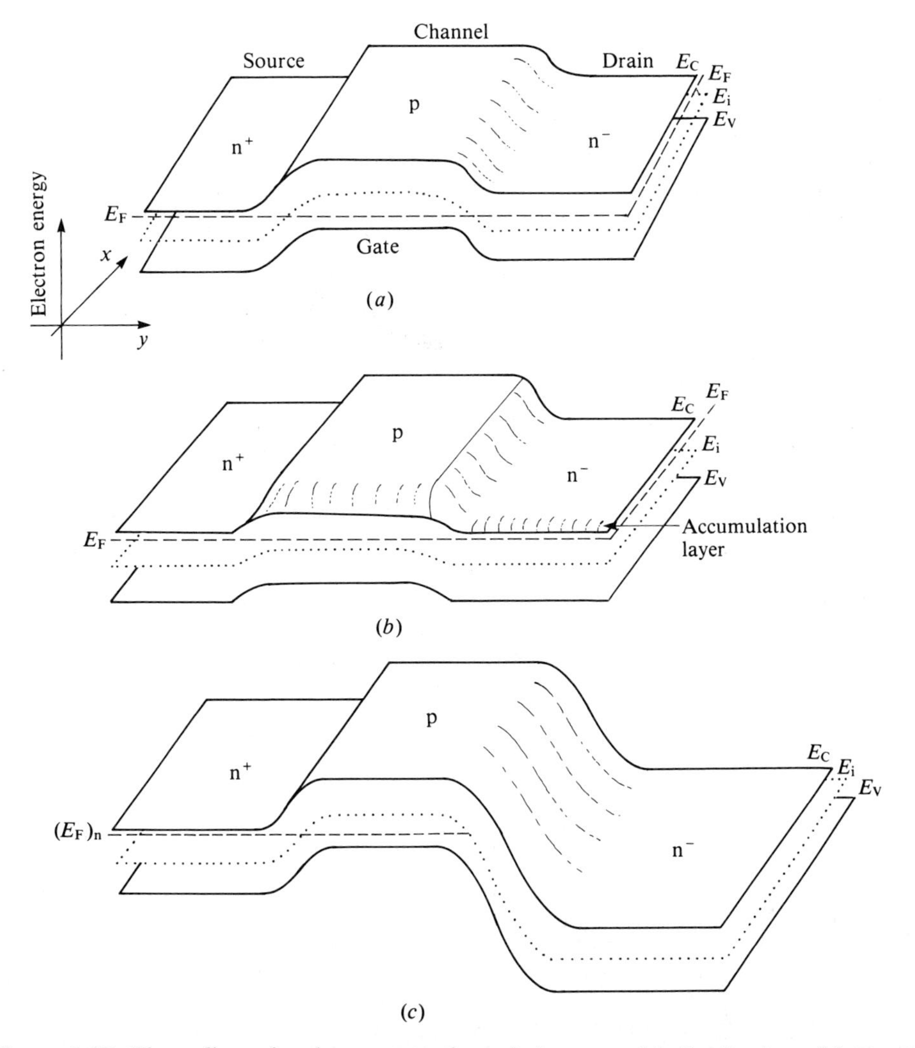

Figure 2.15. Three-dimensional representation of the potential distribution. (*a*) Equilibrium; (*b*) ON (*A*); (*c*) OFF (*B*); (*d*) point *C* in the constant-current region.

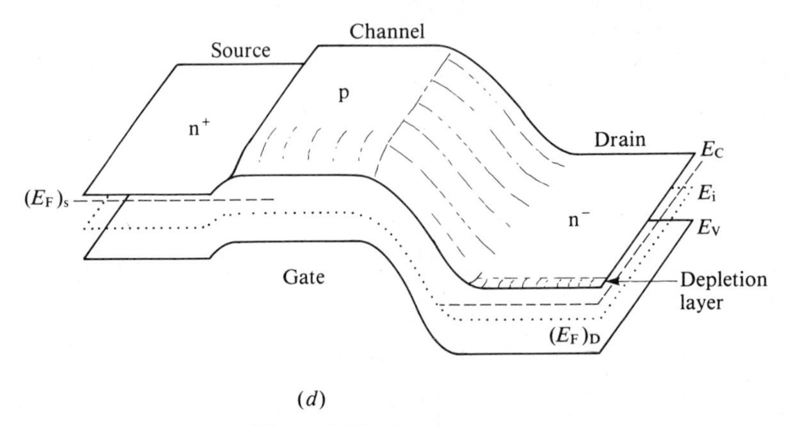

(*d*)

Figure 2.15. (*Continued*)

Figure 2.13*d* represents point *C* on the characteristics. At the source end of the channel, $V_{GB} > V_T$ and an inversion layer forms. However, the voltage drop along the channel is such that at the drain end $V_{GB} < V_T$, so the surface region is depleted, but not inverted. The lateral component of the depletion layer electric field picks up the electrons and sweeps them into the depleted epilayer surface region. They are then accelerated on through the depleted drain drift region to the drain. Changes in V_{DS} merely change the voltage dropped across the depletion layer, and have only a marginal effect on the conduction in the channel in so far as they modulate the length of the inversion layer.

Figure 2.15 is an attempt at a three-dimensional representation of the potential distribution in the bulk semiconductor between source and drain under the different bias conditions shown in Figure 2.13. It is important to recognize that when source, channel, and drain are at the same potential, equilibrium obtains

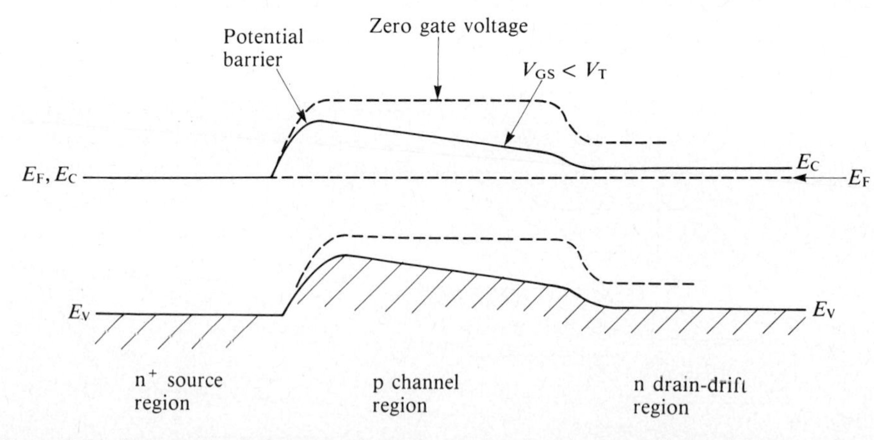

Figure 2.16. Energy-level diagram along the channel with nonuniform doping profile. The built-in potential in the channel results from the doping profile. The potential barrier prevents electrons leaving the source unless $V_{GS} > V_T$, when the channel becomes inverted at the source end. At the drain end, a small applied gate voltage may cause V_{GS} to exceed the local threshold potential, but there is no supply of electrons to maintain equilibrium and form an inversion layer. This is the situation illustrated.

throughout this region and the Fermi energy is at a uniform level. With drain bias applied this is no longer true. As a result, applying a positive drain bias voltage reverse-biases the drain-channel pn junction and sets up a depletion layer in the usual way. This is shown particularly in Figures 2.13*d* and 2.15*d*, where the surface layer in the channel remains inverted only in the regions outside the influence of the drain potential. Even where inversion-layer charge is present in the channel, it is not in thermodynamic equilibrium with the source electrons at points beyond the source potential barrier.

One of the complications caused by the double-diffusion technique is that it produces the impurity concentration gradient along the channel shown in Figure 2.2. This means that there is a built-in potential in the semiconductor and that the local threshold voltage decreases with increasing distance from the source. It is the region of highest doping, next to the source, that determines the actual device threshold voltage, V_T. Until this region has inverted, the device will not turn ON, as electrons from the source cannot surmount the potential barrier. This is shown in Figure 2.16. Even though points further along the channel may have exceeded the local value of threshold voltage, an inversion layer does not form, because there is no ready supply of electrons to fill it. The few electrons generated thermally are accelerated away to the drain.

SUMMARY

For purposes of discussion we consider the active region of the VDMOS FET to be made up of four parts:

Ⓐ, the surface layer of the p diffusion;
Ⓑ, the surface layer of the n epilayer;
Ⓒ, the n regions under the gate oxide, between the p diffusions;
Ⓓ, the main part of the n epilayer, below the level of the p diffusions.

Application of a positive gate voltage greater than a certain threshold value, V_T, gives rise to strong inversion in region Ⓐ, and leads to the formation of a conducting channel between the source and the drain. The theoretical value of the threshold voltage is determined by the doping concentration at the source end of region Ⓐ, by the thickness of the gate oxide, and by the presence of any interface charge at the silicon surface. At room temperature, typical values are in the region of 3.5 V, decreasing at about 6 mV/K as the temperature rises. The threshold voltage normally defined on device data sheets is somewhat lower than this theoretical value. The application of gate voltages above V_T causes a build-up of free electrons in the inversion layer in region Ⓐ, in direct proportion to the excess of the voltage above V_T.

Three device operating regimes can be usefully identified. They are:

A, the conducting, or turned-ON, state;
B, the blocking, or turned-OFF, state;
C, the constant-current condition.

These are shown in Figure 2.12. A physical understanding of the device characteristics can be obtained by considering the electron energy-band distribution along the channel between the source and drain. The flow of current into the channel is essentially controlled by the height of the potential barrier faced by the electrons in the source. Analogy may be made with the fluid model shown in Figure 2.14.

REFERENCES

1. W. L. Engl, H. K. Dirks, and B. Meinerzhagen, "Device modeling." *Proc. IEEE* **71,** 10–33 (1983).
2. S. C. Hu, H. C. Tseng, J. S. Ni, E. A. Wolsheimer, and R. W. Dutton, "Computer-aided design of semiconductor processes and devices." *Philips J Res.* **42,** 533–565 (1987).
3. A. F. Franz and G. A. Franz, "BAMBI—A design model for power MOSFET's." *IEEE Trans. Comput.-Aided Des.* **CAD-4,** 177–189 (1985).
4. S. M. Sze, *Physics of Semiconductor Devices,* 2nd ed. Wiley, New York, 1981. (The standard reference work on the subject.)
5. R. S. Muller and T. I. Kamins, *Device Electronics for Integrated Circuits,* 2nd ed. Wiley, New York, 1986. (A clear presentation of the principles of semiconductor devices, including MOSFETS, designed for VLSI.)
6. B. J. Baliga, *Modern Power Devices.* Wiley, New York, 1987. (A modern and comprehensive review of power semiconductor devices.)
7. M. D. Pocha, A. G. Gonzalez, and R. W. Dutton, "Threshold-voltage controllability in double-diffused-MOS transistrors." *IEEE Trans. Electron Devices* **ED-21,** 778–784 (1974).
8. E. O. Johnson, "The insulated-gate field-effect transistor—a bipolar transistor in disguise." *RCA Rev.* **34,** 80–94 (1973).

CHAPTER 3

Principles of Operation 2: Static Characteristics

3.1. INTRODUCTION

Figures 1.4 and 2.12 are schematic illustrations of typical MOSFET output characteristics. These may be compared with the sets of characteristics of an actual device shown in Figure 3.1. The purpose of the present chapter is to explain these characteristics in terms of the physical properties of the devices. To aid this discussion we divide the total measured drain–source voltage at any instant, V_{DS}, into four separate components. These are:

1. The voltage resulting from the parasitic ohmic resistance of the following parts: the bulk semiconductor regions of the source and substrate; the contacts and the die bonds; the leads and the package pins. We call the total resistance of these parts R_P. The corresponding voltage, $I_D R_P$, is proportional to the drain current.
2. The voltage resulting from the ohmic resistance of the drain drift region in the n^- epitaxial layer. We call this resistance R_{EPI}. A detailed discussion of the steps that may be taken to minimize its value is given in Section 3.5. For convenience we may combine these two ohmic voltages by putting

$$V_R = I_D(R_P + R_{EPI}) \tag{3.1}$$

3. The voltage dropped along the channel inversion layer, V_{Ch}. This is a function of V_{GS} as well as I_D and lies at the heart of the device operation. In particular, V_{Ch} becomes small for large values of V_{GS}. A detailed discussion is given in Section 3.2. We may, if we wish, refer to a "channel resistance", $R_{Ch} = V_{Ch}/I_D$. It is then important to recognise that R_{Ch} is highly nonlinear and depends on both V_{GS} and V_{DS} (or I_D).
4. The voltage V_{SCL}, dropped across the space-charge layer (the depletion layer) that may form around the junction between the channel and the drain drift region. This becomes particularly significant when the transistor is biased into the constant-current part of the characteristics. Note that V_{SCL} may absorb some of the voltage dropped across R_{EPI}.

Provided that we make allowance for any possible interaction between V_{SCL} and

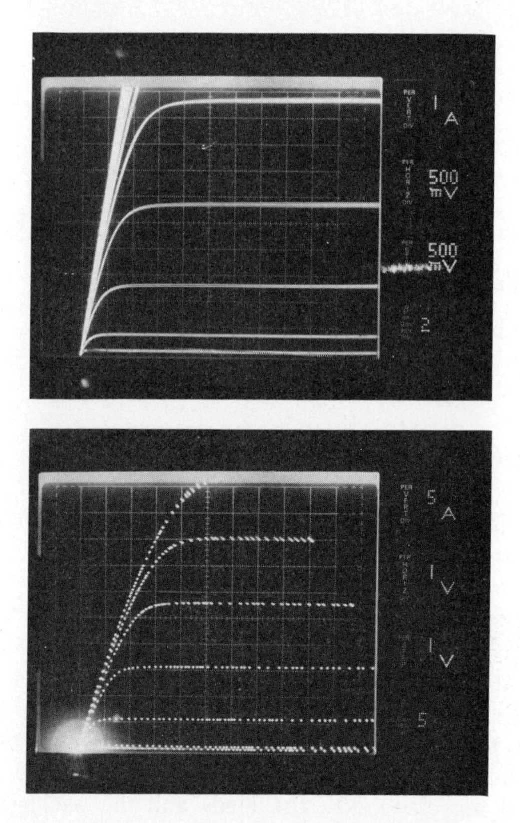

Figure 3.1. Characteristics of drain current versus drain–source voltage for an IRF 540. This is a 100-V, 27-A device with a typical $R_{DS(on)}$ of 0.06 Ω. In the upper set of characteristics, the vertical and horizontal scales are 1 A and 0.5 V per division, and the V_{GS} increments are 0.5 V. In the lower set, the scales are 5 A and 1 V per division, and the increments are 1 V.

R_{EPI}, we may thus write

$$\begin{aligned} V_{DS} &= I_D(R_P + R_{EPI}) + V_{Ch} + V_{SCL} \\ &= V_R + V_{Ch} + V_{SCL} \end{aligned} \tag{3.2}$$

In Figure 3.2, these components of V_{DS} are identified on a typical device characteristic.

The parameter $R_{DS(on)}$ is a most important guide to the efficiency of a power switching transistor. It is simply the ratio V_{DS}/I_D when the device is turned fully ON, as at point A on the characteristics of Figure 3.2. Under these conditions $V_{SCL} = 0$, and, provided that the gate drive is adequate, V_{Ch} is small too. Thus, $R_{DS(on)}$ reduces to the sum $R_P + R_{EPI}$ and so becomes independent of I_D. The curves of Figure 3.1 show this and enable an estimate of $R_P + R_{EPI}$ to be made. This can be seen to be consistent with the value quoted as "typical" in the data sheet, that is, 0.06 Ω. The value of $R_{DS(on)}$ increases rapidly with the required device voltage rating for reasons that are discussed in detail in Section 3.5. It

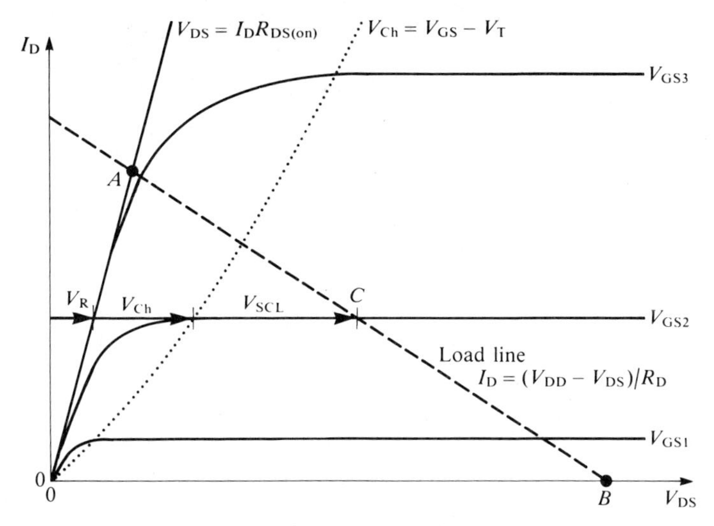

Figure 3.2. An I_D-versus-V_{DS} characteristic showing the various contributions to V_{DS} at an operating point in the constant-current region.

varies inversely with chip area and hence with current rating. Typical values are:

BV_{DSS} (V)	$I_{D(on)}$ (A)	$R_{DS(on)}$ (Ω)
1000	1	10
500	1	8
500	8	0.8
100	8	0.3
100	32	0.06
50	30	0.04
50	51	0.028
50	70	0.018

For reasons discussed in Section 3.6, $R_{DS(on)}$ increases at high current levels.

3.2. CONTROL OF THE DRAIN CURRENT

In this section we deal with the relationship between the drain current I_D, the channel voltage V_{Ch}, and the gate voltage V_{GS} when the gate is biased above threshold. A simple charge-control model is used. This has the merit of giving a clear insight into the basic physical processes which control device operation, something that can easily become obscured in the analytical complexities of more complete theories [1, 2]. The characteristics derived in the simple theory approximate quite well to those observed experimentally, provided that some adjustment is made to the device parameters. They are often quite adequate for device modeling.

When examined in detail, the simple theory reveals some internal inconsistencies. A fuller theory is presented in Appendix 4. This helps to clarify some of the more detailed physical processes involved in the device behavior, but it has to be said that neither theory enables the full range of device characteristics to be predicted accurately from a known structure. Transistor action in a MOSFET is just too complicated for this. Nothing less than a full numerical analysis will do, and even though device-modeling packages are becoming increasingly sophisticated (see Ref. 1 of Chapter 2), this still remains a formidable problem.

We need to consider the distributions of the electric field and the carrier concentration along the channel. For this purpose, we define the set of cartesian coordinates illustrated in Figure 3.3*a*, in which the x-direction is normal to the semiconductor surface, the y-direction lies along the channel from source to drain, and the z-direction is that of the channel width.

Consider first the case illustrated in Figure 3.3*b*, where the gate is biased well above threshold, so that a full inversion layer is formed, but where, also, the drain current and hence the channel voltage are small: $V_{Ch} \approx 0$. Then, the free-electron charge density Q_{IL} is approximately uniform along the whole length of the channel:

$$Q_{IL} = C_{ox}(V_{GS} - V_T) \tag{3.3}$$

with V_T given by Equation (2.10).

The small electric field along the channel, $E_y = V_{Ch}/l$, gives rise to a carrier drift velocity:

$$v_d = \mu_e E_y = \mu_e V_{Ch}/l \tag{3.4}$$

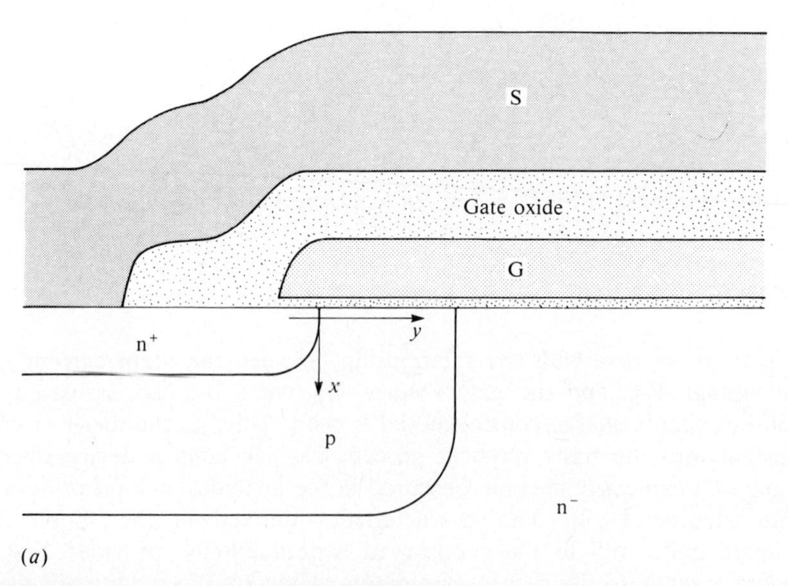

Figure 3.3. The inversion layer. (*a*) Coordinate system; (*b*) very low drain currents; (*c*) before pinchoff; (*d*) after pinchoff.

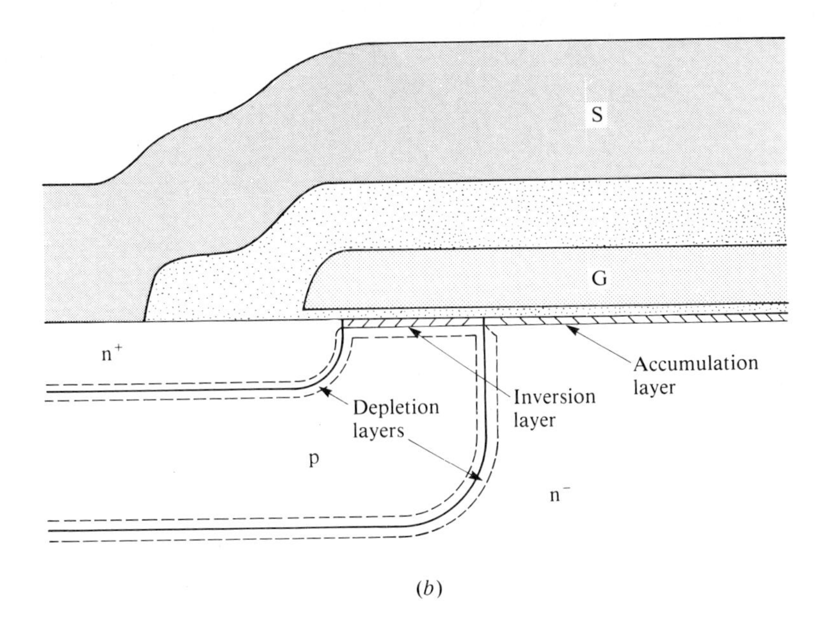

(*b*)

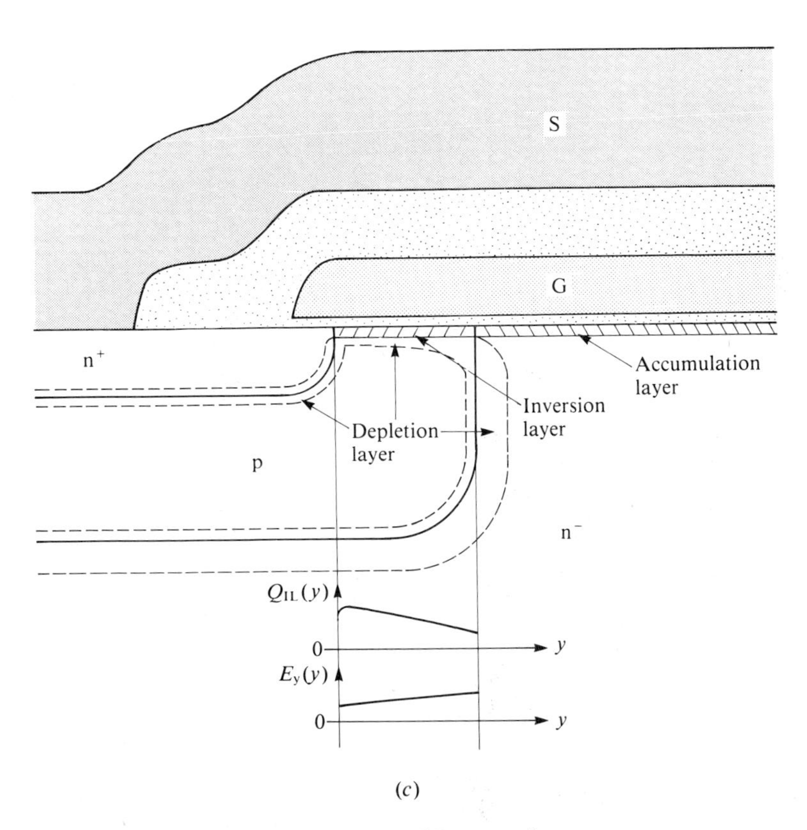

(*c*)

Figure 3.3. (*Continued*)

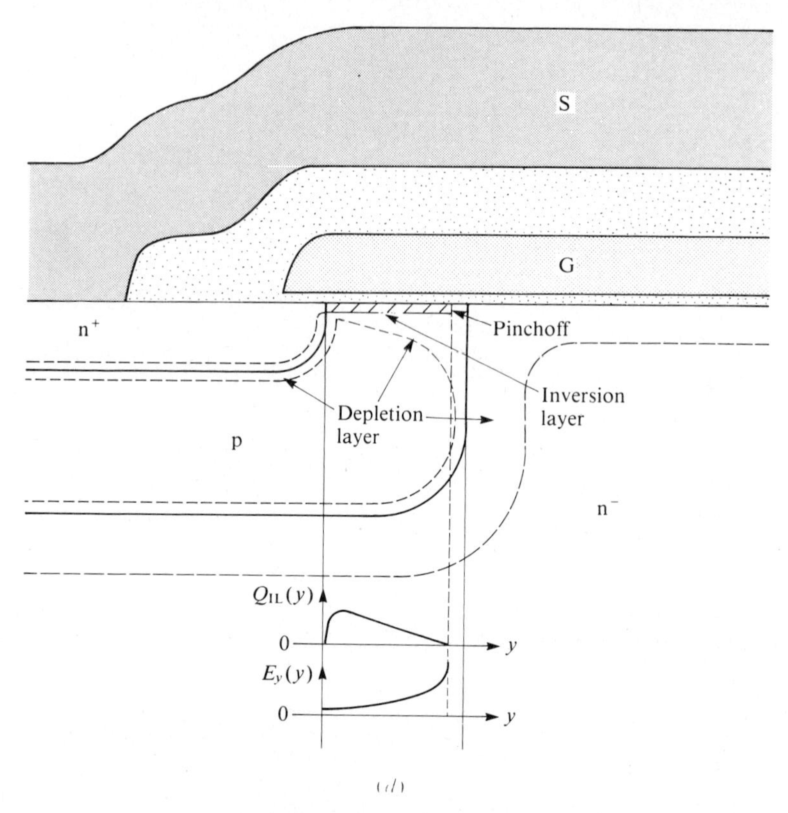

Figure 3.3. (*Continued*)

Here, μ_e is the surface mobility of the electrons. This is lower than the mobility of electrons in the bulk material. Like the bulk mobility it decreases with increasing temperature and varies with crystal orientation. In addition, however, it decreases with increasing normal field E_x. This is not surprising, since E_x forces the electrons into frequent collisions with the surface of the semiconductor. As a typical value, we shall take $\mu_e = 0.05\ \mathrm{m^2/V\,s}$ ($500\ \mathrm{cm^2/V\,s}$).

The total free-electron charge in the channel is $q_e = Q_{IL}wl$, where w is the channel width. The average electron transit time along the channel from source to drain is

$$t_{tr} = \frac{l}{v_d} = \frac{l^2}{\mu_e V_{Ch}} \tag{3.5}$$

The current of free electrons through the channel is thus

$$\begin{aligned} I_D &= \frac{q_e}{t_{tr}} = \frac{Q_{IL}wl}{t_{tr}} \\ &= \mu_e \frac{w}{l} C_{ox}(V_{GS} - V_T)V_{Ch} \end{aligned} \tag{3.6}$$

Note that under these conditions, in which V_{Ch} and I_D are small, the channel behavior is ohmic and $R_{Ch} = dV_{Ch}/dI_D \propto (V_{GS} - V_T)^{-1}$.

At higher values of V_{Ch} and I_D the free-electron surface charge density Q_{IL} and the electric field E_y both vary along the channel, as shown in Figure 3.3*c*. At $y = 0$, the source end, Q_{IL} is unchanged:

$$Q_{IL}(0) = C_{ox}(V_{GS} - V_T) \tag{3.7}$$

But at the drain end, it is reduced as a consequence of the voltage dropped along the channel:

$$Q_{IL}(l) = C_{ox}(V_{GS} - V_T - V_{Ch}) \tag{3.8}$$

If we take $C_{ox}(V_{GS} - V_T - \frac{1}{2}V_{Ch})$ to be the average surface charge density, the total free-electron charge in the channel is reduced to

$$q_e = C_{ox} wl(V_{GS} - V_T - \tfrac{1}{2}V_{Ch}) \tag{3.9}$$

The average drift velocity remains $v_d = \mu_e V_{Ch}/l$, so the transit time is still given by Equation (3.5). The drain current is thus

$$\begin{aligned} I_D = \frac{q_e}{t_{tr}} &= \mu_e \frac{w}{l} C_{ox}[(V_{GS} - V_T)V_{Ch} - \tfrac{1}{2}V_{Ch}^2] \\ &= k[(V_{GS} - V_T)V_{Ch} - \tfrac{1}{2}V_{Ch}^2] \end{aligned} \tag{3.10}$$

The coefficient $k = \mu_e(w/l)C_{ox}$, is often referred to as the *device transconductance parameter.*

When V_{Ch} is increased to the value $V_{GS} - V_T$, $Q_{IL}(l)$ becomes zero. The channel then *pinches off,* and the inversion layer no longer continues through to the drain end. There a depletion layer forms. The situation is illustrated schematically in Figure 3.3*d*. Further increases in V_{DS} do not change V_{Ch} or I_D but merely serve to expand the depletion layer at the pn junction. The increased voltage appears as the voltage V_{SCL} shown in Figure 3.2, which is dropped across this depletion layer. We discussed in Appendix 1 how heavier doping of the p region than the n^- region minimizes any reduction of the channel length caused by the expansion of the depletion layer. As a result, the conditions in the channel are little affected by increases in V_{DS}. The drain current and channel voltage remain sensibly constant. Substituting $V_{Ch} = V_{GS} - V_T$ into Equation (3.10) gives

$$\begin{aligned} I_D = I_{D(Sat)} &= \tfrac{1}{2}\mu_e \frac{w}{l} C_{ox}(V_{GS} - V_T)^2 \\ &= \tfrac{1}{2}k(V_{GS} - V_T)^2 \end{aligned} \tag{3.11}$$

The characteristics shown in Figure 3.1 make it clear that for VDMOS FETS, $I_{D(Sat)}$ remains quite independent of V_{DS} beyond pinch-off. In other types of MOSFET, where a small increase in $I_{D(Sat)}$ does occur, this can best be represented by

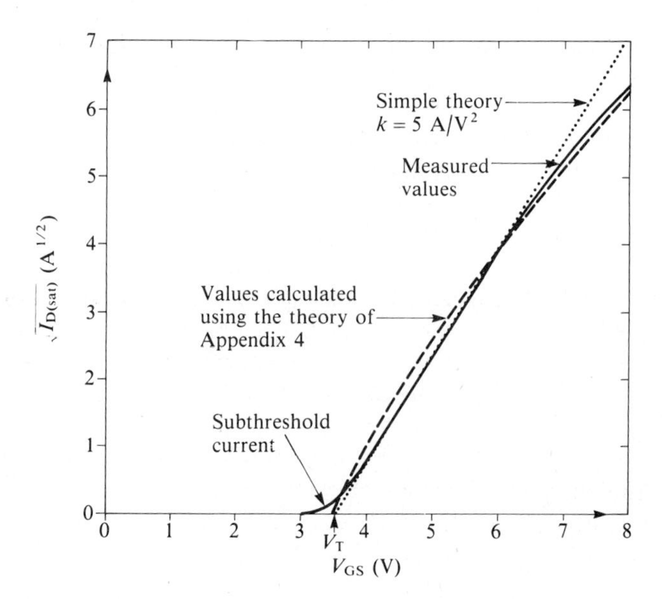

Figure 3.4. A plot of the square root of the saturated drain current versus the gate–source voltage. The data are taken from characteristics like those of Figure 3.1. They show that the behavior of this IRF 540 sample can be represented by $V_T = 3.5$ V, and $k = 5$ A/V^2 for a range of drain currents between 0.2 and 16 A. Deviations occur at higher and lower currents.

introducing a *channel-length modulation parameter,* λ. Then

$$I_D = I_{D(Sat)}(1 + \lambda V_{DS}) \tag{3.12}$$

where $I_{D(Sat)}$ is given by Equation (3.11). For VDMOS devices $\lambda \approx 0$.

In Figure 3.4 a plot of $\sqrt{I_{D(Sat)}}$ versus V_{GS}, taken in part from the curves of Figure 3.1, demonstrates the applicability of Equation (3.11) over a limited range of values. The measured value of k for this device, the IRF 540, is 5 A/V^2. If we assume that $t_{ox} = 100$ nm, so that $C_{ox} = 3.45 \times 10^{-4}$ F/m^2, that the gate width is about 1 m, and that the electron mobility is 0.05 m^2/V s, this value of k implies a channel length in the region of 3.5 μm. This is several times longer than we would expect. Furthermore, the characteristics shown in Figure 3.1 are much steeper near the origin than Equation (3.10) predicts, and they approach the constant-current condition more sharply.

Differentiation of Equation (3.10) with respect to V_{GS} shows that in the linear region, the transconductance $\partial I_D / \partial V_{GS}$ (with V_{Ch} kept constant) should be kV_{Ch}. Differentiating with respect to V_{Ch} (with V_{GS} kept constant) shows that the slope of the I_D versus V_{Ch} characteristics at the origin should be $k(V_{GS} - V_T)$. When allowance is made for $R_{DS(on)}$, the curves of Figure 3.1 indicate a value of k, on this basis, of approximately 24 A/V^2. This implies that the effective channel length is less than 1 μm, which is rather shorter than we would expect.

These discrepancies are not surprising, because there are several assumptions in the simple theory that do not bear close examination. These are discussed in Appendix 4. There it is shown that a physically more self-consistent theory predicts values for I_D and $I_{D(Sat)}$ that fit the measured device characteristics over a much wider range of values of V_{GS} and V_{DS}. We return to these results shortly, but first we consider what happens when higher gate voltages are applied.

The predicted square-law relationship between $I_{D(Sat)}$ and $V_{GS} - V_T$ comes about because first the total quantity of free-electron charge in the pinched-off channel is proportional to $V_{GS} - V_T$, and secondly the average longitudinal electric field in the channel is proporitional to $V_{Ch} = V_{GS} - V_T$ and thus so too is the carrier drift velocity. With the very short channels used in VDMOS FETS this is something that changes at higher values of $V_{GS} - V_T$ and $I_{D(Sat)}$. The electric field along the channel, whose average value is V_{Ch}/l, becomes so large that it causes the electron drift velocity to saturate. This phenomenon is illustrated in Figure 3.5. Only for electric-field strengths less than about 5×10^5 V/m (0.5 V/μm) is there a well-defined, constant electron mobility ($v_d \propto E$). At higher field strengths the drift velocity increases more slowly until, for fields exceeding about 5×10^6 V/m (5 V/μm), v_d becomes constant at the saturated drift velocity, $v_s \approx 9.2 \times 10^4$ m/s. Unlike the electron mobility, v_s does not appear to depend on the normal electric field E_x at the semiconductor surface [3], but it is expected to decrease with increasing temperature.

For $V_{GS} - V_T > E_s l$, where $E_s = v_s/\mu_e$, the carrier drift velocity approaches v_s over an appreciable proportion of the length of the channel before pinchoff occurs. Remember that pinchoff is defined as the point when $Q_{IL}(l) = 0$. The transit time then approaches the limiting value

$$t_{tr} = l/v_s \tag{3.13}$$

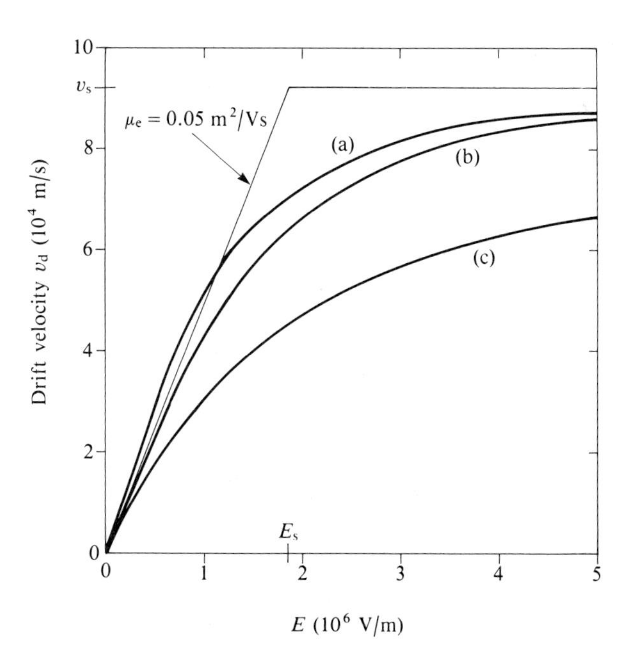

Figure 3.5. Saturation of the carrier drift velocity. Curve (a) shows experimental data based on the results of Ref. 3. Curves (b) and (c) are plots of Equation (A4.22) with $\mu_e = 0.05$ m²/V s, $v_s = 92{,}300$ m/s, and $\alpha = 2$ and 1, respectively.

At pinchoff, $V_{Ch} = V_{GS} - V_T$, so that combining Equations (3.9) and (3.13) gives

$$I_{D(Sat)} = q_e/t_{tr} = \tfrac{1}{2}C_{ox}w(V_{GS} - V_T)v_s \qquad (3.14)$$

At the higher levels of drain current, these effects cause a more even spacing of the I_D-versus-V_{DS} characteristics. This can be seen in the departure from the straight-line relationship in Figure 3.4 at high current levels. When these data are extended and plotted as $I_{D(Sat)}$ versus V_{GS}, the linear relationship expected under velocity saturation conditions is obtained. This is shown in Figure 3.6. Equation (3.14) predicts the gradient of the curve to be $g_{fs} = \tfrac{1}{2}C_{ox}wv_s$. Putting $v_s = 9.2 \times 10^4$ m/s, and using the values suggested previously as appropriate for the IRF 540 (namely $C_{ox} = 3.45 \times 10^{-4}$ F/m^2 and $w = 1$ m), we should expect this slope to be about 16 S. It can be seen in Figure 3.6 that the measured value is 12 S. Thus, once again, the simple theory predicts the general form of the device characteristics quite well, but not the transition to velocity saturation, nor the absolute value of g_{fs} once this is established.

Taking $l \approx 1$ μm, we expect velocity saturation effects to become apparent over the range $0.5\text{ V} < V_{GS} - V_T < 5\text{ V}$. In fact they are seen in Figure 3.6 to be fully established at a gate voltage of 6 V and a drain current of 16 A. This corresponds to $V_{GS} - V_T = 2.5$ V, and the current is well within the maximum continuous current rating of the IRF 540, which is 27 A.

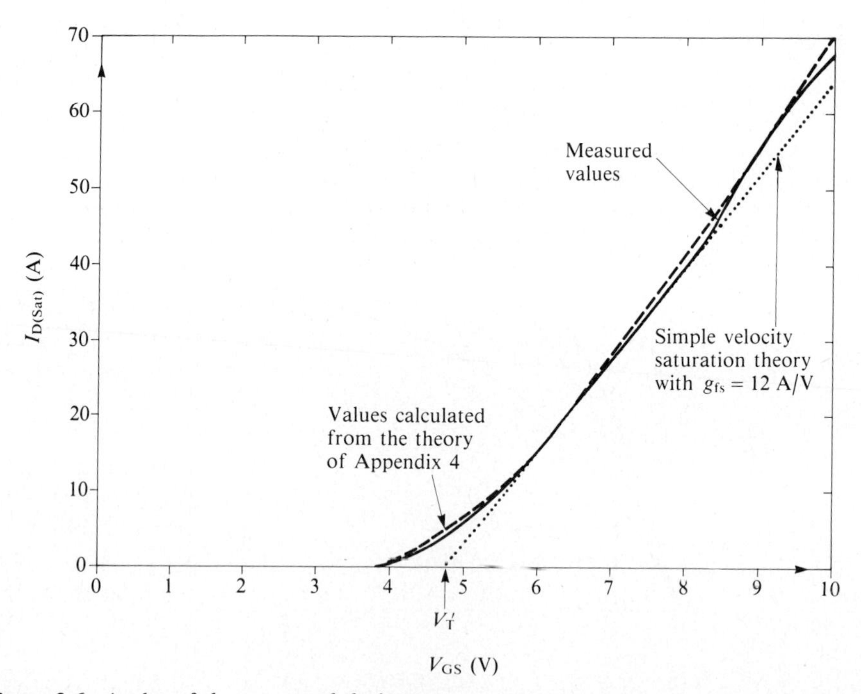

Figure 3.6. A plot of the saturated drain current versus the gate–source voltage. The data are taken from characteristics like those of Figure 3.1. They show that the behavior of this IRF 540 sample can be represented by $g_{fs} = 12$ S and $V'_T = 4.7$ V over the current range 10 to 45 A. Deviations occur at higher and lower currents.

It should be noted that the effective value of the threshold voltage in the velocity saturation region, which we call V'_T, is higher than the effective value, V_T, that applies when the square law [Equation (3.11)] relates I_D and V_{GS}. This can be seen by projecting back the linear parts of the graphs in Figures 3.4 and 3.6 to $I_{D(SAT)} = 0$. This type of behavior would be expected in a more detailed theory, because of the gradual onset of velocity saturation. For purposes of device modelling, V'_T should be used in Equation (3.14), rather than V_T.

Velocity saturation can be a valuable feature of power MOSFETS, even though it causes the drain current to become limited at much lower levels than would otherwise occur. It gives rise to a linear transfer characteristic (g_{fs} = constant) that is beneficial in linear-amplifier applications.

It has become customary to refer to velocity saturation in power MOS devices as a "short-channel" effect. This is unfortunate because to a VLSI engineer, who is designing small MOSFET devices for very large-scale integration, the *short-channel effect* is something quite different and wholly bad. When channel lengths are reduced in the quest for smaller devices, basic device characteristics can be preserved, as long as all linear dimensions and all voltages are scaled in direct proportion and doping densities are varied in inverse proportion [4]. If this is not done, V_{DS} may modulate the channel length, making λ large. It may even cause variations in the height of the potential barrier at the source end of the channel, thereby preventing effective cutoff at the threshold voltage. Clearly, velocity saturation gives rise to a very different kind of short-channel effect in power MOS devices.

When devices are operated at high current levels, two effects may act to reduce the electron mobility and hence the drain current. This may occur either in normal use, or when device characteristics like those of Figure 3.1 are being measured. The first is simply the rise in junction temperature that results from the dissipation. Unless very short pulses at low duty cycle are used, this can cause an apparent negative differential resistance, with the drain current falling as V_{DS} is increased. The second is the increased electric field normal to the surface, E_x, that results from the higher values of V_{GS} required. This is expected to reduce μ_e, but not v_s. The average value of E_x in the inversion layer may be taken to be $(Q_D + \frac{1}{2}Q_{IL})/\epsilon_0\epsilon_{Si}$. It thus varies from about $15\ V/\mu m$ at threshold to over $40\ V/\mu m$ when $V_{GS} = 20\ V$. For device modeling the variation of μ_e is usually approximated by putting $\mu_e = \mu_{e0}/[1 + \theta(V_{GS} - V_T)]$. Values of μ_{e0} and θ may be expected to depend on the doping level and the surface conditions. The values normally used for a typical power VDMOS FET are $0.05\ m^2/V\,s$ and $0.1\ V^{-1}$, respectively. Such a large value for θ includes effects associated with higher gate voltages in addition to the direct effect on the mobility through the normal field E_x.

The more complete theory for the drain-current characteristics presented in Appendix 4 takes into account several effects neglected in the simple theory. These include:

1. the extra charge contained in the depletion layer at the drain end of the channel because of the higher voltage dropped across it;
2. the reduced electron mobility and eventual velocity saturation at high longitudinal field strengths E_y.

Neither of these alters the predicted drain current at very low levels. Both cause I_D to increase more slowly as V_{Ch} rises and the constant-current condition is approached. It is this that gives rise to the discrepancy between the simple theory of Equations (3.10) and (3.11) and the device characteristics.

There remains one more important effect that is neglected in the theory. That is the gradual reduction in the doping concentration of the channel towards the drain end. This leads to a reduction in the charge density in the depletion layer (Q_D reduced), and hence to an increase in the inversion-layer charge (Q_{IL}). As a result, the longitudinal electric field needed to maintain any given drain current is reduced, and the saturation drain current $I_{D(Sat)}$ is increased. It is this that accounts for the unexpectedly low effective channel lengths implied by the characteristics.

The variation of v_d with E_y assumed in the theory of Appendix 4 takes the

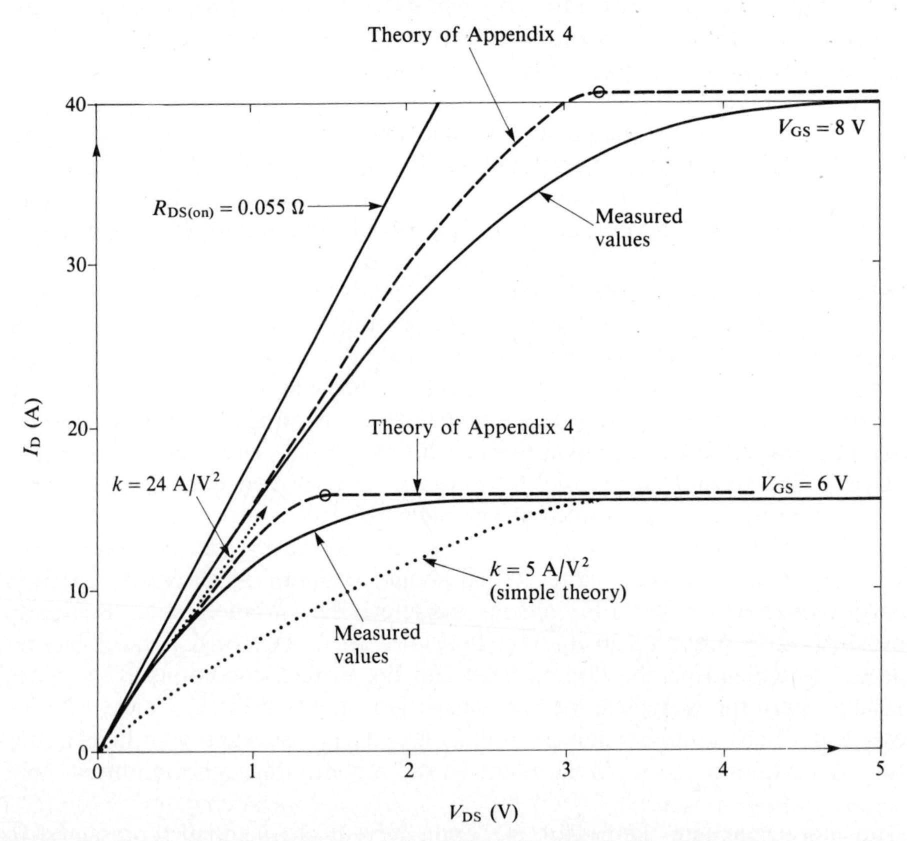

Figure 3.7. Comparison of theoretical and experimental characteristics. Full curves are taken from the experimental characteristics for the IRF 540 shown in Figure 3.1. The dotted curves are plots of Equation (3.10) for $k = 5$ and 24 A/V^2. The lower value fits the constant current region with $V_{GS} = 6$ V; the higher value fits the observed gradient at the origin. The dashed curves are plots of Equation (A4.34), and represent the results of the more sophisticated theory set out in Appendix 4. It can be seen that this theory predicts a sharper approach to the constant-current condition than occurs in practice. In each case, in calculating V_{DS}, a value of 0.055 Ω has been assumed for $R_{DS(on)}$.

form $v_d = \mu_e E_y/(1 + E_y/E_s)$, with μ_e being a function of the gate–source voltage, as has just been mentioned. Then $E_s = v_s/\mu_e$, with $v_s = 9.23 \times 10^4$ m/s. The effect on the theoretical characteristics is simply to increase the effective channel length by V_{Ch}/E_s. Thus k decreases as V_{Ch} and I_D increase. It can be seen in Figure 3.5 that this assumed form for $v_d(E_y)$ underestimates the true drift velocity at any given field.

Results obtained using the theory of Appendix 4 are included in Figures 3.4 and 3.6. The device parameters used previously are assumed: $n_A = 1 \times 10^{23}\,\text{m}^{-3}$, $n_D = 1 \times 10^{26}\,\text{m}^{-3}$, $t_{ox} = 100$ nm, $Q_{SS} = 4 \times 10^{-4}\,\text{C/m}^2$, $w = 1$ m. The effective channel length is taken to be $l = 0.7\,\mu$m, and the mobility parameters used are $\mu_{e0} = 0.06\,\text{m}^2/\text{V s}$ and $\theta = 0.02\,\text{V}^{-1}$.

Equation (3.10) predicts a parabolic variation of I_D with V_{Ch}, with the drain current remaining constant once it reaches its maximum value. In Figure 3.7, the results of this simple theory and the theory developed in Appendix 4 are each compared with two of the experimental characteristics taken from Figure 3.1. In estimating V_{DS}, a value of 0.055 Ω has been assumed for $R_{DS(on)}$. This is a typical value for the IRF 540. It is clear that neither theory provides a close fit to the experimental curves. The simple theory, with a value of k that matches the observed constant current, predicts an initial rise of drain current that is much too low, and a channel length that is much too long. The full theory approaches constant current rather too abruptly, but makes no arbitrary assumptions. All the input parameters can be related in some way to the construction of the device.

3.3. THE SUBTHRESHOLD REGION

Figure 3.4 shows that $I_{D(Sat)}$ departs from the square-law relationship with $V_{GS} - V_T$ at low as well as at high currents. It is clear that $I_{D(sat)}$ does not fall to zero as V_{GS} approaches the threshold voltage in the way predicted by Equation (3.11). The threshold is in fact quite "soft", and a significant current flows even when V_{GS} is less than V_T. This is called the *subthreshold* current. A log–linear plot of $I_{D(Sat)}$ versus V_{GS} shows that in the subthreshold region the drain current varies exponentially with V_{GS}. This can be seen in Figure 3.8, which shows the results of measurements made on an IRF 540. The exponential variation is maintained over more than five decades.

In Section 2.5 we discuss how the drain current in the region of threshold is governed by the height of the potential barrier facing the electrons in the source. This is illustrated in Figure 2.13. It is therefore only to be expected that the transition from the blocking to the conducting state should occur over a range of values of V_{GS}, the drain current varying exponentially with the barrier height. In Appendix 3, Figure A3.2 shows that the inversion-layer surface charge density Q_{IL} does indeed vary exponentially with the surface potential, right through the threshold region.

Because of the very wide gate used in power MOSFETS, the subthreshold current is considerable. On all power-MOS data sheets the threshold voltage is specified at a saturated drain current (with $V_{GS} = V_{DS}$) of 0.25 or 1 mA. This puts it well into the subthreshold region for most devices. Threshold voltages defined in this way

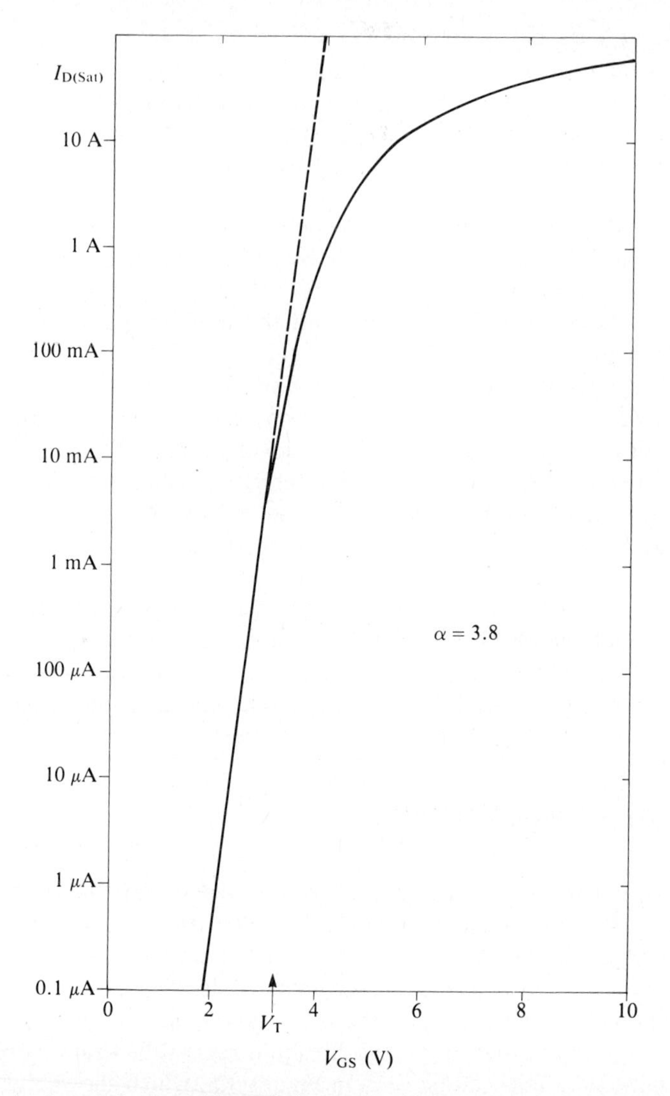

Figure 3.8. A plot of the logarithm of the saturated drain current versus the gate voltage. The data are derived from measured IRF 540 characteristics, including those of Figure 3.1. They demonstrate the exponential variation of the subthreshold current with the gate–source voltage, with $\alpha = 3.8$.

are thus lower than the theoretical values that we define in Chapter 2 and those that are measured from graphs such as Figure 3.4.

It is interesting to estimate the expected saturation drain current, $I_{D(Sat)}$, at threshold. In Appendix 3, it is shown that at threshold, $Q_{IL} \approx 0.1 Q_0$, where $Q_0 = (n_A kT \epsilon_0 \epsilon_{Si})^{1/2}$. Following the discussion of the last section, let us assume that the average inversion-layer charge density along the channel at pinchoff is half this value. Pinchoff occurs at a channel voltage which is sufficient to reduce Q_{IL} to zero at the drain end of the channel. This requires $V_{Ch} = Q_{IL}/C_{ox} =$

$0.1Q_0/C_{ox}$. Thus, at threshold,

$$I_{D(Sat)} = \frac{q_e}{t_{tr}} = \tfrac{1}{2} \times 0.1Q_0 \frac{w}{l} \mu_e V_{Ch} = \tfrac{1}{2}\mu_e \frac{w}{l} \frac{(0.1Q_0)^2}{C_{ox}} \tag{3.15}$$

As an example, assume parameter values based on the IRF 540, for comparison with Figure 3.8. Let $n_A = 1 \times 10^{23}\,\mathrm{m}^{-3}$, so that $Q_0 = 2.08 \times 10^{-4}\,\mathrm{C/m^2}$. Let $w = 1\,\mathrm{m}$, $l = 1\,\mu\mathrm{m}$, and $C_{ox} = 3.45 \times 10^{-4}\,\mathrm{m}^{-2}$, and take $\mu_e = 0.05\,\mathrm{m^2/V\,s}$. Then, at "threshold", $V_{Ch} = 0.06\,\mathrm{V}$ and $I_{D(Sat)} = 30\,\mathrm{mA}$. It can be seen in Figure 3.8 that the corresponding gate voltage is 3.25 V. This is a little lower (0.25 V) than the value of V_T indicated in Figure 3.4.

With $V_{GS} < V_T$, we expect an exponential variation of the drain current with the barrier height, and hence, as observed, with the gate–source voltage. Thus

$$I_{D(Sat)} \propto \exp\left(\frac{e\,\Delta V}{kT}\right) \propto \exp\left(\frac{eV_{GS}}{\alpha kT}\right)$$

The coefficient α relates V_{GS} to the surface potential ΔV at the source end of the channel, where the potential minimum occurs. Its theoretical value can be obtained by applying Equation (2.7):

$$V_{GB} = V_{GS} = V_{ox} + \Delta V - V_{CP} \tag{2.5}$$

$$= \frac{(2e\epsilon_0\epsilon_{Si}n_A\,\Delta V)^{1/2} - Q_{SS}}{C_{ox}} + \Delta V - V_{CP} \tag{3.16}$$

where we have substituted for V_{ox} using Equations (A2.6) and (A2.7). Thus,

$$\alpha = \frac{dV_{GS}}{d(\Delta V)} = \frac{(e\epsilon_0\epsilon_{Si}n_A/2\,\Delta V)^{1/2}}{C_{ox}} + 1$$

$$= \frac{C_D}{C_{ox}} + 1 \tag{3.17}$$

Here, C_D may be thought of as the depletion-layer capacitance,

$$C_D = \frac{dQ_D}{d(\Delta V)} = \left(\frac{e\epsilon_0\epsilon_{Si}n_A}{2\,\Delta V}\right)^{1/2} \tag{3.18}$$

With $n_A = 1 \times 10^{23}\,\mathrm{m}^{-3}$, we have $\Delta V \approx \Delta V_{Th} \approx 0.81\,\mathrm{V}$ and $C_D = 10.2 \times 10^{-4}\,\mathrm{F/m^2}$. With $t_{ox} = 100\,\mathrm{nm}$, $C_{ox} = 3.45 \times 10^{-4}\,\mathrm{F/m^2}$. Together these give $\alpha = 4.0$.

The simplest way of measuring α is to take the change in V_{GS} required to cause a factor-of-ten change in $I_{D(Sat)}$. Call this $(\Delta V_{GS})_{Dec}$. Then, at room temperature,

$$\alpha = \frac{(\Delta V_{GS})_{Dec}}{(kT/e)\ln 10} = 16.9(\Delta V_{GS})_{Dec} \tag{3.19}$$

The graph of Figure 3.8 shows that $(\Delta V_{GS})_{Dec} = 0.225$ V, so the measured value of α is 3.8. This implies an effective doping concentration at the source end of the channel of 0.9×10^{23} m^{-3}. However, in some instances the presence of interface trapping levels may cause Q_{SS} to be a function of ΔV. This would modify the value of α slightly, but the measured value justifies our estimate of n_A very precisely.

The departure from the exponential variation of I_D with V_{GS}, shown in Figure 3.8, starts at 10 mA, when $V_{GS} = 3.05$ V. This is the point at which the inversion-layer charge density becomes big enough to influence the potential at the semiconductor surface significantly.

3.4. DEVICE TRANSCONDUCTANCE

A device parameter of great importance is the transconductance g_{fs} which is defined as:

$$g_{fs} = \left(\frac{\partial I_D}{\partial V_{GS}}\right)_{V_{DS}=\text{constant}} \tag{3.20}$$

This measures the sensitivity of the drain current to changes in the gate–source voltage. It determines the current-carrying capacity of the device.

According to the theory of Section 3.2 we should expect g_{fs} to take on values as follows:

1. Below pinchoff, when $V_{Ch} < V_{GS} - V_T$, differentiation of Equation (3.6) yields

$$g_{fs} = \mu_e \frac{w}{l} C_{ox} V_{Ch} = k V_{Ch} \tag{3.21}$$

2. Above pinchoff but before velocity saturation, that is, for $E_s l > V_{Ch} \approx V_{GS} - V_T$, Equation (3.11) leads to

$$g_{fs} = \mu_e \frac{w}{l} C_{ox}(V_{GS} - V_T) = k(V_{GS} - V_T) \tag{3.22}$$

3. After velocity saturation, $V_{Ch} > E_s l$, Equation (3.14) applies and

$$g_{fs} = \tfrac{1}{2} w C_{ox} v_s \tag{3.23}$$

For practical purposes the specification of g_{fs} for a given device, as on the device data sheet, requires that certain prescribed values of V_{GS} and V_{DS} must be chosen. Usually, V_{DS} is large enough to ensure that the drain current is in the constant-current region, $I_{D(Sat)}$. The choice of V_{GS} usually ensures that this value of $I_{D(Sat)}$ is about half the maximum current rating of the device. Data sheets also display graphs of g_{fs} versus I_D. A typical plot is shown in Figure 6 of the data sheet presented in Appendix 7. These curves demonstrate the increase of g_{fs} with I_D at low values, and its independence of I_D once velocity saturation is established.

For a given value of V_{GS}, the three factors that influence g_{fs} are the gate width w, the channel length l, and the gate oxide thickness t_{ox}. In saying this we are presuming that by choosing silicon as the semiconductor and silicon dioxide as the gate insulator, the values of μ_e (or v_s) and ϵ_{ox} are prescribed.

Decreasing t_{ox} increases g_{fs} through its effect on C_{ox}. The limit is set by the need to preserve the uniformity and integrity of the gate oxide, and for it to support the rated voltage V_{GS}, which is usually ± 20 V. The design of the transistor has to ensure that no greater voltage appears across the gate oxide in any permitted operating condition. With $t_{ox} = 100$ nm (1000 Å), the breakdown voltage of good-quality gate oxide averages about 55 V. Technological improvements have reduced the spread around this value, thereby increasing the margin for reliability.

Decreasing l also increases g_{fs}. The minimum value of l is set by the level of technology, in this case the ability to control the double-diffusion process that determines l. What matters is not what can be achieved in ideal circumstances, but the manufacturability of the device with a reasonable yield within the parameter tolerances specified. This becomes increasingly difficult with $l < 2\ \mu$m.

There is greater scope for increasing g_{fs} by increasing the gate width w, because w varies in direct proportion with the active device area. The parameter determined by technology is the gate width per unit area, w/A. There is an advantage in using a cellular rather than a linear pattern, and several different cellular designs have been proposed and used. Some are illustrated in Figure 3.9. The quantitative theoretical advantages of one pattern over those of another are marginal. They are discussed in Appendix 5. They have been the subject of a number of theoretical studies (e.g. [5]) involving a consideration of the ON-state resistance $R_{DS(on)}$, as well as the transconductance g_{fs}. It has recently been shown [6] that a circular cell geometry offers the best compromise, provided that the full bulk breakdown voltage is realized. The hexagonal structure shown in Figure 3.9*b* approximates this optimum design most closely. Again, though, what really matters is the manufacturability and robustness of the device made by a particular process.

What is clear from theory is that w/A increases as the cell lattice spacing decreases. The limit to this is set in two ways: again by the level of technology, but also by the need to maintain a minimum spacing b between the channel diffusions. This region, Ⓒ in Figure 2.1, acts as a throat which constricts the drain current. The limit on b results from the need to minimize the contribution of this region to $R_{DS(on)}$, bearing in mind that the current density is higher here than in the rest of the drain region. This is discussed in the next section. The problem is exacerbated by the presence here of a parasitic JFET, a matter that is taken up in Section 3.6.

In higher-voltage devices the epitaxial layer is more lightly doped and so has a higher resistivity. This not only increases the resistance of the epilayer, but also causes the channel depletion layers to expand further into the drain drift region, thereby enhancing the JFET action. This can be offset to some extent by increasing the drain doping concentration towards the surface, but the minimum acceptable value of b remains higher in high-voltage devices.

The technological limit is determined by the line width and registration tolerances of the photolithographic processing steps. A key factor is the need to obtain a low-resistance metal contact to the source. The contact area must be sufficient to ensure this after making provision for registration tolerances in the masking processes, which may cause the metal contact to be offset from the center of the cell. The chip area can most readily be filled by a regularly repeated lattice of square or hexagonal cells. Such patterns dominate in commercial devices.

Another limiting factor is the need to radius the channel at each cell apex in order to maintain its length l constant. The more acute the angle of the apex, the more difficult this is, which confers another benefit on the hexagonal cell.

The first range of devices to become widely established commercially in the early 1980s used a cell lattice spacing of 38 μm, giving over 500,000 cells per square inch (approximately 800 per mm^2). The resulting value of w/A is about

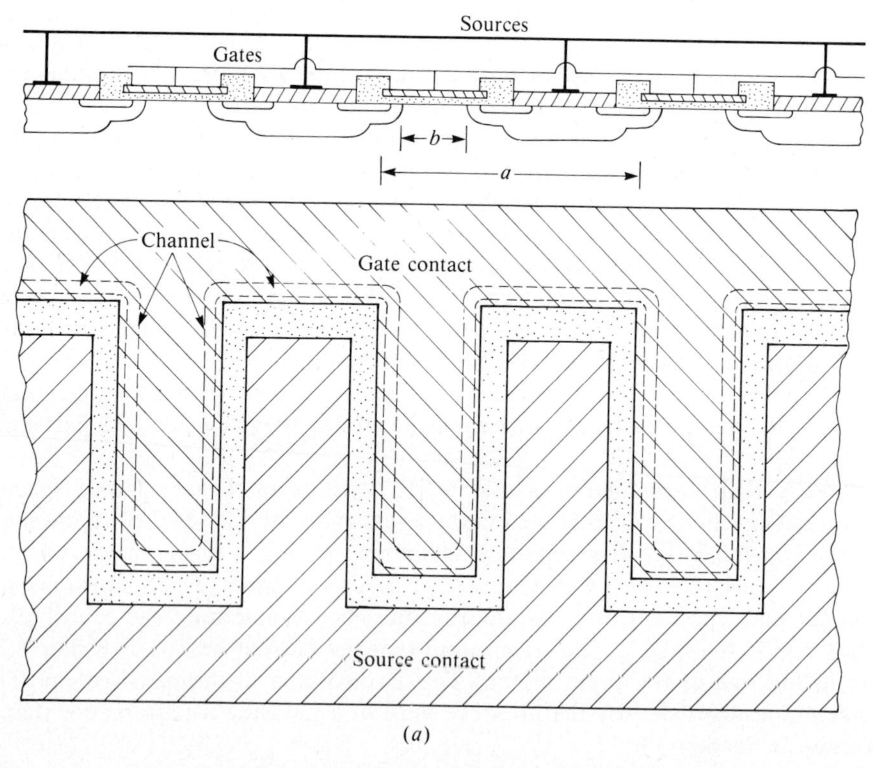

Figure 3.9. Examples of gate patterns: (*a*) Interdigitated. (*b*) Hexagons on hexagonal lattice. The plan view of the poly-Si layer shows the channel positions and the source contact areas. A section through the active region is also given. (*c*) Hexagons on a square lattice. (*d*) Squares on a square lattice. (*e*) Squares on an offset square lattice. (*f*) Squares on a hexagonal lattice. (*g*) Triangles on a hexagonal lattice.

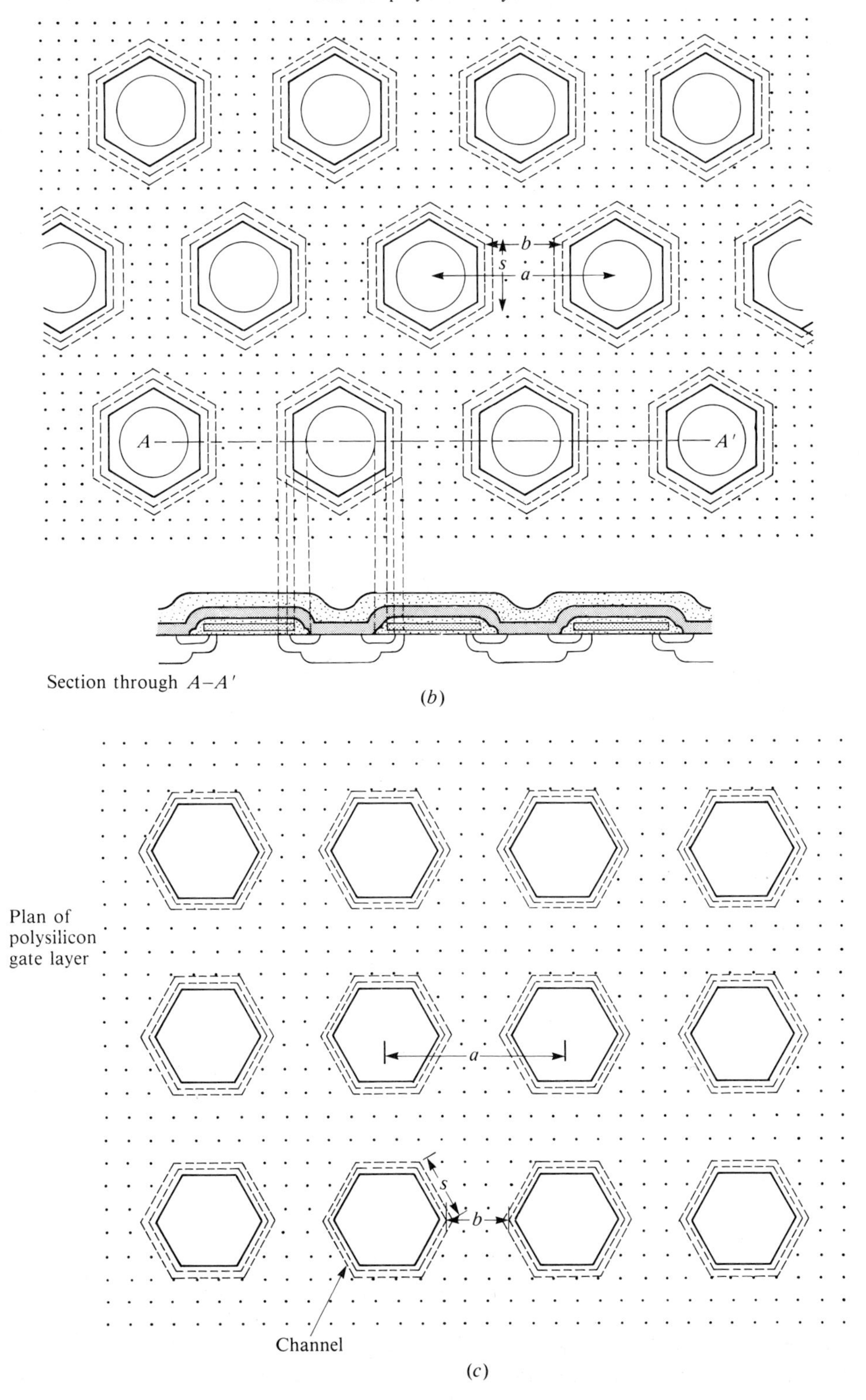

Figure 3.9. (*Continued*)

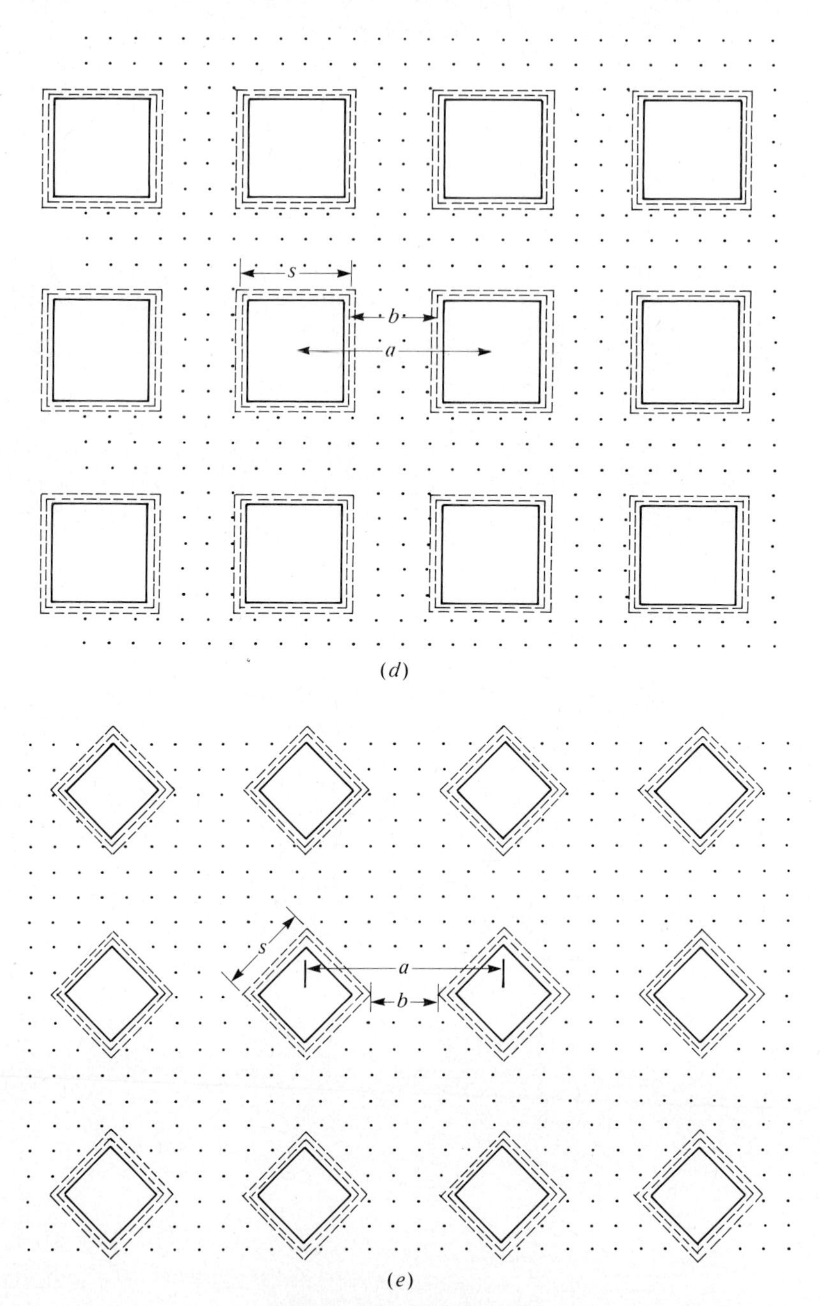

Figure 3.9. (*Continued*)

0.07 m/mm^2. This remains a perfectly satisfactory standard for higher-voltage devices. However, low-voltage MOSFETS have benefited from progressive reductions in the cell lattice spacing to 20 μm and below, with cell densities increasing first to 1,600,000 and more recently to 3,500,000 per square inch (approximately 2500 and 5400 per mm^2, respectively). Corresponding values of w/A are in the

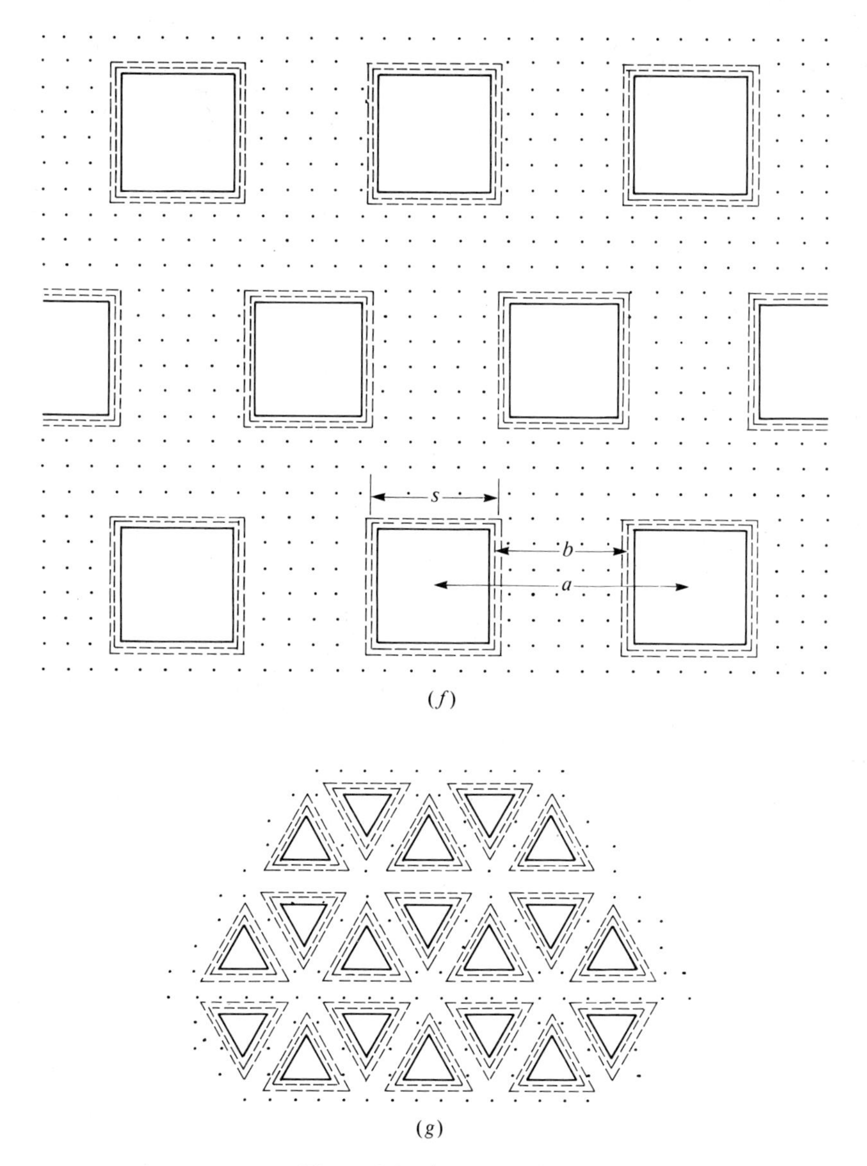

Figure 3.9. (*Continued*)

region of 0.12 and $0.17\,\mathrm{m/mm^2}$, respectively. As a result, these devices can support a given current rating on a chip of smaller area. Since the silicon area required is a major element in the cost of a component, obtaining the maximum performance from a given size of chip is extremely important.

Even so, the proportion of the chip available as active device area A_{A} may be less than 50% of the total area A_{Ch}. For reasons of reliability the source and gate bonding pads are not often laid over active material. The gate vias that distribute the gate current over the area of polysilicon and the field rings used around the edge of the chip to prevent peripheral voltage breakdown all diminish the area

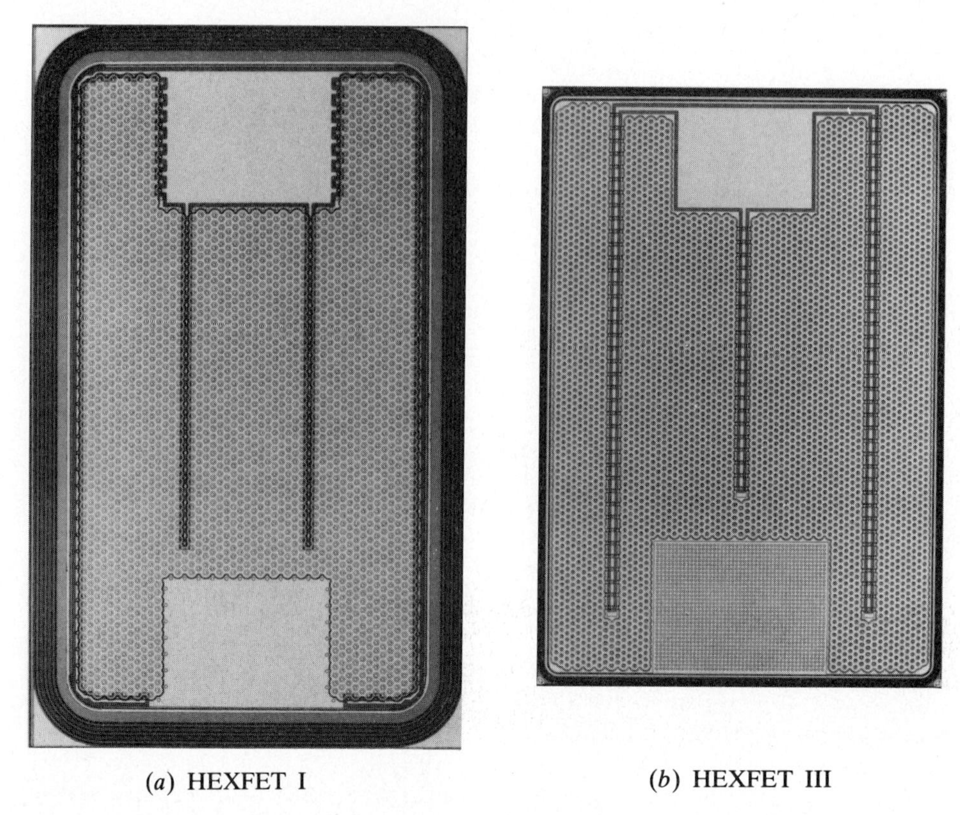

(*a*) HEXFET I (*b*) HEXFET III

Figure 3.10. Photographs of HEXFET™ chips. (*a*) Earlier design, HEXFET™ I; (*b*) more recent design, HEXFET™ III. These plan views show the hexagonal cell pattern, the peripheral field rings, the gate vias, and the bonding pads. In the earlier design, the active area occupies 60% of the total chip area. In the later design, the size of the field rings and the gate bonding pad have been reduced, giving better chip utilization. (Courtesy International Rectifier Corp.)

that can be used for transistor action. The plan views of HEXFET™ chips shown in Figure 3.10 illustrate the type of layout used. Approximately 60% of the chip area is active in the earlier design shown in Figure 3.10*a*. This is increased in the more recent design shown in Figure 3.10*b*.

3.5. THE DRAIN DRIFT REGION

The drain drift region is particularly critical to the design of a power MOSFET. Its principal function is to block and support the full forward voltage held off by the transistor in its turned-OFF state. However, it has also to carry the full forward current in the ON state. It is thus most important that its contribution R_{EPI} to the total ON-state resistance $R_{DS(on)}$ be kept as small as possible. This minimizes dissipation and maximizes the current rating of the device. Thus, the epilayer should be no thicker than it has to be in order to support the maximum blocking voltage. Values range from about 5 μm for a 50-V device to about 50 μm for a 1000-V device.

For purposes of discussion we have subdivided the drain drift region into three parts. In Figure 2.1 these are identified as the surface layer Ⓑ; the part bordered by the source and channel diffusions, Ⓒ; and the main part below the diffused regions, Ⓓ. In the ON state the current should be evenly distributed all across the epilayer. Clearly, this is not possible in the region between the diffusions, where the electron current emerges from the channel regions. The area between the diffusions is a little less than the gate area. Let it be a fraction g of the total active chip area. The even distribution of current here is assisted by the presence of the accumulation layer in the epilayer immediately under the gate oxide, in the turned-on state, and indeed as long as $V_{DS} < V_{GS}$. Under the gate oxide, then, the current density is higher by the factor $1/g$. It is further increased by the presence of the depletion layers at the pn junctions. These expand as the local potential increases at high current levels, and give rise to an effect characteristic of a series junction field effect transistor. This has already been mentioned and is discussed again in Section 3.6.

In high-voltage devices the epilayer is sufficiently thick for the current to be able to spread out to cover the whole of the active chip area fairly evenly by the time it enters the substrate. This is shown in Figure 3.11*a*. With low-voltage devices the chip area can be fully utilized only by further reducing the lateral cell dimensions, as illustrated by Figures 3.11*b* and *c*. This gives an additional incentive for a reduction in the cell pitch a, over and above the increase in w/A that results, as discussed in the last section.

When R_{EPI} is low, as in low-voltage devices, the effect of the other parasitic

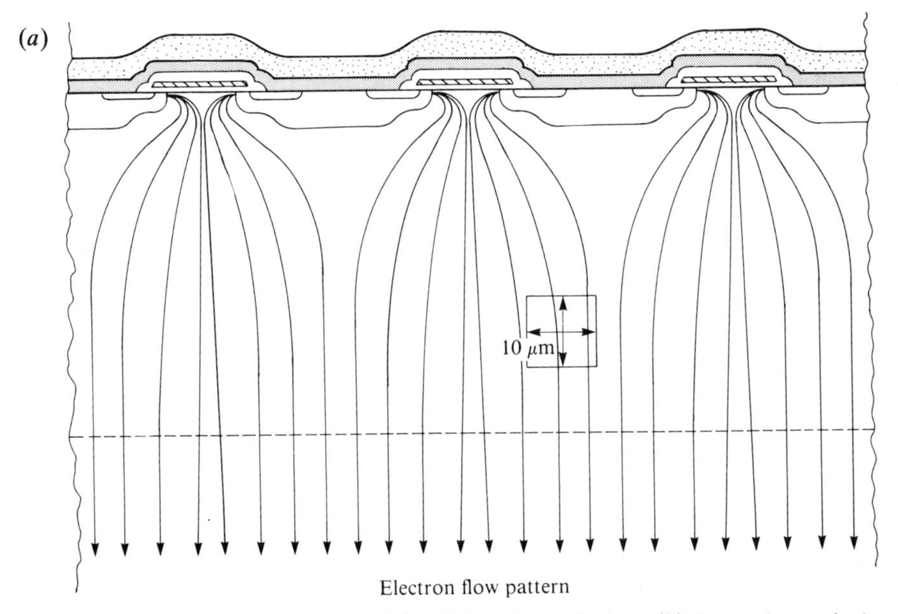

Figure 3.11. Current distribution. (*a*) High-voltage device; (*b*) low-voltage device with same cell pitch; (*c*) low-voltage device with reduced cell pitch.

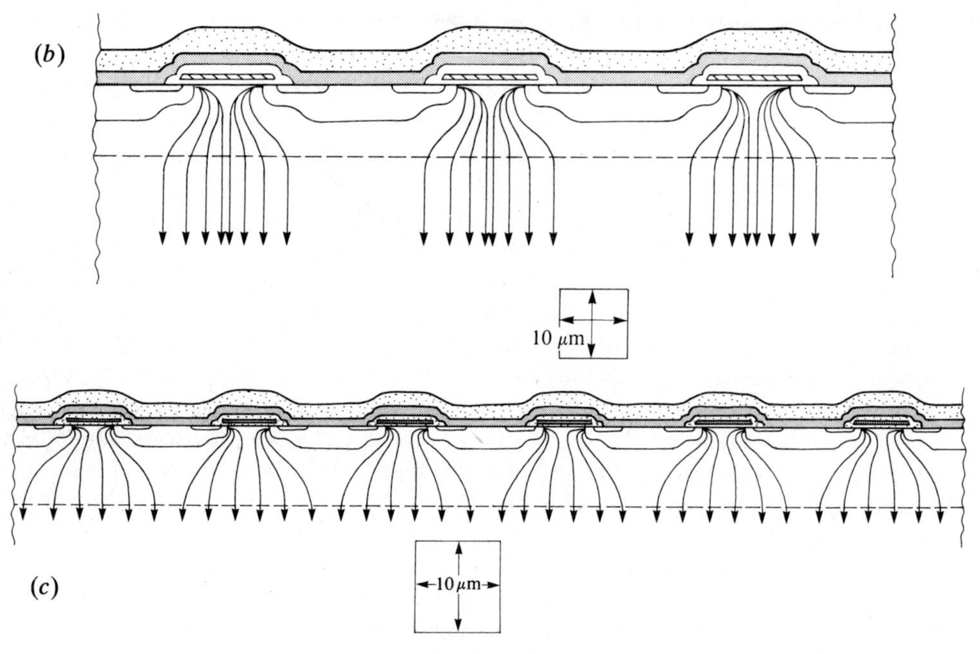

Figure 3.11. (*Continued*)

sources of resistance, R_P, is proportionately greater. A schematic illustration of the relative magnitudes of the different factors contributing to $R_{DS(on)}$ is given in Figure 3.12.

A theory specifying the optimum doping profile and thickness of the epilayer is presented in Appendix 6. It is based on a highly idealized plane-parallel geometry, but it does illustrate the principle of the problem, and it does set into

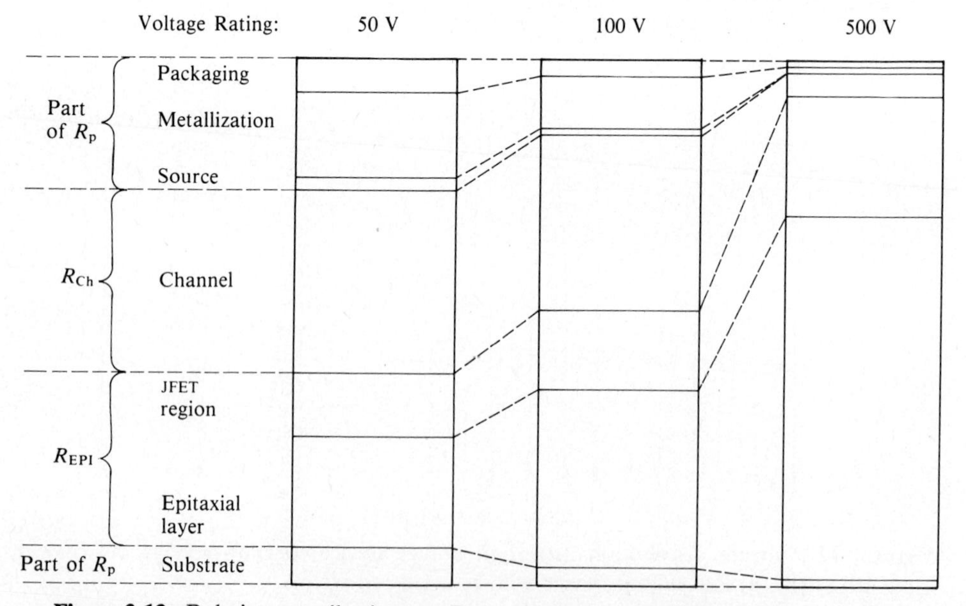

Figure 3.12. Relative contributions to $R_{DS(on)}$ in devices of different voltage rating.

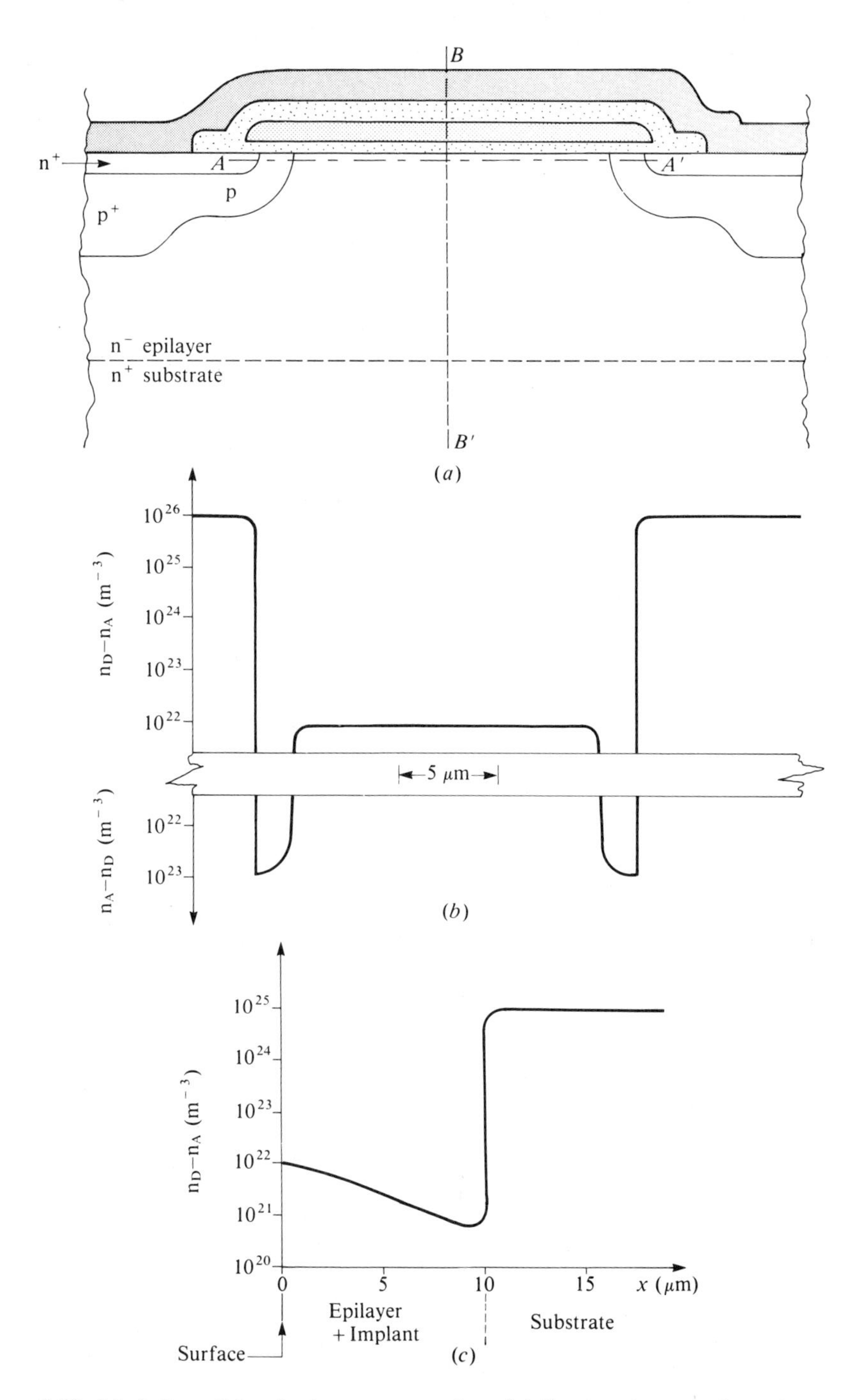

Figure 3.13. Variation of the doping concentration. (*a*) Section through the active region; (*b*) variation parallel to the surface along AA'; (*c*) variation perpendicular to the surface along BB'.

context the much-quoted relationship between the ON resistance and the junction breakdown voltage:

$$R_{DS(on)} \propto V_B^{2.6} \tag{3.24}$$

It is the breakdown voltage V_B that partly determines the voltage rating of the device. The resistance calculated in the theory (R in Appendix 6) is the optimum value of the component R_{EPI} of the ON-state resistance when there is a uniform distribution of current. The lateral nonuniformities of the real device mean that the proposed doping profile of Equation (A6.26) is not appropriate in practice.

Ideally [7], the doping level would follow Equation (A6.26) initially, decreasing from the n^+n^- junction into the epilayer. It would then go through a minimum and increase progressively towards the surface of the chip between the p diffusions. An increase in the epilayer doping concentration in regions Ⓑ and Ⓒ is desirable for two reasons. It helps to minimize the extra contribution to R_{EPI} caused by the higher current density, and more importantly, it reduces the effect of the parasitic JFET. It is found that the breakdown voltage of a well-designed device is not adversely affected. Most benefit is obtained when the doping concentration has lateral uniformity [8], as in the profile shown in Figure 3.13.

Field enhancement would normally occur at the edges and corners of the channel diffusion in the forward-blocking state. But this is reduced if the depletion layers, expanding from the junctions on either side of the throats in the drain drift region beneath the gate oxide, eventually overlap. The breakdown voltage then approaches that of the plane junction between the epilayer and the p^+ diffusion in the centre of the body region under the source contact. In modern devices this junction is designed to withstand avalanche breakdown in a controlled way. These matters are discussed further in the next section.

In Figure 3.14, rated values of $R_{DS(on)}$, normalized against device active area, are plotted against V_B (BV_{DSS}) for some commercially available power MOSFETS. Equations (A6.17) and (A6.28) are also shown. It can be seen that there are significant departures from the theory, especially at low values of V_B. From the earlier discussion this is not surprising, but the reduction of the cell pitch in the more modern designs of low-voltage devices has resulted in their coming nearer to the ideal curves. However, even at the higher values of V_B, the value of $R_{DS(on)}A_A$ is still several times higher than the predicted optimum value RA. One reason is the difficulty of achieving a high fraction of the theoretical value of V_B at the periphery of the chip. Techniques for achieving this are discussed in Chapter 5.

The need to minimize $R_{DS(on)}$ further emphasizes the importance of minimizing the silicon "real-estate overheads" discussed at the end of the previous section. This inevitably leads to compromises in design. Thus, extending the system of field rings at the periphery may help to ensure higher reliability at high voltage, but this is bought only at the expense of a reduced current-handling capacity and an increase in $R_{DS(on)}$. If, in a particular application, this causes the device to run at a higher temperature, its reliability will be degraded.

Some manufacturers have not allowed the bonding pads to overlie active parts of the device. However, an increasing number now have sufficient confidence in

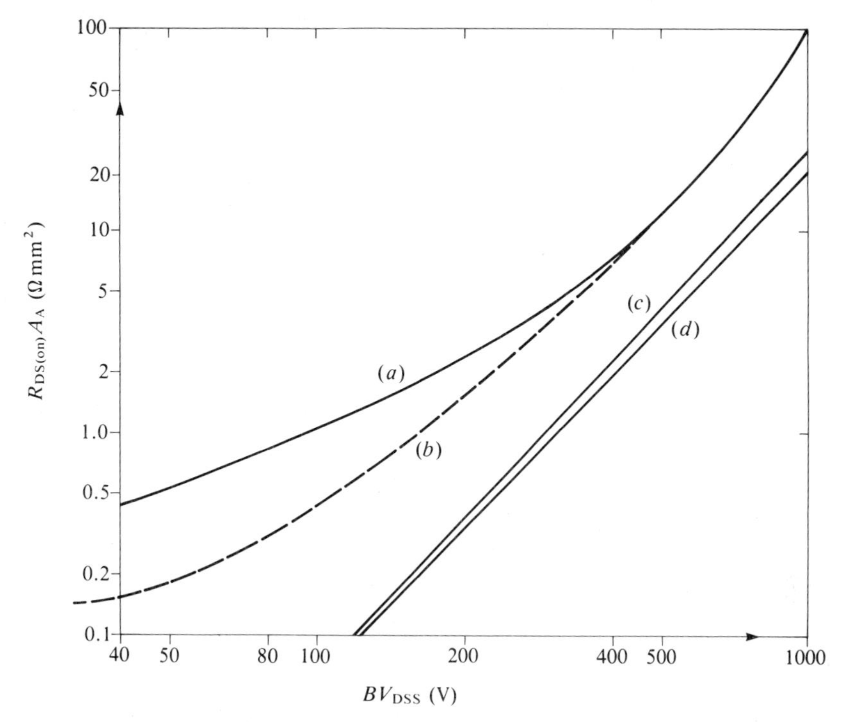

Figure 3.14. Normalized ON resistance versus breakdown voltage. Values of $R_{DS(on)}$ for commercially available devices are compared with the theoretical curves of Appendix 6. Curve (a) shows values typical of standard devices; curve (b) shows values typical of modern low-voltage devices having a high cell density. Curves (c) and (d) represent Equations (A6.17) and (A6.28).

their ability to maintain the integrity of the structure under the bonding pads that they are prepared to extend them over active regions. A final way in which most manufacturers accept an increase in $R_{DS(on)}$ in exchange for extra reliability in the breakdown performance arises from the fact that in general the depletion layer is not permitted to reach through to the substrate. The epilayer is made thick enough to ensure that the field at the n^-n^+ junction is low. Then, for any given voltage, the peak field occurring at the p–n junction is lower than it would be, were reach-through permitted.

3.6. PARASITIC COMPONENTS

3.6.1. Introduction

It will be clear from what has been said already that there are a number of parasitic components inherently associated with the VDMOS FET structure. In addition to the normal resistance of the bulk material and the contacts, the inductance of the leads, and the capacitance of junctions and connections, these parasitic elements include active transistors. Their effects are most pronounced under the transient conditions discussed in Chapter 4. They may severely limit high-frequency operation.

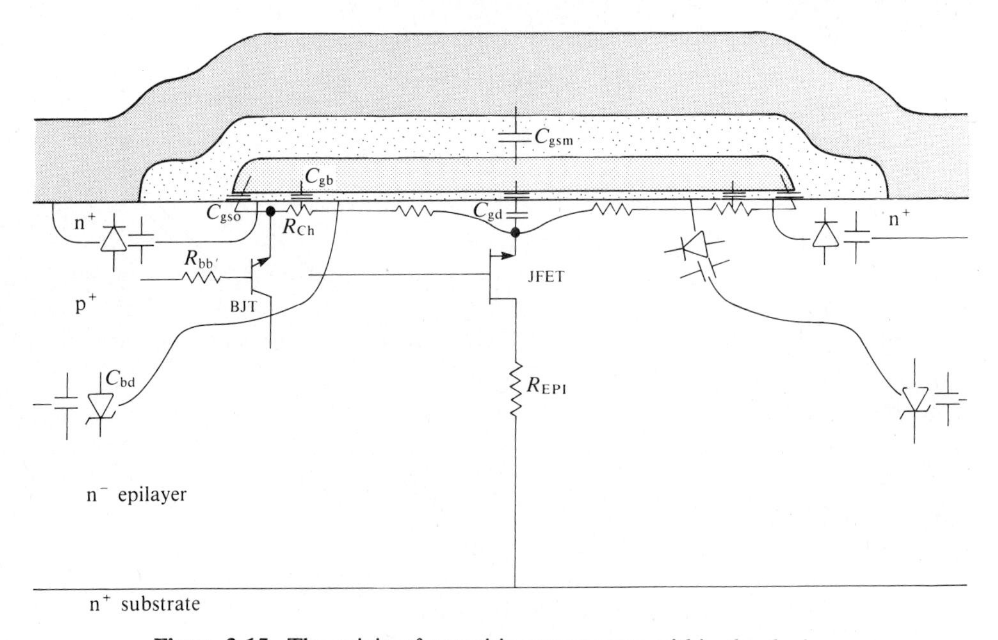

Figure 3.15. The origin of parasitic components within the device.

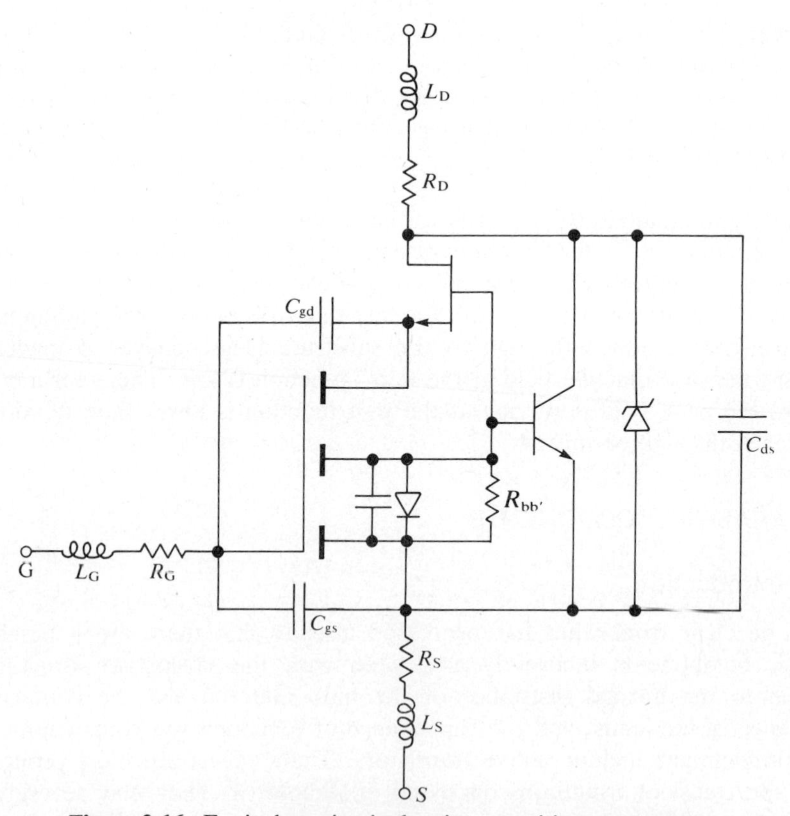

Figure 3.16. Equivalent circuit showing parasitic components.

In Figure 3.15 the various internal parasitic components of a VDMOS FET are shown, drawn onto the usual section through the active region. An equivalent circuit incorporating these and other parasitic elements is shown in Figure 3.16. Passive components include the following:

1. The various resistive elements contributing to $R_{DS(on)}$.
2. The inductance of each of the pin connections.
3. The capacitances between the gate and the source metallization, and between the gate and the source diffusion.
4. The capacitances between the gate and the body region, and between the gate and the epilayer.
5. The capacitances associated with the various pn junctions.
6. The pn junctions show, in addition, their normal diode behavior.

The main active components are:

7. The n-channel junction field-effect transistor (JFET) that forms in the epilayer, in between the channel diffusions.
8. The npn bipolar junction transistor (BJT) formed between the source, the body, and the drain. Under static conditions the base and emitter of the BJT are shorted, leaving only the body–drain diode effective. However, under transient conditions and in avalanche breakdown, the transistor may be activated, and this can seriously degrade the overall performance of the MOSFET.

We consider each of these groups of parasitic components in turn.

3.6.2. Resistance, Inductance and the Parasitic JFET

The causes of parasitic resistance in the path of the drain current have been discussed at length already in this chapter. We will not pursue them further, except in so far as they are influenced by the parasitic JFET, which we deal with next. The potential of this component to affect, adversely, many of the MOSFET static characteristics influences a number of design decisions, and has caused us to mention it several times already. The JFET action occurs in the region between the p diffusions. The p-doped body regions act as the gate, and the throat in the n-type drain drift region as the channel. The effect is more severe in high-voltage devices, because the epilayer is more lightly doped and the depletion layers that form at the body-drain junction extend further into the drain.

In Figure 3.12 the additional resistance caused by the JFET can be seen to account for 25% of the total $R_{DS(on)}$ of a high voltage device. At high ON-state currents, the increased voltage in the drain drift region causes an expansion of the depletion layers and a further increase in $R_{DS(on)}$. This is shown in Figure 3.17. In a poorly designed device, it is possible for pinchoff to occur in the JFET rather than in the MOSFET, under some conditions.

The acceptable spacing b between the channel diffusions is mainly determined by the extent, in the ON state, of these depletion layers. As a result, the maximum

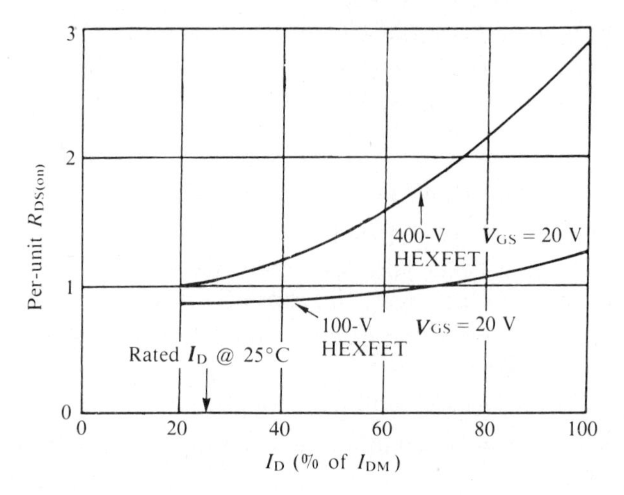

Figure 3.17. Increase in $R_{DS(on)}$ at high drain currents. The current is normalized to I_{DM}, the maximum rated pulse current, which is higher for the lower-voltage device (e.g. 108 A compared to 40 A). The increase in $R_{DS(on)}$ at higher currents is mainly the result of JFET action in the throat region as V_{DS} increases. It can be seen that the increase is much greater in the case of the higher-voltage device, because the epilayer is more lightly doped and the depletion regions expand further into it.

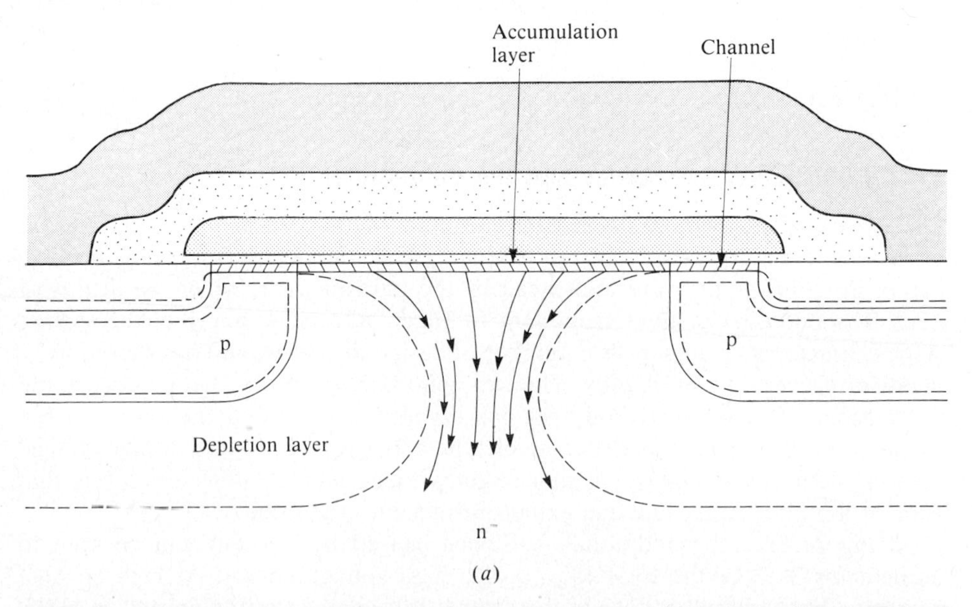

Figure 3.18. Effect of drain implant in reducing JFET action. (*a*) Depletion layers pinch off the channel through the drain neck region in a uniformly doped epilayer. The MOSFET is fully ON. (*b*) Depletion-layer thickness is reduced when the doping level under the gate oxide is increased. (*c*) Under some conditions the potential at the semiconductor surface in the middle of the throat region may exceed V_{GS}. Then a depletion layer forms there too, further constricting the current path.

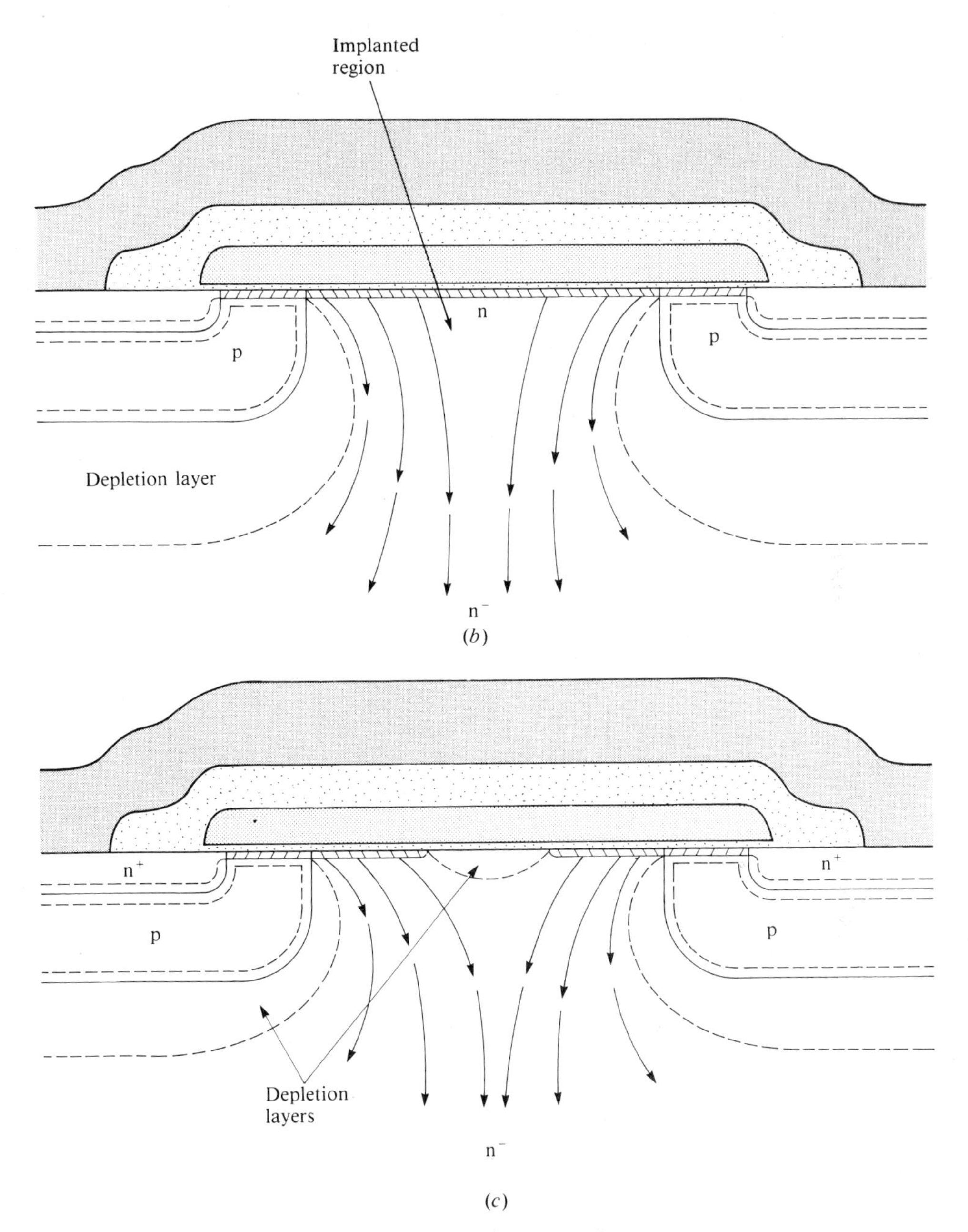

Figure 3.18. (*Continued*)

usable cell density is strictly limited, as discussed in Appendix 5. Increasing the donor doping concentration in the drain throats, as shown in Figure 3.13, has two beneficial effects. The conductivity is improved, and the widths of the depletion layers are reduced, so increasing the cross-sectional area available for conduction. This is illustrated in Figure 3.18. The consequence is that the cell size can be reduced and the resulting benefits of increased g_{fs} and lower $R_{DS(on)}$ can be realized. Of course, the level of the donor concentration in the drain immediately under the gate oxide has to be finely judged. If it is too high, the avalanche breakdown voltage is reduced and channel-length modulation may occur.

Of the lead inductances, that of the gate is not usually significant, but those of the drain, and particularly the source, may have a considerable effect on switching behavior and high-frequency performance. This is discussed in Chapter 4. Values can be minimized by careful encapsulation and design. They lie, typically, in the range 5 to 15 nH.

3.6.3. Capacitance

The physical capacitances shown in Figure 3.15 each contribute to the equivalent-circuit capacitances shown in Figure 3.16. We need to relate them to the terminal capacitances, C_{iss}, C_{oss}, and C_{rss}, that are specified on device data sheets. These are normally measured at 1 MHz, using the circuits shown in Figures A9.4, A9.5, and A9.6 of Appendix 9. Each is a function of the dc drain–source voltage, as shown in Figure 10 of the data sheet of Appendix 7. Values are usually specified for $V_{DS} = 25$ V and $V_{GS} = 0$ V. The three capacitances are related as follows:

1. The input capacitance, C_{iss}, is the parallel sum of C_{gd} and C_{gs}. Thus,

$$C_{iss} = C_{gd} + C_{gs} \tag{3.25}$$

2. The output capacitance, C_{oss}, represents the total capacitance between drain and source with the gate and source shorted together, that is,

$$C_{oss} = C_{ds} + C_{gd} \tag{3.26}$$

   On many early data sheets, an incorrect relationship is given, implying that C_{oss} is the sum of C_{ds} and the series combination of C_{gs} and C_{gd}. The published curves, however, were always obtained using the circuit of Figure A9.5, and are thus correct.

3. The reverse transfer or Miller capacitance, C_{rss}, is simply the gate–drain capacitance with $V_{GS} = 0$. That is,

$$C_{rss} = C_{gd} \tag{3.27}$$

From the discussion in Section 2.4, we can see that C_{gd} is made up of the series combination of the gate oxide capacitance and the capacitance of the drain depletion layer beneath the gate oxide. This latter component exists only when the potential of the drain surface region Ⓑ is more positive than that of the gate. When the gate is more positive, an accumulation layer forms at the semiconductor surface and $C_{gd} \approx C_{ox}A_G$, where $C_{ox} = \epsilon_0\epsilon_{ox}/t_{ox}$, as before, and A_G is the area of the gate. As the drain potential rises above that of the gate, the value of C_{gd} falls quite sharply. It continues to fall progressively as the depletion layer extends deeper into the epilayer. This is shown in Figure 2.11. Depletion-layer capacitance is discussed in Appendix 1. Its variation with the dc bias voltage is shown to

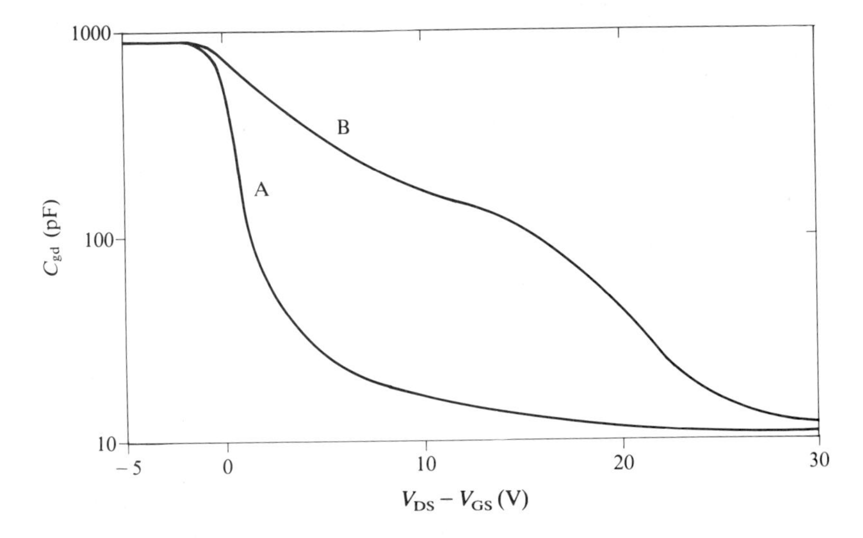

Figure 3.19. Effect of drain implant on the variation of C_{gd} with V_{DS}. Curve A represents a device with a uniformly doped epitaxial layer. Curve B is from one in which the conductivity under the gate oxide has been increased by an implant of 10^{16} phosphorus atoms per m^2 ($10^{12}\ cm^{-2}$).

be a function of the doping profile. The detailed variation of C_{gd} with V_{DS}, given that $V_{GS} = 0$, is therefore a function of the impurity concentration profile in the epilayer. This can be seen in Figure 3.19. Really we are applying the well-known capacitance–voltage (C–V) profiling technique to the whole device.

Physically, the gate–source capacitance C_{gs} comprises the overlap capacitance, the gate–channel capacitance, and the capacitance through the field oxide of the source overlay. It is largely independent of bias.

The principal component of C_{ds} is the capacitance of the junction between the p diffusion and the epilayer. This varies inversely with the square root of the drain–source voltage, as indicated by Equation (A1.18). Thus, at low drain voltages, particularly when $V_{GS} > V_{DS}$, $C_{oss} = C_{ds} + C_{gd}$ becomes quite large.

Because of their considerable variation with the biasing conditions, knowledge of C_{iss}, C_{oss}, and C_{rss} at any one state of bias is of limited value to the circuit designer. Graphs of the three capacitances versus V_{DS} are normally provided, but, unless these are extended to negative values of V_{GD}, they do not show the full extent of the capacitance variation. To overcome this difficulty, it has become normal practice to specify the total amount of charge needed to change the gate voltage from 0 to 10 V while the device switches a prescribed current. The current chosen is normally somewhat greater than the maximum dc rating. The details and implications of this are discussed further in Chapter 4.

3.6.4. The Body–Drain Diode and the Bipolar Junction Transistor

Finally, we come to the bipolar junction transistor structure that is formed by the n^+ source, the p body, and the n^- epilayer. Under normal, static conditions the emitter and the base of the BJT (the source and the body region) are shorted, leaving only the effect of the body–drain diode.

When the MOSFET is reverse biased, the body–drain diode is set into forward conduction. Its current rating is comparable to that of the MOSFET itself. So that this should not lead to excessive dissipation, or to problems in external circuits, its forward voltage ($V_{SD(on)}$) needs to be kept low. Typical values at the full rated current are from 2 to 2.5 V. This reverse-conduction property is valuable in a power switching device. It is particularly necessary in circuits such as dc-to-ac inverters and dc-to-dc choppers, when they are used in applications such as motor speed control, as discussed in Chapter 12.

As well as a low on-state voltage, the body–drain diode must exhibit a rapid, but not too abrupt, recovery waveform, and low reverse recovered charge Q_{RR}. Occasionally, heavy-metal doping (gold or platinum) or electron bombardment have been used to speed up the recovery of the body–drain diode. These are normal techniques used with other types of device. They introduce recombination centers by creating trapping levels in the middle of the band gap, and so kill the carrier lifetime. Unfortunately, this increases both $V_{SD(on)}$ and $R_{DS(on)}$ at room temperature. At elevated temperatures, the trapping levels are no longer effective, and both the lifetime and the ON resistance revert to their normal values. This means that the temperature coefficient of $R_{DS(on)}$ is lower than that of a normal device. The effect on Q_{RR} is discussed in Chapter 13; see Figure 13.22, in particular.

It has been shown [9] that a careful adjustment of the energy and dose of electron irradiation can give valuable reductions in the recovery time and the recovered charge (and so reduce the transient dissipation), without degrading the other parameters significantly. In Ref. 9, 8 Mrad at 3 MeV was reported to be effective. Another adverse effect is a reduction caused to the threshold voltage by the creation of positive charge in the gate oxide. To reduce this charge, it is necessary to anneal the device after irradiation for several hours at about 140°C. This normally restores the threshold voltage to a reasonable value, but leakage currents remain appreciably higher than before irradiation.

The reverse avalanche-breakdown characteristics of the body–drain diode determine the forward voltage rating of the MOSFET, BV_{DSS}. In modern devices, the diode junction is designed to withstand the passage of substantial current under conditions of controlled avalanche breakdown. This enables forward transients to be quenched, and inductive energy to be dissipated, without the use of additional, external clamping and snubber components. To achieve this it is necessary for the breakdown to be very evenly distributed over the p^+ region of the junction and, in particular, to be evenly distributed from cell to cell. This is illustrated in Figure 3.20.

Many devices now have single-shot and repetitive pulsed ratings for the energy (charge × BV_{DSS}) that they can safely dissipate in this way. An example of a device being driven into controlled breakdown by an inductive load is shown in Figure 3.21. The device under test is an IRF 460, rated at 500 V, 24 A. Once the MOSFET is turned on, the current rises approximately linearly with time at a rate determined by the voltage across the inductance (17 A/ms, that is, 100 V across 6 mH). When the transistor is turned off, the voltage across it rises to the breakdown voltage V_B, which is 550 V. This causes the current in the inductance

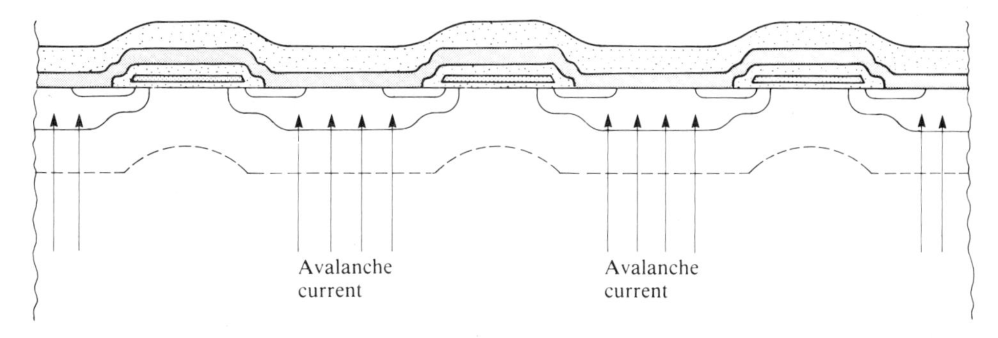

Figure 3.20. Avalanche breakdown of the body–drain diode. The diagram shows the uniform breakdown of the p^+n junction required in a good device. The shielding effect of adjacent cells mitigates the field concentration at the edges and corners of the p diffusion.

to fall at the rate (550 V)/(6 mH) = 90 A/ms. Thus the energy stored in the inductance at the moment of turnoff is dissipated in the transistor during the fall of current. In this case it amounts to about 1.2 J.

The body region has a lateral resistance from the contact to the shoulder that is known as the "pinch-base" resistance, $R_{bb'}$. Current flowing laterally through $R_{bb'}$, as shown in Figure 3.22, causes an ohmic voltage drop. If this voltage exceeds about 0.6 V, it will initiate transistor action in the parasitic BJT. This is because it forward-biases the source–body junction and causes the injection of electrons from the source into the body region. These then become the emitter and base of the parasitic bipolar transistor.

The value of $R_{bb'}$ is minimized by reducing the length of the shoulder linking the channel region to the p^+ diffusion. The depth of the shoulder is determined by the relative diffusion depths of the body and the source, which in turn determine the channel length. With a channel length of about 1 μm, the depth of the p diffusion is rather less than 2 μm. With a maximum boron doping density of about $1 \times 10^{23}\ m^{-3}$ ($1 \times 10^{17}\ cm^{-3}$), the body-region resistivity has a minimum value of about $2 \times 10^{-3}\ \Omega\ m$ (0.2 Ω cm). The surface resistance of the shoulder is

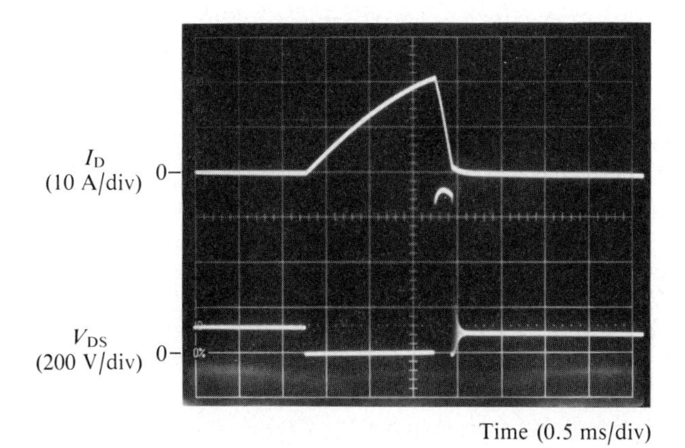

Figure 3.21. Example of controlled avalanche breakdown immediately following turn-off.

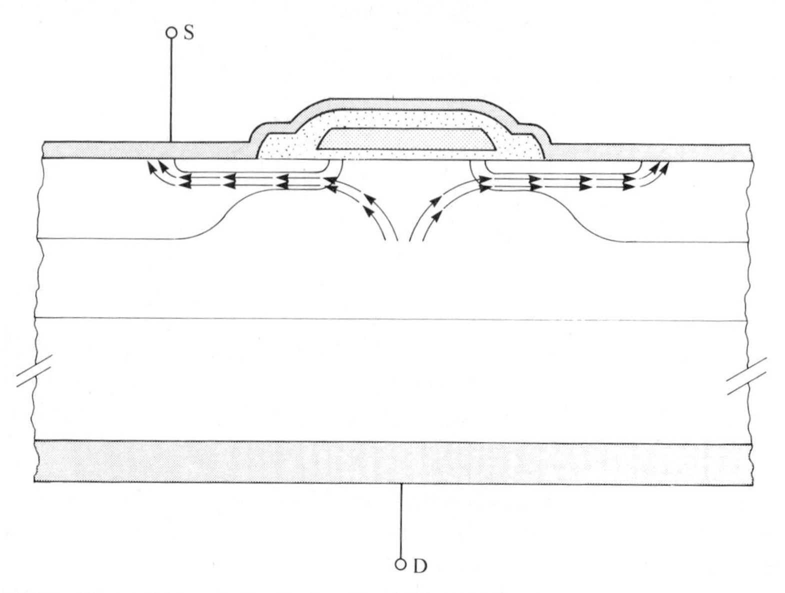

Figure 3.22. Current flow through the pinch-base resistance. This creates the danger that the parasitic BJT may be activated.

Figure 3.23. The avalanche-breakdown behavior of a bipolar junction transistor. (*a*) Typical characteristics; (*b*) common-base configuration (V_{CBO}); (*c*) common-emitter configuration (V_{CEO}); (*d*) emitter and base connected (V_{CER}). Avalanche breakdown is a regenerative process, because some of the electrons and holes created by impact ionization pass back through the high-field region and themselves create more carriers. The breakdown voltage of a junction, V_B, is the voltage at which the carrier avalanche multiplication factor $M = \infty$. As the temperature rises, M decreases, so V_B increases. In a BJT, the breakdown voltage V_{CBO}, is that of the collector–base junction with the emitter open-circuited. A similar value is obtained when the base and emitter are shorted, but it is then called V_{CES}. The breakdown voltage V_{CEO} is that between the collector and the emitter with the base open-circuited. It is significantly lower than V_{CBO}, because of a further process of regeneration. When holes created in the multiplication process at the collector junction pass back across the base and enter the emitter, they may cause the injection of further electrons into the base. A proportion of these reach the collector junction, where they in turn suffer multiplication. Let a fraction α_0 of the current crossing the emitter junction be electrons which reach the collector junction and are multiplied. The current in the transistor can increase without limit when $1 - \alpha_0 M = 0$, that is, when $M = 1/\alpha_0$. With $V_{BE} < 0.6\,V$; $\alpha_0 \approx 0$, but the buildup of carriers in the base, once avalanche multiplication becomes appreciable causes the base–emitter junction to become forward biased. Transistor action is initiated and α_0 increases rapidly. The minimum value of V_{CEO} corresponds to the maximum value of α_0, and is known as $V_{CEO(sus)}$. When a finite resistance connects base and emitter, the breakdown curve (V_{CER}) lies between these two extremes. At low currents α_0 is low and V_{CER} approximates to V_{CBO}. As the current increases, α_0 increases and V_{CER} approaches V_{CEO}. Only if the emitter is open-circuited or shorted to the base can $V_{CBO} = V_{CES}$ be sustained. Further increase in current, if maintained, eventually leads to second breakdown and device destruction.

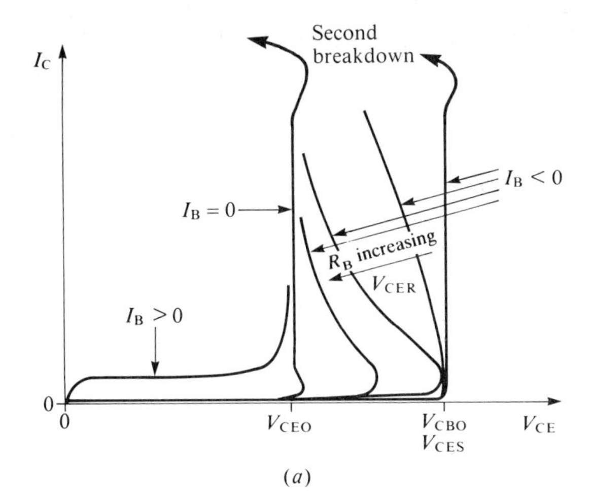

(a)

(b)

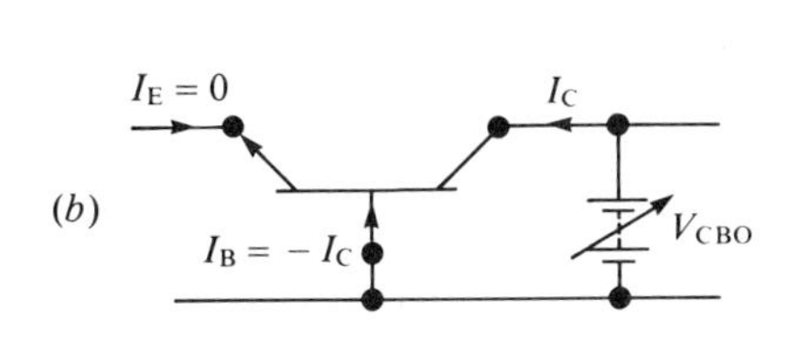

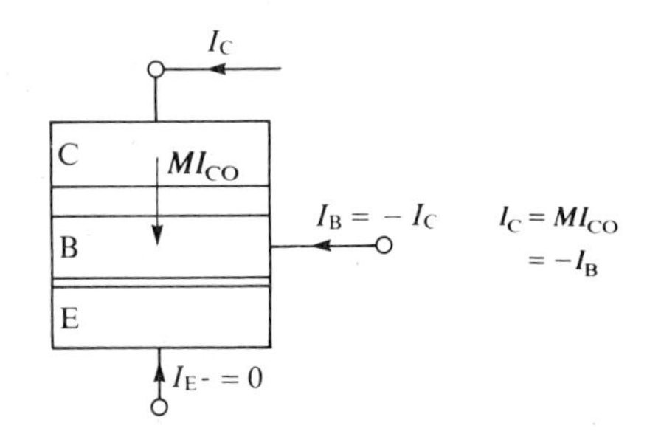

$I_C = MI_{CO}$
$= -I_B$

(c)

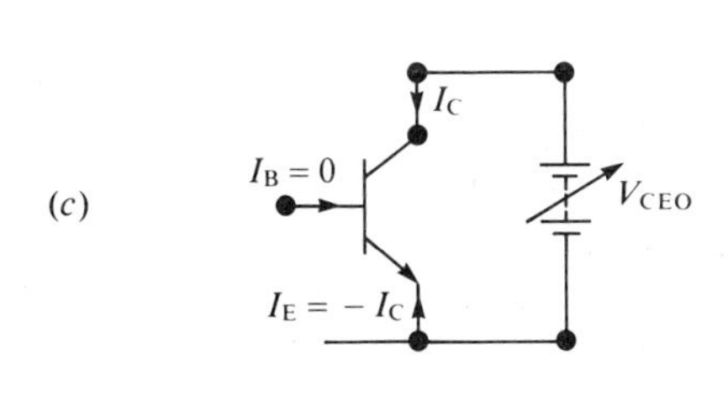

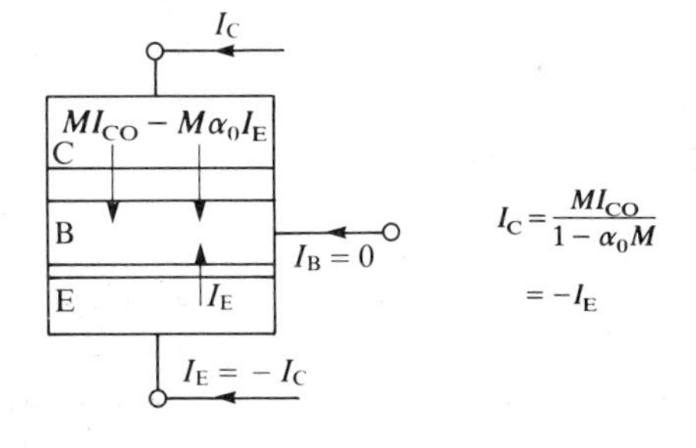

$$I_C = \frac{MI_{CO}}{1 - \alpha_0 M}$$
$$= -I_E$$

(d)

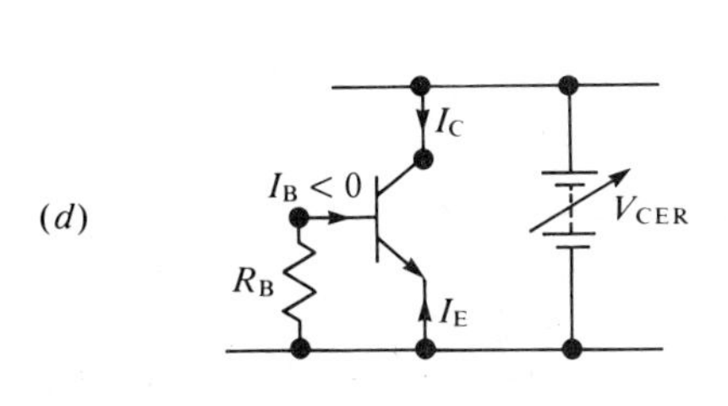

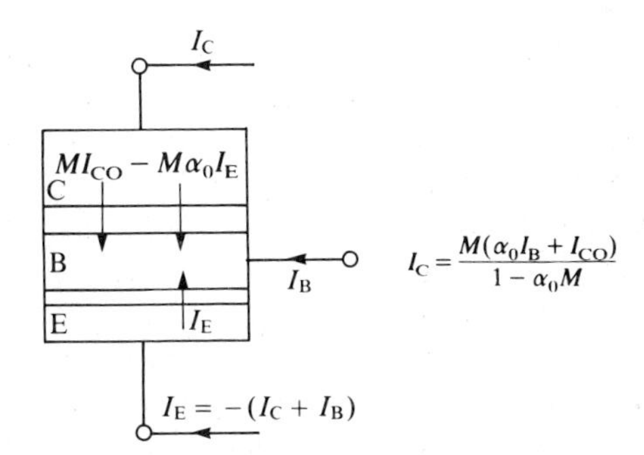

$$I_C = \frac{M(\alpha_0 I_B + I_{CO})}{1 - \alpha_0 M}$$

thus rather more than 1 kΩ/square. With the length of the shoulder about 6 μm and a gate width of about 1 m, the overall value of $R_{bb'}$ is in the region of 6 mΩ, say 6 to 10 mΩ. It would thus require a uniformly distributed current of 60 to 100 A to activate the bipolar transistor. However, any mask misalignment that causes the p^+ diffusion to be displaced also increases the length of the shoulder on one side and so increases $R_{bb'}$ locally. It is important to understand that it only requires the BJT to be turned on in one part of one cell for this region to hog all the current passing through the device. The rise of temperature caused by the resulting dissipation may induce current crowding and lead on to thermal runaway and second breakdown failure. It is a comment on their design quality and process controllability that manufacturers can now guarantee ratings at which such failures do not occur.

The use of the p^+ diffusion to connect the body region back to the source contact can be seen to be an important part of the VDMOS FET design. It reduces $R_{bb'}$ and determines the breakdown voltage of the body–drain diode. Its vertical and lateral doping profiles have to be carefully controlled so that no adverse effects on the breakdown voltage or the properties of the channel region occur.

The presence of the parasitic bipolar junction transistor, with its base and emitter connected by $R_{bb'}$, lowers the breakdown voltage of the body–source junction. This effect is characteristic of bipolar junction transistors in general. It is illustrated in Figure 3.23. There it can be seen that electrons injected into the base (body) from the emitter (source) give rise to an additional regenerative effect. This is serious because field concentration at the edges and corners of the shoulder of the body region would normally reduce the breakdown voltage there anyway. Fortunately, as was mentioned in the previous section and illustrated in Figure 3.20, the proximity of adjacent cells helps to minimize this field concentration. If, as a result of current flowing through $R_{bb'}$, the breakdown voltage is locally reduced to V_{CEO}, and this is below the supply voltage, the MOSFET will be destroyed.

Under transient conditions, the parasitic BJT may be activated in two further ways. First, changes in the potential difference across the body–drain diode require its self-capacitance to be charged and discharged, and the resulting capacitative currents flow through $R_{bb'}$. This could, in principle lead to a forward dV/dt failure. In practice, even taking $C_{ds} = 1$ nF, which would be typical for V_{DS} small, the rate of change of V_{DS} would have to be in the region of 60 to 100 V/ns for $R_{bb'}C_{ds}\,dV_{DS}/dt$ to exceed 0.6 V. This is not a problem with the high degree of cell uniformity achieved in modern devices. The second way of activating the BJT also follows turnoff. Excess carriers left from the previous conduction phase are swept out from the regions either side of the body–drain junction, as the depletion layers expand. This adds greatly to the normal capacitance current. Such carriers are most likely to remain in the region under the gate oxide. From there they are swept into the shoulder of the body region and have to pass the whole of $R_{bb'}$ to reach the source contact. They are, of course, blocked by the source–body junction, until this becomes forward biased—the effect we are seeking to avoid! Even greater care has to be taken over regions at the periphery of the chip and adjacent to the gate vias. Unless special precautions are taken, carriers tend to build up in these "dead spaces" and remain there longer than in the more confined regions of the active parts of the device. Figure 13.24, in

Section 13.13, shows an IRF 530 turning off in an inductive switching circuit. Following the recovery of the body–drain diode, the device is subjected to a fast forward dV_{DS}/dt transient. This is safely clamped at 130 V by a controlled breakdown. More detailed discussions of the origin of the waveforms observed during turn-on and turn-off are given in Chapter 4 and in later chapters in the context of specific applications.

3.7. TEMPERATURE EFFECTS

In Chapter 2 we showed that the threshold voltage would be expected to decrease with an increase in temperature. Carrier mobility and saturation velocity also decrease as the temperature rises, and in the bulk semiconductor regions the resistivity increases. For a given device, in which V_{GS} is kept constant, the net consequence of a rise in temperature resulting from these various effects is as follows:

1. In the subthreshold region ($V < V_T$) the drain current after pinchoff increases. Before pinchoff the increase in the various parasitic resistances tends to reduce the slope of the I_D-versus-V_{DS} characteristic.
2. In the "normal" operating region described by Equations (3.10) and (3.11) in the simple theory, or by their extensions discussed in Appendix 4, the decrease of the carrier mobility more than offsets the consequences of the lowering of V_T. As a result, I_D decreases above and below pinchoff.
3. In the velocity saturation region $I_{D(Sat)}$ again decreases, because of the fall in v_s.

The combination of all these effects on the static characteristics is shown in Figure 3.24. In particular, the transfer characteristics like those of Figure 2 in Appendix 7, which are given in most data sheets, demonstrate how $I_{D(Sat)}$ increases with temperature below a certain current level and decreases with increasing temperature at higher currents. This is important in the bias stabilization of linear amplifiers, and is discussed further in Section 15.2.2. In practice, the normal operating currents of LDMOS FETS lie much further above the threshold voltage than those of VDMOS FETS. The result is that LDMOS FETS are less sensitive to the variation of V_T, and the crossover current occurs at a lower level. This is shown in Figure 15.7, in Section 15.2.2.

The effect of temperature on g_{fs} is shown by Figure 6 of Appendix 7. The reduction in the high-current region, when g_{fs} is constant, is brought about by the lowering of v_s as the temperature rises. The increase in $R_{DS(on)}$ with temperature is shown in Figure 9 of Appendix 7. The effect of this is compounded, because the increase in dissipation that results causes a further temperature rise, and so on.

Operating at elevated temperature increases the junction breakdown voltage, by about 1% for each 10°C rise. It also is likely to reduce the expected device lifetime, as described in Chapters 5 and 17. The device capacitances discussed in the previous section are largely unaffected by changes in temperature.

At higher temperatures, a lower forward bias voltage is needed to support a given current flowing across any of the pn junctions. Thus, $V_{SD(on)}$ is lower.

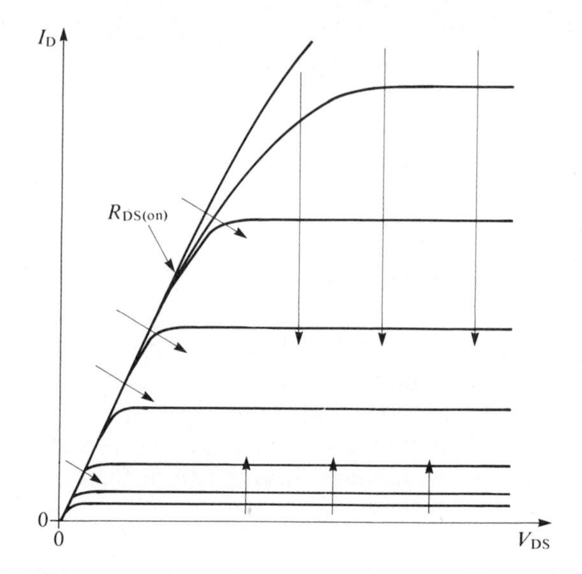

Figure 3.24. Effect of temperature on device characteristics. Arrows indicate the shift of the different parts of the characteristics with an increase in temperature.

Observation of $V_{SD(on)}$, V_T, or $R_{DS(on)}$ permits an indication of the temperature in the active parts of the transistor to be obtained, as discussed in Section 17.2.4.

The parasitic bipolar junction transistor is more easily turned on at high temperatures, because the pinch–base resistance $R_{bb'}$, is increased, and the voltage needed to forward-bias the emitter–base junction is reduced. At the same time the BJT is more rapidly sent into second breakdown failure.

3.8. p-CHANNEL AND OTHER DEVICES

Often considerable advantage in circuit simplicity results when complementary p-channel and n-channel devices are both available. This is as true for power devices as it is for CMOS (in comparison with NMOS) integrated circuits. In silicon, holes have lower mobility as well as a lower saturation drift velocity. As a result, p-channel devices show a higher $R_{DS(on)}$ (about a factor of 2 higher) and a lower value of g_{fs} than n-channel devices having the same silicon area. These important considerations apart, p-channel devices can be made simply by reversing the dopant types, as illustrated in Figure 3.25. Typical characteristics are shown in Figure 3.26.

One advantage of the p-channel device is that the body region is n-type and so has a lower resistivity for a given doping concentration. The pinch–base resistance, $R_{bb'}$, is lower in consequence, and the avalanche characteristics are better because the pnp transistor is less prone to turn on. A further disadvantage, though, is the difficulty of making a good ohmic contact between the source metallization and the body diffusion. Because of the constraints on the body

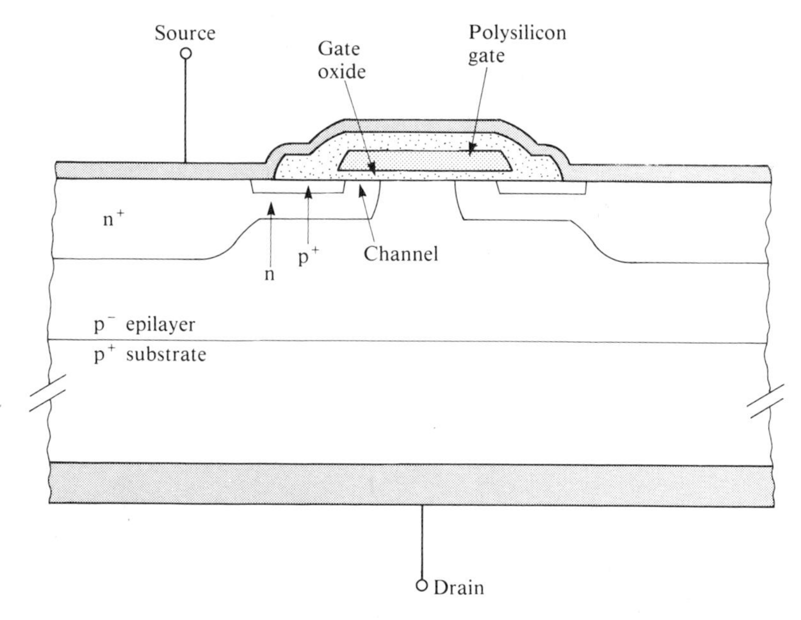

Figure 3.25. A p-channel device.

region doping levels, this contact normally forms a Schottky diode. Its reverse breakdown voltage is in series with the forward voltage of the body–drain diode when reverse current is conducted, so the drain–source voltage drop under these conditions ($V_{SD(on)}$) is high, typically 5 to 6 V.

Many variants on the basic VDMOS FET have been devised and are more or less generally available. These include the IGBT mentioned briefly in Section 1.3, and several other MOS/bipolar devices. They are discussed further in Chapter 20. Current-sensing MOSFETS and radiation-hardened devices are discussed in Chapter 19.

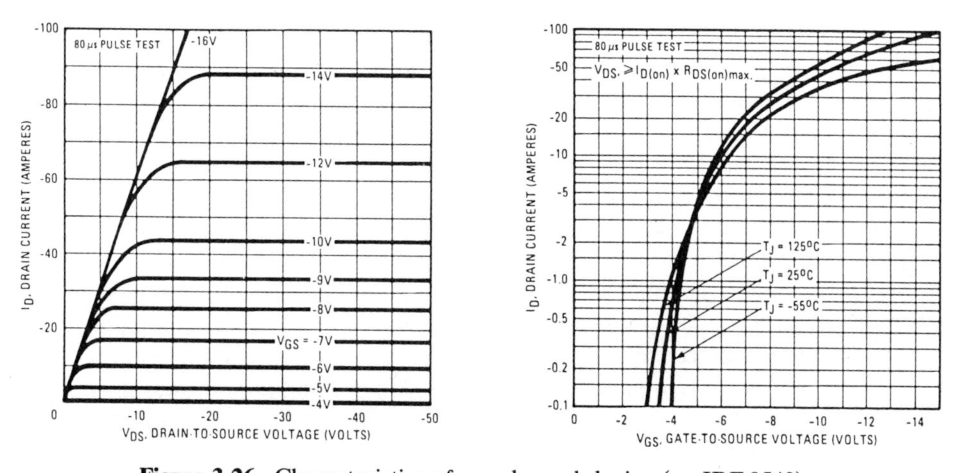

Figure 3.26. Characteristics of a p-channel device (an IRF 9540).

SUMMARY

The total drain–source voltage of the conducting MOSFET comprises the ohmic and the nonlinear voltages illustrated in Figure 3.2. For high-voltage devices in the ON state, the resistance of the epitaxial layer is dominant, and $R_{DS(on)}$ varies approximately as $V_B^{2.5}$. The theory governing these parameters is discussed in Appendix 6. Devices benefit from an increase in the doping level in the drain under the gate oxide. For low-voltage devices it is most important to minimize the contact and lead resistances, and a reduction in the device feature size is beneficial.

A simple charge-control model predicts the following dependence of I_D on V_{GS} and V_{Ch}:

For $V_{GS} > V_T$ and $V_{GS} - V_T > V_{Ch}$,

$$I_D = \mu_e \frac{w}{l} C_{ox}[(V_{GS} - V_T)V_{Ch} - \tfrac{1}{2}V_{Ch}^2] \tag{3.10}$$

For $V_{GS} > V_T$ and $V_{GS} - V_T < V_{Ch}$,

$$I_D = I_{D(Sat)} = \tfrac{1}{2}\mu_e \frac{w}{l} C_{ox}(V_{GS} - V_T)^2 \tag{3.11}$$

At high levels of drain current, electron velocity saturation effects lead to a predicted linear variation of drain current with V_{GS},

$$I_{D(Sat)} = \tfrac{1}{2}C_{ox}w(V_{GS} - V_T)v_s \tag{3.14}$$

Close to threshold, $V_{GS} \approx V_T$, there is an exponential variation of $I_{D(Sat)}$ with V_{GS},

$$I_{D(Sat)} \propto \exp(eV_{GS}/\alpha kT) \tag{3.15}$$

This simple theory can be fitted quite closely to the experimental characteristics of devices, but cannot easily be reconciled with their known physical characteristics. On closer examination it is seen to contain a number of self-contradictions. The more sophisticated theory discussed in Appendix 4 avoids these, but without the introduction of a field-dependent mobility, serious discrepancies with the experimental results remain.

Transconductance is maximized by obtaining the maximum gate width from a given area of silicon. This is inversely proportional to feature size, and is aided by the use of a cellular geometry. A discussion of the implications of some of the different cell patterns that have been used or proposed is given in Appendix 5.

A rise of temperature causes an increase in $R_{DS(on)}$. In the subthreshold region $I_{D(Sat)}$ increases with temperature, because of the reduction in the threshold voltage. At higher current levels the reduction in the electron mobility with temperature offsets this, and above a certain value of $I_{D(Sat)}$, where the temperature coefficient is zero, it decreases. These effects are illustrated in Figure 3.24.

Holes have a lower mobility and a lower saturation drift velocity than electrons, so p-channel devices have a lower transconductance and a higher $R_{DS(on)}$ than otherwise similar n-channel devices. Their voltage drop in reverse conduction is higher, but their reverse breakdown performance is better.

Several parasitic components influence the behaviour of power MOSFETS. As well as the normal parasitic resistance, inductance and capacitance associated with the semiconductor, the leads, and the packaging, several parasitic active components are also inherently present. These include the body–drain diode, the source–body–drain bipolar junction transistor, and a series junction field-effect transistor in the drain drift region. These are illustrated in Figures 3.15 and 3.16. These parasitics limit device performance and set constraints on device design. The body–drain diode is forward biased by any inverse voltage that develops across the transistor and so provides a reverse conduction path. Modern designs of transistors allow the body–drain diode to avalanche safely when reverse biased and so provide a means of dissipating energy stored in inductive switching circuits.

REFERENCES

1. H. C. Pao and C. T. Sah, "Effects of diffusion current on characteristics of metal–oxide (insulator)–semiconductor transistors." *Solid-State Electron.* **9,** 927–937 (1966).
2. R. F. Pierret and J. A. Shields, "Simplified long-channel MOSFET theory." *Solid-State Electron.* **26,** 143–147 (1983).
3. J. A. Cooper, Jr. and D. F. Nelson, "High-field drift velocity of electrons at the Si–SiO_2 interface as determined by a time-of-flight technique." *J. Appl. Phys.* **54,** 1445–1456 (1983).
4. G. Baccarani, M. R. Wordeman, and R. H. Dennard, "Generalised scaling theory and its application to a 1/4 micrometer MOSFET design." *IEEE Trans. Electron Devices* **ED-31,** 452–462 (1984).
5. C. Hu, M.-H. Chi, and V. M. Patel, "Optimum design of power MOSFET's." *IEEE Trans. Electron Devices* **ED-31,** 1693–1700 (1984).
6. H.-R. Chang and B. J. Baliga, "Impact of cell breakdown upon power DMOSFET on-resistance." *IEEE Trans. Electron Devices* **ED-34,** 2360 (1987).
7. V. A. K. Temple, "Ideal FET doping profile." *IEEE Trans. Electron Devices* **ED-30,** 619–626 (1983).
8. A. Lidow and T. Herman, "Process for manufacture of high power MOSFET with laterally distributed high carrier density beneath the gate." U.S. Pat. 4,593,302 (1986).
9. B. J. Baliga and J. P. Walden, "Improving the reverse recovery of power MOSFET integral diodes by electron irradiation." *Solid-State Electron.* **26,** 1133–1141 (1983).

CHAPTER 4

Principles of Operation 3: Transient and High-Frequency Behavior

4.1. INTRODUCTION

The basic action of the MOSFET lies in the control of the drain current by the gate voltage. It may be represented by the very simple equivalent circuit of Figure 4.1*a*. For purposes of transient analysis in this chapter, we make the following simplifying assumptions:

1. $I_D = g_{fs}(V_{GS} - V_T)$, provided that $V_{GS} > V_T$ and $V_{DS} > I_D R_{DS(on)}$.
2. For $V_{GS} < V_T$, $I_D = 0$, and the MOSFET is OFF.
3. When $g_{fs}(V_{GS} - V_T) > V_{DS}/R_{DS(on)}$, the MOSFET is fully ON and $I_D = V_{DS}/R_{DS(on)}$.

This transfer characteristic is illustrated in Figure 4.1*b*. It should be compared with the sets of dc characteristics of actual devices presented in Chapter 3, in particular, Figure 3.6. The simplification can be justified because of the rapid variation of I_D with V_{GS} below threshold. As a result, the transition between the conducting and nonconducting states occurs in a very short time, which can normally be neglected.

Fundamentally, the speed with which I_D can respond to a change in V_{GS} is governed by the time required to establish the corresponding change in the number of carriers in the inversion layer and by the transit time of these carriers along the length of the channel. Under normal operation the carriers are introduced from the source, and removed via the drain. The response time τ thus depends on the length of the channel and on either the mobility of the carriers or on their saturation drift velocity:

$$\tau = \frac{l}{v_d} = \frac{l^2}{\mu_e V_{Ch}} \tag{4.1}$$

or, if $V_{Ch}/l > E_s$,

$$\tau = l/v_s \tag{4.2}$$

In the latter case, taking $l = 1\ \mu m$ and $v_s = 9 \times 10^4$ m/s, we have $\tau = 11$ ps. Then, the upper cutoff frequency may be as high as $f_{co} = 1/2\pi\tau = 14$ GHz. In the fully

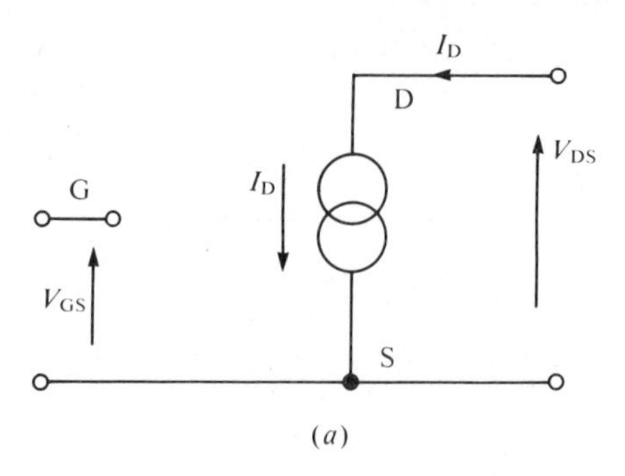

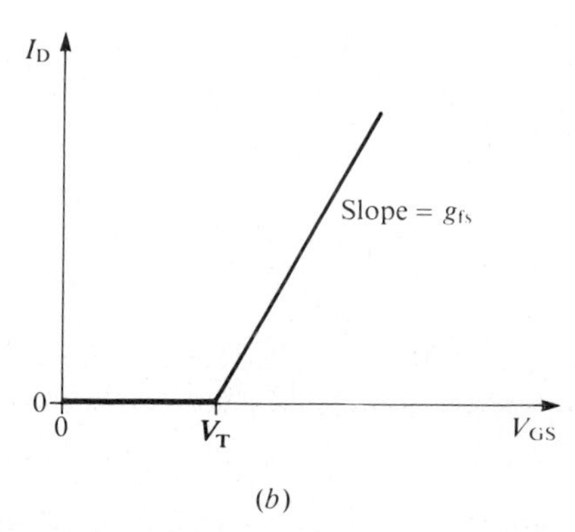

Figure 4.1. MOSFET model used in transient analysis. (*a*) Simple equivalent circuit; (*b*) simplified transfer characteristic:

$I_D = 0$ when $V_{GS} < V_T$ (OFF)

$I_D = g_{fs}(V_{GS} - V_T)$

when $V_{GS} > V_T$ and $g_{fs}(V_{GS} - V_T) < V_{DS}/R_{DS(on)}$ (ACTIVE)

$I_D = V_{DS}/R_{DS(on)}$,

when $V_{GS} > V_T$ and $g_{fs}(V_{GS} - V_T) > V_{DS}/R_{DS(on)}$ (ON)

turned-ON state, when $V_{Ch} < E_s l$, the response time is somewhat longer, as it is when V_{DS} is high and there is a depletion layer of significant width occupying the drain drift region and increasing the transit time proportionally. In either case, however, the basic transistor action is so fast that it is always dominated completely by circuit effects.

In general, in dealing with the transient behavior of electronic components, it is necessary to solve the physical equations governing the device behavior simultaneously with the circuit equations. Under steady-state conditions these are represented by the device characteristics and the load line, respectively. Under transient conditions it is usually extremely difficult to match the nonlinear device characteristics to the terminal characteristics imposed by the circuit. With power

MOSFETS this would be necessary only if microwave frequencies were attainable, but the circuits used and the device parasitics normally limit the operation to much lower frequencies. This greatly simplifies analysis, as we can assume that at any moment the value of the drain current is determined by the instantaneous values of the gate–source and drain–source voltages.

What does determine the switching speed or the frequency response in practice, then, is the time required to establish voltage changes across capacitances and current changes in inductances. At the device level these are the parasitic capacitances and inductances discussed in Section 3.6. Those most relevant to the transient device behavior are shown in the equivalent circuit of Figure 4.2. This should be compared with Figure 3.16. It includes the three interelectrode capacitances, the inductances of the drain and source leads, and the (distributed) resistance of the polycrystalline silicon gate. Of course, each of the electrode leads has parasitic resistance and inductance associated with it, but the components shown are those that are most critical in determining the transient response. Values of the source and drain lead inductances, L_S and L_D, are typically in the region of 5 to 15 nH. The gate resistance R_G is approximately inversely proportional to the active device area A_A. With devices having polycrystalline silicon gates, values of $R_G A_A$ are typically around 20 Ω mm^2. The gate resistance and gate capacitance are distributed together over the device. The implications of this are discussed in Section 4.5.

Transients always cause an increase in dissipation. During turn-on the energy stored in parasitic capacitance has to be absorbed. During turn-off the energy in inductances is dissipated. The circuit-imposed relation between the instantaneous

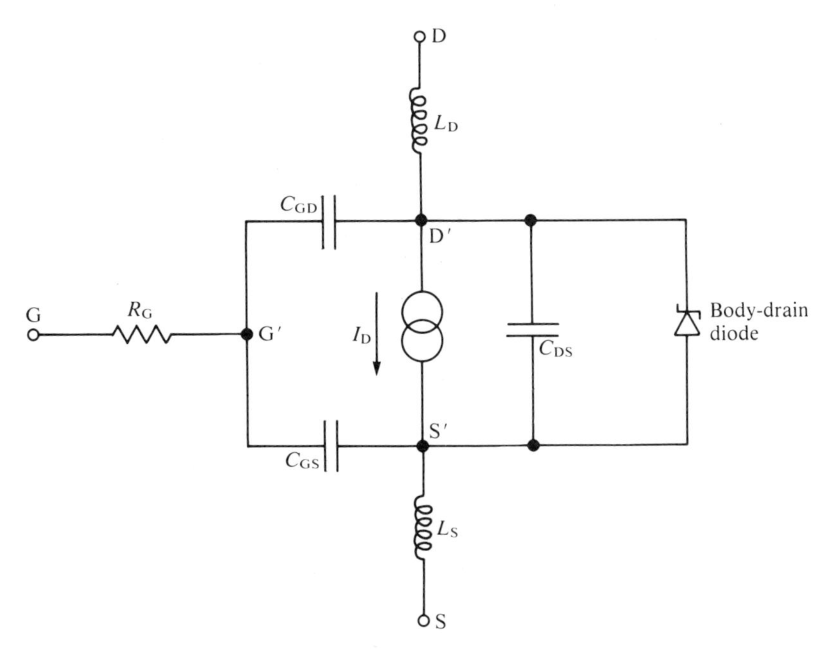

Figure 4.2. Equivalent circuit showing the MOSFET parasitics that have the greatest effect on transient behavior.

values of the device currents and the terminal voltages causes enhanced dissipation during each change of state. Fortunately, as we have just seen, the physical processes governing turn-on and turn-off do not, in practice, slow down these transitions. If they did, the transient dissipation would be further increased.

We shall concentrate on transitions between the fully ON and the fully OFF states, and consider in subsequent sections first turn-on and then turn-off. Under steady-state conditions the power dissipation is simply $I_D^2 R_{DS(on)}$. If the additional energy dissipated during turn-on is E_r, and during turn-off is E_f, and the switching frequency is f, the dissipation becomes:

$$P = I_D^2 R_{DS(on)} + (E_r + E_f)f \tag{4.3}$$

The magnitudes of E_r and E_f depend on the relationsip between the current in the device and voltage across it during the transition, and on the overall transition time, which, as we have said, is determined by the circuit.

For purposes of illustration, we next calculate E_r and E_f using two extreme cases. In each case, we neglect for the moment all the MOSFET parasitics, and simply assume that the changes in the drain current or the drain–source voltage take place linearly, over rise and fall times of τ_r and τ_f, respectively.

In the first example, the drain circuit is taken to be purely resistive, as shown in Figure 4.3. The voltage, current and instantaneous power waveforms are shown in Figure 4.4. For the turn-on transition, they are

$$i = \frac{V_0 t}{R\tau_r}$$

$$v = V_0 - Ri = V_0\left(1 - \frac{t}{\tau_r}\right)$$

$$p = vi = \frac{V_0^2}{R\tau_r}\left(1 - \frac{t}{\tau_r}\right)t \tag{4.4}$$

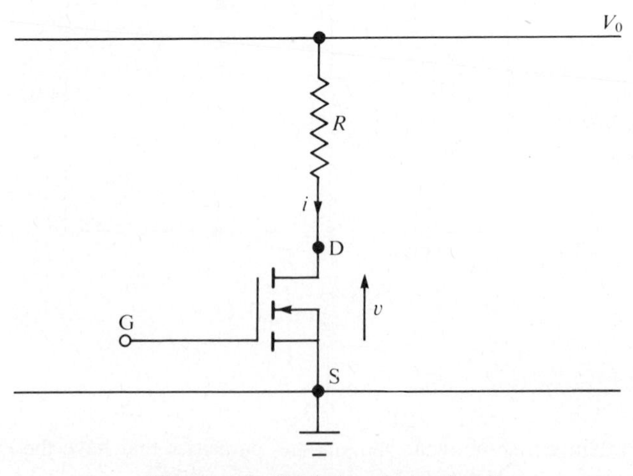

Figure 4.3. Resistive drain circuit.

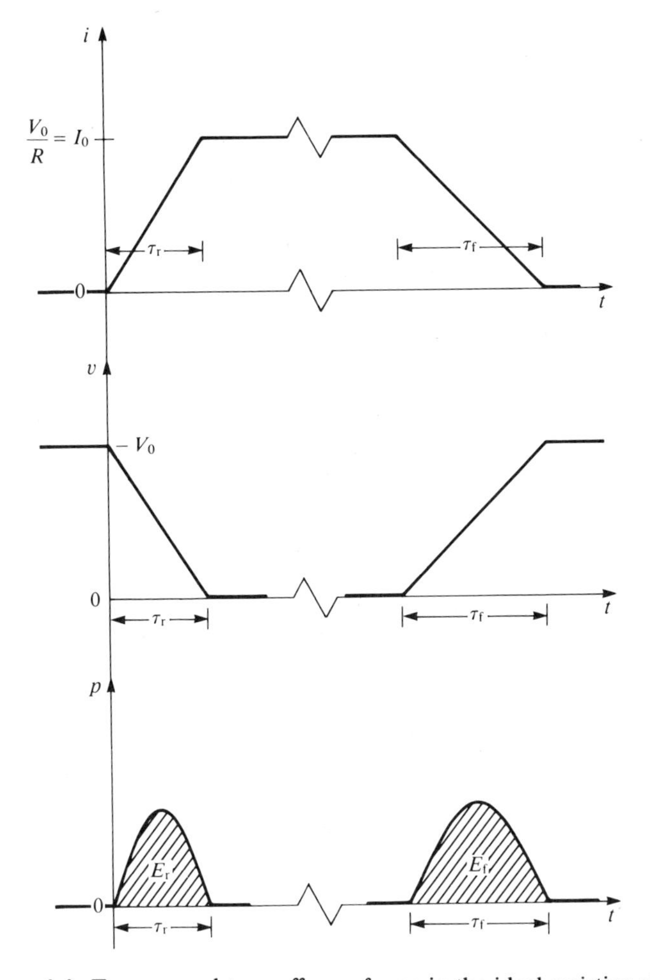

Figure 4.4. Turn-on and turn-off waveforms in the ideal resistive circuit.

Thus, the energy dissipated is

$$
\begin{aligned}
E_r &= \int_0^{\tau_r} p \, dt \\
&= \frac{V_0^2}{R\tau_r}\left[\frac{t^2}{2} - \frac{t^3}{3\tau_r}\right]_0^{\tau_r} \\
&= \frac{V_0^2}{R}\frac{\tau_r}{6} = \frac{V_0 I_0 \tau_r}{6} \qquad (4.5)
\end{aligned}
$$

A similar expression is easily derived for the turn-off transition, so that the total switching energy may be expressed as

$$E_r + E_f = \tfrac{1}{6} V_0 I_0 (\tau_r + \tau_f) \qquad (4.6)$$

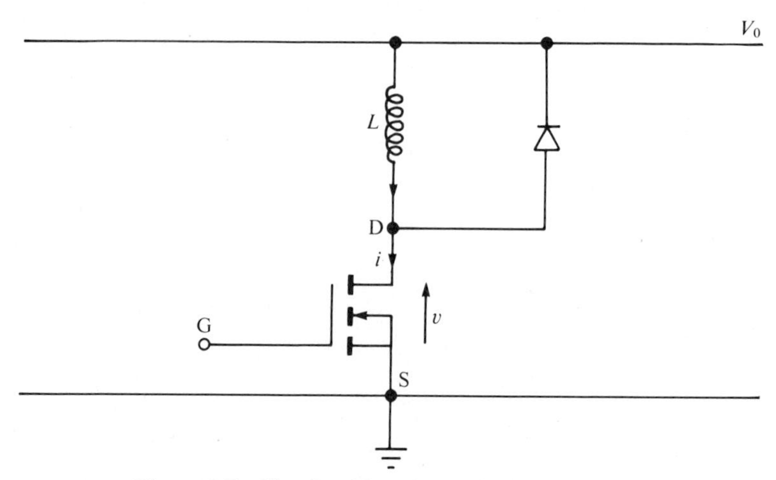

Figure 4.5. Circuit with a clamped inductive load.

The second extreme example is that of the clamped inductive load, as shown in Figure 4.5. This is typical of many applications where the load is a machine or transformer winding, perhaps one in which the current is being controlled using pulse-width modulation. It puts considerably greater stress on the MOSFET switch. Consider what happens when the transistor has been off for some time ($V_{GS} = 0$) and the circuit is completely dead, that is, $I_D = 0$ and $V_{DS} = V_0$. When the MOSFET is first switched on, current builds up in the load at a rate governed by V_0/L, where L is the total inductance. When at some later time the MOSFET is switched off, I_D has reached some value I_0. The inductance maintains this current through the transistor while the drain voltage rises (we assume linearly over a period τ_f) until it reaches $V_0 + 0.7$ V. Only then does the free-wheeling diode become conducting and the current divert from the MOSFET to circulate around the diode. Assume that the current I_0 is still flowing when the MOSFET is turned on again. (In fact it will have decayed to an extent that depends on the losses in the diode circuit.) Now, during turn-on, the voltage across the MOSFET is clamped to $V_0 + 0.7$ V until the drain current has risen to its full value, I_0. Only then can the diode come out of forward conduction, allowing the drain voltage to fall.

The voltage, current, and instantaneous-power waveforms during turn-on and turn-off are shown in Figure 4.6. They are repeated in subsequent switching cycles. A simple calculation shows that $E_r = \frac{1}{2}V_0 I_0 \tau_r$ and $E_f = \frac{1}{2}V_0 I_0 \tau_f$. Thus,

$$E_r + E_f = \tfrac{1}{2}V_0 I_0(\tau_r + \tau_f) \tag{4.7}$$

The clamped inductive load thus increases the transient power dissipation by a factor of three over that of the resistive load.

When an inductive load is not clamped by a free-wheeling diode, the rate of rise of current during each turn-on operation is limited by the inductance and may still be very small when the drain–source voltage has collapsed to its ON-state value. This greatly reduces E_r. However, during each turn-off transition, the energy in the inductance, $\frac{1}{2}LI_0^2$, has to be dissipated, and this adds to E_f. As we

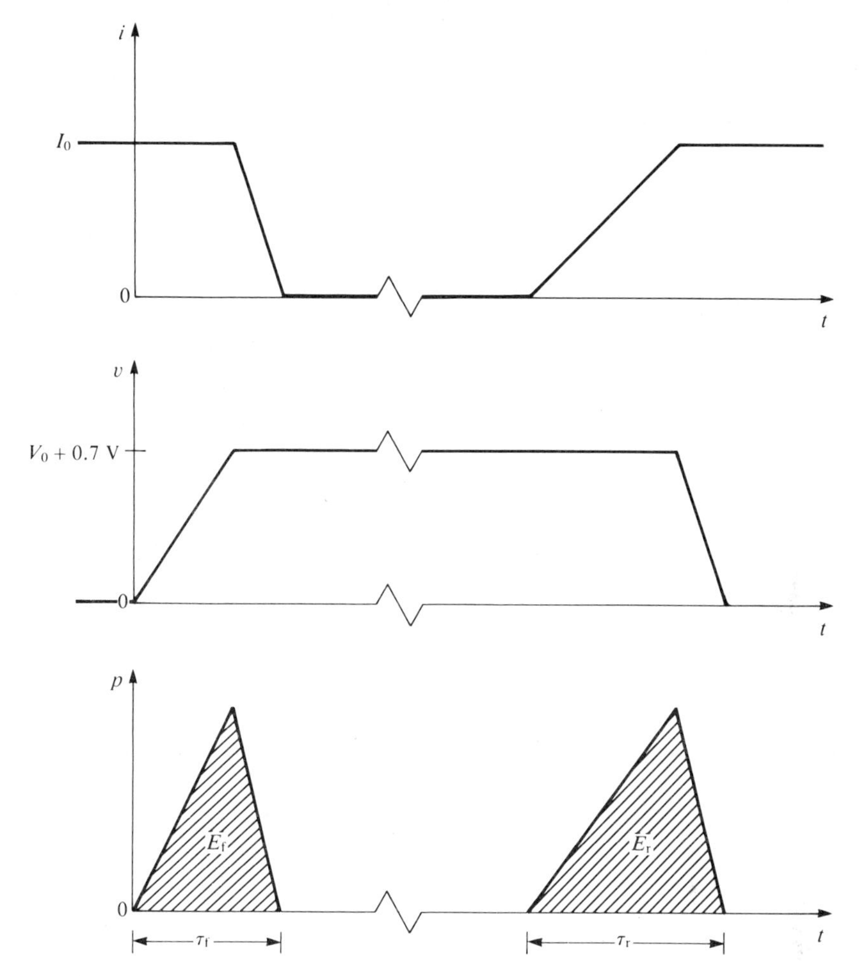

Figure 4.6. Turn-on and turn-off waveforms in an ideal circuit with a clamped inductive load.

shall see in later sections, parasitic drain inductance tends to act in this manner, and may throw the MOSFET into avalanche breakdown. In earlier devices it was necessary to protect against this by means of an external, parallel zener diode which absorbed the current. Modern designs of power MOSFET are able to withstand avalanche breakdown at the body–drain diode junction under these conditions, as was mentioned in Section 3.6.

These simple calculations demonstrate the order of magnitude of the transient dissipation to be expected. In general, of course, the voltage and current waveforms do not vary in the linear manner that we have assumed. However, the actual shape of the controlling voltage or current waveform serves only to modify the precise values ($\frac{1}{2}$ and $\frac{1}{6}$, respectively) of the constant terms derived in Equations (4.6) and (4.7). The essential dependence on V_0, I_0, and the period of the main transition, τ_r or τ_f, remains the same.

In the next two sections we examine in detail the waveforms to be expected, and the factors that determine the switching times, in a particular circuit. Turn-on is dealt with in Section 4.2, and turn-off in Section 4.3. The discussion follows closely the presentations cited in Refs. 1–3. In each case the transition time is divided into a sequence of consecutive intervals during each of which the MOSFET and the other active components are assumed to behave as linear circuit elements. In the transient circuit analysis, the state at the end of any one interval becomes the initial condition for the next. The analysis is carried through for one particular circuit configuration. However, the method can easily be applied to other circuit arrangements that may be more appropriate in other applications.

We have not attempted to make the analysis rigorous, and at each stage assumptions are made about the relative magnitudes of time constants associated with various combinations of circuit elements. The resulting simplifications are valid on the time scales of interest to us for a wide range of relevant component values.

4.2. TURN-ON

4.2.1. The Turn-on Sequence

The circuit we have chosen to illustate MOSFET switching behavior in detail is that of the partially clamped inductive load. This is shown in Figure 4.7, together with the relevant transistor parasitics. It is a frequently used arrangement, in which the

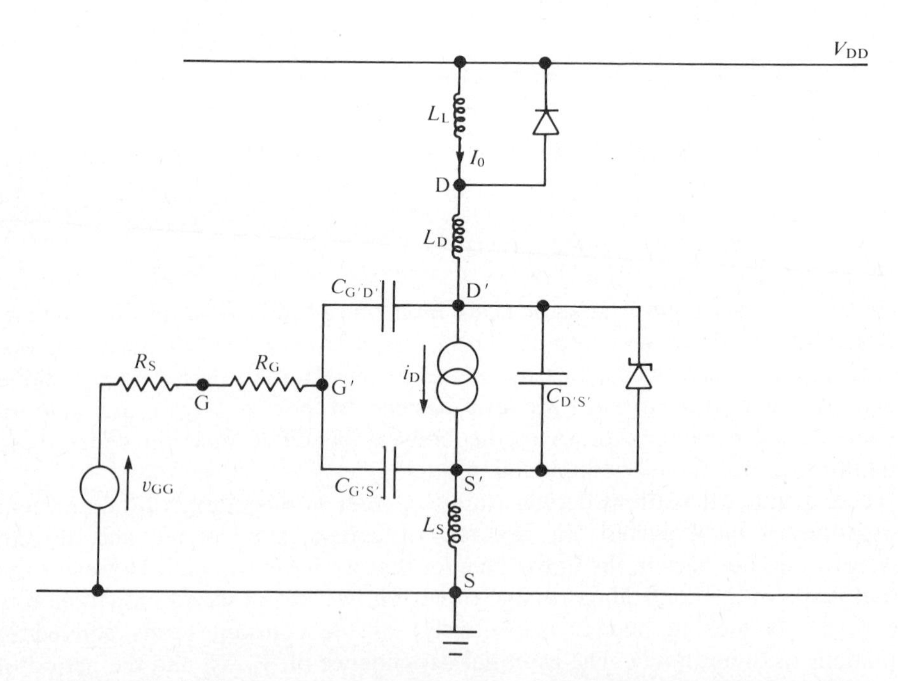

Figure 4.7. Equivalent MOSFET circuit with a partially clamped inductive load.

load is likely to be a machine or transformer winding. The inductance L_D includes not only the packaging inductance of the MOSFET, but also any unclamped portion of the load inductance, for example, the unclamped leakage inductance of a winding. Its value, therefore, may be as much as 0.1 or 1.0 μH.

Initially, the transistor is assumed to be turned off, $v_{G'S'} < V_T$. Thus, $i_D = 0$, and $v_{D'S'} = v_{DS} = V_{DD} + 0.7$ V. The load current circulates around the free-wheeling diode. At time zero, there is a step change in the gate supply voltage, v_{GG}, from V_{GL} ($<V_T$) to V_{GH} ($>V_T$). The subsequent turn-on transition follows at least four distinct phases. These are illustrated schematically in Figure 4.8.

The first period is the delay time τ_D, during which the gate voltage charges up to the threshold voltage V_T. This has no effect in the drain circuit. Next comes the crucial phase, during which there is both a rise in the drain current i_D and a fall in the drain–source voltage $v_{D'S'}$. This phase occupies a time τ_1. It ends *either* when $v_{D'S'}$ has fallen to a low value, so that the transistor is fully ON, *or* when i_D reaches I_0, so that the free-wheeling diode comes out of conduction and the voltage across it can rise. Which of these two possibilities occurs first depends in a rather complicated way on the characteristics of the device and the circuit. This is discussed in detail in Section 4.2.5. In Figure 4.8 we have illustrated the more probable case that the voltage fall is completed before the rise in current. This requires a certain minimum value of the inductance L_D.

In the next phase, which occupies a period τ_2, the drain current completes its rise to I_0 with the MOSFET fully ON, or alternatively, the voltage continues its fall while the current remains at I_0. The final phase sees the gate voltage complete its rise towards V_{GH}, but as the MOSFET is already fully ON, this has negligible effect in the drain circuit. It is assumed throughout the discussion that $V_{GH} - V_T$ is large enough to permit the drain current to rise to its full value of I_0.

4.2.2. The Delay Time

Following the assumed sharp rise in v_{GG}, the resultant gate current, i_G, charges the two gate capacitances, $C_{G'S'}$ and $C_{G'D'}$, which appear in parallel. The effects of L_D and L_S are neglected, and $V_{S'}$ and $V_{D'}$ are assumed to remain constant. The result is that the gate voltage $v_{G'S'}$ charges up exponentially towards V_{GH} with a time constant τ_G given by

$$\tau_G = (R_S + R_G)(C_{G'S'} + C_{G'D'}) \tag{4.8}$$

That is,

$$v_{G'S'} = (V_{GH} - V_{GL})[1 - \exp(-t/\tau_G)] + V_{GL} \tag{4.9}$$

This initial period, the delay time τ_D, lasts until $v_{G'S'} = V_T$. The MOSFET then starts to turn ON. The delay time is given by

$$\tau_D = \tau_G \ln\left(\frac{V_{GH} - V_{GL}}{V_{GH} - V_T}\right) \tag{4.10}$$

Values of $C_{G'S'} + C_{G'D'} = C_{iss}$ typically lie in the range 0.1 to 2.0 nF, as long as

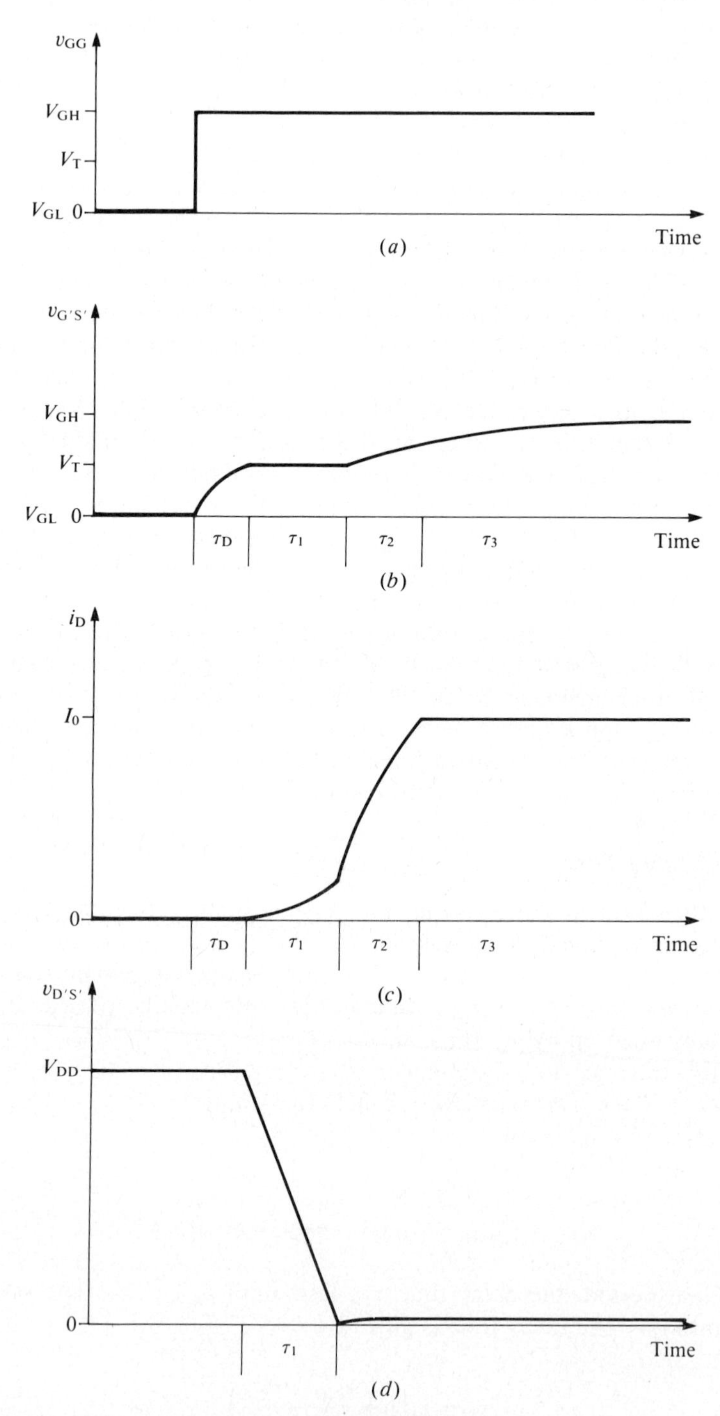

Figure 4.8. Schematic illustration of a typical turn-on sequence. (*a*) Gate supply voltage; (*b*) gate–source voltage; (*c*) drain current; (*d*) drain–source voltage.

$v_{D'S'}$ remains high. In order to keep τ_G somewhere in the range of 5 to 20 ns, $R_S + R_G$ needs to be between 10 and 100 Ω. Larger chips having proportionately higher values of C_{iss} require lower values of gate resistance. A typical gate drive circuit might have $V_{GH} = 10$ V and $V_{GL} = 0$ V. With $V_T \approx 4$ V, the delay time τ_D would approximately equal τ_G.

4.2.3. The Main Transition Period: General Solution

The MOSFET now enters its ACTIVE state, in which $i_D = g_{fs}(v_{GS} - V_T)$. This is rather more difficult to analyze, and really any detailed solutions require computer methods. However, in order to bring out the underlying effects, we shall work through an analytical solution, making simplifying assumptions where necessary.

The first thing to note is that two components give rise to negative feedback from the drain circuit to the gate circuit. These are the source inductance L_S and the gate–drain capacitance $C_{G'D'}$. The former causes a reduction in $v_{G'S'}$ by an amount that depends on the rate of rise of the drain current. The latter acts as a *Miller* capacitance. The rising drain current causes the drain voltage to fall, and this increases $dv_{G'D'}/dt$. As a result, a higher proportion of i_G flows through $C_{G'D'}$, and the rate of rise of $v_{G'S'}$ is much reduced. The rate of rise of i_D is lowered in proportion.

During the main transition period, which occupies the time τ_1, both the drain current and the drain–source voltage vary. However, in most circuits the variation of one dominates that of the other. Whether the drain voltage collapses before there is any appreciable rise in current, or vice versa, depends on several factors which are discussed in detail in Section 4.2.5. Essentially, when L_D is small and $R_S + R_G$ is large, the drain current can reach its maximum value, I_0, before there is any significant fall in the drain–source voltage. This can be avoided by increasing L_D, thereby making the drain circuit "slower", or by reducing $R_S + R_G$, thereby making the gate circuit "faster".

In order to obtain analytical solutions for the period τ_1, without overcomplicating the algebra and so obscuring the essential argument, we neglect at the outset the source lead inductance L_S. During the analysis we also neglect the drain–source capacitance $C_{D'S'}$. We return later to consider the effects of these components, and those of the snubber circuits that are sometimes used to suppress the rate of rise of drain voltage during turn-off. The circuit equations remain third-order. However, the time constant associated with the third-order term is very short, so this term too may be neglected without significantly affecting the solutions.

The simplified equivalent circuit is shown in Figure 4.9. We develop its characteristic equation using s to denote the differential operator, and solve by inspection so as to match the initial conditions. The circuit equations then take the form

$$i_G = sC_{G'D'}v_{G'D'} + sC_{G'S'}v_{G'S'} = \frac{V_{GH} - v_{G'S'}}{R_G + R_S}$$

$$\therefore \quad v_{G'D'} = \frac{(V_{GH} - v_{G'S'})/(R_S + R_G) - sC_{G'S'}v_{G'S'}}{sC_{G'D'}} \tag{4.11}$$

$$v_{D'S'} = V_{DD} - 0.7\text{ V} - sL_D i_D$$

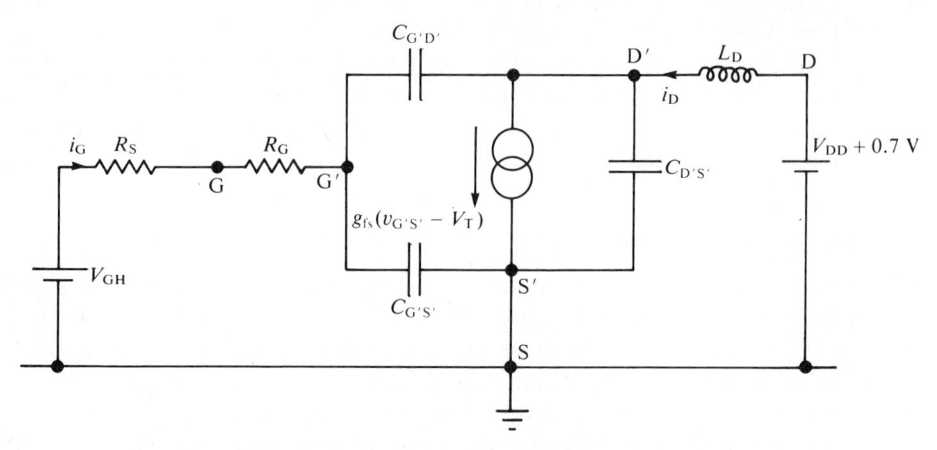

Figure 4.9. Simplified equivalent circuit for the main transition period.

$$i_G + i_D = \frac{V_{GH} - v_{G'S'}}{R_S + R_G} + \frac{V_{DD} - 0.7\text{ V} - v_{D'S'}}{sL_D}$$

$$= g_{fs}(v_{G'S'} - V_T) + sC_{G'S'}v_{G'S'} + sC_{D'S'}v_{D'S'}$$

$$\therefore \quad v_{D'S'} = \frac{V_{DD} - 0.7\text{ V} - sL_D v_{G'S'}[g_{fs} + 1/(R_S + R_G) + sC_{G'S'}]}{1 + sL_D C_{D'S'}} \tag{4.12}$$

$$v_{G'S'} = v_{G'D'} + v_{D'S'}$$

Substituting Equations (4.11) and (4.12) and eliminating all the differentiated dc terms such as $sC_{G'D'}(R_S + R_G)(V_{DD} - 0.7\text{ V})$ gives

$$v_{G'S'} = \frac{V_{GH}}{s^3\tau_{G3}^3 + s^2\tau_{G2}^2 + s\tau_G + 1}$$

where

$$\tau_{G3}^3 = L_D(R_S + R_G)(C_{G'D'}C_{G'S'} + C_{G'S'}C_{D'S'} + C_{D'S'}C_{G'D'})$$

$$\tau_{G2}^2 = L_D\{C_{D'S'} + C_{G'D'}[1 + g_{fs}(R_S + R_G)]\}$$

and τ_G is given by Equation (4.8). Neglecting τ_{G3} and the terms involving $C_{D'S'}$, and assuming $g_{fs}(R_S + R_G) \gg 1$, this reduces to

$$v_{G'S'} = \frac{V_{GH}}{s^2\tau_m\tau_{G'} + s\tau_G + 1} \tag{4.13}$$

where $\tau_m = g_{fs}L_D$, $\tau_{G'} = C_{G'D'}(R_S + R_G)$, and as before $\tau_G = (C_{G'S'} + C_{G'D'}) \times (R_S + R_G)$.

We redefine time zero to coincide with the start of this period, at which point the initial conditions are $v_{G'S'} = V_T$ and $i_D = 0$. As $t \to \infty$, $v_{G'S'} \to V_{GH}$. Equation

(4.13) then solves to give either sinusoidal or exponential solutions, depending on the relative magnitudes of τ_G and $\tau_m\tau_{G'}$. Sinusoidal solutions occur when $\tau_G^2 < 4\tau_m\tau_{G'}$. They take the form:

$$v_{G'S'} = V_{GH} - (V_{GH} - V_T)\exp\left(-\frac{t}{\tau_a}\right)\left(\cos\omega_a t + \frac{1}{\omega_a\tau_a}\sin\omega_a t\right) \tag{4.14}$$

where

$$\tau_a = \frac{2\tau_m\tau_{G'}}{\tau_G} = \frac{2g_{fs}L_D C_{G'D'}}{C_{G'S'} + C_{G'D'}} \tag{4.15}$$

and

$$\omega_a^2 = \frac{1}{\tau_m\tau_{G'}} - \left(\frac{\tau_G}{2\tau_m\tau_{G'}}\right)^2$$
$$= \frac{1}{g_{fs}L_D C_{G'D'}(R_S + R_G)} - \left(\frac{C_{G'S'} + C_{G'D'}}{2g_{fs}L_D C_{G'D'}}\right)^2 \tag{4.16}$$

The drain current and drain voltage waveforms are

$$i_D = g_{fs}(v_{G'S'} - V_T)$$
$$= g_{fs}(V_{GH} - V_T)\left[1 - \exp\left(-\frac{t}{\tau_a}\right)\left(\cos\omega_a t + \frac{1}{\omega_a\tau_a}\sin\omega_a t\right)\right] \tag{4.17}$$

$$v_{D'S'} = V_{DD} + 0.7\text{ V} - g_{fs}L_D\frac{dv_{G'S'}}{dt}$$
$$= V_{DD} + 0.7\text{ V} - g_{fs}L_D\omega_a(V_{GH} - V_T)\exp\left(-\frac{t}{\tau_a}\right)\left[1 + \left(\frac{1}{\omega_a\tau_a}\right)^2\right]\sin\omega_a t \tag{4.18}$$

When $\tau_G^2 > 4\tau_m\tau_{G'}$, exponential solutions are obtained. Then, the equations take the form

$$v_{G'S'} = V_{GH} - (V_{GH} - V_T)\frac{\tau_b\exp(-t/\tau_b) - \tau_c\exp(-t/\tau_c)}{\tau_b - \tau_c} \tag{4.19}$$

$$i_D = g_{fs}(V_{GH} - V_T)\left(1 - \frac{\tau_b\exp(-t/\tau_b) - \tau_c\exp(-t/\tau_c)}{\tau_b - \tau_c}\right) \tag{4.20}$$

$$v_{D'S'} = V_{DD} + 0.7\text{ V} - g_{fs}L_D(V_{GH} - V_T)\frac{\exp(-t/\tau_b) - \exp(-t/\tau_c)}{\tau_b - \tau_c} \tag{4.21}$$

In these equations,

$$\tau_b = \frac{2\tau_m\tau_{G'}}{\tau_G - (\tau_G^2 - 4\tau_m\tau_{G'})^{1/2}} \tag{4.22}$$

$$\tau_c = \frac{2\tau_m\tau_{G'}}{\tau_G + (\tau_G^2 - 4\tau_m\tau_{G'})^{1/2}} \tag{4.23}$$

Although these equations look very different from the sinusoidal ones, the waveforms can be very similar, as we shall see when we examine specific examples.

None of these analytical expressions takes any account of the nonlinearity of the device capacitances. This becomes quite important as the drain voltage falls below the gate voltage. Then, $C_{G'D'}$ increases sharply, as shown in Figure 2.11 in Section 2.4. This slows the rate of fall of $v_{D'S'}$ in the final stages of its collapse, once $v_{D'S'} < v_{G'S'}$. As a result there can be a significant increase in dissipation, because during this period there is usually a substantial drain current flowing.

4.2.4. Device and Circuit Time Constants

The relative magnitudes of the time constants τ_m, τ_G, and $\tau_{G'}$ play a crucial role in determining the analytical form of the turn-on waveforms. We have seen that whether τ_G is greater than or less than $2(\tau_m \tau_{G'})^{1/2}$ determines whether the solutions during the period τ_1 are exponential or sinusoidal. Perhaps surprisingly, this has little bearing on the waveforms themselves. The reason is that the time intervals are usually of quite short duration in comparison with the time constants τ_a, τ_b, τ_c, and $1/\omega_a$. What is much more important is that these parameters help to determine whether it is the current or the voltage transition that is completed first.

The condition for sinusoidal solutions is

$$\tau_m > \tau_G^2 / 4\tau_{G'} \tag{4.24}$$

This may be expanded in terms of the equivalent circuit parameters as

$$g_{fs} L_D > \frac{(C_{G'S'} + C_{G'D'})^2 (R_S + R_G)}{4C_{G'D'}} \tag{4.25}$$

It can be further rearranged to give

$$\frac{L_D}{R_S + R_G} > \frac{(C_{G'S'} + C_{G'D'})^2}{4 g_{fs} C_{G'D'}} \tag{4.26}$$

The term on the left-hand side is a time constant associated with the circuit:

$$\tau_{cct} = \frac{L_D}{R_S + R_G} \tag{4.27}$$

The term on the right-hand side is a time constant associated with the device:

$$\tau_{dev} = \frac{(C_{G'S'} + C_{G'D'})^2}{g_{fs} C_{G'D'}} = \frac{C_{iss}^2}{g_{fs} C_{rss}} \tag{4.28}$$

Because the capacitances and the transconductance are all expected to scale to first order with the active area of the MOSFET, we might think of τ_{dev} as a *process* parameter. It depends on factors such as the oxide thickness, the overlaps, the doping concentrations, and the gate width per unit area. With higher-voltage devices we expect $C_{G'D'}$ to decrease and hence τ_{dev} to increase. For 100-V chips it is typically in the region of 1 ns, whereas for 500-V chips it is more like 3 ns.

4.2.5. The Main Transition Period: Approximate Solutions

The solutions can be simplified under the conditions that L_D is very small or very large. In the first case, $\tau_{cct} \ll \tau_{dev}/4$. As a result, $\tau_b \approx \tau_G \gg \tau_c$, and Equations (4.19) to (4.21) approximate to

$$v_{G'S'} = V_{GH} - (V_{GH} - V_T)\exp\left(-\frac{t}{\tau_G}\right) \tag{4.29}$$

$$i_D = g_{fs}(V_{GH} - V_T)\left[1 - \exp\left(-\frac{t}{\tau_G}\right)\right] \tag{4.30}$$

$$v_{D'S'} = V_{DD} + 0.7\text{ V} - \frac{g_{fs}L_D(V_{GH} - V_T)}{\tau_G}\exp\left(-\frac{t}{\tau_G}\right)$$

$$= V_{DD} + 0.7\text{ V} - \frac{\tau_m}{\tau_G}(V_{GH} - V_T)\exp\left(-\frac{t}{\tau_G}\right) \tag{4.31}$$

When L_D is very large, $\tau_{cct} \gg \tau_{dev}/4$, the second term on the right-hand side of Equation (4.16) is negligible in comparison with the first term. As a result, $\omega_a^2 \approx 1/\tau_m\tau_{G'}$, $\tau_a \gg \tau_G$, and $\omega_a\tau_a \gg 1$. Thus, Equations (4.14), (4.17), and (4.18) approximate to

$$v_{G'S'} = V_{GH} - (V_{GH} - V_T)\cos\omega_a t \tag{4.32}$$

$$i_D = g_{fs}(V_{GH} - V_T)(1 - \cos\omega_a t) \tag{4.33}$$

$$v_{D'S'} = V_{DD} + 0.7\text{ V} - g_{fs}L_D\omega_a(V_{GH} - V_T)\sin\omega_a t$$

$$= V_{DD} + 0.7\text{ V} - \omega_a\tau_m(V_{GH} - V_T)\sin\omega_a t \tag{4.34}$$

A much more important approximation can be made, provided that $g_{fs}(V_{GH} - V_T) \gg I_0$. Then, τ_1 is sufficiently short to permit the expansion of the exponential, sine, and cosine terms of both the sinusoidal and the exponential solutions. Under these conditions (which are not compatible with L_D very small) both sets of equations (4.14), (4.17), (4.18) and (4.19) to (4.21) reduce, to first order, to

$$v_{G'S'} = V_T + (V_{GH} - V_T)t^2/2\tau_m\tau_{G'} \tag{4.35}$$

$$i_D = g_{fs}(V_{GH} - V_T)t^2/2\tau_m\tau_{G'}$$

$$= (V_{GH} - V_T)t^2/2L_D\tau_{G'} \tag{4.36}$$

$$v_{D'S'} = V_{DD} + 0.7\text{ V} - (V_{GH} - V_T)t/\tau_{G'} \tag{4.37}$$

Note that the time τ_v taken for the drain voltage to fall from V_{DD} to zero is then

$$\tau_v = \frac{V_{DD}}{dv_{D'S'}/dt} = \frac{V_{DD}\tau_{G'}}{V_{GH} - V_T} \tag{4.38}$$

In the final stage of the voltage collapse, when $v_{G'D'} < 0$, the rate of fall of $v_{D'S'}$ can decrease by as much as a factor of 10, because of the increase in $C_{G'D'}$ and hence of $\tau_{G'}$.

Substitution into Equation (4.36) shows that the time τ_i required for the drain current to rise from zero to I_0 is

$$\tau_i = \left(\frac{2I_0L_D\tau_{G'}}{V_{GH} - V_T}\right)^{1/2} \tag{4.39}$$

Thus the condition that the voltage should complete its fall before the current has risen to I_0 is

$$k = (\tau_i/\tau_v)^2 > 1$$

where

$$k = \frac{2(V_{GH} - V_T)}{V_{DD}} \frac{L_D I_0}{V_{DD}\tau_{G'}}$$

$$= \frac{2(V_{GH} - V_T)}{V_{DD}} \frac{L_D}{R_0\tau_{G'}} \tag{4.40}$$

$$= \frac{2(V_{GH} - V_T)}{V_{DD}} \frac{\tau_{Dr}}{\tau_{G'}} \tag{4.41}$$

We have substituted $R_0 = V_{DD}/I_0$, and put $\tau_{Dr} = L_D/R_0$, representing respectively an effective drain-circuit impedance and an effective drain-circuit time constant.

Early in the analysis, in Section 4.2.3, we neglected the drain–source capacitance $C_{D'S'}$. It can be seen from Equation (4.37) that the rate of fall of the drain voltage is $(V_{GH} - V_T)/\tau_{G'}$. If we take as typical values $V_{GH} - V_T = 5$ V, $\tau_{G'} = 1$ ns, and $C_{D'S'} = 0.2$ nF, the current required to discharge this capacitance is 1 A. The capacitative current in the MOSFET is thus quite significant in comparison with the external current i_D. As a result we must expect the rate of fall of the drain voltage to be reduced below the values predicted by the simplified theory. The energy stored in $C_{D'S'}$, $\frac{1}{2}C_{D'S'}V_{DD}^2$, is dissipated in the transistor during the turn-on period.

We have also left the source inductance L_S out of the analysis so far. This is justifiable when the drain inductance is large and $k \gg 1$. But with reduced drain inductance its effect can be considerable. According to Equation (4.36), the maximum rate of rise of drain current occurs at the end of τ_1. Differentiating this equation and substituting $t = \tau_1 = \tau_v$ from Equation (4.38) gives a peak value of V_{DD}/L_D for di_D/dt. This would introduce a voltage $V_{DD}L_S/L_D$ into the gate drive circuit. Only when $L_S/L_D \ll V_{GH}/V_{DD}$ can this be discounted.

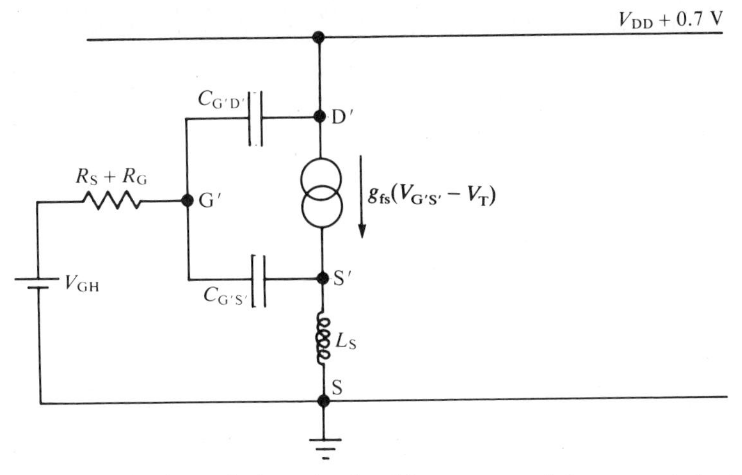

Figure 4.10. A simple equivalent circuit taking account of the source inductance.

When $k \ll 1$, an approximate analysis which takes account of L_S can be made, on the assumption that there is very little change in $v_{D'S'}$ during the rise of current. We are therefore neglecting the effects of L_D and $C_{D'S'}$, and can put $dv_{G'D'}/dt = dv_{G'S'}/dt$. The equivalent circuit that then applies is shown in Figure 4.10. The circuit equations become

$$v_{S'S} = sL_S(sC_{G'S'}v_{G'S'} + g_{fs}v_{G'S'})$$

and

$$\begin{aligned} V_{GH} &= v_{S'S} + v_{G'S'} + (R_S + R_G)(sC_{G'S'}v_{G'S'} + sC_{G'D'}v_{G'D'}) \\ &= v_{G'S'}[s^2L_SC_{G'S'} + s\{(R_S+R_G)(C_{G'S'}+C_{G'D'}) + g_{fs}L_S\} + 1] \\ &= v_{G'S'}(s^2\tau_s^2 + s\tau_f + 1) \end{aligned} \tag{4.42}$$

where

$$\tau_s = \sqrt{L_SC_{G'S'}} \tag{4.43}$$

and

$$\tau_f = (R_S+R_G)(C_{G'S'}+C_{G'D'}) + g_{fs}L_S = \tau_G + g_{fs}L_S \tag{4.44}$$

Usually $\tau_f \gg \tau_s$, so that the term $s^2\tau_s^2$ in Equation (4.42) can be neglected on the time scale of interest to us. Solutions then take the form given in Equations (4.29)–(4.31), where L_S was neglected, except that now τ_f replaces τ_G as the time constant in the exponential function. The presence of source inductance "slows down" the gate circuit. Another way of looking at its effect is to say that it appears to introduce a further resistance R_{LS} into the gate circuit, given by

$$R_{LS} = \frac{g_{fs}L_S}{C_{G'S'} + C_{G'D'}} \tag{4.45}$$

A proper analysis becomes very complicated, but the addition of R_{LS} to $R_S + R_G$, when evaluating the solutions given in Section 4.2.3 and earlier in this section, represents a reasonable first approximation. In Equations (4.19)–(4.23), τ_f should be substituted for τ_G, and τ_s^2 for $\tau_m \tau_{G'}$. Certainly we cannot expect di_D/dt to exceed the maximum value predicted by this analysis, namely, $(V_{GH} - V_T)/L_S$, at $t = 0$. With $V_{GH} - V_T = 5$ V and $L_S = 10$ nH, this is 0.5 A/ns.

4.2.6. The Periods Following the Main Transition

At the end of the main transition period, τ_1, one of two situations obtains. Either $v_{D'S'} \approx 0$ and $i_D = I_{D*}$, or $i_D = I_0$ and $v_{D'S'} = V_{D*}$. We discuss each case in turn.

The more common situation is for L_D to be large enough and $R_S + R_G$ to be small enough to allow the drain voltage to complete its fall, at least to $v_{G'S'}$, while i_D is still rising. During the subsequent period, τ_2, the drain voltage remains low while the drain current rises from I_{D*} to I_0. As $v_{D'S'}$ is then relatively constant, the Miller effect no longer operates, and the input circuit reverts to that of the initial switch-on period τ_D. The gate current supplies $C_{G'S'}$ and $C_{G'D'}$ in parallel, and $v_{G'S'}$ resumes its exponential approach towards V_{GH}. However, the time constant is now substantially longer than it was during the delay period, because $C_{G'D'}$ is now much larger. To distinguish it from τ_G, we shall write it as $\tau_{G''}$. Thus,

$$v_{G'S'} = (V_{GH} - V_{G*})[1 - \exp(-t/\tau_{G''})] + V_{G*} \tag{4.46}$$

with $V_{G*} = V_T + I_{D*}/g_{fs}$.

Because the MOSFET is now fully ON, the drain current rises at a rate determined only by the drain inductance, that is, as long as $i_D < g_{fs}(v_{G'S'} - V_T)$. With $v_{D'S'} \approx 0$, V_{DD} is entirely dropped across L_D and $di_D/dt \approx V_{DD}/L_D$. Thus,

$$i_D = I_{D*} + \frac{V_{DD}}{L_D}t \tag{4.47}$$

The current rises linearly until it reaches I_0, which marks the end of τ_2. Thus,

$$\tau_2 = L_D \frac{I_0 - I_{D*}}{V_{DD}} \tag{4.48}$$

An equivalent circuit which includes the source inductance and applies in period τ_2 is shown in Figure 4.11*a*. It can be further reduced to the circuit of Figure 4.11*b* on the assumption that $R_{DS(on)}$ is negligibly small. As long as $L_S < (V_{GH}/V_{DD})L_D$, the approximation of Equation (4.47) remains valid. If this condition is not satisfied, it is clear from the figure that the gate supply voltage cannot maintain the transistor in its ON state. It then remains in the ACTIVE

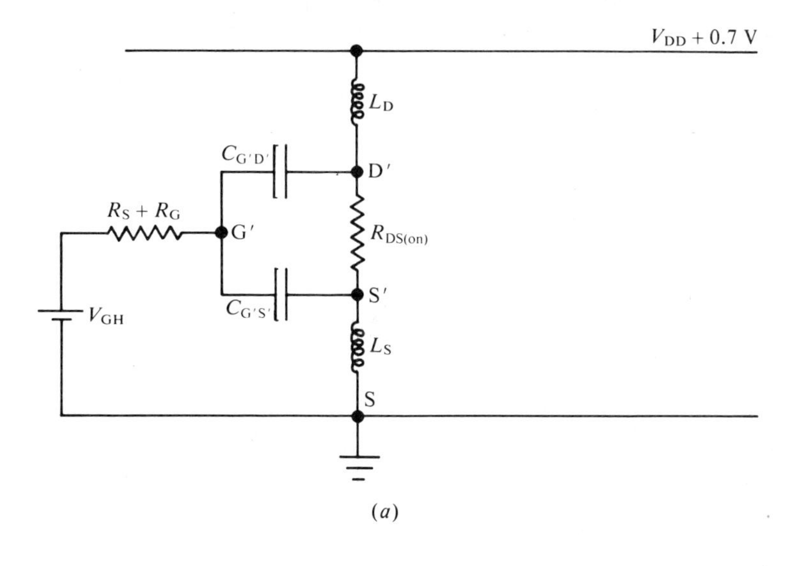

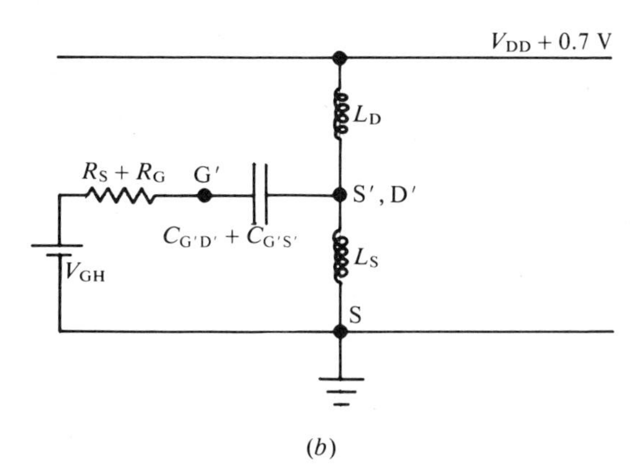

Figure 4.11. Equivalent circuits, including the source inductance, for the period of the rise of current following the collapse of the drain voltage. (*a*) Including $R_{DS(on)}$; (*b*) reduced equivalent circuit neglecting $R_{DS(on)}$.

condition until the rise of current is complete. This could be thought of as the result of the effective resistance R_{LS} in the gate circuit.

We next deal with the situation that applies when $k < 1$, and the drain current reaches I_0 before $v_{D'S'}$ becomes small. Having already reached I_0, the drain current i_D and the gate–source voltage $v_{G'S'}$ remain constant. As a result, $C_{G'S'}$, L_S, and L_D can be disregarded during the remainder of the transition period, τ_2. The corresponding equivalent circuit is shown in Figure 4.12. The gate current serves to discharge $C_{G'D'}$. It is constant, so the drain voltage falls at a constant rate. Summing currents at the node D', we obtain

$$g_{fs}(v_{G'S'} - V_T) = I_0 + i_G - C_{D'S'}\frac{dv_{D'S'}}{dt}$$

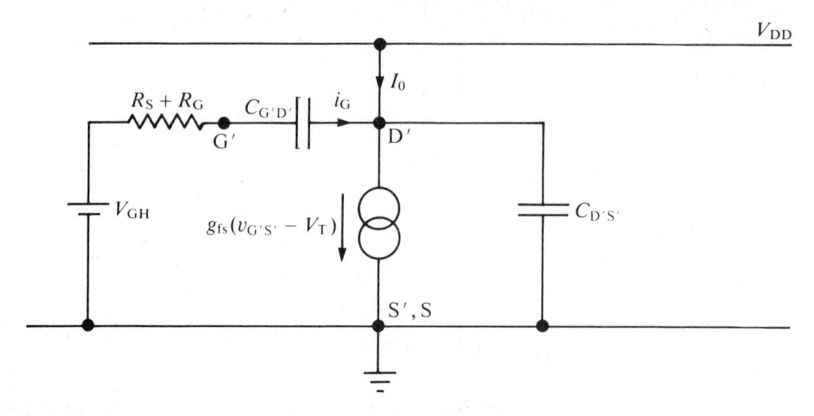

Figure 4.12. Equivalent circuit for the fall of the drain voltage after the drain current has reached I_0.

But

$$i_G = C_{G'D'} \frac{dv_{G'D'}}{dt}$$

and

$$v_{G'S'} = V_{GH} - (R_S + R_G)i_G$$

$$\therefore \quad I_0 + C_{G'D'} \frac{dv_{G'D'}}{dt} - C_{D'S'} \frac{dv_{D'S'}}{dt} = g_{fs}\left((V_{GH} - V_T) - (R_S + R_G)C_{G'D'} \frac{dv_{G'D'}}{dt}\right)$$

Also, $dv_{G'D'}/dt = -dv_{D'S'}/dt$, so that

$$-\frac{dv_{D'S'}}{dt} = \frac{g_{fs}(V_{GH} - V_T) - I_0}{C_{D'S'} + [1 + g_{fs}(R_S + R_G)]C_{G'D'}}$$

$$\therefore \quad v_{D'S'} = V_{D*} - Kt \tag{4.49}$$

where

$$K = \frac{g_{fs}(V_{GH} - V_T) - I_0}{C_{D'S'} + [1 + g_{fs}(R_S + R_G)]C_{G'D'}} \tag{4.50}$$

The duration of period τ_2 is thus V_{D*}/K. Note, however, that the increase in $C_{G'D'}$ as $v_{G'D'}$ goes through zero causes the value of K to be sharply reduced at the end of this phase. It is sensible, then, to seperate τ_2 into two parts. Period τ_{21} ends when $v_{D'S'} = v_{G'S'}$. There is then a further period, τ_{22}, during which $v_{D'S'}$ falls more slowly from $v_{G'S'} = V_T + I_0/g_{fs}$ to zero. We can denote the values of K in periods τ_{21} and τ_{22} by K_1 and K_2, respectively.

Of course, the observed value of $v_{D'S'}$ falls not to zero but to $I_0R_{DS(on)}$. However, for the purposes of the transient analysis we regard this as a steady-state voltage which has to be superimposed onto any additional transient voltages. We therefore neglect it here. Inside the MOSFET the voltage $I_0 R_{DS(on)}$ is dropped mainly in the drain drift region, regions Ⓒ and Ⓓ of Figure 2.1. When the transistor approaches its turned-on state, any additional component of $v_{D'S'}$ is dropped in the channel region, Ⓐ in Figure 2.1. In particular, before an accumulation layer can form in region Ⓑ, the potential there must be less than the gate potential. It is thus reasonable to decouple these two contributors to the observed value of v_{DS}.

4.2.7. Worked Examples

At this point it is worthwhile working through some numerical examples, to illustrate the different forms taken by the analytical functions that we have derived. We start with a set of parameters, representing a typical low-voltage circuit of moderate drain current. It uses a "fast" gate circuit having a total resistance of 10 Ω. We examine the effect of changing drain inductance, starting with a relatively high value. The chosen parameters are:

$$V_{DD} = 50\text{ V}, \quad I_0 = 10\text{ A}, \quad g_{fs} = 5\text{ S}, \quad L_D = 1\ \mu\text{H}$$

$$V_{GL} = 0\text{ V}, \quad V_{GH} = 10\text{ V}, \quad V_T = 5\text{ V}, \quad R_S + R_G = 10\ \Omega$$

$$C_{G'S'} = 0.6\text{ nF}, \quad C_{G'D'} = 0.1\text{ nF}, \quad C_{D'S'} = 0.2\text{ nF}$$

$$\text{when} \quad v_{D'G'} < 0, \quad C_{G'D'} \to 1\text{ nF}, \quad L_S = 10\text{ nH}$$

The various time constants are

$$\tau_{cct} = 100\text{ ns}, \quad \tau_{dev} = 1\text{ ns} \quad (\therefore \text{ sinusoidal solutions})$$

$$k = 40 \quad (\therefore \text{ voltage collapses first})$$

$$\tau_m = 5\ \mu\text{s}, \quad \tau_G = 7\text{ ns}, \quad \tau_{G'} = 1\text{ ns}, \quad \tau_{G''} = 16\text{ ns}$$

$$\tau_a = 1.4\ \mu\text{s}, \quad \omega_a = 1.4 \times 10^7\text{ rad/s}, \quad \omega_a \tau_a = 20$$

$$\tau_f = 57\text{ ns}, \quad \tau_s = 2.45\text{ ns}$$

The waveforms drawn in Figure 4.13 with a solid line have been derived from these parameters using Equations (4.14), (4.17), and (4.18) for the period τ_1, and Equation (4.47) for τ_2.

The delay time $\tau_D = 5$ ns, the drain voltage fall time $\tau_1 = 10$ ns (19 ns if the reduced rate of fall at low values is included), and the current rise time $\tau_2 = 195$ ns. The overall transition time, $\tau_r = \tau_1 + \tau_2$, is therefore some 215 ns.

At the end of the main voltage collapse period, the gate voltage is 5.05 V, that is, 0.05 V above threshold, and the drain current is 0.25 A. The period during which the drain voltage falls from 5 V to zero may be a further 10 ns, as a result of the increased value of $C_{G'D'}$. It thus overlaps the rise of current in a rather complicated way and significantly increases the dissipation. On the basis of the waveforms shown, the dissipation is about 20 nJ. However, the energy stored in $C_{D'S'}$ has to be added, and with the values given this represents a further 250 nJ. We have not included the dissipation in $R_{DS(on)}$ in the analysis, on the grounds

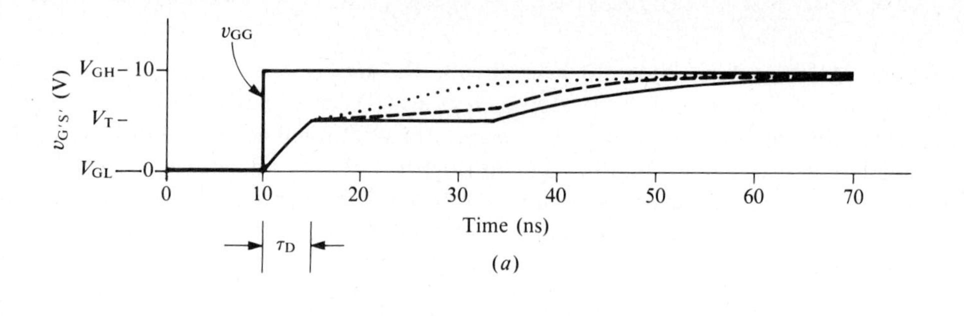

v_{GG}
V_{GH} – 10
V_T –
V_{GL} — 0
$v_{G'S'}$ (V)
0 10 20 30 40 50 60 70
Time (ns)
τ_D

(a)

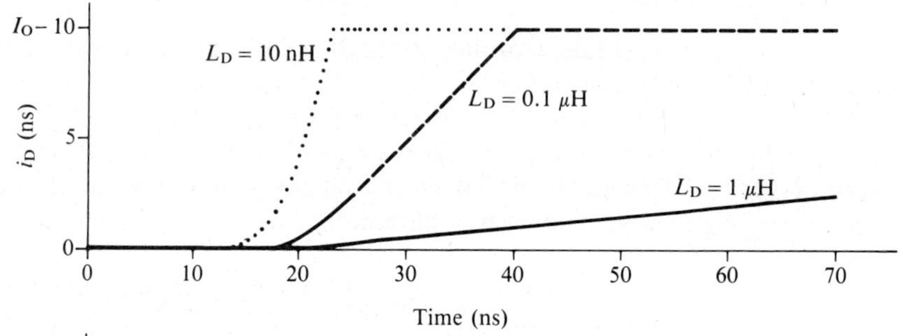

I_O – 10
5
0
i_D (ns)
L_D = 10 nH
L_D = 0.1 μH
L_D = 1 μH
0 10 20 30 40 50 60 70
Time (ns)

(b)

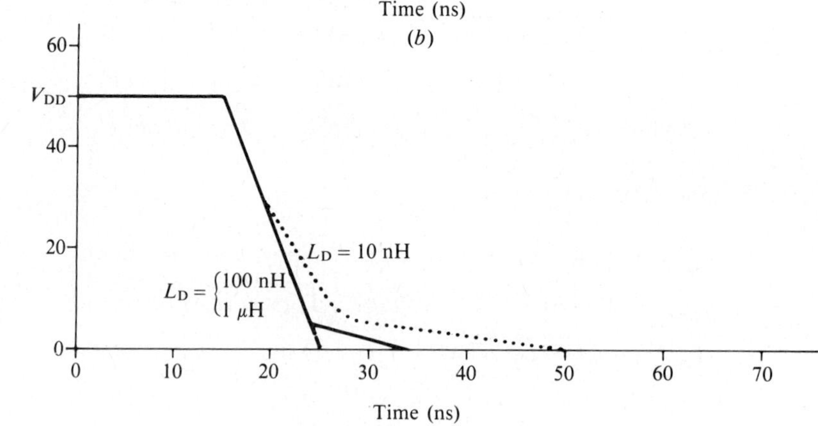

60
V_{DD}
40
20
0
$v_{D'S'}$ (V)
L_D = 10 nH
L_D = {100 nH, 1 μH
0 10 20 30 40 50 60 70
Time (ns)

(c)

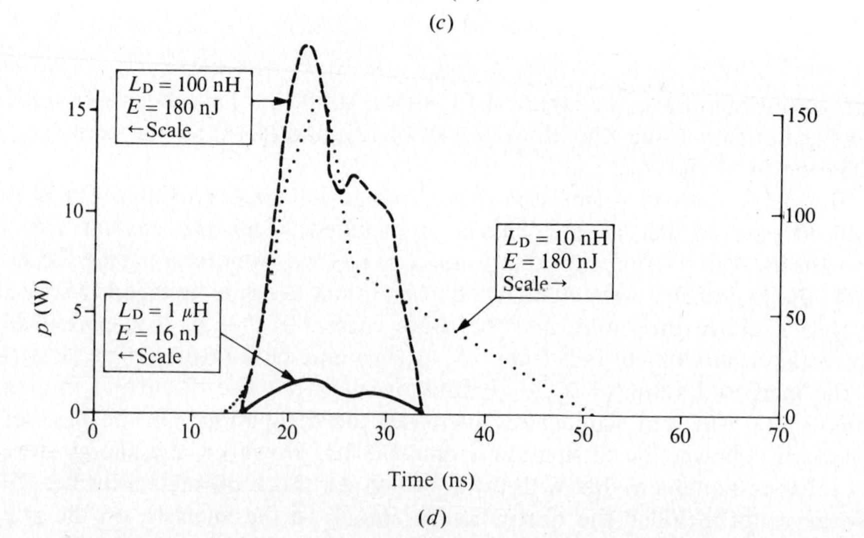

L_D = 100 nH
E = 180 nJ
←Scale
L_D = 10 nH
E = 180 nJ
Scale→
L_D = 1 μH
E = 16 nJ
←Scale
p (W)
15
10
5
0
150
100
50
0
0 10 20 30 40 50 60 70
Time (ns)

(d)

that it is a steady-state loss and adds to the transient losses that we have calculated.

Decreasing L_D from 1 μH to 0.1 μH causes only the drain-current waveform to be modified to any significant extent. This is shown by the dashed curves in Figure 4.13. The time constants now are

$$\tau_{cct} = 10\text{ ns}, \quad \tau_{dev} = 1\text{ ns} \qquad (\therefore \text{ sinusoidal solutions})$$

$$k = 4 \qquad (\therefore \text{ voltage collapses first})$$

$$\tau_m = 500\text{ ns}, \qquad \tau_G = 7\text{ ns}, \qquad \tau_{G'} = 1\text{ ns}, \qquad \tau_{G''} = 16\text{ ns}$$

$$\tau_a = 143\text{ ns}, \qquad \omega_a = 4.4 \times 10^7\text{ rad/s}, \qquad \omega_a \tau_a = 6.3$$

$$\tau_f = 57\text{ ns}, \qquad \tau_s = 2.45\text{ ns}$$

The current now climbs to 2.5 A during the voltage collapse, and the transient dissipation increases to about 180 nJ, in addition to the 250 nJ of $C_{D'S'}$. The transition times become $\tau_D = 5$ ns, $\tau_1 = 10$ ns, $\tau_2 = 15$ ns. Thus, $\tau_r = 25$ ns. The source inductance has been neglected. A value of 10 nH would have a significant effect on these waveforms, extending the transition times and thereby increasing the dissipation.

The effect of a further decrease in L_D, to 10 nH, is also shown in Figure 4.13, by the dotted waveforms. Again L_S has been neglected. Although the solutions are still sinusoidal, the current rise is now completed while the drain voltage is still falling. The time constants are

$$\tau_{cct} = 1\text{ ns}, \quad \tau_{dev} = 1\text{ ns} \qquad (\therefore \text{ sinusoidal solutions})$$

$$k = 0.4 \qquad (\therefore \text{ current completes rise during voltage collapse})$$

$$\tau_m = 50\text{ ns}, \qquad \tau_G = 7\text{ ns}, \qquad \tau_{G'} = 1\text{ ns}, \qquad \tau_{G''} = 16\text{ ns}$$

$$\tau_a = 14\text{ ns}, \qquad \omega_a = 1.2 \times 10^8\text{ rad/s}, \qquad \omega_a \tau_a = 1.75$$

$$\tau_f = 57\text{ ns}, \qquad \tau_s = 2.45\text{ ns}$$

The transient dissipation is about 2000 nJ, and the transition times are $\tau_D = 5$ ns, $\tau_1 = 8$ ns, $\tau_2 = 4 + 24 = 28$ ns, giving $\tau_r = 36$ ns. However, the rate of rise of current, 5 A/ns at the end of τ_1, is incompatible with a typical stray source inductance of 10 nH.

An indication of how the waveforms might be modified by 10 nH of source inductance is given in Figure 4.14, using the approximate theory set out at the end of Section 4.2.5. The rise of current is slowed, but more significantly, the drain voltage scarcely falls at all during this period. The transition times are increased to $\tau_1 = 29$ ns, $\tau_2 = 15 + 23 = 38$ ns, $\therefore$ $\tau_r = 67$ ns. The dissipation rises to

---

Figure 4.13. Calculated turn-on waveforms for a MOSFET switching 10 A at 50 V. (*a*) $v_{G'S'}(t)$; (*b*) $i_D(t)$; (*c*) $v_{D'S'}(t)$; (*d*) $p(t)$. $R_S + R_G = 10\,\Omega$, $g_{fs} = 5$ S, $C_{G'S'} = 0.6$ nF, $C_{G'D'} =$ 0.1 nF. $L_D = 1\,\mu$H (solid line), 100 nH (dashed line), 10 nH (dotted line).

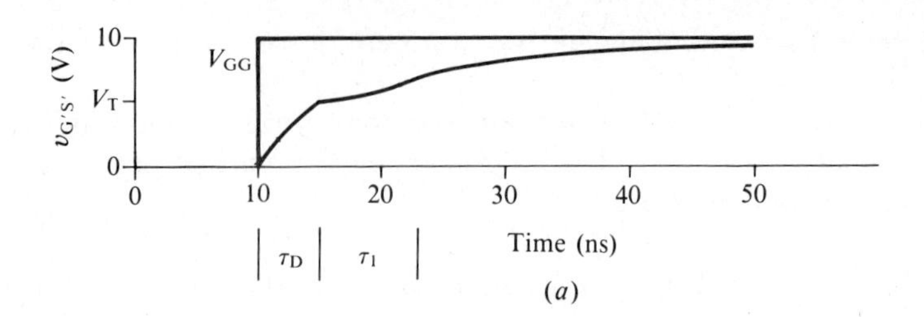

$v_{G'S'}$ (V)
10
V_{GG}
V_T
0
0
10
20
30
40
50
τ_D
τ_1
Time (ns)

(a)

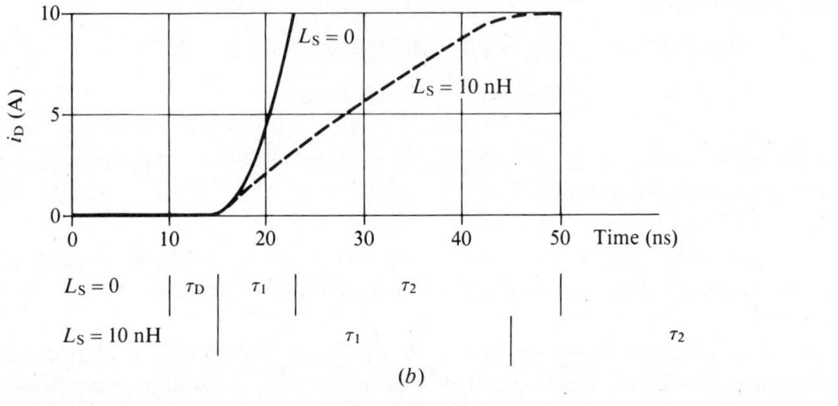

i_D (A)
10
5
0
$L_S = 0$
$L_S = 10$ nH
0
10
20
30
40
50
Time (ns)
$L_S = 0$
τ_D
τ_1
τ_2
$L_S = 10$ nH
τ_1
τ_2

(b)

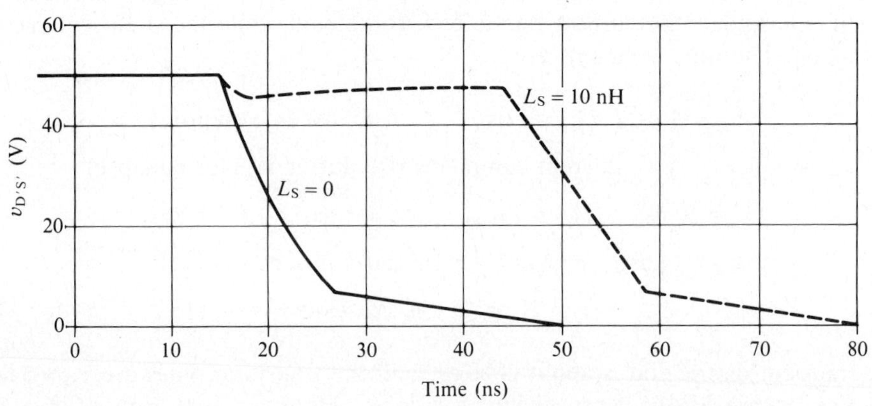

$v_{D'S'}$ (V)
60
40
20
0
$L_S = 10$ nH
$L_S = 0$
0
10
20
30
40
50
60
70
80
Time (ns)

(c)

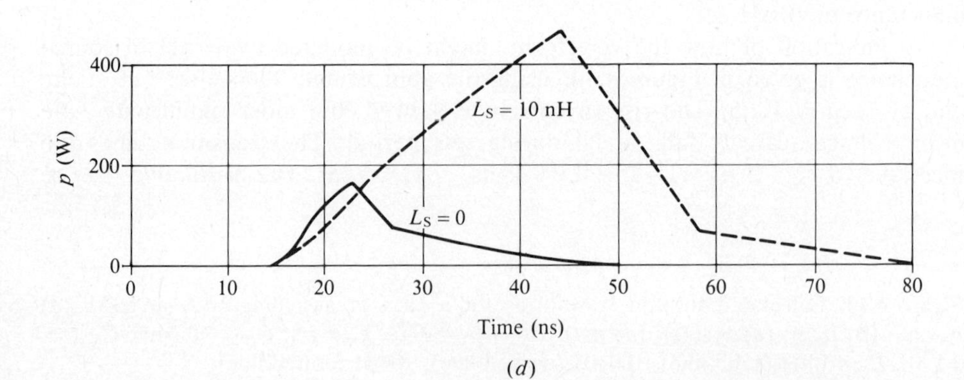

p (W)
400
200
0
$L_S = 10$ nH
$L_S = 0$
0
10
20
30
40
50
60
70
80
Time (ns)

(d)

nearly 12,000 nJ. The importance of keeping the source inductance to an absolute minimum in a fast circuit is brought out very clearly.

We next examine some examples of a higher-voltage circuit, in which a higher gate resistance of 50 Ω is also used. The gate circuit is thus rather "slow". As a result, the drain current rises before the drain voltage has completed its fall, even in the presence of considerable drain inductance. Again we start with $L_D = 1\ \mu H$:

$$V_{DD} = 400\ V, \quad I_0 = 4\ A, \quad g_{fs} = 3\ S, \quad L_D = 1\ \mu H$$

$$V_{GL} = 0\ V, \quad V_{GH} = 10\ V, \quad V_T = 5\ V, \quad R_S + R_G = 50\ \Omega$$

$$C_{G'S'} = 0.6\ nF, \quad C_{G'D'} = 40\ pF, \quad C_{D'S'} = 0.1\ nF$$

$$\text{when} \quad v_{D'G'} < 0, \quad C_{G'D'} \to 1\ nF, \quad L_S = 10\ nH$$

Hence

$$\tau_{cct} = 20\ ns, \quad \tau_{dev} = 3\ ns \quad (\therefore \text{ sinusoidal solutions})$$

$$k = 0.125 \quad (\therefore \text{ voltage collapses after the rise of current})$$

$$\tau_m = 3\ \mu s, \quad \tau_G = 32\ ns, \quad \tau_{G'} = 2\ ns, \quad \tau_{G''} = 80\ ns$$

$$\tau_a = 375\ ns, \quad \omega_a = 1.3 \times 10^7\ rad/s, \quad \omega_a \tau_a = 4.7$$

$$\tau_f = 62\ ns, \quad \tau_s = 2.45\ ns$$

These parameters lead to the waveforms shown as solid curves in Figure 4.15. The delay time remains $\tau_D = 5$ ns, but the transition times are much increased: $\tau_1 = 61$ ns, $\tau_2 = 146 + 86 = 232$ ns, giving a total τ_r of almost 300 ns. The dissipation is correspondingly high at 100 μJ.

Reducing L_D to 0.1 μH speeds up the rise of current, reducing τ_1 to 22 ns. However, the rate of fall of drain voltage is virtually unchanged, and so is τ_r. The dissipation increases to 140 μJ. Further reduction of L_D continues this trend. At $L_D = 10$ nH we have $\tau_1 = 11$ ns, and the dissipation is 160 μJ. The key time constants are

$$L_D = 100\ nH$$

$$\tau_{cct} = 2\ ns, \quad \tau_{dev} = 3\ ns \quad (\therefore \text{ sinusoidal solutions, just!})$$

$$k = \frac{1}{80} \quad (\therefore \text{ voltage collapses after the rise of current})$$

$$\tau_a = 37.5\ ns, \quad \omega_a = 3.1 \times 10^7\ rad/s, \quad \omega_a \tau_a = 1.16$$

Figure 4.14. The effect on the turn-on waveforms of 10-nH source inductance. (*a*) $v_{G'S'}(t)$; (*b*) $i_D(t)$; (*c*) $v_{D'S'}(t)$; (*d*) $p(t)$. The circuit parameters are the same as those used in Figure 4.13, with $L_D = 10$ nH, and $L_S = 0$ (solid line), 10 nH (dashed line).

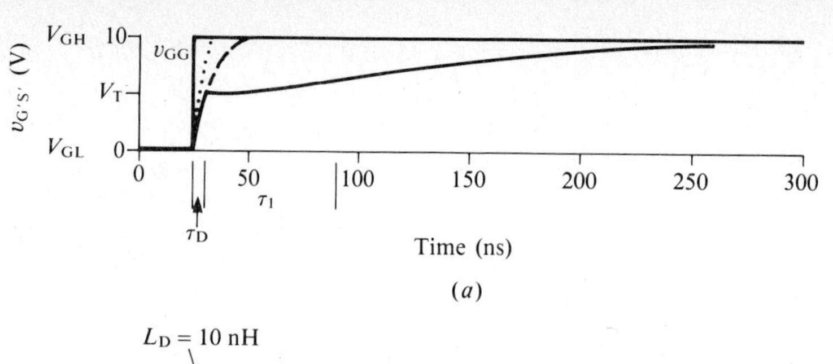

V_{GH}
V_T
V_{GL}
$v_{G'S'}$ (V)
v_{GG}
τ_1
τ_D
Time (ns)

(*a*)

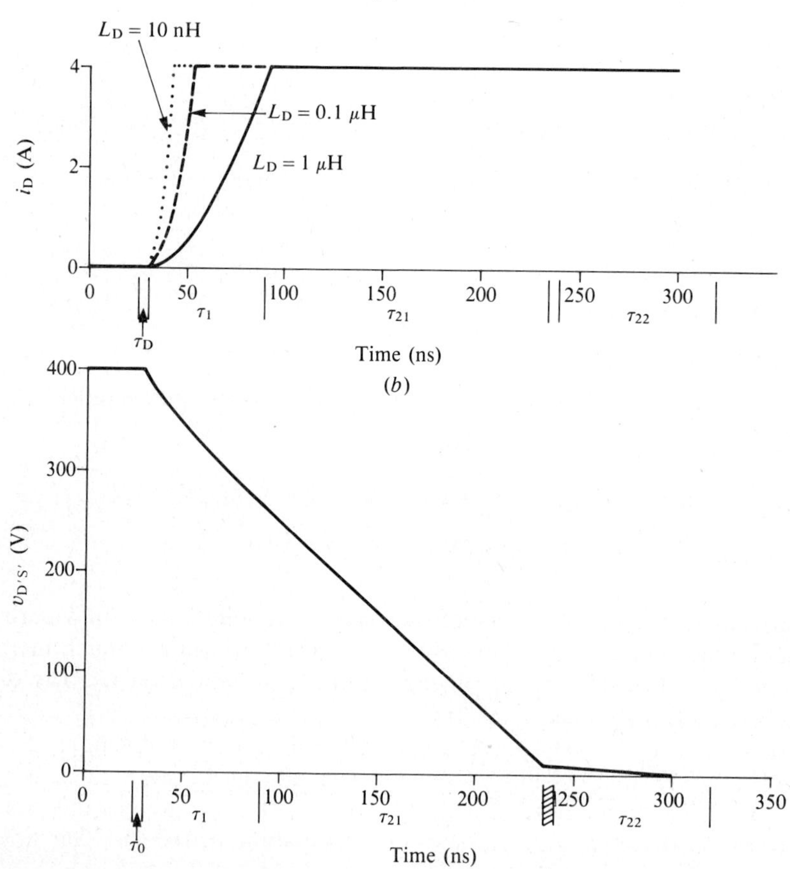

$L_D = 10$ nH
$L_D = 0.1\ \mu$H
$L_D = 1\ \mu$H
i_D (A)
τ_1
τ_{21}
τ_{22}
τ_D
Time (ns)

(*b*)

$v_{D'S'}$ (V)
τ_1
τ_{21}
τ_{22}
τ_0
Time (ns)

(*c*)

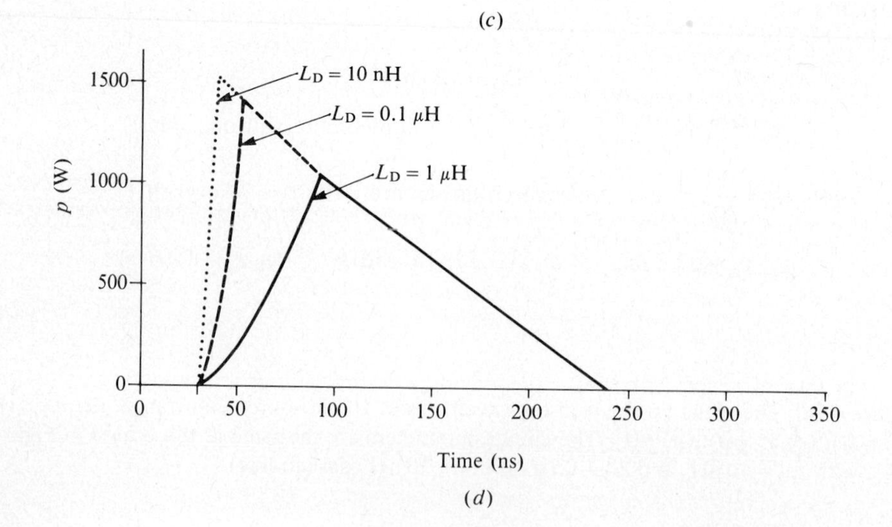

$L_D = 10$ nH
$L_D = 0.1\ \mu$H
$L_D = 1\ \mu$H
p (W)
Time (ns)

(*d*)

and

$$L_D = 10\text{ nH}$$

$$\tau_{cct} = 0.2\text{ ns}, \quad \tau_{dev} = 3\text{ ns} \quad (\therefore \text{ exponential solutions})$$

$$k = \frac{1}{800} \quad (\therefore \text{ voltage collapses after the rise of the current})$$

$$\tau_m = 30\text{ ns}, \quad \tau_b = 30\text{ ns}, \quad \tau_c = 2\text{ ns}$$

Both current waveforms are shown in Figure 4.15. The effect of 10-nH source inductance is shown in Figure 4.16, using the same approximation as before. Its significance is now much less.

With such a slow gate circuit, a very large unclamped drain inductance is needed if the current is not to rise to its full value while the drain voltage is still high. The final illustration in this sequence shows the effect of 10-μH drain inductance:

$$\tau_{cct} = 200\text{ ns}, \quad \tau_{dev} = 3\text{ ns} \quad (\therefore \text{ sinusoidal solutions})$$

$$k = 1.25$$

$$\tau_m = 30\ \mu\text{s}, \quad \tau_a = 3.75\ \mu\text{s}, \quad \omega_a = 4.1 \times 10^6\text{ rad/s}, \quad \omega_a\tau_a = 15.3$$

The drain current now rises to 3.5 A during the initial fall of voltage. Overall this takes $175 + 85 = 260$ ns. Dissipation is reduced to 20 μJ. The waveforms are shown in Figure 4.17.

These examples demonstrate clearly that turn-on dissipation is greatly reduced by ensuring that τ_{cct} is large. But, as we shall find in the next section, making τ_{cct} large by increasing L_D merely transfers the dissipation from the turn-on transition to the turn-off period. The key to ensuring low transient power consumption is to achieve a fast gate circuit by using a low value of R_S. It is then important to be sure that source inductance is minimal.

4.3. TURN-OFF

4.3.1. The Turn-off Sequence

In order to illustrate the turn-off process we once again use the circuit in which an inductive load is partially clamped by means of a free-wheeling diode. The equivalent circuit of Figure 4.7 still applies. But now the starting conditions are that the load current I_0 flows through the fully ON transistor, with $v_{GG} = v_{G'S'} = V_{GH}$. At some point the gate supply voltage v_{GG} drops instantaneously to V_{GL}.

Figure 4.15. Calculated turn-on waveforms for a MOSFET switching 4 A at 400 V with a "slow" gate circuit. (*a*) $v_{G'S'}(t)$; (*b*) $i_D(t)$; (*c*) $v_{D'S'}(t)$; (*d*) $p(t)$. $R_S + R_G = 50\ \Omega$, $g_{fs} = 3$ S, $C_{G'S'} = 0.6$ nF, $C_{G'D'} = 0.04$ nF. $L_D = 1\ \mu$H (solid line), 100 nH (dashed line), 10 nH (dotted line).

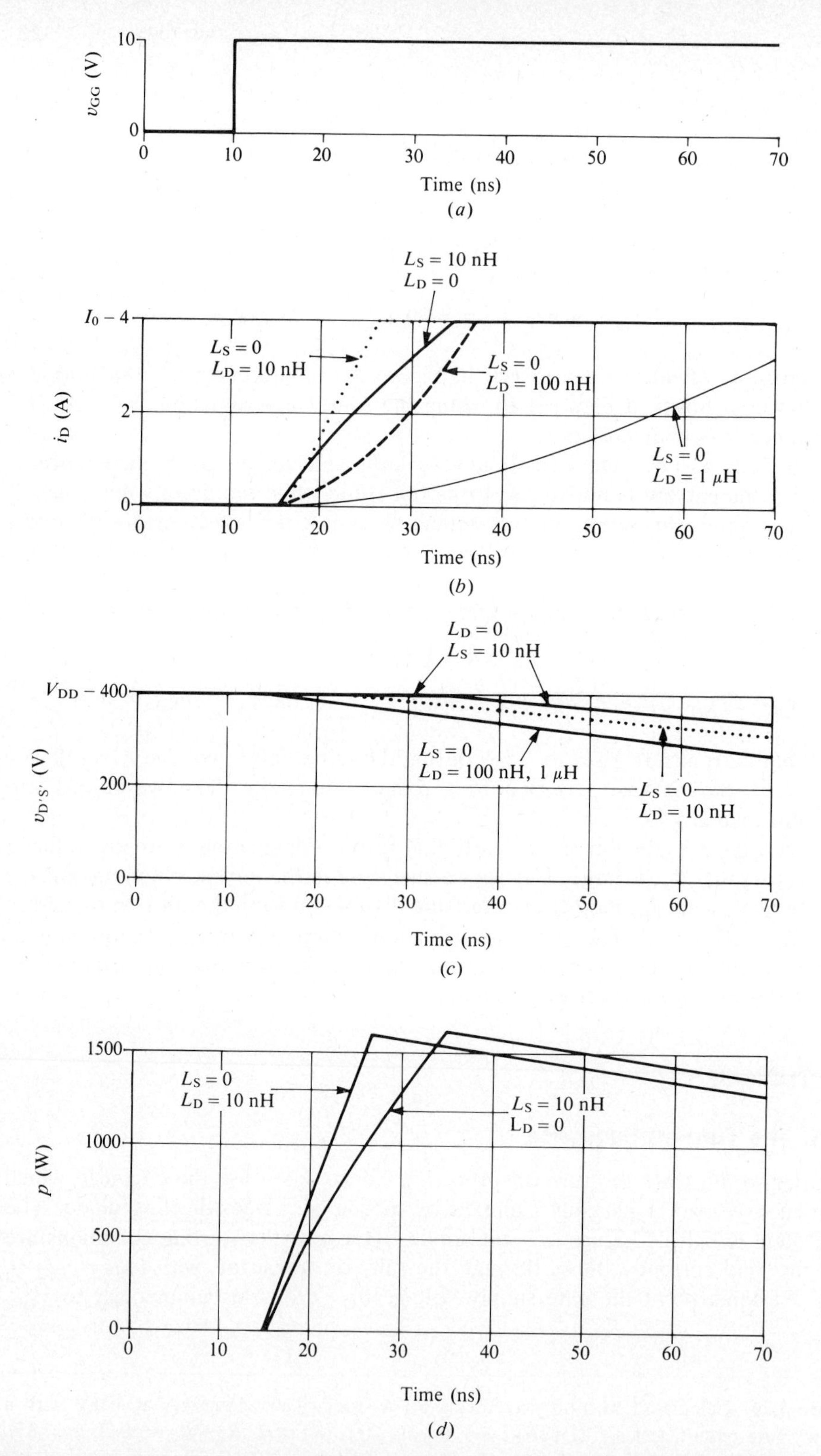

Figure 4.16. Effect of 10-nH source inductance on the turn-on waveforms of Figure 4.15. (a) $v_{GG}(t)$; (b) $i_D(t)$; (c) $v_{D'S'}(t)$; (d) $p(t)$.

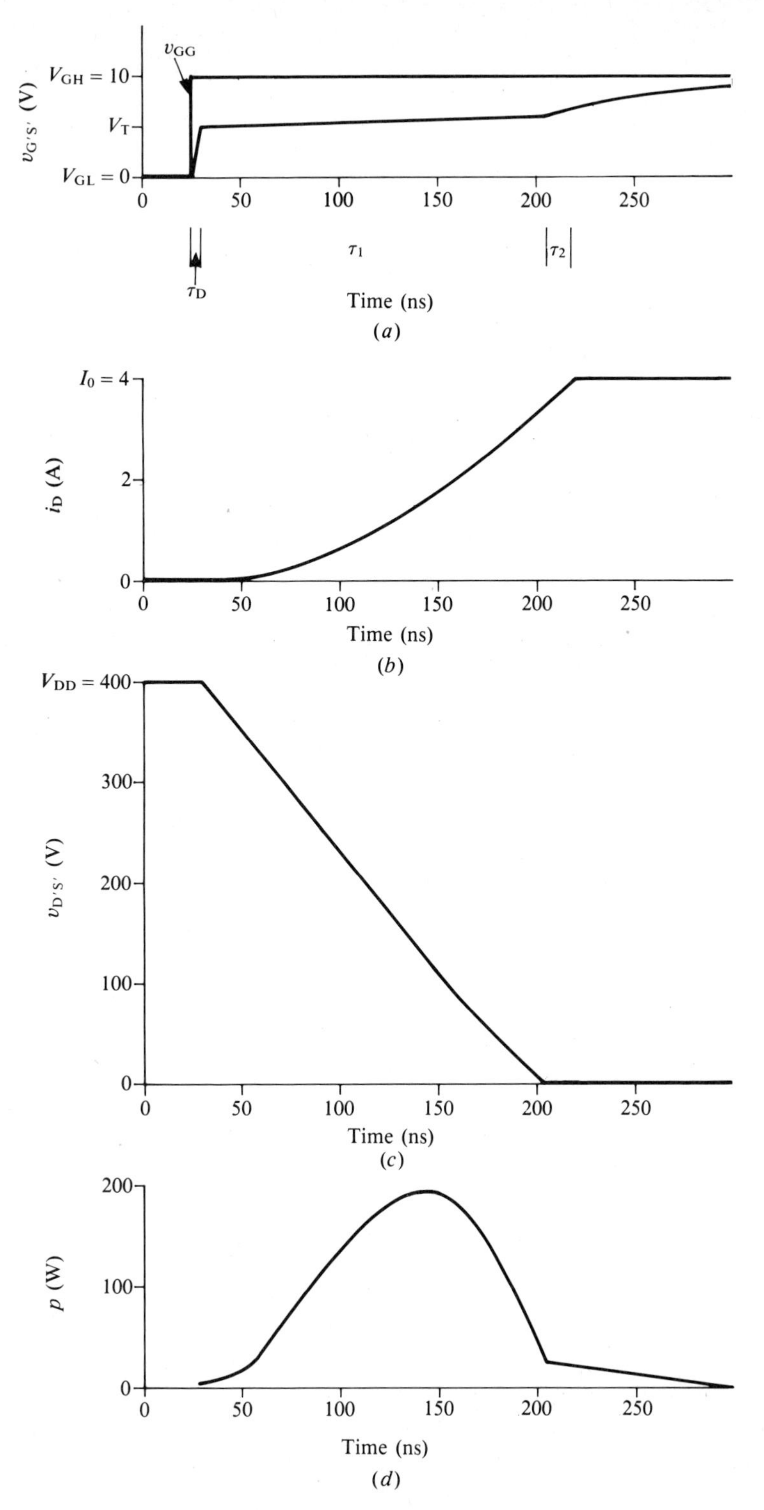

Figure 4.17. Turn-on waveforms with slow gate circuit and $L_D = 10\,\mu$H. (*a*) $v_{G'S'}(t)$; (*b*) $i_D(t)$; (*c*) $v_{D'S'}(t)$; (*d*) $p(t)$.

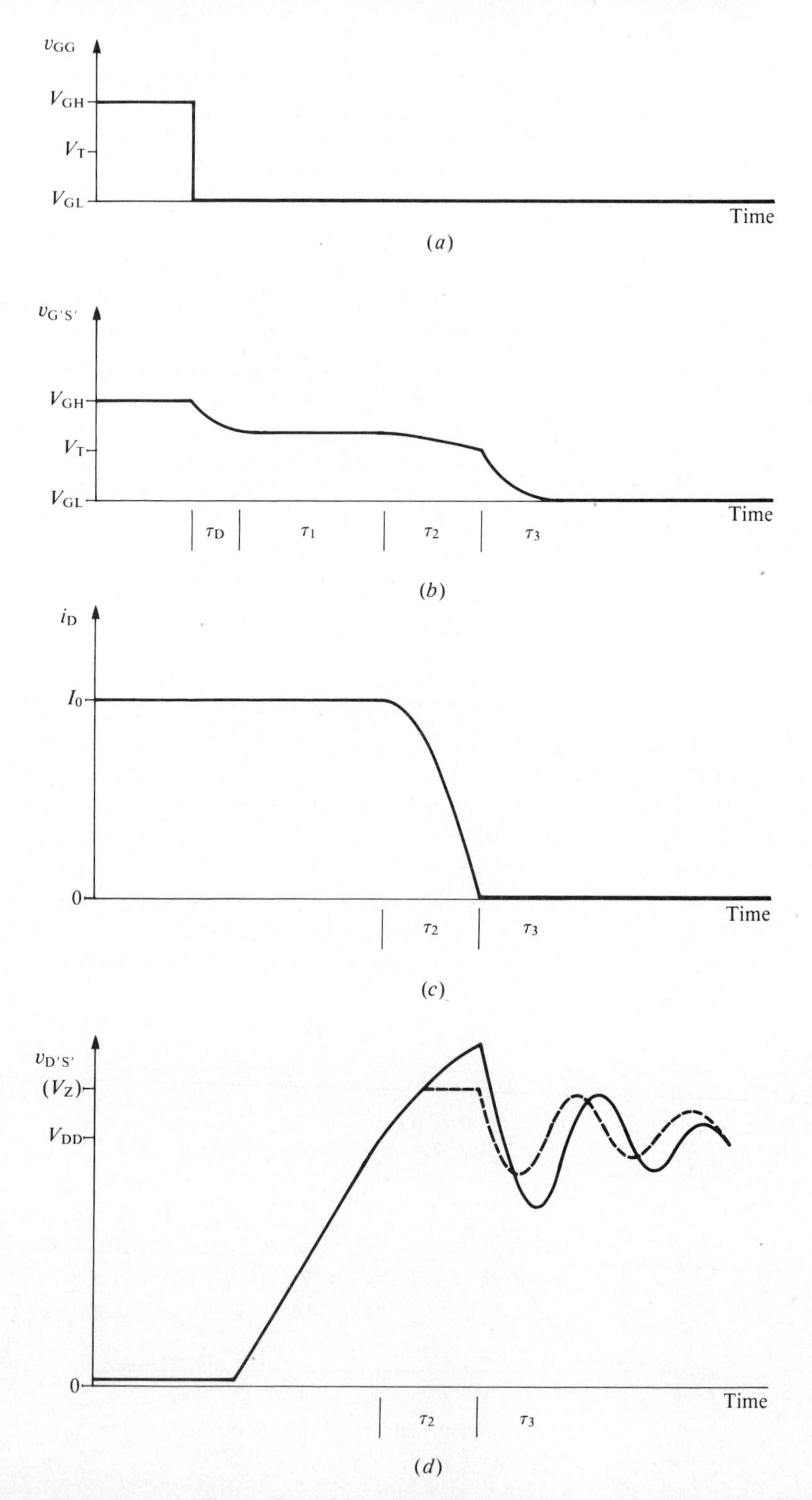

Figure 4.18. Schematic illustration of a typical turn-off sequence. (*a*) v_{GG}; (*b*) $v_{G'S'}(t)$; (*c*) $i_D(t)$; (*d*) $v_{D'S'}(t)$.

The gate voltage then starts to fall, thereby discharging the device capacitances $C_{G'S'}$ and $C_{G'D'}$. As long as $v_{G'S'} \geq V_T + I_0/g_{fs}$, the load current I_0 continues to flow through the MOSFET unchanged, and the drain circuit is unaffected. The time taken for $v_{G'S'}$ to fall to $V_T + I_0/g_{fs}$ is the turn-off delay time τ_D. It is the first of four stages of the turn-off sequence, which is illustrated by the typical set of waveforms shown in Figure 4.18.

During the next phase of turn-off, the drain voltage rises up to $V_{DD} + 0.7$ V and continues beyond this value. While $v_{D'S'}$ is below $V_{DD} + 0.7$ V, the full load current continues to flow into the MOSFET, because the free-wheeling diode remains in reverse bias and therefore nonconducting. The drain current i_D stays sensibly constant, and with it so does $v_{G'S'}$. This assumes that we can neglect the small currents required to charge up the capacitances $C_{G'S'}$ and $C_{D'S'}$. This phase of turn-off occupies the period τ_1.

When $v_{D'S'}$ reaches and rises above $V_{DD} + 0.7$ V, we enter the third phase of turn-off. There is then a voltage across L_D which opposes I_0 and enables an increasing fraction of this current to be diverted through the free-wheeling diode. The MOSFET current thus starts to decrease. To reduce it to zero requires a voltage–time integral across L_D equal to $L_D I_0$. That is,

$$\int [v_{D'S'} - (V_{DD} + 0.7 \text{ V})] \, dt = L_D I_0 \tag{4.51}$$

In practice, the zener diode which connects the drain and the source, as shown in Figure 4.7, effectively clamps $v_{D'S'}$ at some value V_Z. Many modern power MOSFETS are designed so that the body–drain diode junction provides a controlled avalanche capability within the device itself, as described in Section 3.6.4. Otherwise it is important that an external zener diode be connected as closely as possible to the transistor, to prevent it from suffering an overvoltage. The third phase of turn-off may therefore subdivide into two parts. In the first part, the drain current and the drain voltage both vary. In the second part, should it occur, the drain voltage is clamped while the drain current continues to fall. In either case, this phase of turn-off ends when the drain current reaches zero, that is, after a time τ_2.

In the final phase the drain and gate circuits are to some extent decoupled. The gate capacitance discharges, and $v_{G'S'}$ exponentially approaches V_{GL} with a time constant τ_G. At that same time, a "ring" is set up in the drain circuit. At the start of this period $v_{D'S'} > (V_{DD} + 0.7 \text{ V})$ and the transistor is OFF. An oscillation is set up between the total drain capacitance $C_{D'S'} + C_{D'G'}$ and the drain inductance L_D. This "ring" decays with a time constant that depends on the total resistance in the circuit—a factor we have hitherto neglected. The oscillation may be observed in the gate circuit by virtue of the ac current that flows through $C_{G'D'}$.

4.3.2. The Delay Time

During the first phase of turn-off, the equivalent circuit shown in Figure 4.19 applies. Only the gate–source voltage varies, diminishing as

$$v_{G'S'} = (V_{GH} - V_{GL}) \exp(-t/\tau_{G''}) + V_{GL} \tag{4.52}$$

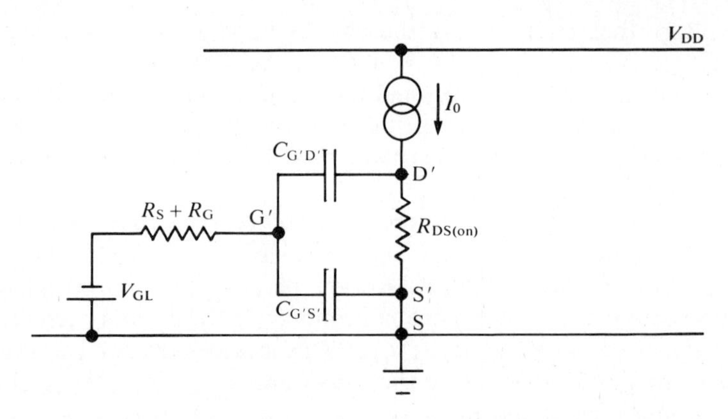

Figure 4.19. Equivalent circuit for the first phase of turn-off.

where $\tau_{G''} = (C_{G'S'} + C_{G'D'})(R_S + R_G)$, and the value of $C_{G'S'} + C_{G'D'} = C_{iss}$ is the enhanced value that applies when $v_{G'D'} > 0$.

This phase ends when $v_{G'S'} = V_T + I_0/g_{fs}$. It thus takes a time τ_D given by

$$\tau_D = \tau_{G''} \ln \frac{V_{GH} - V_{GL}}{V_T + I_0/g_{fs} - V_{GL}} \tag{4.53}$$

Typically, $\tau_D \approx \tau_{G''}$.

4.3.3. The Rise of the Drain Voltage

Once $v_{G'S'}$ reaches the value $V_T + I_0/g_{fs}$, it is held there by the demands of the transistor and the circuit, while the drain voltage rises to $V_{DD} + 0.7$ V. During this period the equivalent circuit is the same as that shown in Figure 4.10, but with the gate supply voltage equal to V_{GL}. An analysis similar to that given in Section 4.2.6 predicts a linear rise of drain voltage:

$$v_{D'S'} = Kt \tag{4.54}$$

where

$$K = \frac{I_0 + g_{fs}(V_T - V_{GL})}{C_{D'S'} + [1 + g_{fs}(R_S + R_G)]C_{G'D'}} \tag{4.55}$$

This phase of turn-off ends when $v_{D'S'} = V_{DD}$. Thus,

$$\tau_1 = V_{DD}/K \tag{4.56}$$

However, it must be remembered that $C_{G'D'}$ decreases as $v_{D'S'}$ rises. We should therefore expect K to increase sharply as $v_{D'S'}$ exceeds $v_{G'S'}$, and to continue rising as $v_{D'S'}$ gets bigger. A better approximation to τ_1 is obtained by subdividing the period into two parts, in which K takes the constant values K_1 and K_2, respectively. Then,

$$\tau_1 = \frac{V_T + I_0/g_{fs}}{K_1} + \frac{V_{DD} - (V_T + I_0/g_{fs})}{K_2} \tag{4.57}$$

4.3.4. The Fall of the Drain Current

The drain voltage continues to rise above $V_{DD} + 0.7\ \text{V}$. Usually it is clamped at some higher value V_Z. For the moment we neglect this possibility and assume that $v_{D'S'}$ can rise as the circuit dictates. The equivalent circuit is then that of Figure 4.9, except that the gate supply voltage is now V_{GL} ($<V_T$) rather than V_{GH}. Analysis similar to that given in Section 4.2.3 leads to a differential equation equivalent to that of Equation (4.13). With the new boundary conditions, the solutions are as follows. If

$$\tau_{cct} > \tau_{dev}$$

then

$$v_{G'S'} = \left(\frac{I_0}{g_{fs}} + V_T - V_{GL}\right) \exp\left(-\frac{t}{\tau_a}\right)\left(\cos \omega_a t + \frac{1}{\omega_a \tau_a} \sin \omega_a t\right) + V_{GL} \tag{4.58}$$

where τ_a and ω_a are defined, as before, by Equations (4.15) and (4.16). Then

$$\begin{aligned} i_D &= g_{fs}(v_{G'S'} - V_T) \\ &= [g_{fs}(V_T - V_{GL}) + I_0] \exp\left(-\frac{t}{\tau_a}\right)\left(\cos \omega_a t + \frac{1}{\omega_a \tau_a} \sin \omega_a t\right) - g_{fs}(V_T - V_{GL}) \end{aligned} \tag{4.59}$$

$$\begin{aligned} v_{D'S'} &= V_{DD} + 0.7\ \text{V} - g_{fs} L_D \frac{dv_{G'S'}}{dt} \\ &= V_{DD} + 0.7\ \text{V} + [g_{fs}(V_T - V_{GL}) + I_0] L_D \omega_a \exp\left(-\frac{t}{\tau_a}\right)\left[1 + \left(\frac{1}{\omega_a \tau_a}\right)^2\right] \sin \omega_a t \end{aligned} \tag{4.60}$$

If

$$\tau_{cct} < \tau_{dev}$$

then

$$v_{G'S'} = \left(\frac{I_0}{g_{fs}} + V_T - V_{GL}\right) \frac{\tau_b \exp(-t/\tau_b) - \tau_c \exp(-t/t_c)}{\tau_b - \tau_c} + V_{GL} \tag{4.61}$$

and

$$i_D = [g_{fs}(V_T - V_{GL}) + I_0]\frac{\tau_b \exp(-t/\tau_b) - \tau_c \exp(-t/\tau_c)}{\tau_b - \tau_c} - g_{fs}(V_T - V_{GL}) \quad (4.62)$$

$$v_{D'S'} = V_{DD} + 0.7\text{ V} + [g_{fs}(V_T - v_{GL}) + I_0]L_D \frac{\exp(-t/\tau_b) - \exp(-t/\tau_c)}{\tau_b - \tau_c} \quad (4.63)$$

For small arguments the sine, cosine, and exponential terms can be expanded, as in Section 4.2.5. This is usually valid. Then, both the sinusoidal and the exponential solutions approximate to

$$v_{G'S'} = \left(\frac{I_0}{g_{fs}} + V_T - V_{GL}\right)\left(1 - \frac{t^2}{2\tau_m \tau_{G'}}\right) + V_{GL} \quad (4.64)$$

$$i_D = [g_{fs}(V_T - V_{GL}) + I_0]\left(1 - \frac{t^2}{2\tau_m \tau_{G'}}\right) - g_{fs}(V_T - V_{GL}) \quad (4.65)$$

$$v_{D'S'} = V_{DD} + 0.7\text{ V} + \left(\frac{I_0}{g_{fs}} + V_T - V_{GL}\right)\frac{t}{\tau_{G'}} \quad (4.66)$$

This phase ends when $i_D = 0$. Its duration is thus given by

$$\tau_2 = \left(\frac{2\tau_m \tau_{G'} I_0}{g_{fs}(V_T - V_{GL}) + I_0}\right)^{1/2} = \left(\frac{2L_D C_{G'D'}(R_S + R_G)I_0}{I_0/g_{fs} + V_T - V_{GL}}\right)^{1/2} \quad (4.67)$$

At $t = \tau_2$,

$$v_{D'S'} = V_{D^*} = V_{DD} + 0.7\text{ V} + \left(\frac{I_0}{g_{fs}} + V_T - V_{GL}\right)\frac{\tau_2}{\tau_{G'}} = V_{DD} + 0.7\text{ V} + \left(\frac{2L_D I_0(I_0/g_{fs} + V_T - V_{GL})}{\tau_{G'}}\right)^{1/2} \quad (4.68)$$

If $v_{D'S'}$ is clamped to a voltage V_Z less than V_{D^*}, this phase of the turn-off also divides into two parts. The rise to V_Z takes a time τ_Z given by

$$\tau_Z = \tau_{G'} \frac{V_Z - (V_{DD} + 0.7\text{ V})}{I_0/g_{fs} + V_T - V_{GL}} \quad (4.69)$$

The drain current at this time, I_{D^*}, is obtained by substituting τ_Z into Equation (4.65). Once $v_{D'S'}$ is clamped, the equivalent circuit takes the form shown in Figure 4.20. With the voltage across L_D constant, the MOSFET current falls to zero

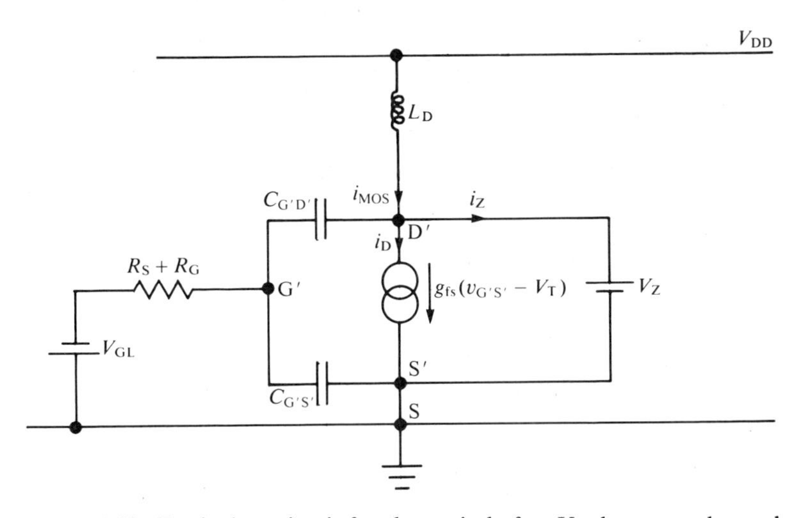

Figure 4.20. Equivalent circuit for the period after V_{DS} becomes clamped.

at a constant rate:

$$i_{MOS} = i_{D*} - \frac{V_Z - V_{DD}}{L_D} t \tag{4.70}$$

The current divides between that flowing normally through the device, i_D, and that flowing in parallel through the zener diode, i_Z:

$$i_{MOS} = i_D + i_Z \tag{4.71}$$

The normal drain current follows the decay of $v_{G'S'}$ towards V_{GL} as the gate capacitance discharges. This is exponential:

$$v_{G'S'} = [v_{G'S'}(0) - V_{GL}] \exp(-t/\tau_G) + V_{GL} \tag{4.72}$$

and it continues through into the final stage of turn-off. When $v_{D'S'}$ is clamped, $t = 0$ and $v_{G'S'}(0) = V_T + I_{D*}/g_{fs}$ at the moment that $v_{D'S'} = V_Z$. The drain current is thus given by

$$i_D = [g_{fs}(V_T - V_{GL}) + I_{D*}] \exp(-t/\tau_G) - g_{fs}(V_T - V_{GL}) \tag{4.73}$$

and it falls to zero in the time

$$\tau_Z = \tau_G \ln\left(1 + \frac{I_{D*}}{g_{fs}(V_T - V_{GL})}\right) \tag{4.74}$$

When $v_{D'S'}$ is not clamped, Equation (4.72) applies from the end of the period τ_2, with $v_{G'S'}(0) = V_T$.

The effect of any source inductance L_S is to force the potential of the source itself (S′) to go negative during this phase of turn-off. There are several possibilities here, depending on whether the drain voltage is clamped, and if so, whether the zener diode is internal to the MOSFET or external. We examine just one case which gives a particularly simple analytical result. This is with an internal zener, where we assume that all the current, $i_{MOS} = i_Z + i_D$, passes through L_S. The rate of fall of i_{MOS} can be seen from Equation (4.70) to be simply $(V_Z - V_{DD})/L_D$, so that

$$v_{S'S} = (V_Z - V_{DD})L_S/L_D \tag{4.75}$$

It is most important that this voltage should not cause the absolute maximum gate–source voltage rating to be exceeded.

4.3.5. The Turn-off Ring

During the final stage of turn-off, as the gate voltage returns to V_{GL} in the way just described, an oscillation of the drain voltage usually occurs. We put $v_{D'S'} = V_{D^*}$, at the start of this period. If the drain voltage has been clamped, $V_{D^*} = V_Z$. Otherwise, V_{D^*} is given by Equation (4.68). The equivalent circuit is shown in Figure 4.21. We have included the component R_D to represent the resistive losses in the drain circuit. This is a simple LCR circuit, in which the drain voltage oscillates as

$$v_{D'S'} = (V_{DD} + 0.7\text{ V}) + [V_{D^*} - (V_{DD} + 0.7\text{ V})] \exp(-t/\tau_d) \cos \omega_d t \tag{4.76}$$

where

$$\tau_d = 2L_D/R_D \tag{4.77}$$

and

$$\omega_d = \left[\frac{1}{L_D C_D} - \left(\frac{R_D}{2L_D}\right)^2\right]^{1/2} \tag{4.78}$$

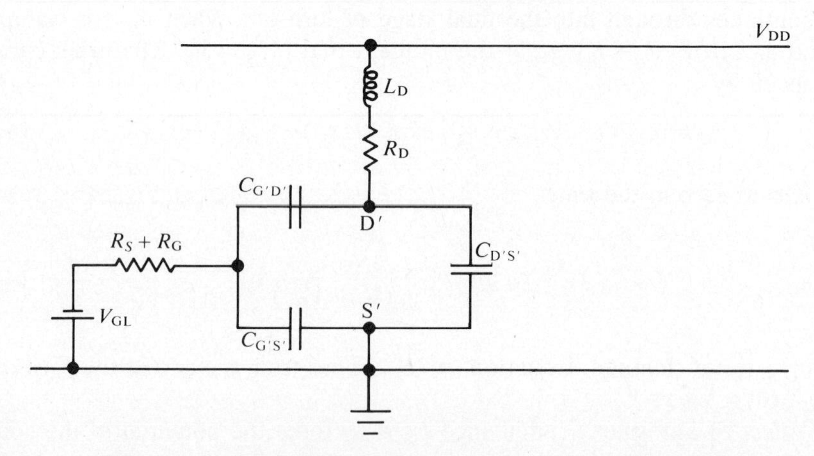

Figure 4.21. Equivalent circuit for the final phase of turn-off.

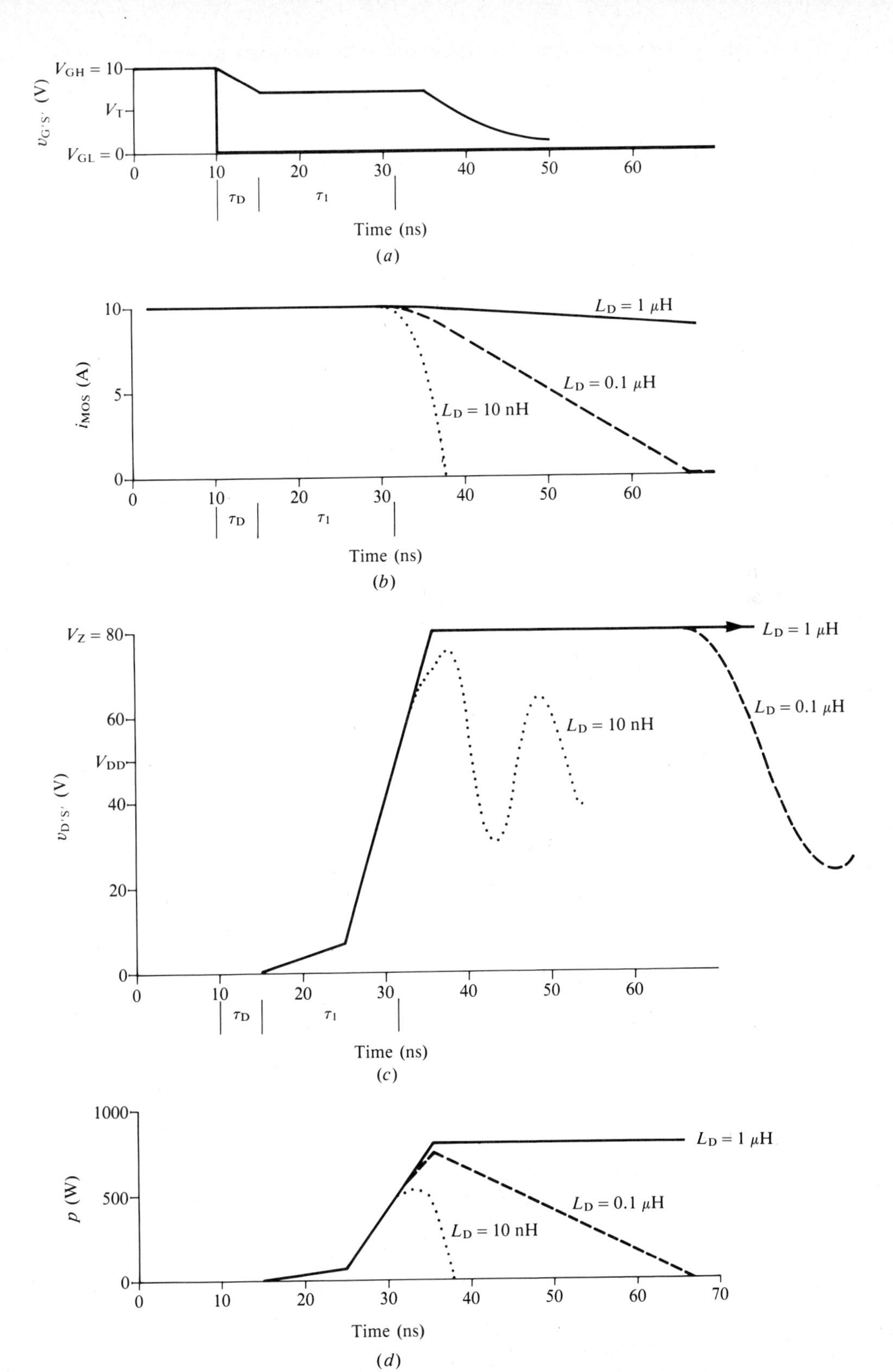

Figure 4.22. Turn-off waveforms for "fast" gate circuit. (*a*) $v_{G'S'}(t)$; (*b*) $i_{MOS}(t)$; (*c*) $v_{D'S'}(t)$; (*d*) $p(t)$. $L_D = 1\ \mu H$ (solid line), 100 nH (dashed line), 10 nH, (dotted line).

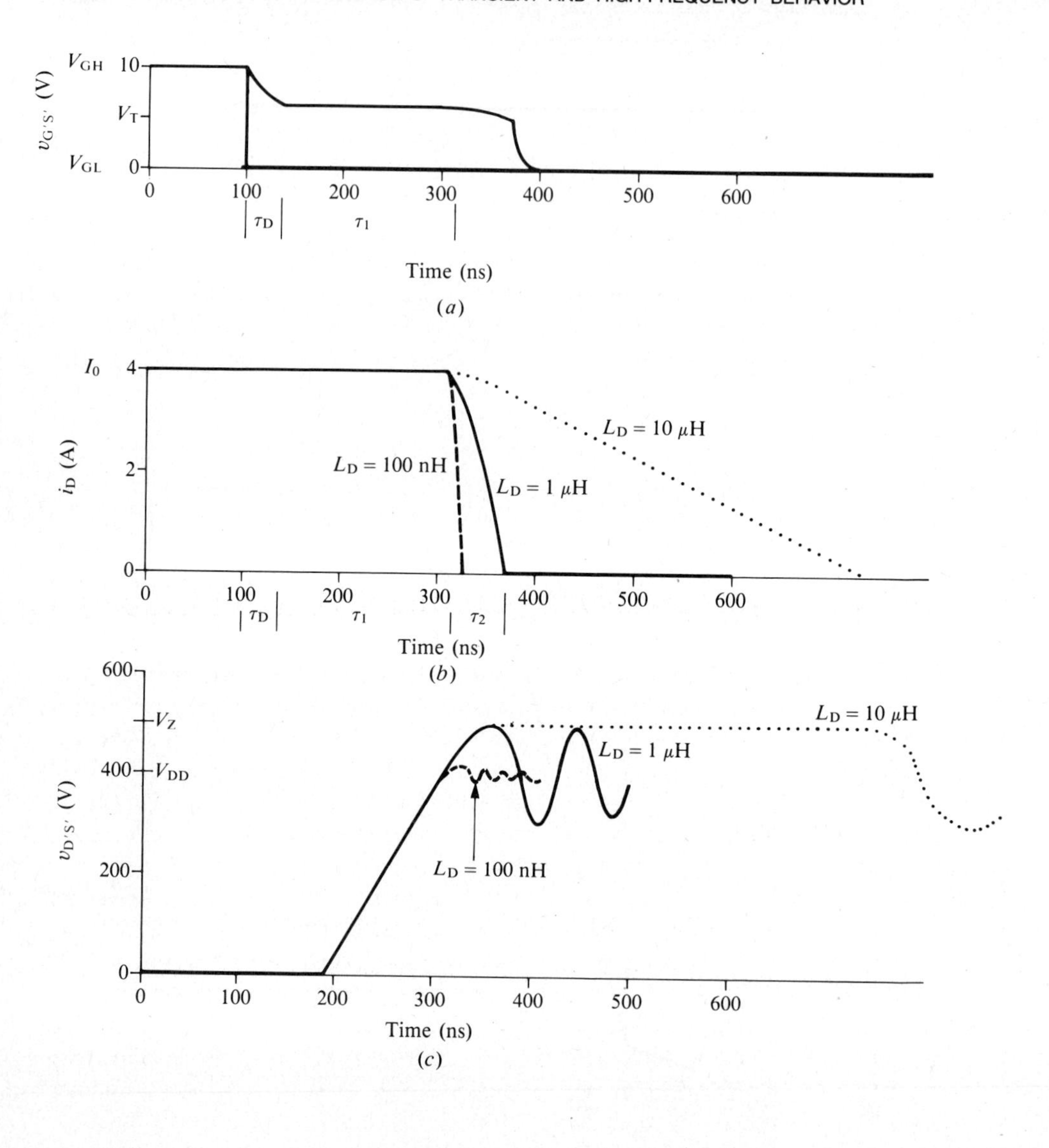

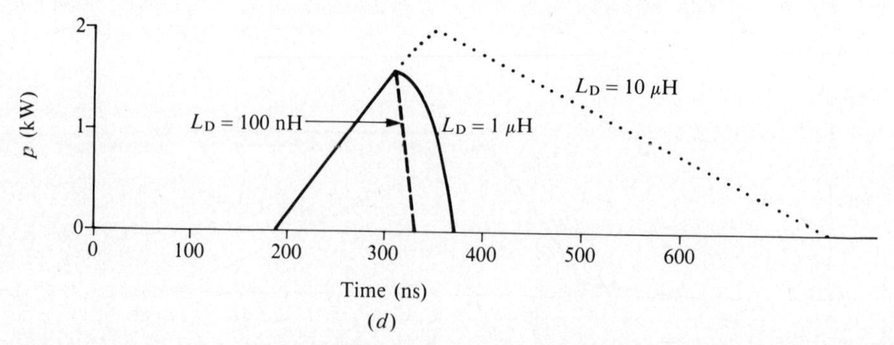

Figure 4.23. Turn-off waveforms for "slow" gate circuit. (a) $v_{G'S'}(t)$; (b) $i_{MOS}(t)$; (c) $v_{D'S'}(t)$; (d) $p(t)$. $L_D = 100$ nH (dashed line), 1 μH (solid line), 10 μH (dotted line).

with

$$C_D = C_{D'S'} + C_{G'D'} \tag{4.79}$$

4.3.6. Worked Examples

Waveforms corresponding to the first three examples given in Section 4.2.7 are plotted in Figure 4.22. It is assumed that the drain voltage is clamped at 80 V. The value of L_D has no effect on the delay time ($\tau_D = 5$ ns) or on the rise of the drain voltage to V_{DD} ($\tau_1 = 10\text{ ns} + 6\text{ ns} = 16$ ns). Its effect on the fall of the current, on the other hand, is considerable.

With $L_D = 1\ \mu$H, we have $\tau_2 \approx 340$ ns, and the total energy dissipated is about 140 μJ. Of this dissipation, 50 μJ represents the energy stored in the drain inductance ($\frac{1}{2}L_D I_0^2$). Note that this compares with 25 nJ dissipated during turn-on (275 nJ when the energy in $C_{D'S'}$ is included).

With $L_D = 0.1\ \mu$H, we have $\tau_2 \approx 35$ ns, and the dissipation is reduced to about 14 μJ, of which some 5 μJ is the energy initially in L_D. Note that the turn-on energy in this case is 180 nJ plus the 250 nJ of $C_{D'S'}$. Turn-off dissipation still dominates.

With $L_D = 10$ nH, $v_{D'S'}$ does not reach the zener breakdown voltage of 80 V, and there is a marked difference in its waveform above V_{DD}. The current now falls in some 6 ns, giving rise to about 4 μJ of dissipation, of which 0.5 μJ derives from L_D. The turn-on dissipation was estimated to be 2 μJ. In this case the "ring" on the drain voltage following the end of the conduction period is illustrated. In practice, we must expect the waveforms to be modified by source inductance during turn-off as well as during turn-on.

Results of similar calculations, based on the higher-voltage circuit of Section 4.2.7 with the slower gate drive supply, are shown in Figure 4.23. Now, $\tau_D = 37$ ns and $\tau_1 = 50 + 125 = 175$ ns. With $L_D = 10$ nH there is only a 5-V overshoot of V_{DD} and no ring. The current falls in 9 ns, and the dissipation is 90 μJ. With $L_D = 100$ nH there is a small ring, the current falls in 19 ns, and the dissipation is 110 μJ. With $L_D = 1\ \mu$H the drain voltage reaches the clamping voltage of 500 V in 34 ns when the current is 2.3 A. It falls to zero in a further 23 ns. The dissipation is about 160 μJ. With $L_D = 10\ \mu$H, the total turn-off time extends to nearly 600 ns, and the dissipation increases to some 560 μJ. Of this, $\frac{1}{2}L_D I_0^2 = 80\ \mu$J. These and the earlier results are summarized in Table 4.1.

4.4. GATE CHARGE

The nonlinearity of the parasitic capacitances, and the often incomplete data on their variation over the full range of relevant voltages, make gate circuit design by conventional methods exceedingly difficult. To overcome this problem it has become standard practice to specify the total gate charge Q_g that has to be supplied in order to establish a particular drain current under given test conditions. Data sheets normally divide the total charge into that required to charge the gate–source capacitance, Q_{gs}, and that required to supply the gate–drain or Miller capacitance, Q_{gd}. The merit of the gate-charge parameter is

TABLE 4.1. Turn-on and Turn-off Parameters

50-V, 10-A Circuit with Fast Gate, $R_S + R_G = 10\ \Omega$

		Turn-on		Turn-off		
L_D	L_S	Time	Energy[a]	Time	Energy[b]	Total Energy
1 μH	0	215 ns	270 nJ	356 ns	90 + 50 μJ	140 μJ
0.1 μH	0	25 ns	430 nJ	52 ns	9 + 5 μJ	14.5 μJ
10 nH	0	(36 ns	2.2 μJ)	22 ns	4 + 0.5 μJ	(6.7 μJ)
10 nH	10 nH	67 ns	12 μJ	22 ns	4 + 0.5 μJ	16.5 μJ

400-V, 4-A Circuit with Slow Gate, $R_S + R_G = 50\ \Omega$

	Turn-on		Turn-off		
L_D	Time	Energy[c]	Time	Energy[b]	Total Energy
10 μH	293 ns	36 μJ	600 ns	480 + 80 μJ	596 μJ
1 μH	293 ns	116 μJ	232 ns	152 + 8 μJ	276 μJ
0.1 μH	293 ns	156 μJ	194 ns	109 + 0.8 μJ	266 μJ
10 nH	293 ns	176 μJ	184 ns	90 + 0.08 μJ	266 μJ

[a] Includes 250 nJ stored in $C_{D'S'}$ at the start of turn-on.
[b] The second term is the energy stored in L_D at the start of turn-off.
[c] Includes 16 μJ stored in $C_{D'S'}$ at the start of turn-on.

that it is relatively insensitive to the drain current and the precise circuit conditions used, and quite independent of temperature. It permits a very simple design methodology for obtaining any desired switching time, and it enables the total charge and the total energy required to be estimated easily. The resulting mean current and mean power needed from the gate circuit can then be obtained simply by multiplying by the operating frequency.

A schematic representation of the test circuit is given in Figure 16 of Appendix 7. The use of effectively constant-current sources for the supplies to the gate and drain greatly simplifies the turn-on transition, as can be seen in Figure 11 of Appendix 7 and Figure 4.24. In particular, the inductances L_D and L_S in the equivalent circuit for the device under test can be neglected. Initially the device under test is OFF with the full line voltage applied across drain and source. At time zero, the gate current is switched into the device. With I_G maintained constant, C_{GS} and C_{GD} are charged at a constant rate,

$$\frac{dv_{GS}}{dt} = \frac{I_G}{C_{GS} + C_{GD}} \tag{4.80}$$

Since v_{DS} is high, usually 80% of its maximum rated value, $C_{GD} \ll C_{GS}$, and this initial period is essentially characteristic of C_{GS}.

When v_{GS} reaches V_T, the MOSFET under test enters its active region and its

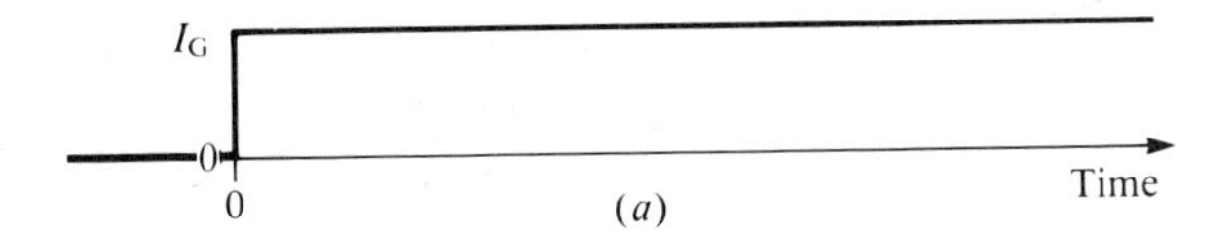

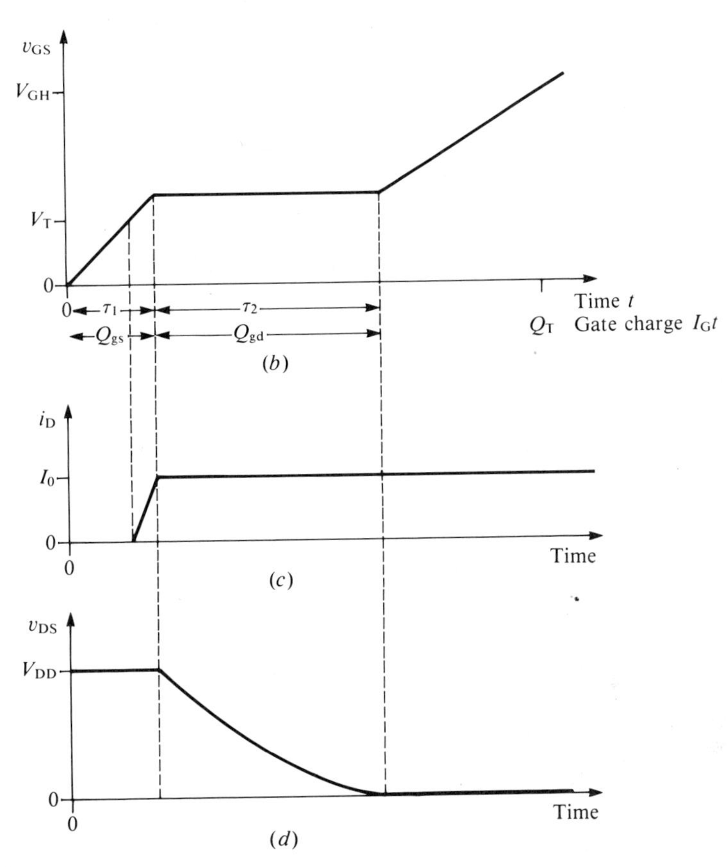

Figure 4.24. Waveforms during gate-charge test. (*a*) constant gate current; (*b*) gate–source voltage; (*c*) drain current; (*d*) drain–source voltage.

drain current starts to rise, with v_{GS}, and therefore also at a constant rate:

$$i_D = g_{fs}(v_{GS} - V_T) \tag{4.81}$$

The rise of v_{GS} is unaffected and continues until $i_D = I_0$, the value set by the control MOSFET. At this point the drain voltage of the device under test starts to fall. With $i_D = I_0 =$ constant, $v_{GS} = V_T + I_0/g_{fs} =$ constant. So the constant gate current discharges the Miller capacitance C_{GD} at a rate given by:

$$\frac{dv_{DS}}{dt} = -\frac{I_G}{C_{GD}} \tag{4.82}$$

As v_{DS} falls, C_{GD} increases and dv_{DS}/dt decreases. This can be seen in Figure 4.24.

Once the drain voltage has fallen to its ON-state value, the gate voltage resumes its constant upward rise for as long as the gate current continues. The rate of rise is lower than it was at first, because of the higher value of C_{GD}.

Because I_G is constant, the plot of gate voltage against time shown in Figure 4.24*b* is identical to a plot of gate voltage against gate charge, $Q_g = I_G t$. This is indicated along the horizontal axis. If $I_G = 1$ mA, each microsecond corresponds to 1 nC of gate charge. The period τ_1 represents the gate–source charge, $Q_{gs} = I_G \tau_1$, and τ_2 the gate–drain charge, $Q_{gd} = I_G \tau_2$.

Of course, in real gate-drive circuits the current is not constant, but this has only a second-order effect on the overall gate charge Q_T needed in any particular case to raise V_{GS} from V_{GL} to V_{GH}. This normally includes a margin over the minimum charge Q_g needed to switch on the transistor. Likewise, Q_g and Q_T are only slightly sensitive to the initial value of v_{DS}. The effects of different, constant drain currents and initial voltages are shown in Figure 4.25, where it can be seen that changes in I_0 affect Q_{gs} and changes in V_{DD} influence Q_{gd}.

The circuit designer can develop a drive circuit that deposits the total charge Q_T onto the gate within the required switching time. The total energy delivered

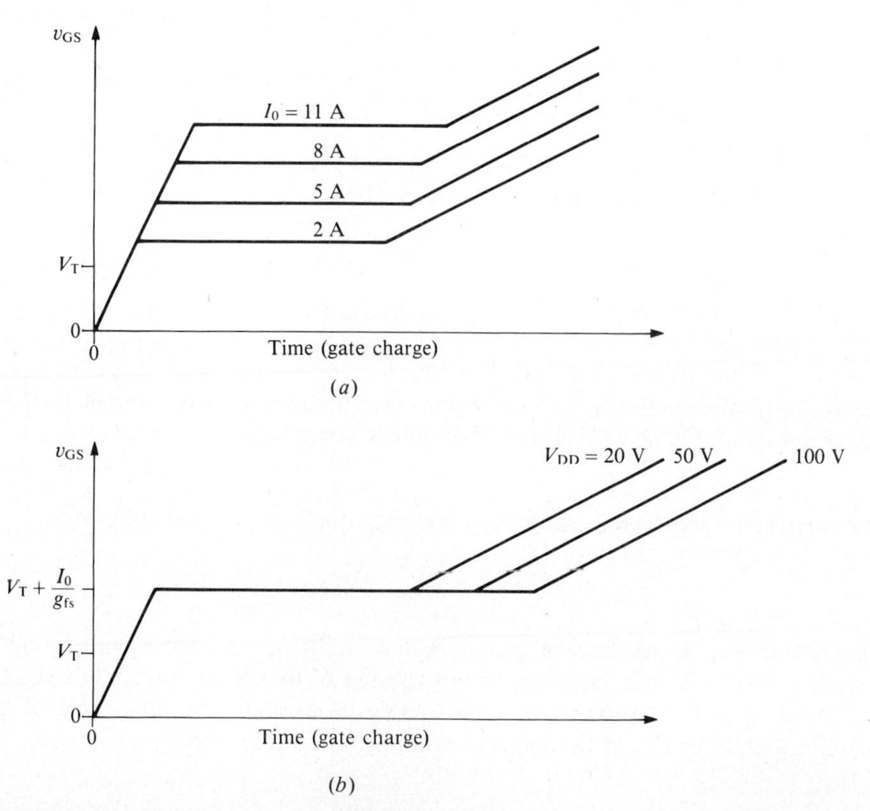

Figure 4.25. Variation of gate voltage with gate charge (time). (*a*) Effect of drain current; (*b*) effect of initial drain–source voltage.

by the gate drive supply is $V_{GH}Q_T$, so the gate power consumption is $V_{GH}Q_Tf$, where f is the switching frequency.

4.5. LIMITATIONS OF THE POLYCRYSTALLINE SILICON GATE AT HIGH FREQUENCY

Rapid voltage transients applied to the gate contact have to charge and discharge the gate capacitance. The charging and discharging currents have to flow through the significant sheet resistance of the polycrystalline silicon layer. As a result, they are subject to propagation delays and degradation as they travel to the more remote regions of the gate. This means that some care has to be taken over the layout of the gate vias, in order to ensure that no part of the gate is more than some minimum distance from the nearest via. Otherwise the turn-on of the more remote cells would be delayed significantly.

The complex shape of the gate web makes precise calculation difficult, but it is instructive to consider what happens in a long, rectangular length of poly-Si, having a low-impedance contact along one end. The situation is illustrated in Figure 4.26. The resistance per unit length of the poly-Si is given by $r = \rho/td = R_t/d$, where ρ is the bulk resistivity, d the width, and t the thickness of the layer. A typical value of the sheet resistance, $R_t = \rho/t$, is 30 Ω/square. The capacitance of the gate is assumed to be dominated by the capacitance through the gate oxide to the silicon. The capacitance per unit length, c, is given by $c = \epsilon_0\epsilon_{ox}d/t_{ox} = C_{ox}d$, where $\epsilon_0\epsilon_{ox}$ is the permittivity of the gate oxide and t_{ox} is its thickness. Thus, for $t_{ox} = 100$ nm (1000 Å), $C_{ox} = 3.45 \times 10^{-4}$ F/m^2 and

$$rc = R_tC_{ox} = 30 \times 3.45 \times 10^{-4} = 0.01 \text{ s/m}^2 \tag{4.83}$$

For a length $l = 1$ mm, the time constant $l^2rc = 10$ ns. Note that this is independent of the width d. It is not, however, the characteristic charging and discharging time constant of that length of gate. That would imply that all the capacitance charging current had to flow through all the resistance, an overpessimistic assumption.

A fluctuating voltage, applied to the end of the poly-Si, propagates along it according to the diffusion equation. The situation is identical to the variation of temperature along an insulated rod, heated at one end. Consider an elementary section of the poly-Si, of length δx, at position x, as shown in Figure 4.26. It may be modeled by the equivalent circuit shown in Figure 4.26*c*. The relationships between the voltages and the currents are:

$$\frac{\partial v}{\partial x}\,\delta x = -i(x)r\,\delta x \tag{4.84}$$

$$\frac{\partial i}{\partial x}\,\delta x = -\frac{\partial v}{\partial t}c\,\delta x \tag{4.85}$$

Therefore,

$$\frac{\partial i}{\partial t} = \frac{1}{rc}\frac{\partial^2 i}{\partial x^2} \tag{4.86}$$

and

$$\frac{\partial v}{\partial t}=\frac{1}{rc}\frac{\partial^2 v}{\partial x^2} \tag{4.87}$$

For an infinite length of poly-Si, excited from one end, the solutions are straightforward. With the voltage initially zero everywhere, a step change from zero to V_0 at $x=0$ at $t=0$ leads to a subsequent distribution given by

$$v(x, t)=V_0 \operatorname{erfc}\left(\frac{x}{2(t/rc)^{1/2}}\right) \tag{4.88}$$

where $\operatorname{erfc}(y)=(2/\sqrt{\pi})\int_y^\infty \exp(-t^2)\,dt$ is the complementary error function. This step response is shown in Figure 4.26*d*. There is an initial delay, a relatively rapid

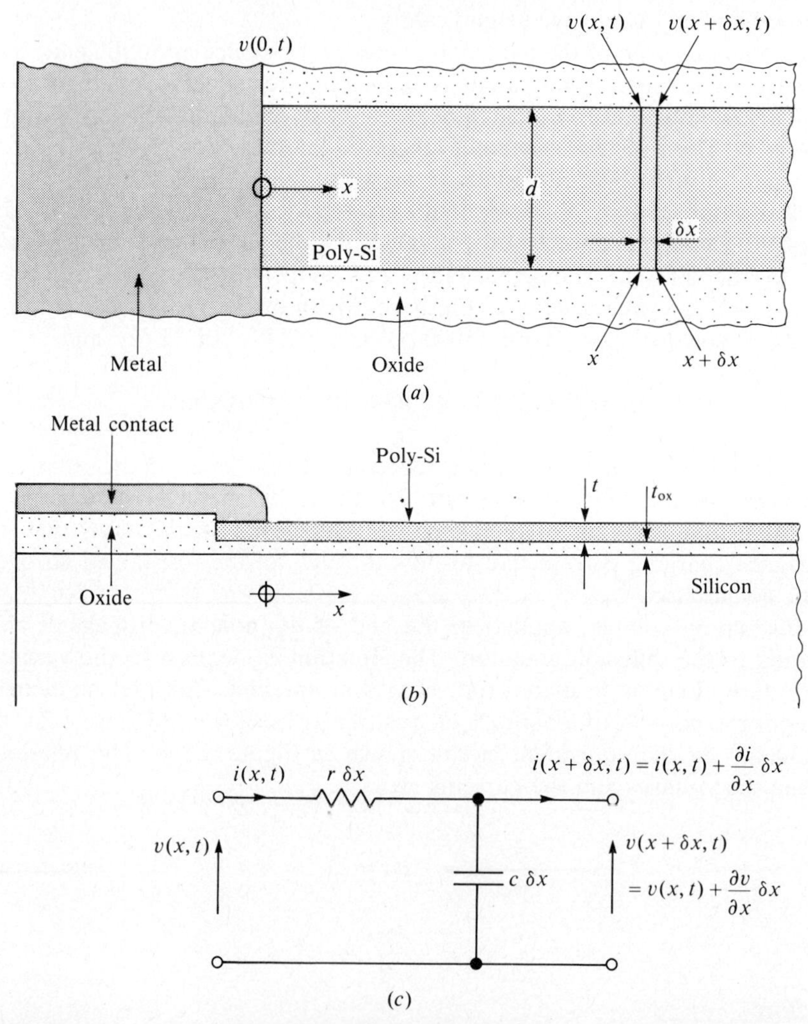

Figure 4.26. Propagation of voltage along a polycrystalline silicon gate finger. (*a*) Plan view of gate finger; (*b*) section; (*c*) equivalent circuit; (*d*) step propagation; (*e*) sine-wave attenuation.

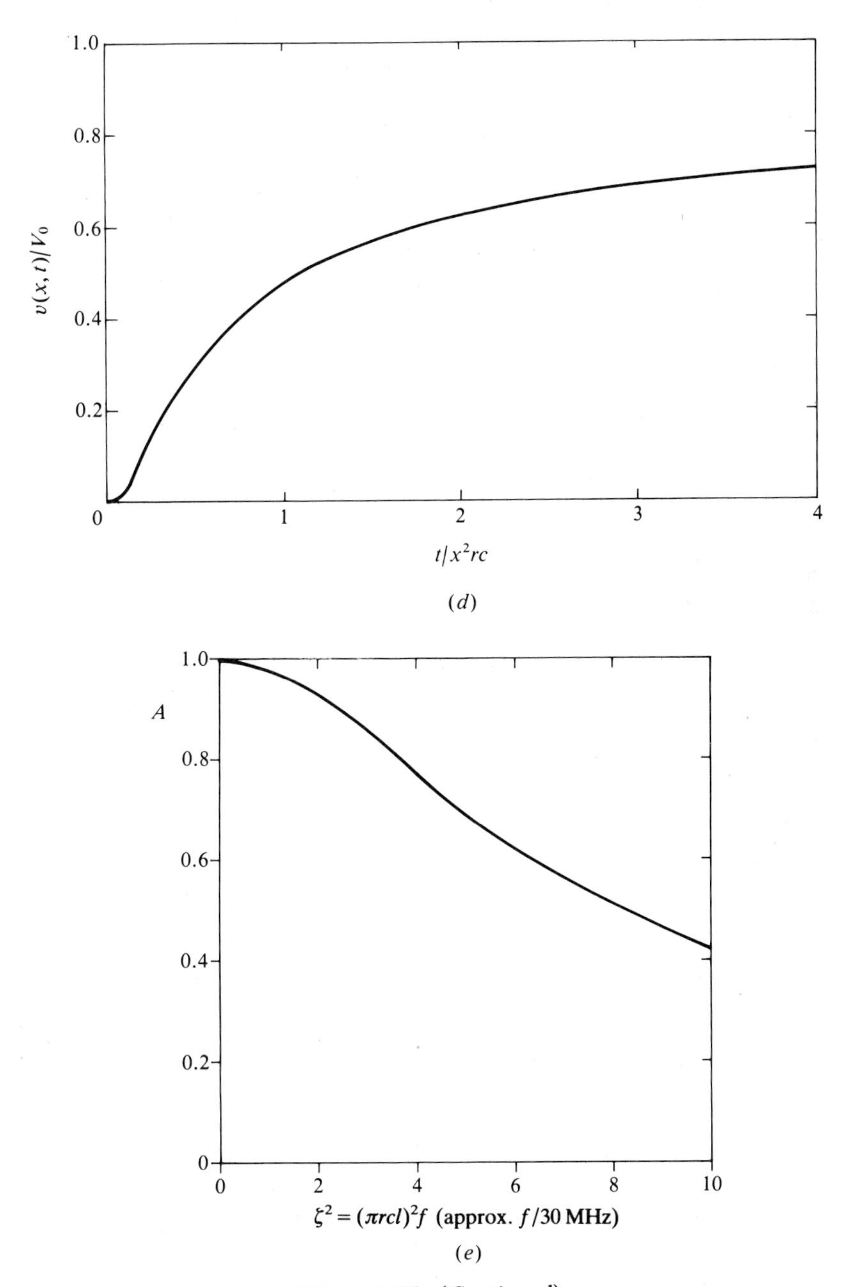

Figure 4.26. (*Continued*).

rise during which the gate voltage normally crosses the threshold voltage, and lastly, a very gradual approach to the final value V_0.

When a sinusoidal voltage of amplitude $V(0)$ and frequency f is applied to the end at $x = 0$, the voltage fluctuations decay with distance as

$$\left|\frac{V(x)}{V(0)}\right| = \exp[-x(\pi rcf)^{1/2}] \tag{4.89}$$

The amplitude is reduced by a factor of $\sqrt{2}$ when $x(\pi rcf)^{1/2} = 0.35$. With the values quoted in Equation (4.83), this corresponds to a distance of 1 mm at a frequency of 11 MHz.

In practice, in most cases, the polysilicon is excited from both ends. This can be seen in the plan view of the HEXFET™, shown in Figure 3.10. The gate vias are rather less than 1 mm apart. The actual spacing ought to be multiplied by $2/\sqrt{3}$ to allow for the angled path between the hexagons, so an effective value of 1 mm for the spacing l between the ends is reasonable. Solutions are rather more complicated [4], but the ratio of the amplitude at the midpoint to that at the ends when a sinusoidal voltage of frequency f is applied is

$$A = \left| \frac{V(\frac{1}{2}l)}{V(0)} \right| = \sqrt{2}(\cosh \zeta + \cos \zeta)^{-1/2} \tag{4.90}$$

where

$$\zeta = l(\pi rcf)^{1/2} \tag{4.91}$$

This frequency response is shown in Figure 4.26*f*. The voltage at the centre is less than the applied voltage by the factor $1/\sqrt{2}$ when $\zeta = 2.2$. For small ζ, A varies as $1 - (\zeta^4/48)$. Thus, the points furthest from the gate vias rapidly cease to be effective as ζ exceeds 2, that is, as f exceeds $4/\pi l^2 rc$. With the values $l = 1$ mm, $R_t = 30\ \Omega$/square, and $C_{ox} = 3.45 \times 10^{-4}\ \mathrm{F/m^2}$, this corresponds to frequencies exceeding about 100 MHz.

SUMMARY

The transit-time-limited switching speed of VDMOS FETS is less than 0.1 ns. In practice, switching is limited by circuit and parasitic capacitance and inductance. During the turn-on and turn-off transitions, the transistor may be assumed to be in one of three states: ACTIVE, ON, or OFF, and can be modeled by the very simple transfer characteristic shown in Figure 4.1. Additional energy, proportional to the transition time, is dissipated in each turn-on and turn-off operation. The additional power dissipated is proportional to the switching frequency.

Each turn-on and turn-off transition may be divided into four periods. They are illustrated in Figures 4.8 and 4.18. For turn-on there is an initial delay while the gate voltage rises to V_T. This is followed by periods when the drain current rises and the drain–source voltage falls, when either I_D completes its rise or V_{DS} its fall, and when V_{GS} completes its rise to V_{GH}. During turn-off there is a delay while V_{GS} falls to the level needed to maintain the drain current. In a clamped inductive circuit the drain voltage then rises above the supply voltage. Next I_D falls while V_{DS} remains above V_{DD}, possibly clamped by avalanche breakdown in the body–drain diode or by an external zener diode. Finally, there is usually a damped drain-circuit oscillation while the gate voltage decays back to its lower value.

Turn-on and turn-off are both speeded by the use of a "fast" gate circuit that is, one in which the gate supply impedance has a low value. A "slow" drain circuit, in which L_D is increased, reduces the turn-on time, but increases the energy dissipated during turn-off.

Analysis of the transitions is complicated by the nonlinearities of $C_{G'S'}$ and $C_{G'D'}$ as v_{DS} varies. However, the gate charge needed to turn on the transistor is insensitive to the current being switched and to the precise waveforms. This parameter is given on data sheets and makes gate-drive circuit design quite straightforward.

A layout of the gate vias that minimizes the polysilicon path lengths is crucial if the whole transistor is to be activated in high-frequency operation. For efficient operation above 50 MHz, the gate vias must be less than 1 mm apart.

REFERENCES

1. S. M. Clementi, B. R. Pelly, and A. Isidori, "Understanding power MOSFET switching performance." *IEEE Ind. Appl. Symp.* **IAS81,** Pap. 32B, 763–776 (1981).
2. S. M. Clementi, B. R. Pelly, and A. Isidori, "Analysis and characterization of power MOSFET switching-interval performance in power converters." *Proc. Powercon* **8,** Pap. H2, 1–11 (1981).
3. S. M. Clementi, B. R. Pelly, and A. Isidori, "Understanding HEXFET® switching performance." *Int. Rectifier Appl. Note* **AN-947** (1984).
4. H. S. Carslaw and J. C. Jaeger, *Conduction of Heat in Solids,* 2nd ed. Oxford Univ. Press, London and New York, 1959 (see, in particular, Sects. 20 and 38).

CHAPTER 5

Fabrication and Reliability

5.1. FABRICATION

5.1.1. A Standard Process for n-Channel Devices

Most VDMOS FETS are fabricated in processes involving several deposition, implantation, and diffusion stages, and some five or six separate masking steps. The photolithography should permit a 2- to 3-μm minimum line width to be defined. This means that whole-wafer proximity printing may be used, together with chemical (wet) etching methods. A number of more modern device designs utilize a smaller cell size, and so demand a finer geometry. This is particularly true for devices intended for low-voltage applications. For these, the difficulty of maintaining the required definition and registration tolerances over the whole slice is overcome by the use of a step-and-repeat masking process. A mask covering a single device or a few devices only, is projected onto the photoresist, usually with a $\times 10$ reduction, and progressively stepped across the whole wafer. This technique has to be used when larger diameter slices (6 in. rather than 4 in.) are used to increase throughput. Likewise, plasma (dry) etching techniques are then needed in order to avoid undercutting and lateral spread during etching. For a comprehensive review of all aspects of fabrication technology, the reader is referred to Ref. 1.

As was made clear in Chapter 1, VDMOS FETS are fabricated on $\langle 100 \rangle$-oriented slices. The slice thickness has to be chosen with some care. Heat dissipated in the active parts of the device has to be removed through the full thickness of the substrate. The thicker the slice, the greater is the thermal resistance of the final device. However, too thin a slice will lead to breakages during manufacture and a lower yield, particularly with larger-diameter slices. Typically, slice thicknesses between 250 and 500 μm (10–20 mils) are used. The slice thickness also adds to $R_{DS(on)}$. However, heavy doping with phosphorus ensures that the resistivity is low. Typical values exceed 10^{24} atoms/m^3 (10^{18} cm^{-3}), giving 2×10^{-4} Ω m (0.02 Ω cm), or less.

After the normal cleaning and polishing routines, the first step is to grow the n-type epitaxial layer in which the device will be formed. Ideally this should be defect-free. Its thickness and doping level are determined by the intended voltage rating of the transistors. For 50-V devices, as little as 10 μm of material may be deposited. The resistivity would then typically be about 0.01 Ω m (1 Ω cm), corresponding to a phosphorus doping level of about 5×10^{21} m^{-3} (5×10^{15} cm^{-3}). The thickness and resistivity of the epitaxial layer increase as the voltage rating increases. For 500-V devices, the thickness is more likely to be 50 μm and the resistivity 0.2 Ω m (20 Ω cm). The corresponding phosphorus concentration is 2×10^{20} m^{-3} (2×10^{14} cm^{-3}).

The surface is oxidized for the first masking process. This defines the p-type wells that in the final device connect the channel regions back to the source contacts. It also defines the p-type regions that will underlie the gate vias and the bonding pads. Similar p-type regions form the guard rings for the high-voltage structure around the periphery of the chip. Windows are etched through the oxide, and some 10^{18} boron atoms per square meter (10^{14} cm^{-2}) are implanted at low energy (<10 kV) and then diffused in to form a junction at a depth of about 5 μm. In an optimized process, some compensation has to be made for the doping level of the epilayer. The sheet resistance is typically about 200 Ω/square. During the high-temperature treatment, some outdiffusion of dopant from the substrate into the epilayer occurs.

The surface is left oxidized for the second masking sequence, which is the reverse of the first. All the oxide, *except* that covering the p-type wells, is removed. The slice is then ready for a high-energy (>100 kV) implant of phosphorus, everywhere except into the p-type wells. The situation is illustrated in Figure 5.1(a). Some 10^{16} to 10^{17} atoms per square meter are implanted (10^{12} to 10^{13} cm^{-2}). This gives enhanced conductivity to what will become the drain drift regions under the gate oxide.

Growth of the gate oxide is the next stage. The surface is cleaned, except for the thicker layer of oxide that covers the p-type wells. It is then subjected to a dry thermal oxidation, giving a uniform film of high-integrity silicon dioxide, some 50 to 100 nm thick. Onto this, some 0.5 μm of polycrystalline silicon, heavily doped with phosphorus, is deposited. This will form the gate web. Its sheet resistance is usually in the region of 30 Ω/square. Next, a layer of oxide is deposited over the whole surface by low-pressure chemical vapor deposition. This will form the mask for the subsequent, self-aligned diffusions. A cross-section through the device at this stage is shown in Figure 5.1*b*.

The third masking process defines the pattern of cells, hexagonal or square, that covers the active area of the device. The cells are centered on the previously formed p-type wells, which still carry their caps of oxide. The regions within the cell areas are etched down to the silicon surface, except for the thicker oxide covering the p-type wells. At the same time, fingers of the polycrystalline silicon layer are defined for the gate vias, as are annular rings around the periphery of the chip which will form part of the guard ring structure. This stage is shown in Figure 5.1*c*.

The slice is now ready for an implant of about 10^{18} boron atoms per square meter (10^{14} cm^{-2}). This is normally carried out at about 50 kV, and is followed by a drive-in diffusion for up to 1 or 2 hours at around 1200°C. This forms a pn junction with the n-type epitaxial layer at a depth of some 2 to 3 μm. The junction extends about 2 μm under the edge of the mask, because of lateral diffusion. It is this region immediately under the gate oxide that forms the channel region of the FET. Note that the high-temperature treatment drives the earlier phosphorus implant deeper into the epitaxial layer, and causes further outdiffusion of the substrate dopant.

Any thin coating of oxide formed on the exposed silicon surface is cleaned off,

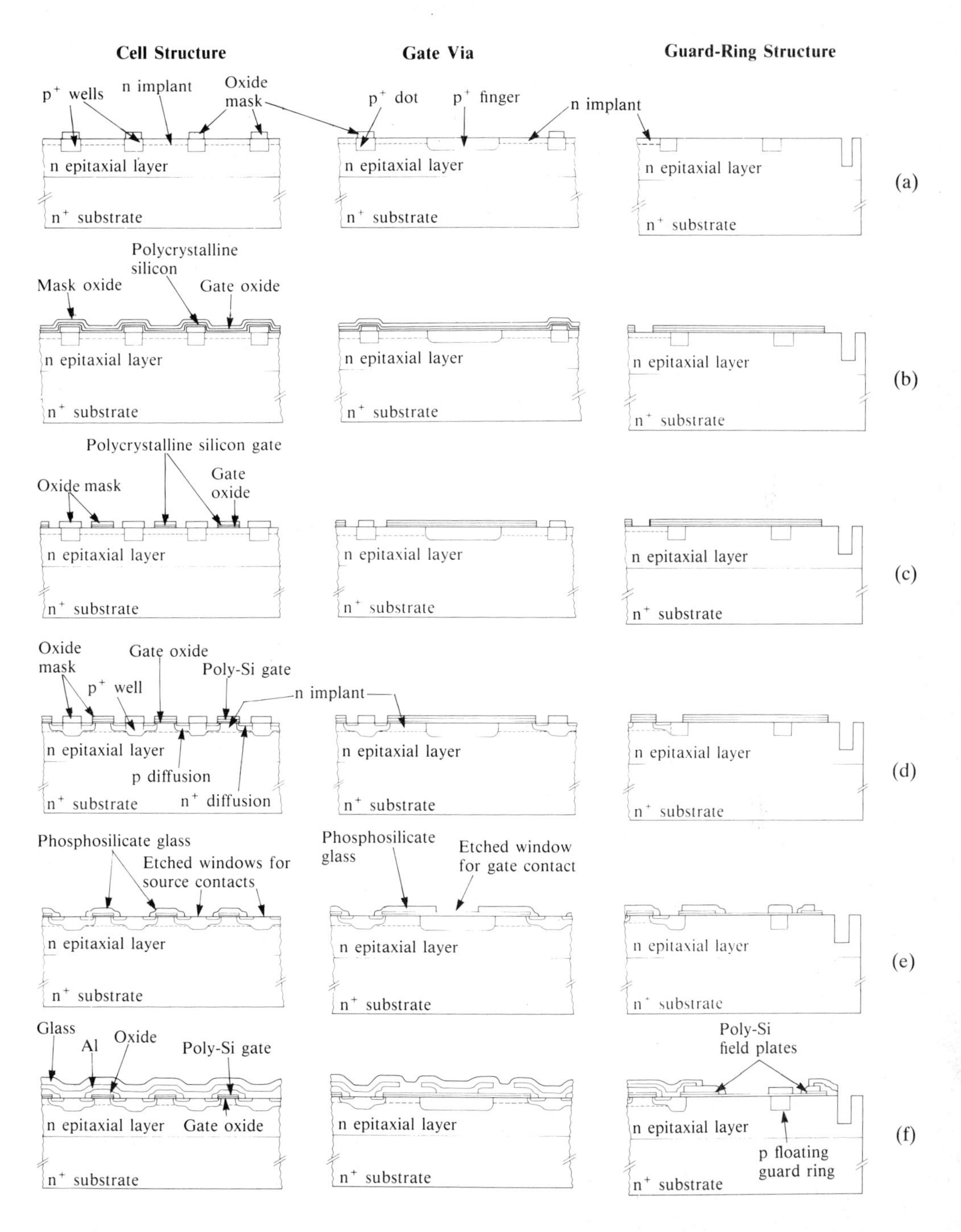

Figure 5.1. A typical processing sequence for a power MOSFET. (*a*) Section through the device after epitaxial growth and p and n implants; (*b*) following the growth of the gate oxide and the deposition of the poly-Si gate; (*c*) following the etching of the cells and the definition of the gate web; (*d*) after two self-aligned diffusions have defined the channel region; (*e*) following the etching of the contact windows; (*f*) the finished device.

and the slice is exposed to a short phosphorus diffusion, typically for about 30 minutes at about 950°C, through the same windows. This forms what will become the source regions. The surface concentration is high, and a relatively abrupt junction is formed at a depth of about 1μm. The junction extends rather less than 1 μm under the edge of the oxide mask.

We have now formed the basic transistor. The channel is defined by the two pn junctions, each following the edge of the mask. The fact that both ends of the channel are defined using the gate structure as a mask ensures precise alignment of gate and channel and accurate control of the channel length. This is the essence of the self-aligned process. The situation at this stage is shown in Figure 5.1*d*.

All the oxide formed during the previous processing (except the gate oxide, which is masked by the poly-Si) is removed using hydrofluoric acid, and a new, thin layer of oxide is grown over the whole surface. Onto this a thicker layer of phosphorus-doped silicon dioxide is deposited. This can easily be made to reflow and so form smooth glassy contours over the edges of the polycrystalline silicon. This aids the continuity of the subsequent metallization stages.

The fourth mask defines windows for the source contacts. These windows are etched down to the silicon surface, as shown in Figure 5.1*e*, and aluminum is evaporated over the entire top surface of the wafer. At the same time windows are etched down to the polycrystalline silicon layer over the gate vias, so the aluminum contacts there as well. The fifth mask enables the aluminum to be etched, so as to separate the main source contact from the gate contact and also from the field rings at the chip periphery. After removal of the photoresist, a final layer of phosphosilicate glass is deposited over the whole wafer for protective purposes. A section through the device at this stage is shown in Figure 5.1*f*. A sixth mask is used to define windows which are etched through the glassy layer down to the aluminum of the source and gate bonding pads.

The individual MOSFETS are now fully formed. A drain contact is made, for example by evaporating chromium, nickel, and silver layers onto the rear of the wafer. The individual chips are then each probe-tested. The wafer is scribed and broken, and the chips separated, cleaned, die-bonded, and connected.

5.1.2. Process Variations

In the previous section, we have described what may be thought of as a standard process for the fabrication of VDMOS FETS. Each manufacturer has proprietary variations. Some processes are more complex than the one described, making use of the full range of techniques developed for VLSI, in particular, plasma etching and step-and-repeat photolithography. They may employ as many as eight masks. Others aim at greater simplicity and lower cost. One manufacturer has introduced [2] a four-mask process with an interdigitated (noncellular), aluminum-on-polysilicon gate. This has been optimized for high-voltage (500 V), high-frequency applications. Another manufacturer has described [3] a very different interdigitated design, intended to maximize gate width and minimize ON resistance in a low-voltage device. This process applies the technology of VLSI to a development of the VVMOS device. We give a brief account, next, of each of these very different designs.

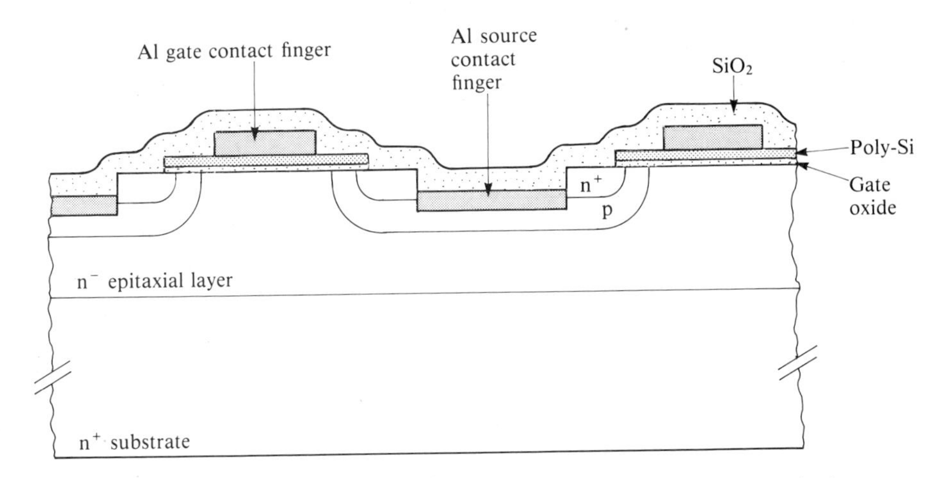

Figure 5.2. Cross-section through the active region of a metal gate device.

The first process aims at achieving a high yield from a simple process with low masking tolerances and a reduced sensitivity to defects. It achieves the required ratings by employing a larger chip area. While this is more costly in raw material, it brings the immediate benefits of lower ON resistance, lower thermal resistance, and lower current density. A cross-section through the active region is shown in Figure 5.2. The two main novel features can be clearly seen. The source–body contact fingers lie at the bottom of trenches etched through the source diffusion and into the body region. This removes the need for the p^+ diffusion, while minimizing the effect of the parasitic bipolar transistor. Because the source metal does not overlie the gate, C_{GS}, and therefore C_{iss}, are reduced. The polycrystalline silicon gate fingers are overlaid with aluminum metallization. This reduces the gate turn-on propagation delay, and ensures that at high frequency all parts of the transistor are fully activated. Only one of the masking steps is critically aligned, compared with the conventional process where the first four masks are each critical in their alignment. The electrical characteristics of the device are: $BV_{DSS} = 500$ V, $I_D = 16$ A, $R_{DS(on)} = 0{\cdot}3\ \Omega$, $C_{iss} = 2$ to 2.5 nF, $Q_g = 55$ to 75 nC.

The second design is shown in plan and cross-section in Figure 5.3, and an outline of the process sequence is illustrated by Figure 5.4. The devices are formed in an 8-μm-thick, phosphorus-doped epilayer of resistivity 0.006 Ω m (0.6 Ω cm). This is grown on an antimony-doped substrate of resistivity less than $6 \times 10^{-5}\ \Omega$ m (6 mΩ cm). A boron implant and diffusion forms the body region, with the junction at a depth of about 4 μm. This is followed by an implant and diffusion of arsenic to form the source with a junction depth of 0.4 μm. Thin films of silicon dioxide and silicon nitride are deposited, then a thicker layer of oxide, as shown in Figure 5.4*a*. Rectangular grooves are then cut to a depth of 2.5 μm, by reactive-ion etching, and after cleaning the surface, the 50-nm-thick gate oxide is grown, as in Figure 5.4b. A polycrystalline silicon layer 0.5 μm thick and heavily doped with phosphorus is deposited next, and its surface is oxidized. This is followed by a deposition of a second layer of poly-Si, this time undoped, which fills the grooves and leaves a planar surface, as seen in Figure 5.4*c*. This is etched

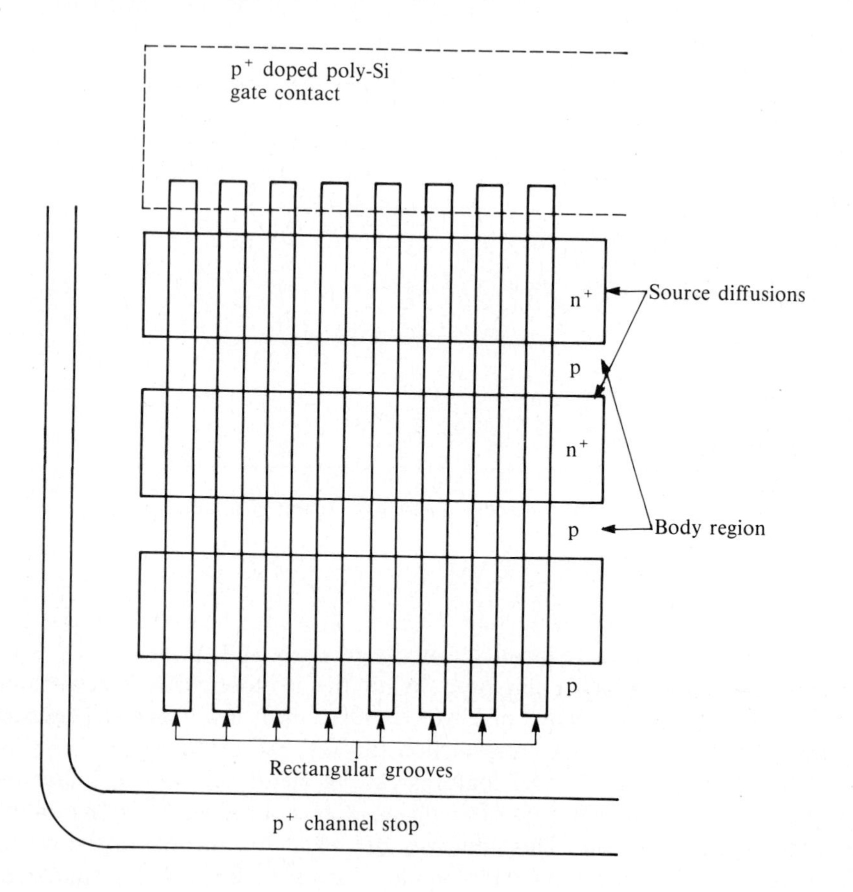

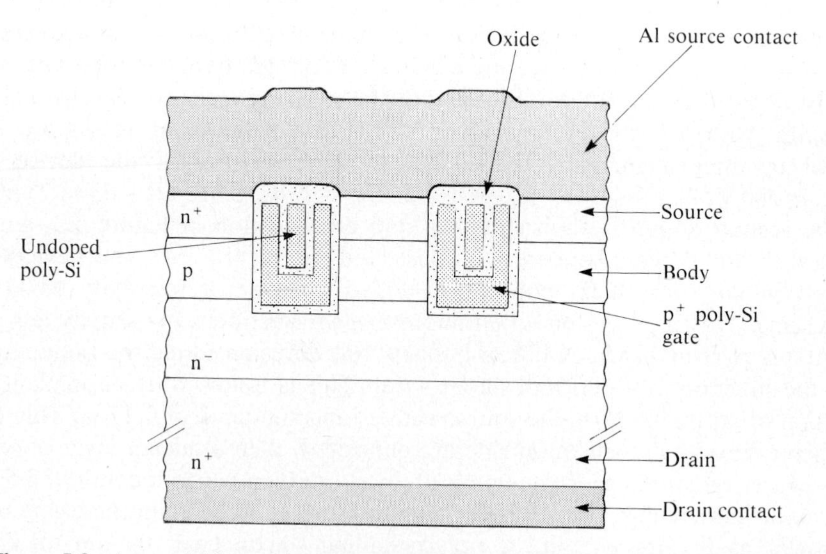

Figure 5.3. A very high-packing-density device. (*a*) Plan view; (*b*) cross-section.

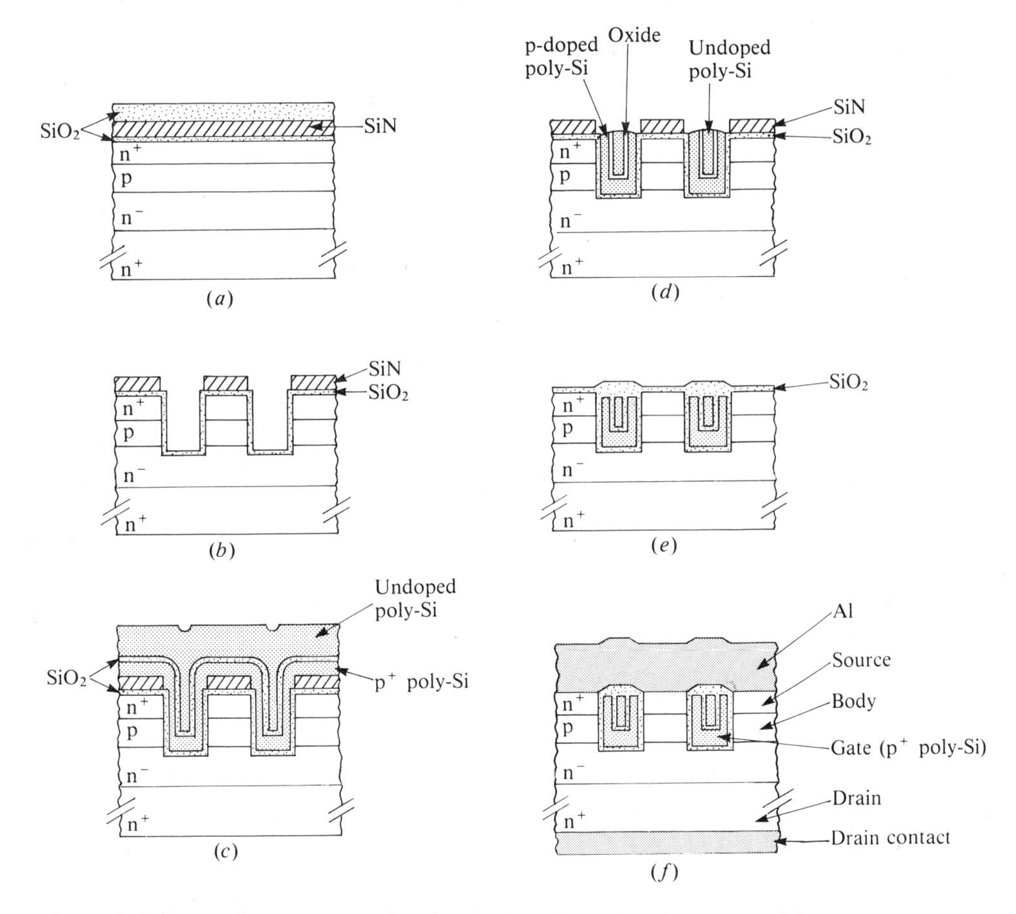

Figure 5.4. Processing sequence for the device shown in Figure 5.3. (*a*) After epitaxial growth, p and n diffusions, and the deposition of insulating layers; (*b*) etched rectangular grooves; (*c*) poly-Si infilling; (*d*) poly-Si etch; (*e*) local oxidation; (*f*) the final device.

back to the nitride layers, again using a reactive-ion etch. These then act as a mask for a local oxidation of the poly-Si of the gate fingers, as in Figure 5.4*d* and *e*. The nitride and gate oxide are removed from the surface of the source, and an aluminum contact layer is evaporated everywhere, as shown in Figure 5.4*f*. The entire gate and channel structure is self-aligned with a single mask. As a result it is possible to fabricate the gate grooves on a pitch of 4 μm. This gives a gate width per unit area of 0.5 m/mm^2 and a value of 0.14 Ω mm^2 for $R_{DS(on)}A$. The electrical characteristics of a device of total area 3.8 × 4.0 mm are $BV_{DSS} = 32$ V, $R_{DS(on)} = 9$ mΩ, $g_{fs} = 62$ S, $C_{iss} = 6.4$ nF, $t_r = 50$ ns, $t_f = 200$ ns.

5.2. QUALITY AND RELIABILITY

5.2.1. Introduction

The VDMOS FET has been designed from the outset as a highly reliable component with a high manufacturing yield. We distinguish the terms *quality* and *reliability* as follows:

Quality is inversely related to the proportion of defective or out-of-specification parts at the time of shipment. Statistical sampling methods are used to enable the supplier to monitor the outgoing quality level (OQL). Power MOSFETS can achieve an assured electrical outgoing quality level with fewer than 40 defective parts per million (AOQL < 40 ppm). This is obtained by rigorous monitoring and control of the incoming materials and the process and assembly at every stage, and by the careful screening of individual components. At this level of AOQL, individual device inspection by the purchaser is likely to be counterproductive, in that the extra handling involved may well introduce more defective components than it finds.

Reliability measures the capability of the device to perform as specified over an extended operational lifetime. In common with suppliers of other semiconductor components, power-MOSFET manufacturers have attempted to quantify the reliability of their products under the conditions likely to be experienced in normal operation. For this to be of use to the design engineer, reliable data on failure rates have to be presented in such a way that they can readily be related to a wide range of possible operating conditions. In order to achieve this, the most likely modes of failure have to be identified, and a precise definition of what constitutes a failure has to be made. This is usually quite straightforward, and involves a particular parameter such as leakage current or $R_{DS(on)}$ going outside specification. To obtain these data on a reasonable time scale and at reasonable cost, some means has to be found to accelerate the degradation and failure processes. Otherwise, the information would take so long to acquire that, before it was available, the device would likely to be obsolete. For this information to be reliably extrapolated to normal operating conditions, the physical causes of each mode of failure have to be understood. This enables one or more significant stress parameters, such as temperature or electric field, to be isolated. Next, a physical law relating the failure rate to a particular stress parameter has to be established. Finally, the validity of the law is tested, and the data required by the user obtained in accelerated life tests carried out under overstressed conditions. For the results to be statistically meaningful, tests have to be made on a large number of devices. Then confidence levels can be assigned to each predicted lifetime.

In the next section we give a brief review of the terms used in assessing reliability. Later we go on to examine the principal causes of VDMOS failure. Practical aspects of the reliability of power MOSFETS are discussed further in Chapter 17.

5.2.2. Basic Reliability theory

A plot of the failure rate against time, for many components, follows the classic *bathtub* curve. This is shown in Figure 5.5. The instantaneous failure rate is the sum of three components. At early times the failure rate is high. This is caused by devices that are slightly defective in some way, but which still meet the device

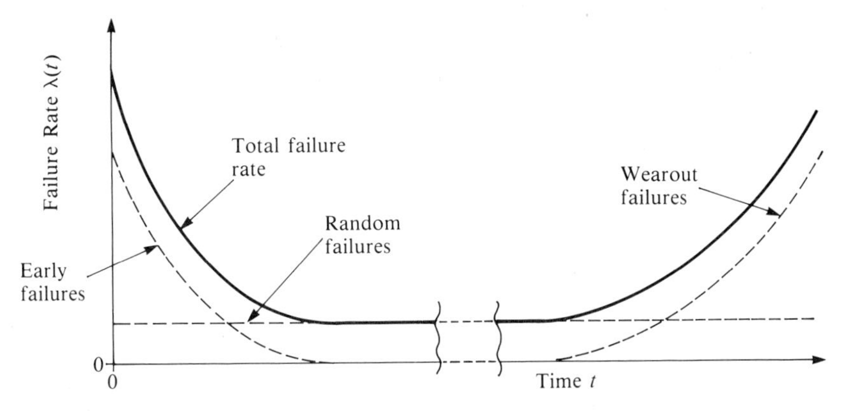

Figure 5.5. The "bathtub" curve.

specification initially. They are thus sent out as good devices, but the defect causes a rapid change in some parameters, so that they are soon out of specification. They are known as *infant mortalities.* They can often be isolated by subjecting all components to a period of *burn-in,* that is, a short test at high temperature and sometimes with other imposed stresses such as high voltage or high current. It is, of course, important that the burn-in should not itself induce new defects and significantly accelerate the ultimate failure of good devices. One of the great benefits of having readily available and reliable life-test data on components is that the equipment manufacturer is relieved of the trouble and cost of running a burn-in schedule, with all its attendant risks.

Throughout the life of the component, failures may occur randomly, for no correlated or clearly ascribable reasons. It is often assumed that during the main period of the life of many components this is the dominant failure mode. The failure rate is taken to be constant and to depend on the level of stress imposed. Under conditions of low stress it may be very low indeed, and then, this period may last for many millions of hours. Really, a random failure phase is most likely to be observed in a complex system in which failures result from many independent and individually unlikely causes. Most causes of failure in a single component, after infant mortalities have been screened out, are associated with a wearout mechanism. Any observed period of apparently constant failure rate is likely to be at best an approximation, and coincidental.

Following any period of random failure comes the final period of *wearout* failure. The failure rate rises as physical and chemical degradation processes accelerate, until finally no good components remain.

It is necessary to define some terms which can be applied to a population of components subjected to life test. The *reliability* $R(t)$ represents the probability that a component survives to the time t. The probability that the component fails sometime during the period t to $t + dt$ is $f(t)\,dt$, where $f(t)$ is known as the *failure probability density function.* The probability that the component fails sometime during the period t to $t + dt$, *given that it survives to the time t,* is $\lambda(t)\,dt$, where $\lambda(t)$ is known as the *failure rate.* It is $\lambda(t)$ that is plotted in Figure 5.5. Clearly,

$$f(t) = \lambda(t)R(t) \tag{5.1}$$

Other relationships between these parameters may easily be derived. The probability of failure during the whole of the period from the start of life to time t is given by

$$\int_0^t f(t)\,\mathrm{d}t = 1 - R(t) = F(t) \tag{5.2}$$

where $F(t)$ is known as the *cumulative failure probability function.* Note that $F(0) = 0$, $R(0) = 1$, $F(\infty) = 1$, and $R(\infty) = 0$. Differentiation of Equation (5.2) shows that

$$f(t) = \frac{-\mathrm{d}R}{\mathrm{d}t} \tag{5.3}$$

And substituting Equation (5.3) into Equation (5.1) gives

$$\lambda(t) = \frac{-\mathrm{d}R/\mathrm{d}t}{R(t)} \tag{5.4}$$

Integration of Equation (5.4) gives:

$$R(t) = \exp\left(-\int_0^t \lambda(t)\,\mathrm{d}t\right) \tag{5.5}$$

The relationship between these parameters for a component that follows the "bathtub" curve is illustrated in Figure 5.6.

During any random failure period, when the failure rate λ is constant, $R(t) = e^{-\lambda t}$, $F(t) = 1 - e^{-\lambda t}$, and $f(t) = \lambda e^{-\lambda t}$. It is only during a period when λ is constant that it is helpful to the user for the failure rate of a component to be specified. When this is possible, the failure rate is usually measured in units representing one failure per 10^9 device hours. These are known as failure units, or FITS. Sometimes percentage failures per 1000 hours are quoted. Note that a failure rate of 1% per 1000 hours corresponds to 10^4 FITS.

The *lifetime* $\langle t \rangle$, or the *mean time to failure* MTTF, is defined as

$$\langle t \rangle = \int t f(t)\,\mathrm{d}t \tag{5.6}$$

and the standard deviation σ about the mean lifetime, is given by

$$\sigma^2 = \int (t - \langle t \rangle)^2 f(t)\,\mathrm{d}t = \int t^2 f(t)\,\mathrm{d}t - \langle t \rangle^2 \tag{5.7}$$

It is easily shown that when λ is constant, $\langle t \rangle = \sigma = 1/\lambda$. The term *mean time between failures* (MTBF) is often used for $\langle t \rangle$, although really this is best applied to a system containing many subsystems or components that are repaired or replaced on failure.

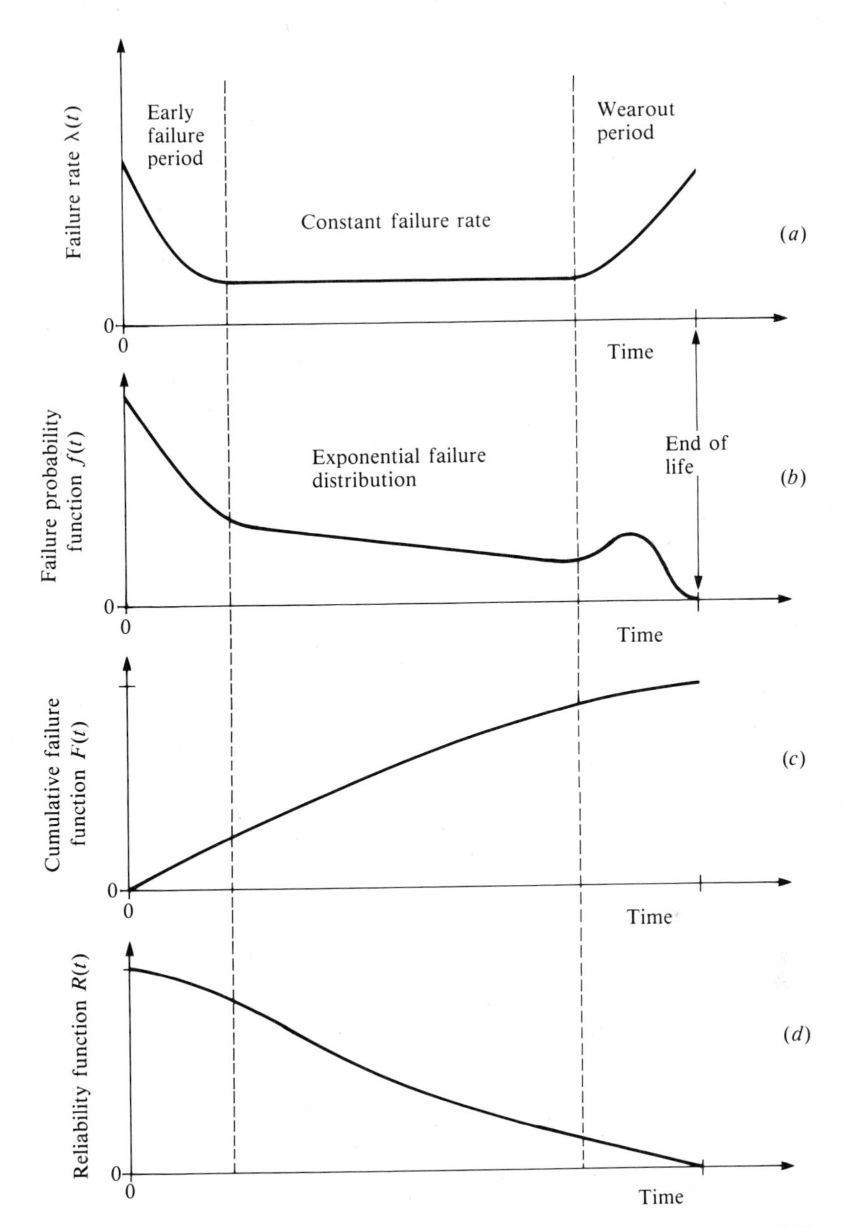

Figure 5.6. Variation of the failure parameters during life when failure follows the "bathtub" curve. (*a*) Failure rate $\lambda(t)$; (*b*) failure probability density function $f(t)$; (*c*) cumulative failure probability function $F(t)$; (*d*) reliability function $R(t)$.

In many instances, the cumulative failure function for semiconductor devices follows a *lognormal* distribution. The implication of this is that failure is the random consequence of the *product* of a number of independent factors. It is perhaps better to think of it as occurring with a probability proportional to the exponential function of the *sum* of many independent events. As these events result from the operation of the component, this is essentially a wearout mechanism. The failure probability distribution function takes the form

$$f(t) = (2\pi)^{-1/2}(ts)^{-1}\exp\left(\frac{-[\ln(t/t_m)]^2}{2s^2}\right) \tag{5.8}$$

In Equation (5.8), t_m is the geometrical mean of the failure times and is also the *median* lifetime, that is, the time at which 50% of the components will have failed. The factor s is sometimes called the dispersion. It represents the range of lifetimes about their mean value $\langle t \rangle$, which is given by:

$$\langle t \rangle = t_m \exp(s^2/2) \tag{5.9}$$

In Figure 5.7, we have plotted the various failure functions for a typical lognormal failure distribution. In Figure 5.7*c*, it can be seen that a graph of $f(t)$ versus $\ln t$ gives a normal or Gaussian distribution. The integral of this distribution, shown in Figure 5.7*d*, can be transformed into a straight line by the use of the so-called probability scale for the ordinate, as shown in Figure 5.7*e*. The validity of the lognormal distribution function can be established by recording the times of successive failures in this way. The fraction of the total population of devices under test that have failed is recorded on a normal probability scale, against time plotted on a logarithmic scale. Once the straight line is established, the median lifetime and the dispersion can easily be determined.

When the life test is carried out on a small sample from the total population of components, the supposition is that the distribution of the samples is representative of that of the whole population. However, let us say that we test a very small sample of ten components. When the first one fails, we might at first sight presume that this time is representative of the time to failure of 10% of the population. This is unduly pessimistic, because it is most unlikely that *all* 10% of the population would have failed by this time, had they all been put on test. There are methods available to correct for this. One that is quite accurate and easy to apply is known as Bernard's method. The ith failure out of a sample of n is regarded as representing the fraction $(i - 0.3)/(n + 0.4)$, rather than i/n, of the total population. This is what should be plotted on the ordinate scale of Figure 5.7*e*.

In general, knowledge of the median lifetime of a component is of little value to an equipment designer. Much more important is to know the likely time to the first failure among the many components used in a given system. When a lognormal failure distribution is established for a sample of these components, an estimate can be made of the likely time to failure of, say, the first 0.1% of components. Furthermore, depending on the sample size used, confidence limits, expressing the probability that such an estimate is representative of the whole population, can be calculated.

Many of the degradation processes in semiconductors involve the formation of dislocations and other crystal defects, or the diffusion and aggregation of impurities. All of these are thermally activated processes, having a characteristic

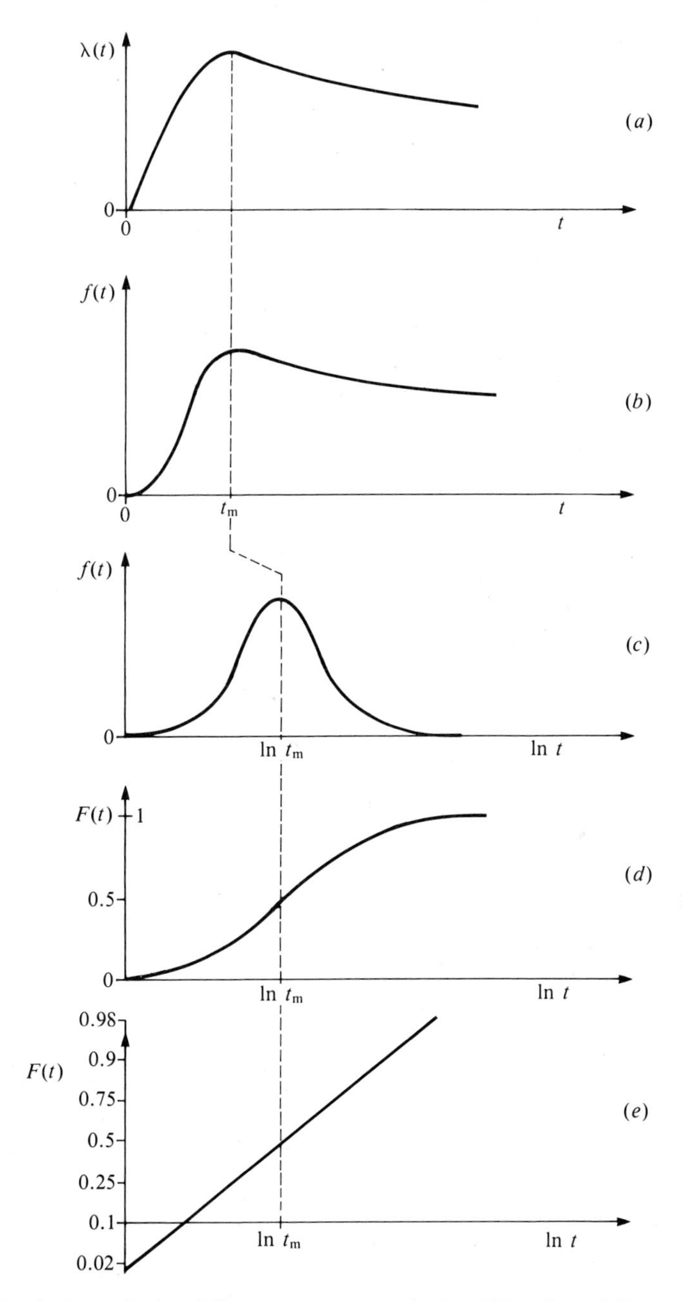

Figure 5.7. Variation of the failure parameters during life when failure follows the lognormal distribution. (*a*) Failure rate $\lambda(t)$; (*b*) failure probability density function $f(t)$; (*c*) $f(t)$ versus time plotted on a logarithmic scale; (*d*) cumulative failure probability function $F(t)$ versus time plotted on a logarithmic scale; (*e*) $F(t)$ plotted on a normal probability scale, versus time plotted on a logarithmic scale.

activation energy, E_a. They are enhanced by an increase in temperature, according to the law of Arrhenius [4]:

$$\text{rate of activation} \propto \exp(-E_a/kT) \tag{5.10}$$

The time required for a sufficient number of such events to produce a condition of failure is thus proportional to $\exp(E_a/kT)$, and it follows the lognormal distribution function. This can be established by plotting the median lifetime (or some other characteristic time) on a logarithmic scale versus the reciprocal of the absolute

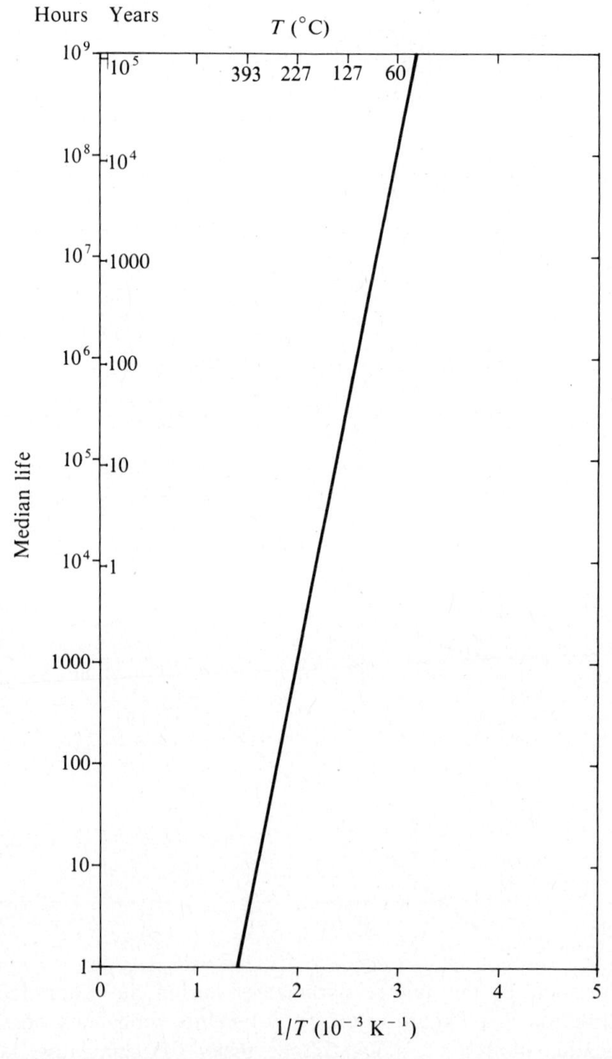

Figure 5.8. An Arrhenius plot: a graph of the median time to failure, plotted on a logarithmic scale, against the reciprocal of the absolute temperature. The graph shown corresponds to the example given in the text, namely, the activation energy is 1 eV and the median life at 227°C is 1000 hours.

temperature, as shown in Figure 5.8. The slope of the line is governed by the activation energy, which in some instances can be determined independently.

Establishing the value of E_a with a degree of confidence is the basis of the use of high temperatures to accelerate life testing. To give an example, if E_a is 1 eV, the lifetime is proportional to $\exp(e/kT) = \exp[(11{,}602\text{ K})/T]$. The expected life of a device run at 500 K (227°C) is shorter than that of one run at 400 K (127°C) by a factor of $\exp[11{,}602(1/400 - 1/500)]$, that is, by a factor of $\exp(5.8) = 330$. Provided that the same failure mechanism is dominant at both temperatures, a median lifetime of 1000 hours measured at the higher temperature leads to a predicted median lifetime of 330,000 hours ($37\frac{2}{3}$ years) for a device run at the lower temperature.

5.2.3. The Main Causes of VDMOS Failure

Device failure, as with most semiconductor devices, is usually the result of faulty circuit design causing one or more of the device operating limits to be exceeded. Particularly vulnerable periods occur during the switching of inductive circuits, when overvoltage or overrapid reapplication of voltage (excess dV/dt during diode recovery) may occur. The most common causes of failure resulting from such misuse are:

1. Gate short, usually caused by excessive voltage applied to the gate, either in circuit or during handling (electrostatic discharge).
2. Safe-operating-area failure, caused by one or more excessive thermal transients, or by poor heat-sinking.
3. Avalanche failure, when excessive avalanche current is drawn during unclamped inductive switching.
4. Fused leads, following excessive transient overload current.

Failures that may develop during long periods of normal use at or near to the specified limits of the device operating conditions usually take one of the following forms:

1. Gate failure. The gate leakage current increases, and eventually the transistor becomes stuck in the ON state, leading to a catastrophic circuit failure.
2. Drain–source leakage increase. The transistor becomes overheated and thermal runaway ensues.
3. Connection failure. Thermomechanical fatigue or corrosion at a wire bond, or at the die bond, leads to a rise in the thermal impedance R_{Th} or in $R_{DS(on)}$, and hence in dissipation.

In each case the most likely symptom of approaching failure is a gradual increase in the device temperature, with the attendant risk of thermal runaway during onerous operating conditions.

The underlying physical causes of these progressive failures mechanisms are conveniently divided into those associated with the die and those associated with the packaging. Two die defects are the most common. The first is the electric field distortion brought about by the accumulation of polar molecules and ionic contaminants in the passivation of the high-field region, in particular the field rings around the periphery of the chip. The second is the growth of defects in the gate oxide. These usually take the form of voids or inclusions. Several mechanisms that cause failure in other types of semiconductor devices do *not* arise in practice in properly constructed VDMOS FETS. These include the development of slow trapping levels in the active region, microcracking of the aluminum metallization, and electromigration of the metallization (which can occur at high current densities). Packaging failures include thermal fatigue of the die bond brought about by differential expansion between the silicon and the header, wire-bond fatigue, and corrosion, usually caused by adsorbtion of water (OH^-) onto the chip surface. We deal with each of these failure mechanisms in turn, examining how they may be enhanced for the purposes of accelerated life-testing.

Field Distortion. This manifests itself as a gradual increase in the drain–source leakage current I_{DSS}. Allowed to progress far enough, it can lead to thermal runaway. Accelerated life tests are carried out at 150 and 175°C, with 80% of the rated drain–source breakdown voltage (BV_{DSS}) applied, and $V_{GS} = 0$ V. These are known as high-temperature reverse-bias (HTRB) tests. Tests carried out at different temperatures and different bias levels have established that temperature acceleration follows the Arrhenius law with an activation energy of 1 eV, and that bias acceleration increases the failure rate by the factor $e^{V/(311\text{V})}$. Published data have usually been treated as though a constant failure rate were to be expected, and the results are quoted with an upper confidence level of 60% in equivalent FITS at a temperature of 90°C. See for example Figure 19 of Appendix 7. Note that as this failure mechanism is essentially one of wearout, burn-in is not capable of reducing the failure rate.

Oxide Defects. The accumulation of oxide defects causes a gradual increase in the gate–source leakage current I_{GSS}, leading eventually to a possible rupture of the gate oxide and a gate–source short circuit. Accelerated gate stress life tests are carried out at 150 and 175°C with $V_{DS} = 0$ V and $V_{GS} = 20$ V, the maximum rated value. Again, even though this is essentially a wearout mechanism, it is usually assumed that a constant failure rate occurs. Thermal acceleration follows the Arrhenius law, with $E_a = 0.4$ eV. The electric field in the gate oxide, E_{ox}, which is proportional to V_{GS}, enhances the failure rate by the factor $\exp(-E_{ox}/E_c)$, where $E_c = 10.8\ \text{V}/\mu\text{m}$. Again, burn-in is not effective. The estimated time to 1% failure probability is shown in Figure 18 of Appendix 7.

Die-Bond and Wire-Bond Fatigue. Differential thermal expansion between the silicon die and the header, and thermomechanical and current pulsing stresses, cause fatigue and degradation of the die bond. The thermal impedance is very sensitive to this effect and soon increases, leading to an increase in the junction temperature. The wire bonds may degrade in a similar manner, causing $R_{DS(on)}$ to rise and become unstable. The vulnerability of devices to these effects is tested in two ways. In the power cycling test the device is switched on for some two or three minutes, during which time heat generated by the dissipation in the die raises the case temperature from an ambient of 30°C to 100 or 130°C. The transistor is then switched off and forced-air-cooled back to 30°C. These cycles are continued to failure. In the temperature cycling test the devices are taken from −55 to +150°C and back over a period of about 30 minutes. No bias is applied. Lifetests usually extend for at least 10,000 cycles. In this case degradation results from differential thermal expansion. Plastic encapsulated (TO-220) devices, which have to use hard-soldered bonds, are likely to be more vulnerable than those that are hermetically sealed (TO-3), which can use soft-soldered connections. Transient temperature gradients created in the power cycling test give rise to further stresses. In each case failure times have been found to vary according to the Arrhenius law, with an activation energy of about 0.3 eV.

Surface Corrosion. The ingress of water molecules into non-hermetically-sealed packages leads to device deterioration in two ways. First, ingress into the field termination region distorts the electric field distribution and degrades the voltage breakdown characteristics, causing leakage currents, as described previously. Secondly, hydroxyl ions penetrating to the regions of the bonding pads stimulate cathodic corrosion of the aluminum contact. This leads to an increase in the values of $R_{DS(on)}$ and V_{SD}, both of which become unstable. Drain bias accelerates this effect. However, the increased working temperature of an operating device helps to exclude the water vapor. Samples of non-hermetically-sealed devices are subjected to a *temperature–humidity–bias* test of at least 1000 hours at 85°C and 85% relative humidity with $V_{GS} = 0$ V and $V_{DS} = 10$ V or at the full rating (BV_{DSS}). A variant on the test that is sometimes used is the *highly accelerated temperature and humidity stress test* (HAST). This is a high-pressure (0–40 psig), high-humidity (50–80% relative humidity) test normally carried out for several tens or hundreds of hours with $V_{DS} = 8$ V. Similar failure mechanisms occur.

Avalanche Failure. A test has been introduced recently to detect any degradation brought about by subjecting devices to repetitive avalanche breakdown. The test circuit is similar to that shown in Figure 14 of Appendix 7.

SUMMARY

Power MOSFETS are fabricated by different manufacturers using a range of processes of differing sophistication and complexity. The processes use all the techniques normally associated with MOS integrated-circuit fabrication.

A particularly important part of the design is that of the field rings around the periphery. These are formed at the same time as the main device. They are critical for the achievement of the voltage rating, especially when the body–drain diode is required to sustain controlled avalanche breakdown.

Power MOSFETS have been designed from the outset as high-quality and highly reliable components. Device failures that are not the result of faulty circuit design usually follow a lognormal distribution function. This indicates wearout, rather than random failures, and means that burn-in is of limited benefit. The degradation processes are accelerated by increased temperature. The rate of degradation often follows the law of Arrhenius, indicating an activation-energy-controlled process. Humidity and applied electric field also act as accelerating stress factors for many of the degradation mechanisms.

REFERENCES

1. S. M. Sze (Ed.), *VLSI Technology*. McGraw-Hill, New York, 1983.
2. T. Daly, "New technology makes power MOSFETS faster, more efficient." *Power Convers. Intell. Motion* (*PCIM*) pp. 14–18 (1988).
3. D. Ueda, H. Takagi, and G. Kano "An ultra-low on-resistance power MOSFET fabricated by using a fully self-aligned process." *IEEE Trans.* on *Electron Devices* **ED-34,** 926–930 (1987).
4. S. Arrhenius, "Über die Dissociationswärme und den einfluss der temperatur auf den Dissociationsgrad der Elektrolyte" ("On the heat of dissociation and the influence of temperature on the degree of dissociation of an electrolyte"). *Z. Phys. Chem.* **4,** 96–116 (1889).

CHAPTER 6

Dissipation and Heatsink Design

6.1. INTRODUCTION

For the purposes of sizing the heatsink, or calculating the junction temperature if the heatsink size is already fixed, a knowledge of the dissipation in the power MOSFET is essential. The total power dissipation P_T is the sum of the switching losses P_S, conduction losses P_C, internal gate losses P_G, and leakage current losses P_L [1].

6.2. SWITCHING LOSSES

Power-MOSFET switching is analyzed in detail in Chapter 4. Switching losses can generally be estimated to sufficient accuracy for the purposes of heatsink design by using simplified voltage and current waveforms. The switching waveforms for a clamped inductive load and a resistive load may be represented by the straight-line approximations shown in Figures 4.6 and 4.4 respectively. The power loss resulting from turn-on and turn-off for a clamped inductive load will then be given by

$$P_S = \frac{V_{DS(max)} I_{D(max)}}{2} t_s f_s \tag{6.1}$$

where t_s is the average switching time and f_s is the switching frequency. Alternatively the switching loss per cycle may be calculated from Equation (4.7). For a resistive load the power loss will be

$$P_S = \frac{V_{DS(max)} I_{D(max)}}{6} t_s f_s \tag{6.2}$$

For a resistive load the energy loss per cycle is given in Equation (4.6). If the waveforms are not reasonably close to either of these approximations, the energy loss during switching will have to be determined by piecewise multiplication and integration of the current and voltage waveforms.

In practice the drain voltage waveform may be substantially modified by the drain–source capacitance and case–sink capacitance as shown in Figure 6.1. As well as crossover losses there are losses associated with the discharge of

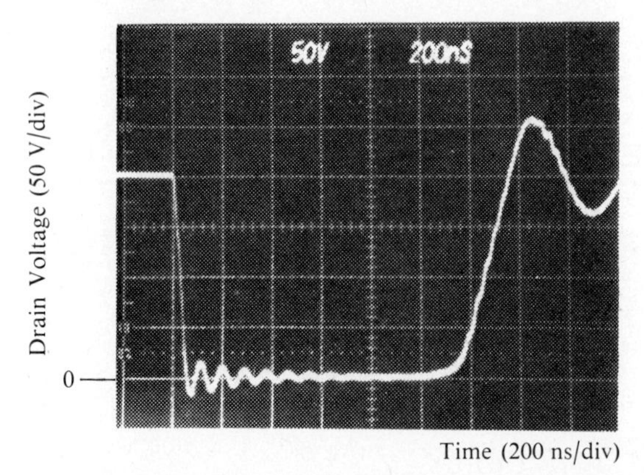

Figure 6.1. Drain–source capacitance slows the rise of drain voltage at turn-off. Compare this with the rate of fall at turn-on.

drain–source parasitic capacitance during turn-on. At high frequencies these losses may not be negligible. At turn-off the rise in drain voltage may be regulated by the charging of the drain–source capacitance by the load current. If losses are being calculated using an estimated switching time based on data-sheet parameters and the gate-drive design, a rough value of the energy loss per cycle can be obtained by adding to the calculated switching losses the losses associated with discharging this capacitance. The power dissipated in discharging the drain capacitance is given by

$$P_{SC} = \tfrac{1}{2} C_{DS} V_{DS}^2 f_s \tag{6.3}$$

where C_{DS} includes the case–sink capacitance and V_{DS} is the drain voltage at turn-on.

To enhance MOSFET reliability a snubber may be used to reduce dissipation in the device and allow the MOSFET to operate at a lower junction temperature. Clearly, the procedure followed to calculate the switching losses will then depend on the type of snubber circuit used [2].

6.3. LEAKAGE-CURRENT LOSSES

The losses due to drain–source leakage current can usually be discounted even in high-voltage MOSFETS. However, the leakage current may not be insignificant when the off-state gate-drive voltage approaches the threshold voltage. Such a situation can arise if the gate drive is derived from a logic integrated circuit with a pullup resistor on the output and if the MOSFET has a low threshold voltage. Losses due to leakage currents are given by

$$P_L = I_{DSS} V_{DS} (1 - \delta) \tag{6.4}$$

where δ is the ON-time duty cycle.

6.4. GATE DISSIPATION

A proportion of the gate drive power will be dissipated in the internal resistance of the gate. At frequencies of the order of 100 kHz these losses will be negligible, but they become significant, at least for polysilicon gate devices, when frequencies in the 10-MHz range are considered. The internal gate losses are given by

$$P_{G(int)} = V_{GS} Q_G f_s \frac{R_{G(int)}}{R_S + R_{G(int)}} \tag{6.5}$$

where Q_G is the gate charge transfer during switching, $R_{G(int)}$ is the internal gate resistance, and R_S is the external gate resistance. $R_{G(int)}$ is typically 0.5 to 2 Ω.

6.5. CONDUCTION POWER LOSS

The conduction power loss is given by

$$P_C = I^2_{D(rms)} R_{DS(on)} \tag{6.6}$$

The rms value of a waveform is given by

$$I_{rms} = \left(\frac{1}{T} \int [I(t)]^2 \, dt \right)^{1/2} \tag{6.7}$$

If the waveform can be approximated by a combination of waveforms of known rms value, then the rms value of the whole is given by

$$I_{rms} = \sqrt{I^2_{RMS(1)} + I^2_{RMS(2)} + \cdots + I^2_{RMS(N)}} \tag{6.8}$$

The rms values of a selection of common waveforms are given in Figure 6.2 [3].

Next the appropriate value of $R_{DS(on)}$ must be found. The difficulty with this step is that $R_{DS(on)}$ increases significantly with temperature, thereby increasing the dissipation and the junction temperature. Therefore the solution has to be found either by a series of iterations or by a graphical method.

In the iterative approach, a junction temperature is assumed and $R_{DS(on)}$ calculated from the data-sheet graph of normalized $R_{DS(on)}$ versus temperature (see Appendix 7). The losses are then calculated and, using the appropriate value of thermal resistance, a new junction temperature is calculated. The process is repeated until the junction-temperature calculations have converged sufficiently. The initial value of $R_{DS(on)}$ is usually based on the maximum guaranteed value, since the usual object of the calculation is to obtain the peak possible junction temperature within the allowed manufacturing spread of the characteristics. Some devices show a significant increase in $R_{DS(on)}$ with drain current, and this may also need to be incorporated into the calculation of the operating value of $R_{DS(on)}$. A low value of V_{GS} can also result in an increase in $R_{DS(on)}$. However, if the device

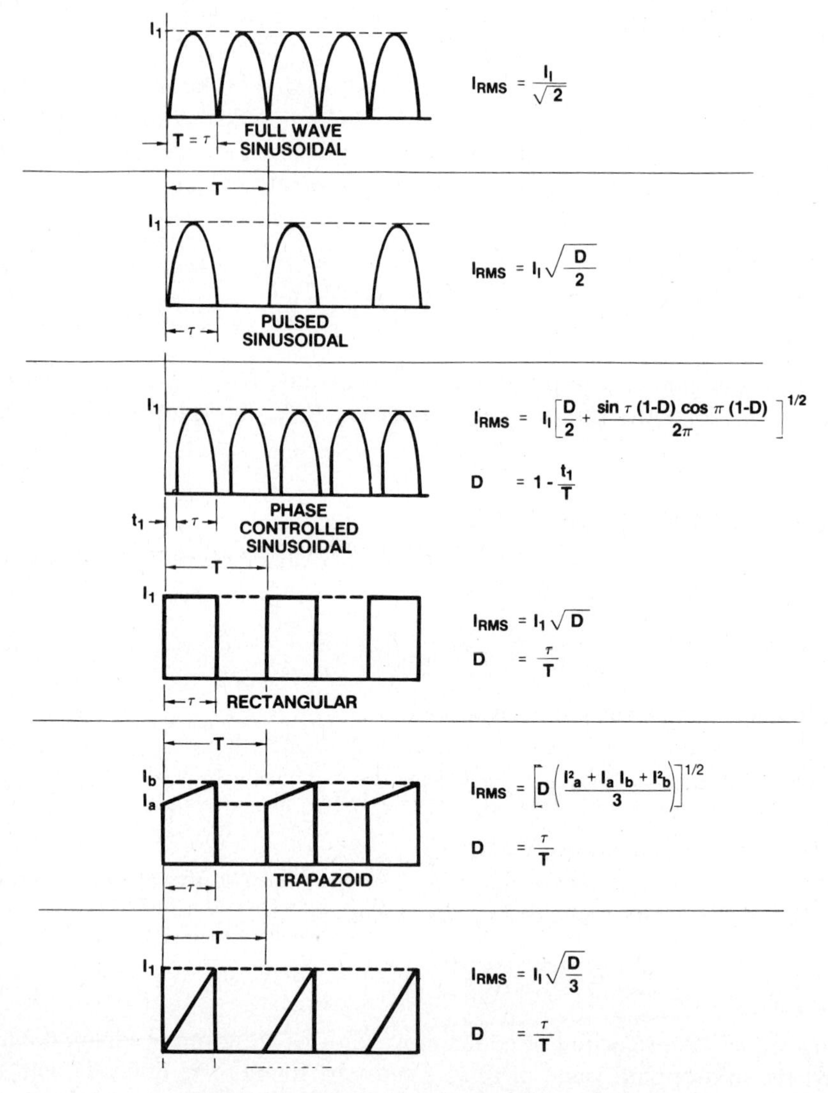

Figure 6.2. RMS values of common waveforms.

is not fully turned on, conduction losses are best calculated using the transfer characteristics and multiplying V_{DS} by I_D.

The graphical solution is illustrated in Figure 6.3. A graph is plotted of total dissipation of the MOSFET against junction temperature. The total dissipation can be expressed as

$$P_T = P_1 + I^2_{D(rms)} R_{DS(on)} \tag{6.9}$$

where

$$P_1 = P_S + P_L + P_G$$

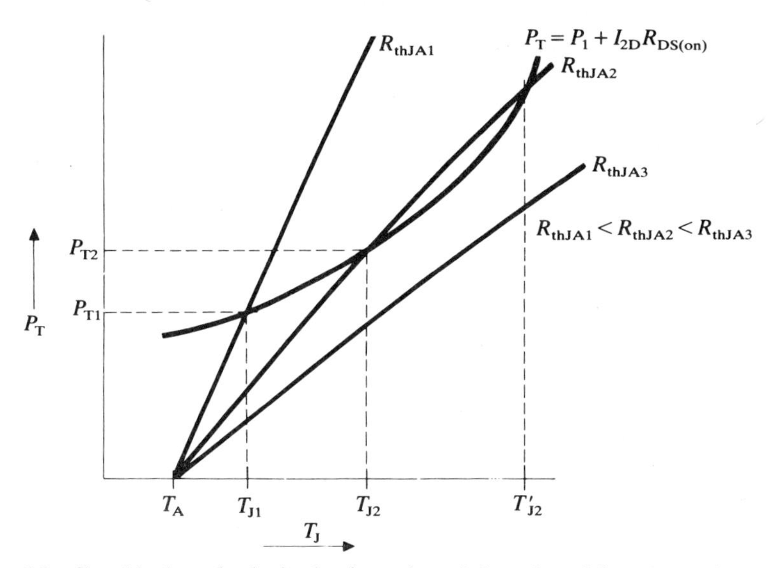

Figure 6.3. Graphical method of selecting a heatsink and avoiding thermal runaway.

A line representing the thermal characteristics of the case and the heatsink is added to the graph. The intersection of these two lines gives the junction operating temperature. Three different heatsink arrangements are illustrated in Figure 6.3, represented by the three lines, R_{thJA1}, R_{thJA2} and R_{thJA3}.

Stable equilibrium can only exist if the heat input P_T from the MOSFET equals the heat dissipated by the heatsink. The heatsink represented by R_{thJA3} has too high a thermal resistance for thermal equilibrium to be established. The curves do not cross and thermal runaway will occur. The heatsink represented by R_{thJA1} has a low enough thermal impedance for the curves to cross, and the temperature will stabilize at temperature T_{J1}. The heatsink represented by R_{thJA2} is conditionally stable at temperature T_{J2}. However, should the temperature ever exceed T'_{J2}, the system enters an unstable region and thermal runaway results. The heatsink must therefore have a sufficiently low thermal resistance that an unstable condition is never reached.

6.6. STEADY-STATE THERMAL RESISTANCE

Heat flows from the silicon die to the cooling medium through a series of thermal resistances. In the usual situation in which a heatsink is attached to the power semiconductor device and the cooling medium is air, a number of thermal resistances are involved: junction-to-case (R_{thJC}), case-to-sink (R_{thCS}), and sink-to-ambient (R_{thSA}). Thermal circuits may be treated in a manner analogous to electrical circuits, with thermal resistance the equivalent of electrical resistance, temperature the equivalent of voltage, and power transfer (Pd) the equivalent of current flow. Thus

$$P_d = \frac{T_J - T_A}{R_{thJA}} \tag{6.10}$$

and

$$T_J = T_A + P_d R_{thJA} \tag{6.11}$$

The total thermal resistance is found by summing the individual resistances through which the heat flows, so that

$$R_{thJA} = R_{thJC} + R_{thCS} + R_{thSA} \tag{6.12}$$

Parallel paths for heat flow may be treated in the same manner as parallel paths for current flow in electric circuits. The commonly used units of thermal resistance are degrees centigrade per watt (°C/W) or degrees Kelvin per watt (K/W). These units are generally used interchangeably. In steady-state calculations P_d is the average power dissipation and the thermal resistances are steady-state values.

The designer should be prepared for a broad spread in the values of R_{thJC}, R_{thCS}, and R_{thSA}. The semiconductor device manufacturer will specify a maximum (worst-case) value for R_{thJC} which has to take into account variation in the quality of the die mounting and variation in silicon wafer thickness. R_{thCS} is very dependent on the manner in which the device is attached to the heat sink. The use of an insulating washer, the use of heat-sink compound, the presence of dirt in the interface, and the pressure used in mounting can all have very large effects on the thermal resistance. Furthermore, allowance may need to be made for an increase in thermal resistance with the elapse of time as the quality of the interface deteriorates due to corrosion, the evaporation of heat-sink compound, changes in mounting pressure, etc. In the interests of reliability it is desirable to assign a high value to R_{thCS}, but a larger heatsink will then be required to give a low value of R_{thSA} in compensation. The heatsink manufacturer will specify a maximum value for R_{thSA}, but it should be borne in mind that R_{thSA} is very dependent on the ability of air to circulate around the heatsink and the orientation of the heatsink. Very often the junction temperature which any particular cooling arrangement produces will not be known accurately until a prototype has been built and tested.

6.7. TRANSIENT THERMAL IMPEDANCE

When a transistor is carrying a pulsed current waveform, the power dissipation in the junction will also occur in pulses. The junction temperature will therefore fluctuate during the cycle. If the pulse frequency is sufficiently high, it may be possible to ignore fluctuations in the die temperature and to use only steady-state calculations to assess the die temperature. If, however, the interval between pulses is long with respect to the thermal time constant of the die, then an

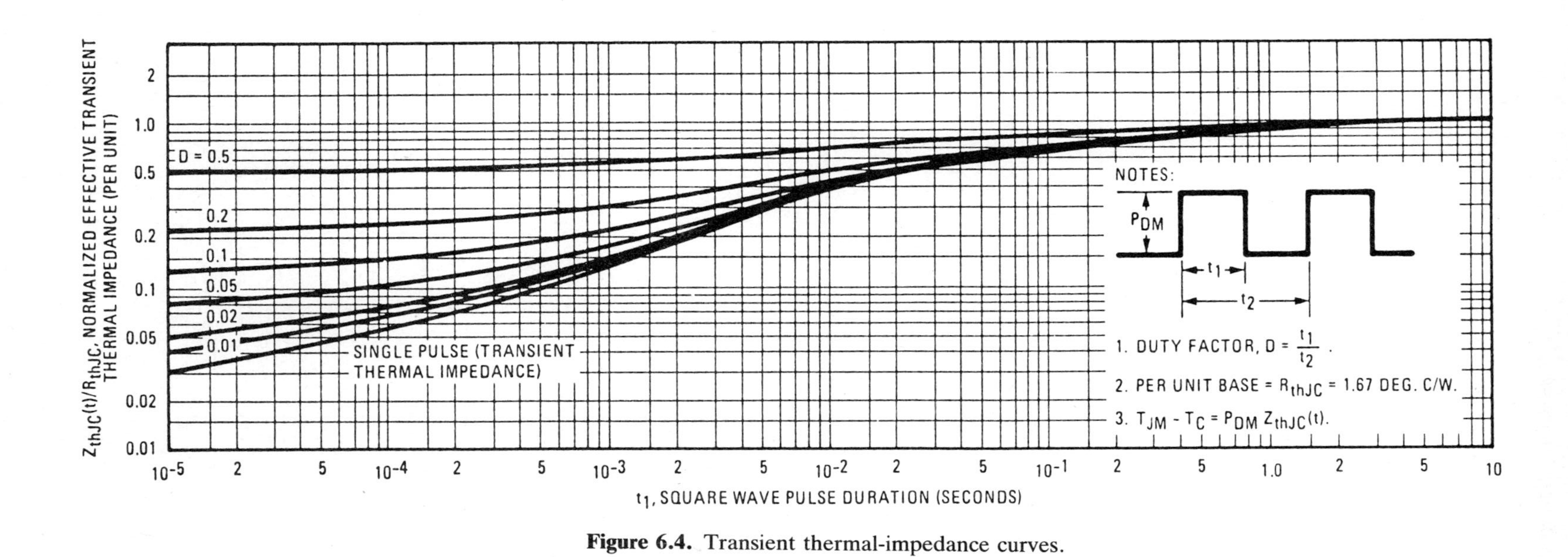

Figure 6.4. Transient thermal-impedance curves.

alternative means of calculating the die temperature must be used. To meet this need, device manufacturers generally supply transient thermal-impedance characteristics for a rectangular current waveform. Figure 6.4 shows an example of such a characteristic. The transient thermal impedance Z_{thJC} is normalized with respect to the thermal resistance R_{thJC}, so that

$$Z_{thJC} = r(D, t_1)R_{thJC} \tag{6.13}$$

Z_{thJC} can be obtained from the graph using the appropriate values of the case temperature T_C, conduction time t_1, and duty cycle D. The peak junction temperature can then be calculated from T_C, Z_{thJC}, and the peak power dissipation P_{dm} as follows:

$$T_{J(max)} = T_C + P_{dm}Z_{thJC} \tag{6.14}$$

For nonrectangular pulse waveforms, an estimate of the peak temperature can be obtained by equating the pulses with rectangular pulses that produce the same dissipation in the junction.

6.8. FORCED COOLING

In situations where forced air cooling is available, the effective thermal resistance of the heatsink is reduced. The heatsink manufacturers' data are usually available to enable the new thermal resistance to be found. Some manufacturers' data may also be available on the effect of forced cooling on the thermal resistance of power MOSFET packages, particularly where the device is not commonly used with a heatsink. For example, Figure 6.5 shows the relationship between air velocity and thermal resistance of a power MOSFET in a *dual in-line package* (DIP) [4].

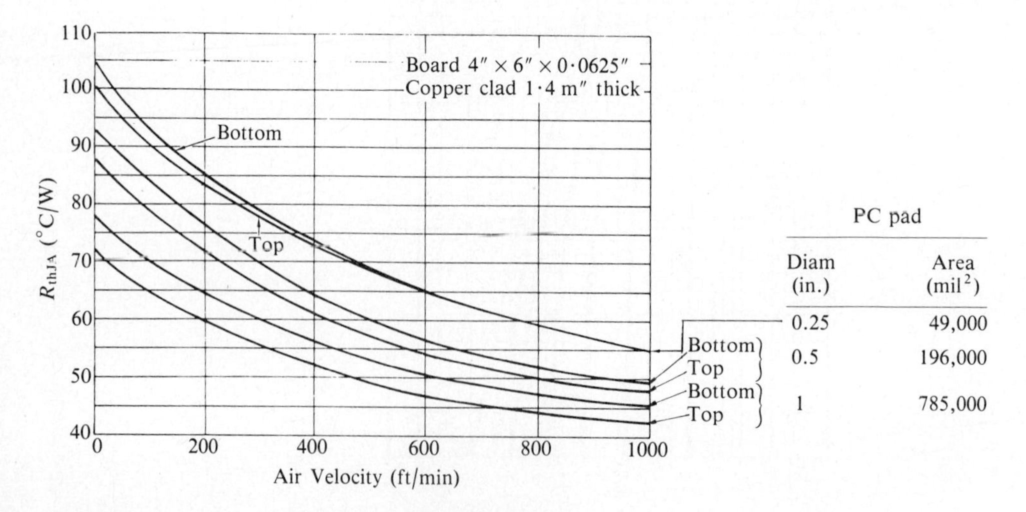

Figure 6.5. Effect of forced cooling on the thermal resistance of a DIP-packaged MOSFET.

REFERENCES

1. R. Severns, "Simplified HEXFET power dissipation and junction temperature calculation speeds heatsink design." *Int. Rectifier Appl. Note* **AN-942** (1981).
2. P. J. Carlson, "Use snubbers and clamps to improve power MOSFETS' performance and reliability." *GE Appl. Note* (1985).
3. S. Clemente, B. R. Pelly, and R. Ruttonsha, "Current ratings, safe operating area, and high frequency switching performance of power HEXFETs." *Int. Rectifier Appl. Note* **AN-949A** (1984).
4. S. Brown and G. Xenakis, "More power from HEXDIPs." *Int. Rectifier Appl. Note* **AN–953** (1983).

CHAPTER 7

Parallel and Series Operation

7.1. PARALLEL OPERATION

7.1.1. Introduction

There are a number of reasons for paralleling power MOSFETS:

- *Higher I_D rating.* To obtain the necessary current rating when a single device of adequate rating is not available.
- *Lower $R_{DS(on)}$.* To reduce $R_{DS(on)}$ so that the forward voltage drop is below a certain value.
- *Improved thermal performance.* Distributing the sites of dissipation over a heatsink or a header reduces the effective thermal resistance.
- *Lower cost.* The cost of manufacturing a die increases rapidly as its size goes up, due to a reduction in yield. Two smaller dice in parallel may be cheaper than one single die with the same total silicon area.

Power MOSFETS have a distinct advantage over bipolar transistors in the ease with which they can be paralleled. The temperature coefficient of the forward voltage drop of a saturated bipolar transistor is usually negative. This makes the bipolar transistor susceptible to thermal runaway in parallel operation, since the hottest device tends to attract more load current. The MOSFET has a positive temperature coefficient of ON-state voltage, since $V_{DS} = I_D \times R_{DS(on)}$, and $R_{DS(on)}$ increases with temperature due to a reduction in carrier mobility. The problem of thermal runaway does not arise when power MOSFETS are operated in parallel in the switching mode on an adequately sized heat sink. However, thermal runaway is a possibility during operation in the linear regime. For this reason it is necessary to consider parallel connection for the switching mode and linear modes of operation separately since different considerations apply. Paralleling for switched operation is dealt with in Sections 7.1.2 to 7.1.7, and paralleling for linear operation in Section 7.1.8.

Equalization of the currents flowing through the individual devices, in both static and dynamic conditions, is often seen as the criterion for successful parallel operation. However, equalization of device heating, and even more important, equalization of junction temperatures, can be more appropriate objectives. In the steady state all these aims are served by matching the $R_{DS(on)}$ of the devices where this is possible and by ensuring good thermal coupling of the paralleled devices. A mismatch in $R_{DS(on)}$ is the dominant cause of a static temperature differential.

Threshold voltage and transconductance mismatch will have only a very minor effect, since the gate voltage is usually well above the minimum value required to fully enhance the device.

Good dynamic current sharing is less easily achieved. The switching performance of a MOSFET depends on a number of device parameters, such as the threshold voltage, the transconductance, and the various parasitic capacitances. Variations in these device parameters can cause paralleled devices to switch at different times and at different rates. This leads to poor dynamic current sharing and a momentary overload of some of the transistors. Fortunately the MOSFET has a good overload capability, so that the consequence of poor dynamic current sharing is usually a disparity in junction temperatures rather than the immediate destruction of a device.

7.1.2. Static Conditions

At one time it was a commonly held belief that the prescription for successful paralleling of power MOSFETS under static conditions was to thermally isolate the devices from each other. The MOSFET that carries the most current dissipates the most heat and therefore has the highest junction temperature. The rise in junction temperature increases the $R_{DS(on)}$ of the device, thereby limiting the share of the load current that it carries. However, thermal isolation of the MOSFETS does not give the best result when the total current rating of the parallel combination is considered. The criterion by which a parallel arrangement should be judged is the disparity in junction temperature for a given load current. The total current must not be greater than that value which produces the maximum allowable temperature in the hottest junction. Keeping all junctions at approximately the same temperature allows all devices to be operated close to the chosen T_{Jmax} and gives a minimum value of total resistance. If some junctions are cooler than others, the cooler transistors are not being utilized to the full. In order to keep temperature variations between junctions to a minimum, the MOSFETS should be as closely coupled thermally as possible. Usually this means mounting the devices on the same heatsink.

The value of $R_{DS(on)}$ for a particular device increases with increasing drain current. The effect is most prevalent in high-voltage devices (Figure 7.1). This is due to the lighter doping of the epitaxial drain region in high-voltage devices, which causes the body–drain depletion region to be broader for a given value of channel current. The broader depletion region produces greater "throttling" of the drain current as it passes through the JFET region between cells (see Section 3.5). Since $R_{DS(on)}$ increases with drain current, the resistance of the MOSFET that carries the greatest current will increase the most, thereby reducing the disparity between the $R_{DS(on)}$-values of the transistors. However the nonlinearity in $R_{DS(on)}$ is not generally sufficient to produce a significant improvement in the current imbalance. For example, the current imbalance resulting from the paralleling of two dissimilar devices of the type illustrated in Figure 7.1*b* would be reduced by approximately 6% by the nonlinearity of $R_{DS(on)}$.

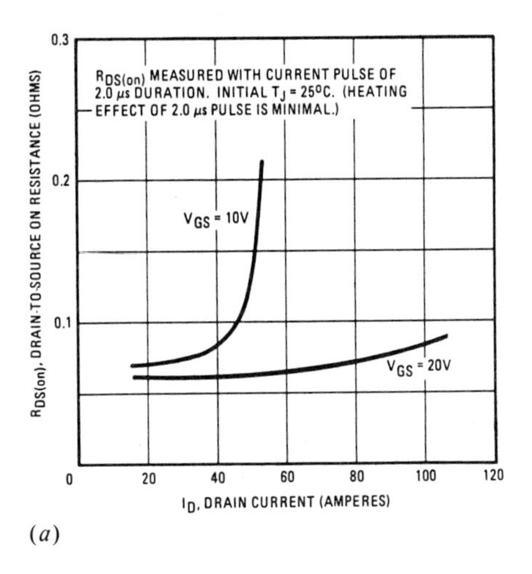

(*a*)

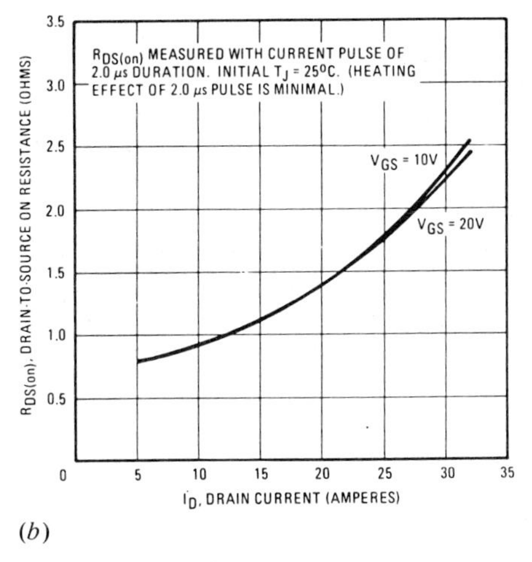

(*b*)

Figure 7.1. Variation of $R_{DS(on)}$ with drain current. (*a*) 100-V device (IRF 540); (*b*) 500-V device (IRF 840). The steep rise in $R_{DS(on)}$ of the IRF 540 for $V_{GS} = 10$ V is due to the device entering the constant current region.

7.1.3. Thermal Model for Static Conditions

Figure 7.2 shows the thermal model for two similar MOSFETS mounted on a common heatsink. (In this section thermal resistances are identified by an alphabetic subscript and electrical resistances are denoted by a numeric subscript.) The location of the thermal resistance between the transistors depends on whether they are mounted on the same heatsink or on the same header. In both cases the diagram can be simplified as shown in Figure 7.3, where R_a equals the thermal impedance before the coupling path and R_d equals the thermal impedance after the coupling path. An analysis of this model [1] with two MOSFETS

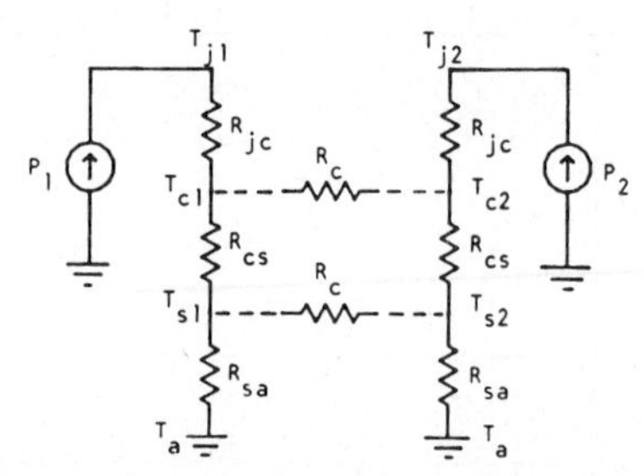

Figure 7.2. Static thermal model for two MOSFETS in parallel. From Kassakian [1]. By permission of the Institute of Electrical Engineers of Japan.

with ambient temperature values of $R_{DS(on)}$ of 0.12 Ω and 0.16 Ω gave the results shown in Figure 7.4.

These results show that the least temperature difference between the two junctions is obtained when $R_c = 0$, that is, with the closest possible thermal coupling of the two transistors. It can be seen from Figure 7.4 that the imbalance in drain currents due to the mismatch in the $R_{DS(on)}$ of the devices is corrected only a small amount by the variation of $R_{DS(on)}$ with temperature, whatever the degree of thermal coupling between the two devices.

An alternative method of analysis is to compute the individual junction temperatures in an iterative fashion. Junction temperatures are assumed for the first iteration to enable the power dissipation in each of the devices to be calculated. From these results a new value of the die temperature is calculated and the cycle repeated until the algorithm converges on a solution. Results obtained by this method [2] corroborate the finding that close thermal coupling between devices gives lowest junction temperatures.

7.1.4. Dynamic Current Sharing

During the switching process power MOSFETS pass through the linear region. Current sharing in this regime is influenced by factors such as threshold voltage and transconductance which have little effect on static current sharing in the fully conducting state. The effect of a 2-V mismatch in threshold voltage is illustrated in the SPICE-model results shown in Figure 7.5. The effect of a 20% mismatch in transconductance is shown in Figure 7.6. Other device parameters which influence switching speed, and therefore dynamic current balance, include

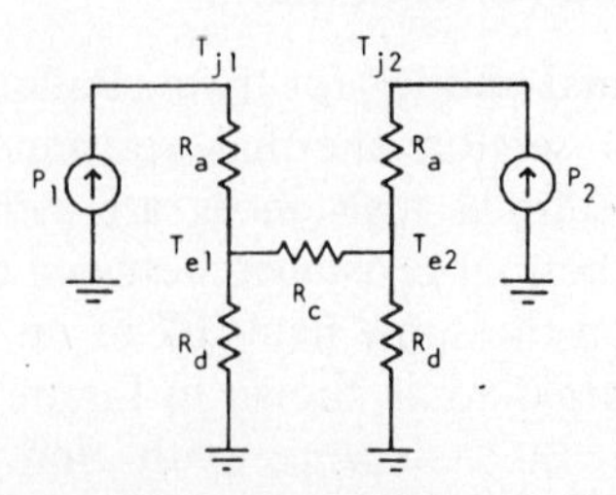

Figure 7.3. Generalized static thermal model. From Kassakian [1]. By permission of the Institute of Electrical Engineers of Japan.

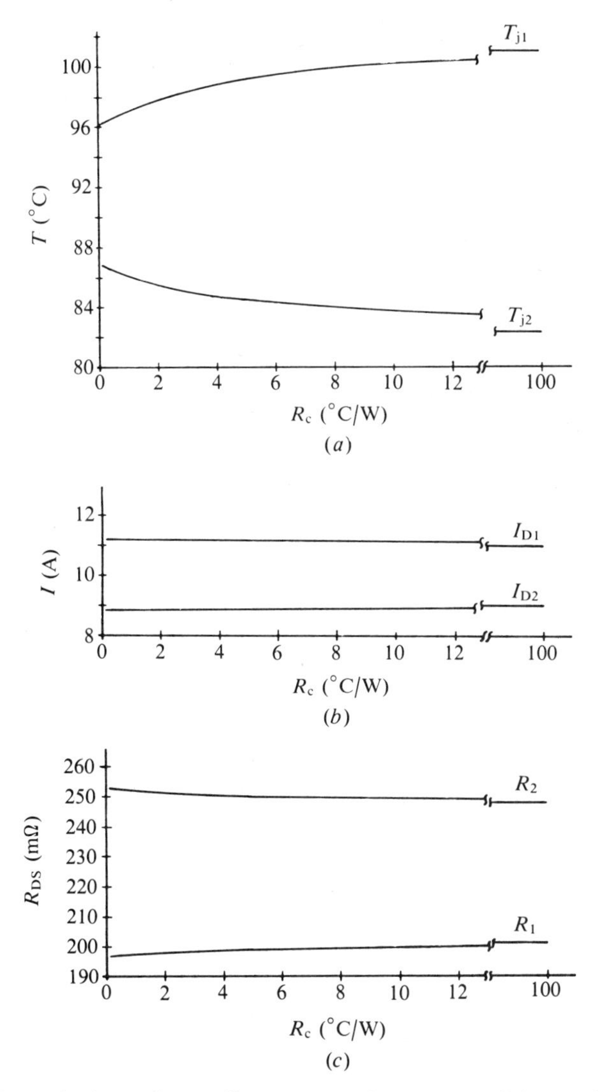

Figure 7.4. Effect of thermal coupling on steady-state conditions of two paralleled MOSFETS. (*a*) Junction temperature; (*b*) drain current; (*c*) $R_{DS(on)}$. From Kassakian [1]. By permission of the Institute of Electrical Engineers of Japan.

gate–source capacitance, drain–source capacitance, and source inductance. Since many factors are involved in shaping the switching waveforms of power MOSFETS, it is a practical impossibility to match devices in all respects. However, a number of measures can be taken to improve dynamic current sharing [2–6]:

1. Source inductances should be as equal as possible. At high switching speeds the source inductance becomes a crucial factor in determining the switching speed of a MOSFET and therefore in determining the division of the load current. The layout of paralleled devices should therefore be as symmetrical as possible to equalize source inductances. Providing these are equal, current sharing is aided

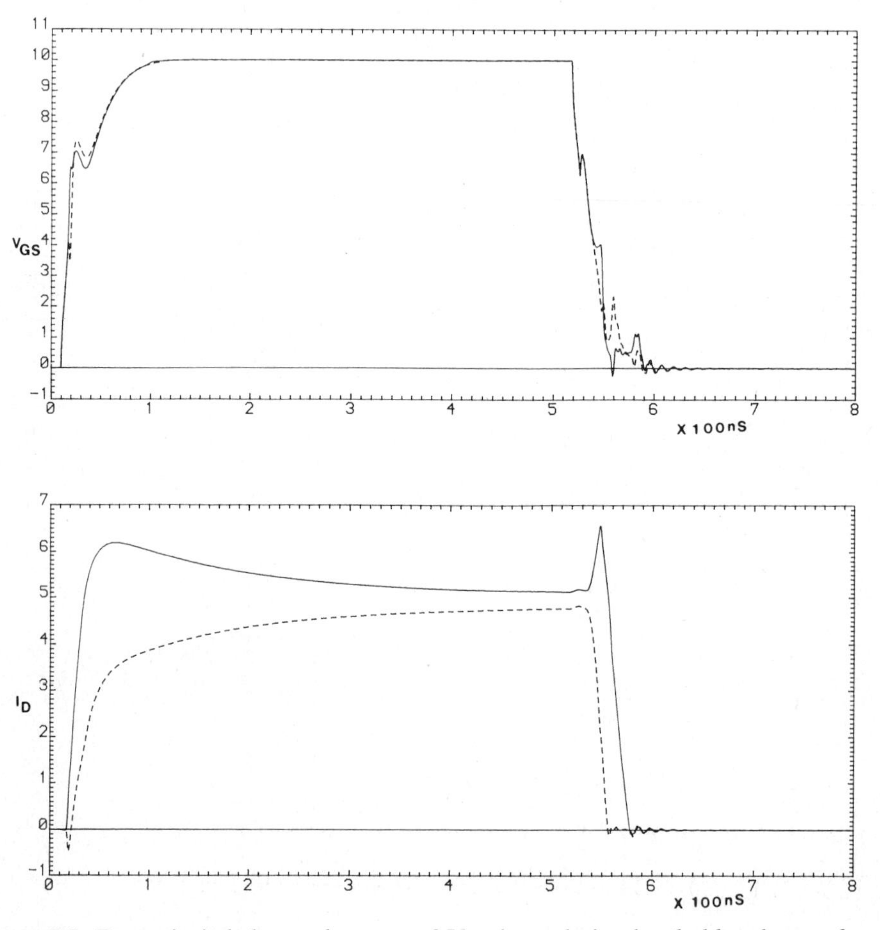

Figure 7.5. Dynamic imbalance due to a 2 V mismatch in threshold voltage of two MOSFETS connected in parallel. Solid line, $V_T = 2$ V; dashed line, $V_T = 4$ V.

by large values of source inductances, since their effect swamps the effect of other mismatched parameters.

2. Devices may be selected to match threshold and transconductance characteristics. An appropriate but simple way of doing this is to measure the value of I_D produced by a given value of V_{GS} (in the constant current region).

3. Decoupling of the gates with series gate resistors assists dynamic current sharing, so that the gate resistors included to suppress parasitic oscillations will also have a beneficial effect on the dynamic current balance.

Fortunately power MOSFETS have a high transient overload capability, so that unequal die heating in high-frequency applications is likely to be the only effect of dynamic imbalance. However, it is conceivable that in situations where many devices are paralleled to carry a very high current, the current carried by the first device to turn on or the last device to turn off could be of sufficient magnitude to cause damage.

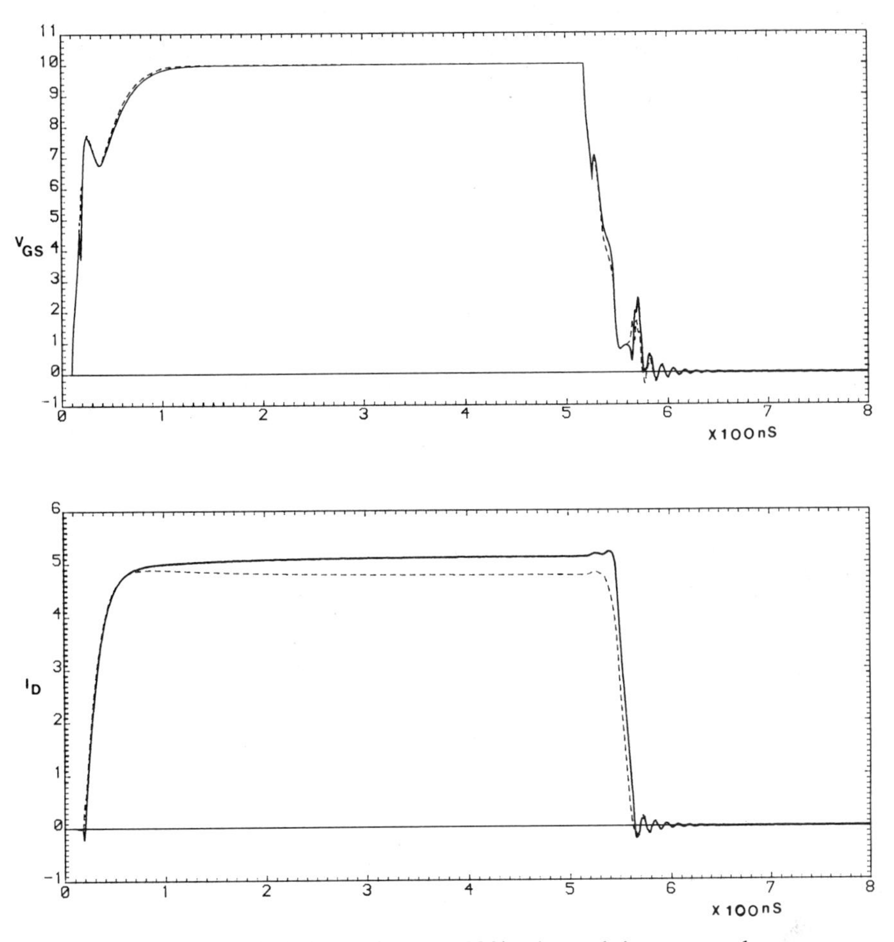

Figure 7.6. Dynamic imbalance due to a 20% mismatch in transconductance.

7.1.5. Oscillations During the Switching of Paralleled MOSFETS

When MOSFETS are paralleled, there is a possibility of oscillation during switching as the MOSFETS pass through the linear region. The oscillations that occur in parallel operation are generally attributed to differential-mode instability [7, 8]. This means that a circuit with perfect dynamic balance may be unstable. A pair of MOSFETS can operate as a local oscillator in the push–pull mode, the tuned circuit being composed of the parasitic inductance of the interconnecting wire and the parasitic capacitances of the MOSFETS. Feedback is provided by the parasitic capacitances of the MOSFETS. Parasitic oscillations may cause paralleled power MOSFETS to fail for no apparent reason. The frequency of the oscillations is often in the range of 100 to 250 MHz, and an appreciation of the extent of the problem can only be gained using an oscilloscope with a bandwidth of this order. The peak gate voltage generated by such oscillations can be sufficiently large to break down the gate oxide.

Oscillation can usually be prevented by the inclusion of resistors in series with the individual gate connections. The resistors provide differential-mode damping and ensure the stability of any loops formed as a result of paralleling transistors. The gate resistors slow down the switching of the MOSFETS, and a conflict may arise in applications where fast switching is required. Good component layout then becomes critical.

The stability of two MOSFETS in parallel has been analyzed by Kassakian and Lau [7]. In the linear region the circuit is represented by the incremental model shown in Figure 7.7. The analysis assumes that the devices are identical. Since all nodes on the plane of symmetry of the incremental model are at incremental ground potential, the model can be simplified as shown in Figure 7.8. The gain provided by the current generator $g_m V$ and the feedback provided by C_2 make this circuit potentially unstable. The stability of the circuit can be determined by analyzing the roots of its characteristic equation. To simplify the analysis, two separate situations were considered. In the first case L_S was zero, while in the second L_D was zero. This gives rise to the models shown in Figure 7.9*a* and *b*.

The stability of the circuits may be determined by applying the Routh–Hurwitz criterion, the essence of which is that the circuit will be unstable if there are any right–half-plane (RHP) zeros. Kassakian and Lau used this method to analyze the behavior of two 100-V, 10-A MOSFETS in parallel. The values of the parasitic

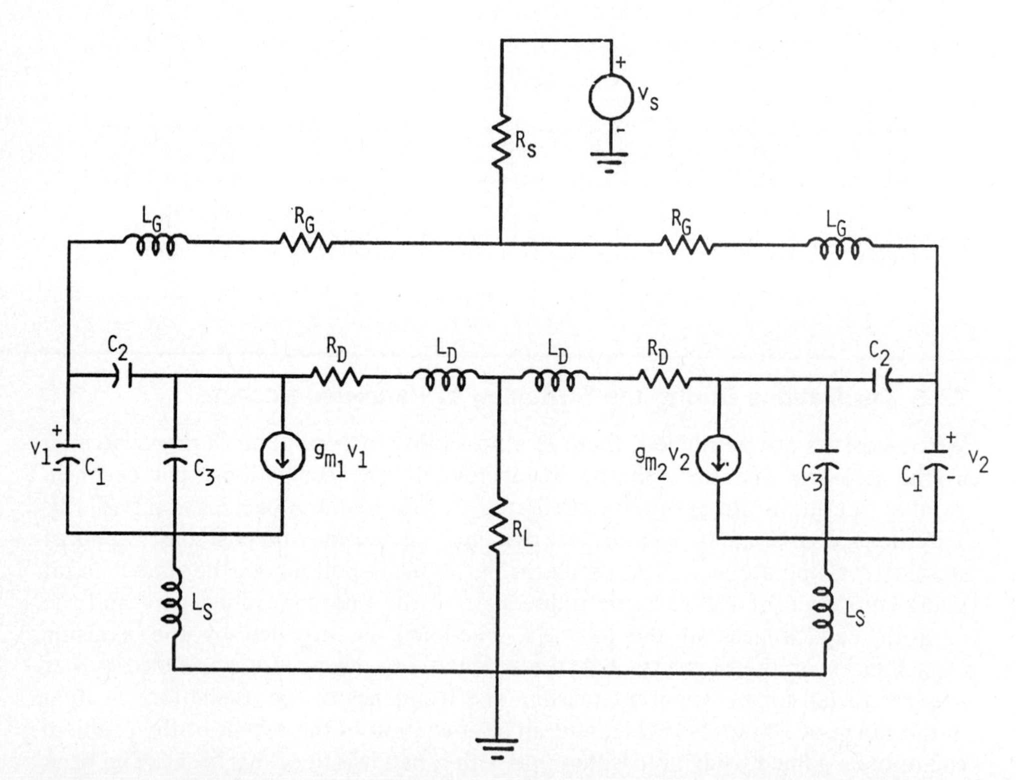

Figure 7.7. Incremental model, including parasitic elements, for the linear-region operation of paralleled MOSFETS. From Kassakian and Lau [7]. © IEEE 1984.

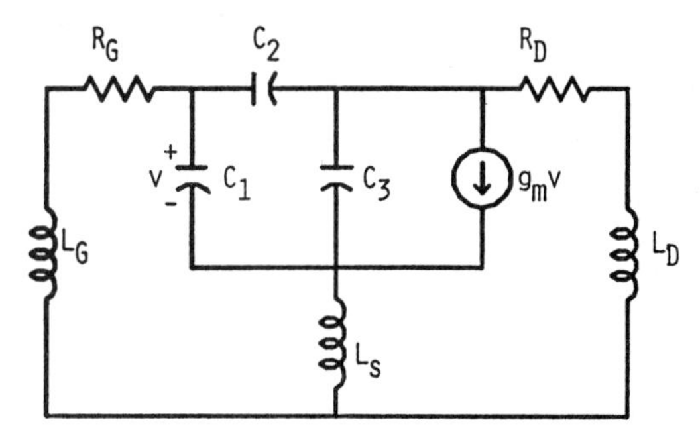

Figure 7.8. Differential-mode equivalent circuit for paralleled MOSFETS. From Kassakian and Lau [7]. © IEEE 1984.

capacitances used were 700 pF for C_1, 100 pF for C_2, and 300 pF for C_3. The Routh–Hurwitz criterion was used to determine the stability of the circuit for values of g_m between 0.1 and 10 S, for parasitic inductance values between 0.5 and 50 nH, and for parasitic differential gate resistance (parasitic plus external) between 0.01 and 100 Ω. This range of parameters is adequate to cover values typically encountered when discrete devices are paralleled and when chips are paralleled on a common header.

The results are shown in Figures 7.10 and 7.11. These graphs show the minimum value of R_G necessary for stability for various values of L_D and L_G in

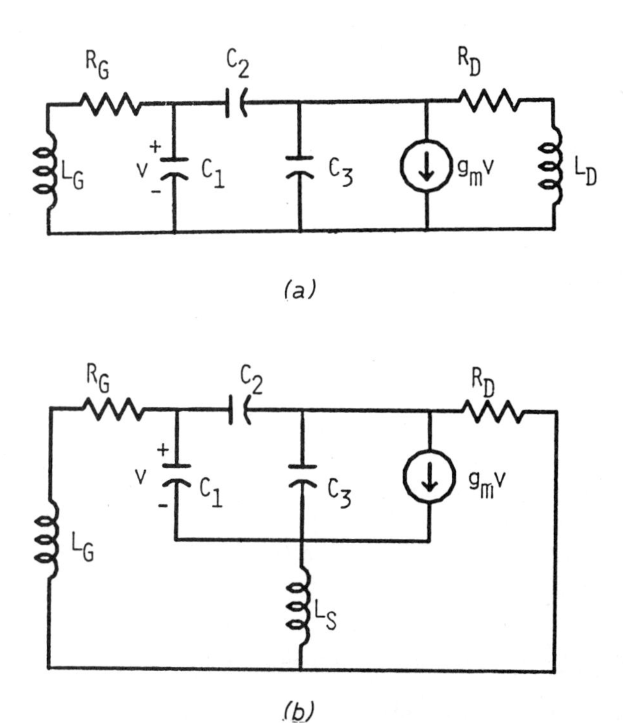

Figure 7.9. Differential-mode equivalent circuits for paralleled MOSFETS. (*a*) With $L_S = 0$; (*b*) with $L_D = 0$. From Kassakian and Lau [7]. © IEEE 1984.

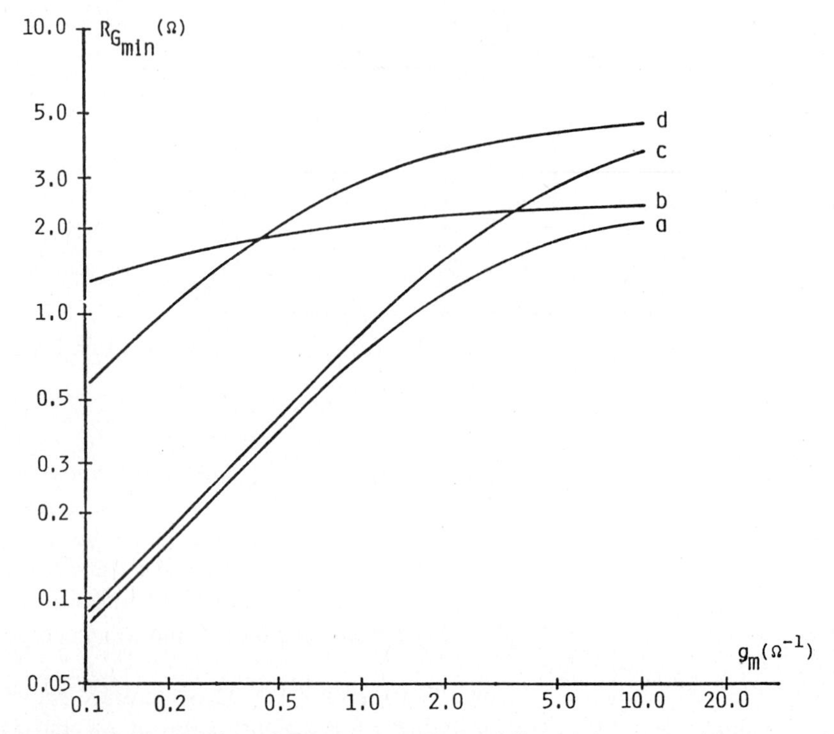

Figure 7.10. Minimum value of R_G necessary for stability of the model (Figure 7.9*a*) with $R_D = 0$ as a function of g_m for various values of L_D and L_G. *a*, $L_D = 5$ nH, $L_G = 5$ nH; *b*, $L_D = 20$ nH, $L_G = 5$ nH; *c*, $L_D = 5$ nH, $L_G = 20$ nH; *d*, $L_D = 20$ nH, $L_G = 20$ nH. From Kassakian and Lau [7]. © IEEE 1984.

Figure 7.10 and various values of L_S and L_G in Figure 7.11. The area below the curves represents a region of unstable operation. It is clear from these results that instability is a possibility in a wide range of practical situations. These results support the popular rule of thumb that a resistor of 10 Ω in series with each gate will prevent parasitic oscillation in most situations where MOSFETS are paralleled. Furthermore, Kassakian and Lau found that no RHP roots existed when L_G/L_S was greater than approximately 2.6. This lends justification to the practice of putting ferrite beads on the individual gate leads to suppress oscillation.

The results shown in Figures 7.10 and 7.11 were obtained with R_D equal to zero. The effect of R_D on stability can be gauged from Figure 7.12, which shows the minimum value of R_G required for stability for the case in which $L_G = L_S = 20$ nH and $L_D = 0$. R_D clearly has a damping effect on the circuit, and there is, in fact, a value of R_D above which no RHP roots exist when $R_G = 0$. The danger of instability is therefore greatest when low-resistance devices are paralleled.

7.1.6. Effect of Gate-Driver Source Impedance

If paralleled MOSFETS oscillate due to instability in the linear region of the switching cycle, the oscillations will grow at a rate determined by the Q of the circuit. If the time spent in the linear regime during switching is short compared

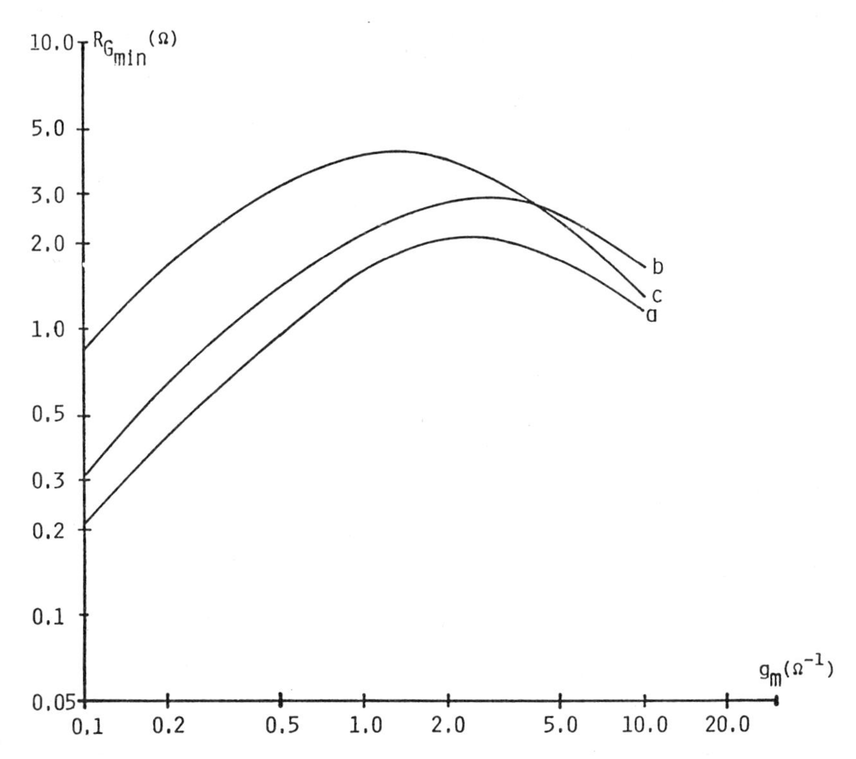

Figure 7.11. Minimum value of R_G necessary to assure stability of the model (Fig. 7.9*b*) with $R_D = 0$ as a function of g_m for various values of L_G and L_S. *a*, $L_G = 5$ nH, $L_S = 5$ nH; *b*, $L_G = 5$ nH, $L_S = 20$ nH; *c*, $L_G = 20$ nH, $L_S = 20$ nH. From Kassakian and Lau [7]. © IEEE 1984.

with the time constant governing the buildup of oscillations, then the oscillations may not attain a significant amplitude and the fact that the system is unstable will be of no consequence. In this case, R_S, the gate-driver source impedance, should be kept as low as possible so that the MOSFETS pass through the linear region as quickly as possible. In the case of chip-paralleled devices in which the parasitic inductances will be low, the oscillation could grow very quickly even for very fast switching times. These oscillations may not be observable outside the package, since all elements of the oscillatory circuit are contained within the package. It would therefore seem advisable when paralleling MOSFETS at chip level to include resistors in series with the gates or by some other means to ensure that the series gate resistance is sufficiently large for stability.

Kassakian and Lau found some discrepancy between the predictions of the model and the experimental results, particularly in the range of g_m over which oscillations occur. The range of instability found in practice is somewhat less than predicted. They suggest a variety of possible causes for these discrepancies, including the damping effect of parasitic loss mechanisms within the MOSFETS, variation in the value of parasitic capacitances due to drain–source voltage variations, and a difference between the data-sheet values of parasitic capacitance, measured at 1 MHz, and the actual values at the frequency of oscillation, which could be in excess of 100 MHz.

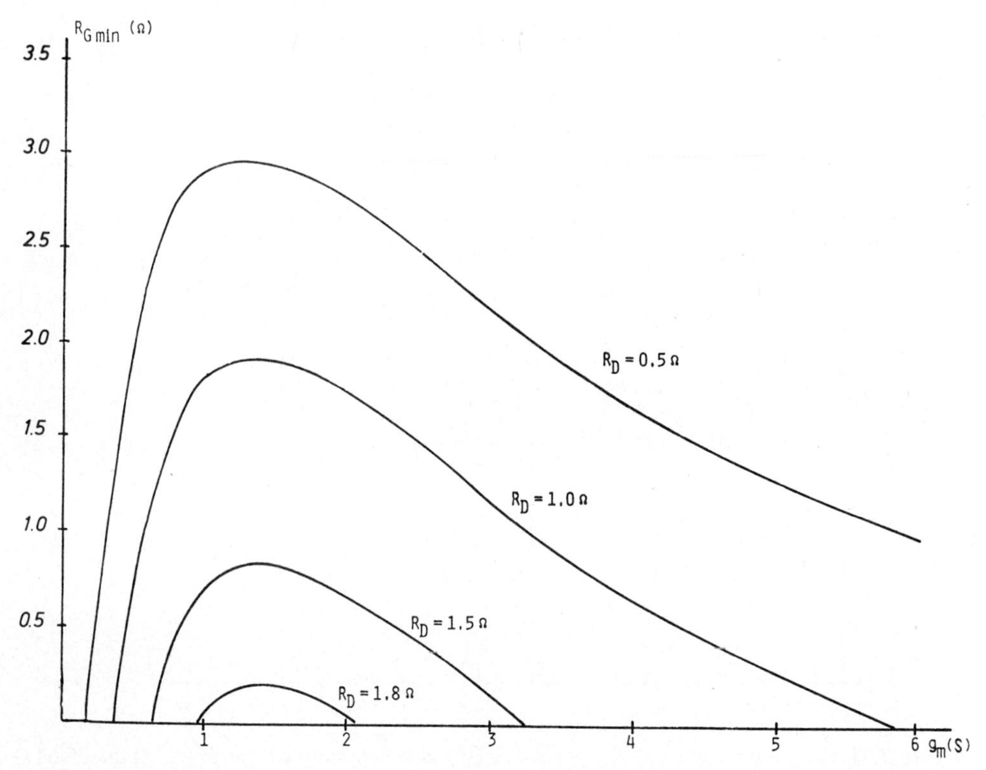

Figure 7.12. Minimum value of R_G necessary to assure stability of the model (Fig. 7.9*b*) as a function of g_m for various values of R_D. From Kassakian and Lau [7]. © IEEE 1984.

7.1.7. Stability of Parallel Packaged Devices

Because of the differential nature of the oscillations produced by instability in paralleled MOSFET dice, it is possible for these oscillations to occur inside a multiple-die package without clear indications at the device terminals of what is occurring within the package. This could have been the cause of mysterious failures said to have occurred when some MOSFET manufacturers first introduced multiple-die devices [1]. It has now become standard practice in the manufacture of parallel-die power MOSFETS either to include individual gate resistors for each die or to make the sheet resistance of the polysilicon gate high enough to provide the required degree of differential mode damping. A resistance of an ohm or two is usually adequate to ensure stability.

Verification of the absence of differential-mode oscillation in parallel packaged devices is difficult, since there may be no evidence of the oscillation in the common mode signals. Inserting current-measuring instruments in the wiring of the device is not possible without substantially altering the prevailing conditions. One method of detecting oscillations which causes minimum alteration of the operating conditions is to open the package, locate a search coil near the wire bonds, and look for evidence of oscillation using a high-frequency spectrum analyser [8].

7.1.8. Paralleling for Linear Operation

Although paralleled power MOSFETS are not subject to current hogging when operated in the switching mode, this is not necessarily the case when operating in the linear mode. This is because in the linear region the temperature coefficient of forward voltage drop may be negative below a certain value of drain current. Figure 15.7*a* shows the transfer characteristics of a device for which a rise in junction temperature produces a fall in drain–source voltage when the gate voltage is held constant at some value below the zero-temperature-coefficient point. This means that a device that carries more current will get hotter, V_{DS} will tend to fall, and the device will consequently carry even more current. Close thermal coupling of the junctions reduces this tendency, since forced elevation of the temperature of the junction carrying the least current will encourage it to conduct more. Current sharing in the linear regime can be promoted by matching devices not only for threshold voltage but also for transconductance. A match of one of these parameters does not necessarily guarantee a match in the other, and ideally the transconductance curves should be matched, although matching at one important point on the curve may be adequate.

Current sharing may be forced by source-resistor ballasting as shown in Figure 7.13. Resistors of a few ohms in series with the source leads will insert a negative feedback voltage into the gate–source loop, so that an increase in source current reduces the gate voltage, thus tending to reduce the drain–source current. The penalty for this improvement in stability is a reduction in the gain of the stage and extra losses due to the presence of the source resistor. The gain of each transistor when ballasted becomes a function of the source resistor [4], so that

$$g'_{m} = \frac{g_{m}}{1 + Rg_{m}} = \frac{1}{R + 1/g_{m}} \tag{7.1}$$

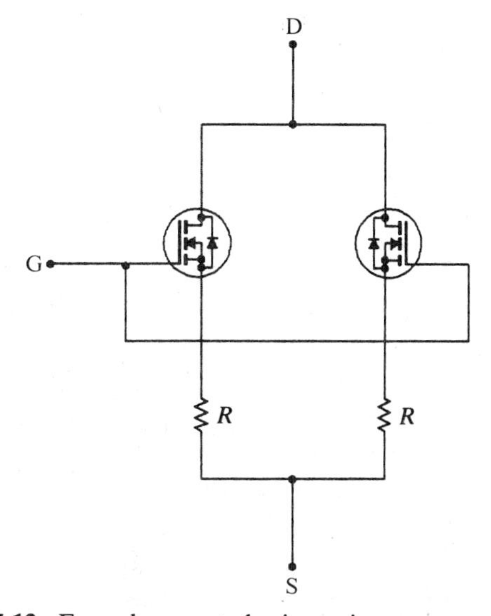

Figure 7.13. Forced current sharing using source resistors.

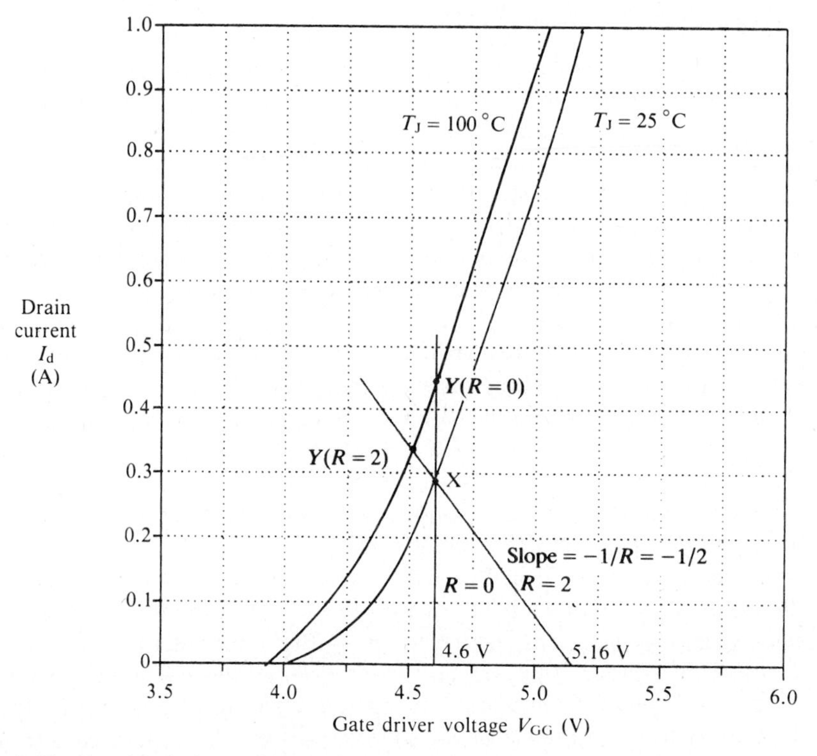

Figure 7.14. Graphical determination of the value of the source resistor required for stabilization of the quiescent operating point.

If $R \gg 1/g_m$, the effective transconductance of each paralleled device is independent of g_m. Values of $1/g_m$ typically range from 0.1 to 1 Ω.

The value of the source resistance needed to produce the required operating conditions can be determined graphically [9] as illustrated in Figure 7.14. The quiescent current points for the two extremes of the operating temperature range are chosen on the basis of the deviation that is considered acceptable. A line drawn through these two points is projected to the point where it cuts the horizontal axis. This gives the new quiescent point gate voltage. The value of R can be determined from the slope of the line, which is $1/R$.

7.2. INSTABILITY DURING SWITCHING OF A SINGLE MOSFET

Closely allied to the problem of oscillation of parallel devices is the tendency of single-device circuits to ring when switched. While switching between the ON and OFF states a power MOSFET passes through the linear region. During the time the device is in the linear region, instability can occur if care is not taken with the layout of the circuit, particularly the gate drive circuit. This instability can result in a burst of oscillation, which can cause excessive gate voltage, leading to oxide damage and the destruction of the device. Causes of instability in single devices operating in the switching mode are discussed next. Instability in parallel-connected MOSFETs is discussed in Section 7.1.5.

The power MOSFET is susceptible to instability because of its high bandwidth and high input impedance. The gain of a power-MOSFET stage can still be greater than unity at 100 MHz, and at this frequency a tuned circuit can easily be formed by parasitic capacitance in the MOSFET and a few centimeters of PCB track. The high input impedance makes it easy for feedback signals to couple into the gate circuit.

The most obvious step to take to avoid instability is to avoid magnetic coupling between the drain and gate circuits, since this can provide the positive feedback necessary to sustain oscillation. The area of the loop formed by the gate track and the return path from the source should be kept as small as possible and as far away as possible from any conductors which carry the drain current. However, when this cause of oscillation is totally eliminated, a power MOSFET circuit can still oscillate if the damping is inadequate. The situation has been analyzed by Giandomenico, Kuo, Hu, and Choi [10] using the small-signal circuit model of the type shown in Figure 7.15. Using the Routh–Hurwitz stability criterion, they showed that the condition for stability is:

$$L_G < (C_{GD} + C_{DS})^2 R_G / C_{GD} g_m \tag{7.2}$$

where L_G is the gate circuit inductance. Their results show that a gate inductance of only one or two nanohenrys can be sufficient to cause instability. The clear conclusion is that not only does the gate loop need to be small to prevent coupling with the drain circuit, but it must also be small in order to reduce the inductance in series with the gate. Gate drivers should be located as close as possible to the MOSFET, and twisted pairs should be used for wiring. Stability was

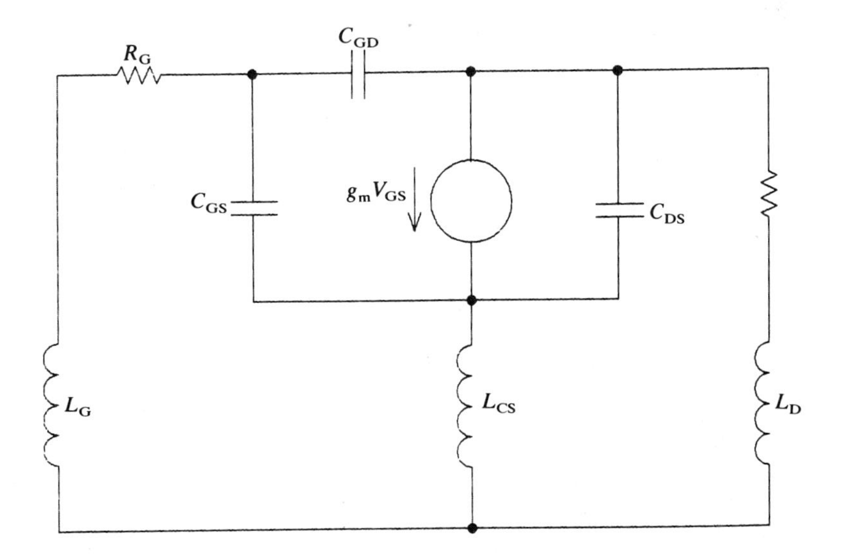

Figure 7.15. Small-signal circuit model of a power MOSFET operating in the linear region.

shown to decrease as the die area (and current rating) increases, so that greater care is required with larger devices. A few ohms of gate resistance can be adequate to ensure stability. The gate resistor should be located as close as possible to the gate lead, so that the maximum amount of parasitic gate inductance is decoupled. Common source inductance also has a damping effect.

Parasitic oscillation in power MOSFETS can often be cured by locating a ferrite bead over the gate lead. This can be a convenient retrofit cure for oscillation, since it need not involve a change in PCB layout. The ferrite bead may simply be slipped over the gate lead of the MOSFET. Given the results of the above analysis an increase in gate inductance would appear to make oscillation more likely and indeed in some cases the addition of the ferrite bead can make the situation worse. However it has been shown that when MOSFETS are paralleled, ferrite beads on the gate can enhance stability (see Section 7.1.5), and ferrite beads can undoubtedly be beneficial in single MOSFET circuits. There are a number of possible explanations for the elimination of oscillation in this way, including the damping effect of the ferrite losses, decoupling of a feedback signal from the drain, and lowering of the resonant frequency of the tuned circuit.

7.3. SERIES OPERATION

Power MOSFETS may be chained simply in the source-follower configuration as shown in Figure 7.16. Static voltage sharing is aided by parallel resistors, and

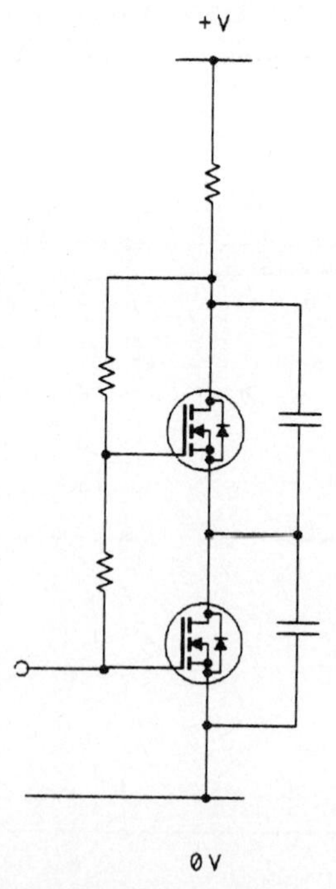

Figure 7.16. Series operation of MOSFETS

dynamic voltage sharing by parallel capacitors. This method of series operation has the disadvantages that a high-voltage gate control signal is required, that capacitors are needed across each device to ensure dynamic voltage sharing, and that if the gate pullup resistors are made large to reduce dissipation, switching is slow. The presence of external capacitors further increases the losses produced by the charging and discharging of C_{DS}.

An arrangement which avoids these disadvantages at the expense of increased component count is shown in Figure 7.17. The individual gates are transformer-driven. Transient overvoltage of any one device is prevented by zener clamping, but if the MOSFETS possess adequate avalanche capability, the zener clamps may be dispensed with. The arrangement is capable of switching several kilovolts in a few tens of nanoseconds.

The increasing availability of MOSFETS with voltage ratings of over 1000 V has reduced the need to operate devices in series. Although the $R_{DS(on)}$ per unit area of silicon in devices of this voltage rating is high, the cost of gate driving arrangements for MOSFETS in series is also high, and it is unusual for devices to be operated in series when a device of the required voltage rating is available. MOSFETS with voltage ratings of up to 1200 V will find application in motor drives operating from full-wave-rectified 440-V three-phase supplies. Beyond this

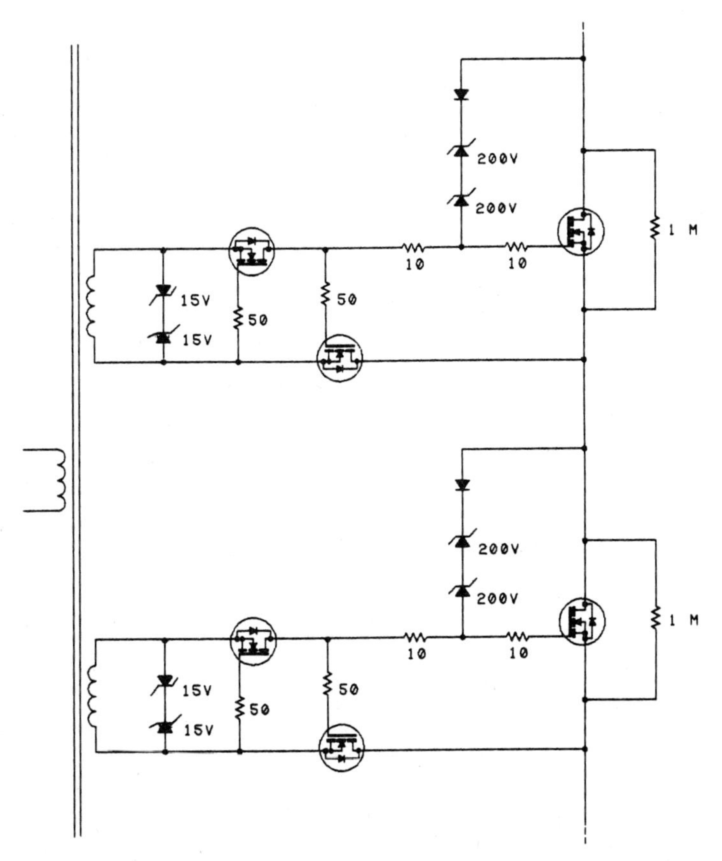

Figure 7.17. Series operation of MOSFETS using transformer-coupled gate drivers.

voltage the market for MOSFETS thins out, and manufacturers are unlikely to develop a broad range of MOSFETS with higher voltage ratings. Therefore series operation will be necessary for such higher-voltage applications.

REFERENCES

1. J. G. Kassakian, "Some issues related to the behavior of multiply paralleled power MOSFETS." *Proc. Int. Power Electron. Conf.* Tokyo 324–335 (1983).
2. K. Gauen, "Paralleling power MOSFETS in switching applications". *Motorola Appl. Note* **AN-918** (1984).
3. J. Forsythe, "Paralleling of power MOSFETS for higher power output". *IEEE, Proc. Ind. Appl. Soc. Conf.* pp. 777–786 (1981).
4. R. Severns, "Parallel operation of power MOSFETS." *Siliconix Tech. Article.* **TA84-5** (1984).
5. J. Forsythe, "Techniques for controlling dynamic current balance in parallel power MOSFET configurations." *Proc. Powercon* **8,** G.3.1–G.3.11 (1981).
6. S. Clemente and B. Pelly, "A chopper for motor speed control using parallel connected power HEXFETS." *Int. Rectifier Appl. Note* **AN-941** (1982).
7. J. G. Kassakian and D. Lau, "An analysis and experimental verification of parasitic oscillations in paralleled power MOSFETS." *IEEE Trans Electron Devices* **ED-31** (7), 959–963 (1984).
8. D. Giandomenico, D.-S. Kuo, and C. Hu, "Oscillations in multichip power MOSFETS." *Power Convers. Intell. Motion* (*PCIM*) August, pp. 74–78 (1985).
9. K. Gauen, "Paralleling power MOSFETS in linear applications." *Proc. Power Convers. Int. Conf.* April, pp. 144–151 (1984).
10. D. Giandomenico, D.-S. Kuo, C. Hu, and J. Choi, "Analysis and prevention of power MOSFET anomalous oscillations." *Proc. Powercon* **11,** H.4.1.–H.4.9 (1984).

CHAPTER 8

Gate Drive Circuits

8.1. BASIC PRINCIPLES

To switch a MOSFET from the nonconducting state to the conducting state, the gate–source voltage must be raised from below to above the threshold voltage by a transfer of charge into the gate. Turn-off is the reverse of this procedure. The manner in which the transition between the two steady-state levels of gate voltage is made determines the switching performance of the device. The theory of MOSFET switching and the influence of device and circuit parameters on the process are discussed in Chapter 4.

A gate drive circuit may generally be represented by a voltage source in series with a resistance (Figure 8.1). (Where the gate drive circuit is asymmetric, two voltage sources and corresponding resistances are required to represent the turn-on and turn-off conditions.) The resistance may be introduced intentionally to control the MOSFET switching speed, or it may represent the impedance of the voltage source. This simple equivalent circuit forms a convenient basis for comparing the performance of gate drive methods.

8.2. CHOOSING THE GATE-DRIVER VOLTAGE

In the steady state, since the gate-to-source resistance will generally be many orders of magnitude greater than the gate-driver output impedance, the MOSFET gate voltage will be equal to the gate-driver open-circuit voltage. A steady-state gate voltage lower than the open-circuit driver voltage is an indication of gate oxide damage. Oxide damage does not always result in an immediate gate-to-source short circuit, and a MOSFET may continue to operate for a time with a leaky gate provided the gate driver can supply adequate current to raise the gate voltage to a level sufficient to turn the device on.

In a switching application the open-circuit gate-driver voltage has two steady-state values, representing the ON and OFF conditions. The open-circuit gate-driver voltage will influence the switching time of the MOSFET, but generally the gate driver voltage is chosen primarily to give the required static conditions. Or it may be chosen because it is a level which can be conveniently supplied. It is often possible to choose the gate-driver voltages for convenience, since there is usually a broad range of possible gate voltage between the minimum voltage required to turn the MOSFET on and the maximum gate voltage allowed by considerations of gate oxide lifetime. The required switching time is then obtained by adjusting the impedance in series with the gate.

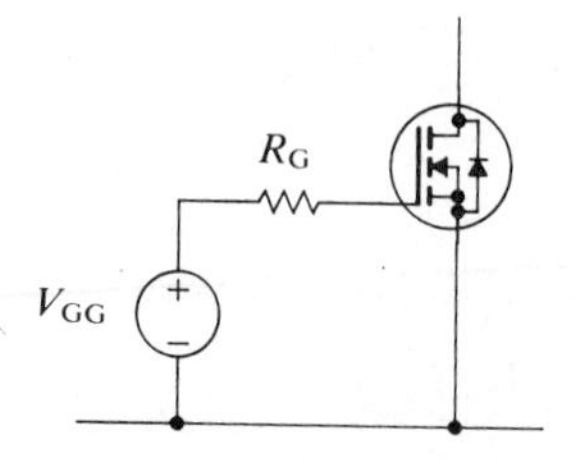

Figure 8.1. Gate driver equivalent circuit.

8.3. OFF-STATE GATE VOLTAGE

The OFF-state gate voltage must be sufficiently below the threshold voltage to provide adequate noise immunity. An OFF-state gate voltage of zero (source potential) usually provides sufficient noise immunity. The noise immunity also depends on the gate-driver impedance, and a low driver impedance will help ensure good noise immunity.

An OFF-state gate voltage of zero may not provide sufficient noise immunity in situations in which the drain voltage rises rapidly, for example in inverters. As Figure 8.2 shows, the drain voltage of Q2 can rise rapidly as the current commutates to Q1 after circulating through D2. This rapid rise in drain voltage causes a rise in gate voltage of Q2 due to the parasitic capacitance between the drain and the gate. If as a result of this the gate voltage rises above the threshold voltage, Q2 will turn on, resulting in a short circuit between the power rails of the inverter. The rise in gate voltage is restrained by the gate–source capacitance and by the gate-driver circuit. A low gate-driver impedance therefore is beneficial in restraining the rise in gate voltage. However, even when R_G is zero and the gate is effectively connected to the source, there remains the resistance of the gate structure itself, which in the case of a polysilicon gate can be equivalent to one or two ohms of series resistance. Furthermore, the lead inductance of the MOSFET package and the inductance of the external wiring will also reduce the ability of

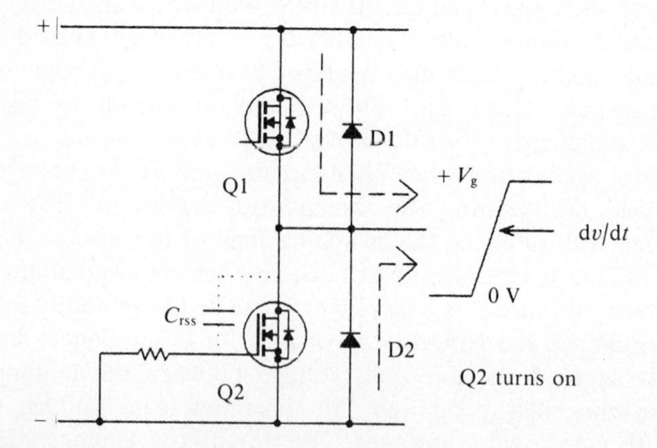

Figure 8.2. dv/dt-induced turn-on due to drain–gate capacitance.

the gate driver to hold down the gate voltage. Under conditions of very high drain dv/dt it may therefore be necessary to apply a negative bias to the gate to prevent spurious turn-on of the MOSFET. A negative gate bias is not usually used unless absolutely necessary, since it adds to the complexity of the gate drive circuit and increases the turn-on delay time.

8.4. ON-STATE GATE VOLTAGE

The minimum ON-state gate voltage required to ensure that the MOSFET remains fully-on when carrying maximum drain current can be determined from the forward transfer characteristics of the device (see Figure 8.3). The gate voltage should be greater than this minimum value by an amount that allows for manufacturing spread in component values and for supply-voltage variation.

The gate voltage may also be raised above the minimum required by other considerations in order to ensure a low value of ON-state drain–source voltage. As Figure 8.4 shows, increasing the gate voltage once saturation is achieved can further reduce the ON-state drain–source voltage. This can be particularly important in applications where a low ON-state voltage drop is essential to efficiency, such as when operating from a low-voltage supply. An increase in gate voltage results in a reduction in channel resistance due to the channel being inverted to a greater depth, and to an increase in the depth of the accumulation layer in the drain region under the gate. These effects are most significant in

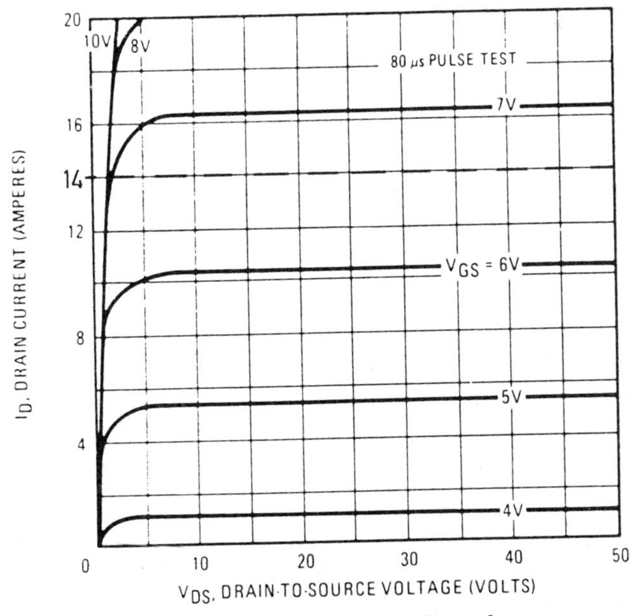

Figure 8.3. Calculating the minimum required gate voltage from MOSFET output characteristics. A gate voltage of at least 7 V is required if this MOSFET is to remain fully turned on while carrying the rated drain current of 14 A.

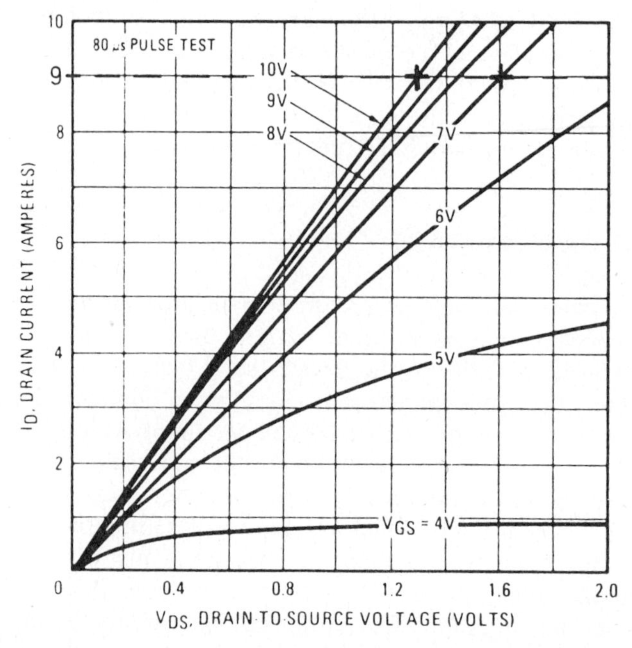

Figure 8.4. Typical MOSFET saturation characteristics. This expanded version of the output characteristics of Figure 8.3 shows that the ON-state voltage drop can be reduced by an increase in V_{GS}.

low-voltage MOSFETS, in which the channel and accumulation layer resistances form a more important part of $R_{DS(on)}$.

However, there are disadvantages and dangers in using an elevated ON-state gate voltage. The turn-off delay time increases as the ON-state gate voltage increases. Also, reliability of the MOSFET can be undermined by use of an excessive gate voltage. The possibility of a MOSFET failing in service becomes significant as the gate voltage approaches the maximum allowable value (see Figure 17.11). Therefore there should be an adequate margin between the maximum ON-state gate voltage and the absolute maximum allowable gate-voltage rating. For standard-process power MOSFETS the maximum desirable gate voltage is typically 15 V.

There are thus a number of conflicting requirements which influence the selection of the ON-state gate voltage. In practice, the choice often owes more to the availability of a particular supply voltage than to any other consideration. A compromise is then made between MOSFET performance and system cost. When the gate is driven through a transformer, this conflict does not arise, since the number of turns on the secondary winding may be chosen to give the optimum gate voltage.

8.5. GATE-DRIVER IMPEDANCE

The gate-driver impedance comprises two elements—an inescapable component involved in generating the gate voltage, and a component deliberately introduced to tailor the response of the MOSFET. The usual method of modifying the MOSFET switching time once the gate-driver voltage has been chosen is to insert resistance in series with the gate.

In bipolar-transistor base drive circuits a speed-up capacitor is sometimes connected in parallel with the base current-limiting resistor to ensure fast switching of the transistor. There is little point in putting a speed-up capacitor in parallel with the gate resistor of a MOSFET, since the gate only draws current during switching, and the series resistor should not be bypassed during the switching period, since its presence is required to regulate the switching speed. While switching speed is increased by the use of a bypass capacitor, a cleaner switching waveform is generally obtained if the switching speed is increased by a reduction in the value of the series resistor.

8.6. ASYMMETRIC GATE DRIVE CIRCUITS

Different turn-on and turn-off times can be achieved by bypassing the series gate resistor with a diode, as shown in Figure 8.5. Asymmetry may be inherent in the

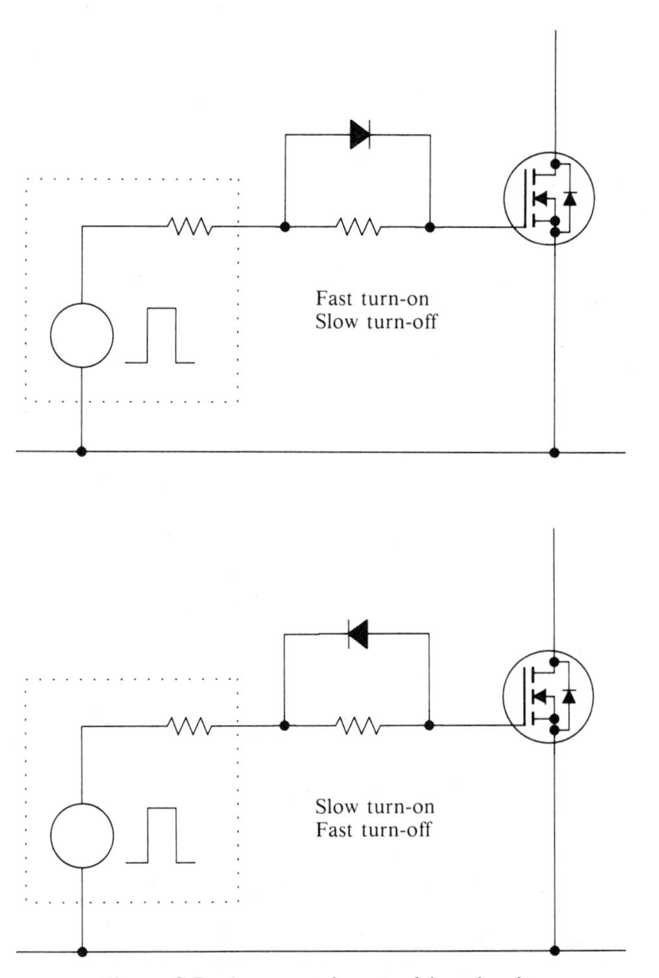

Figure 8.5. Asymmetric gate drive circuits.

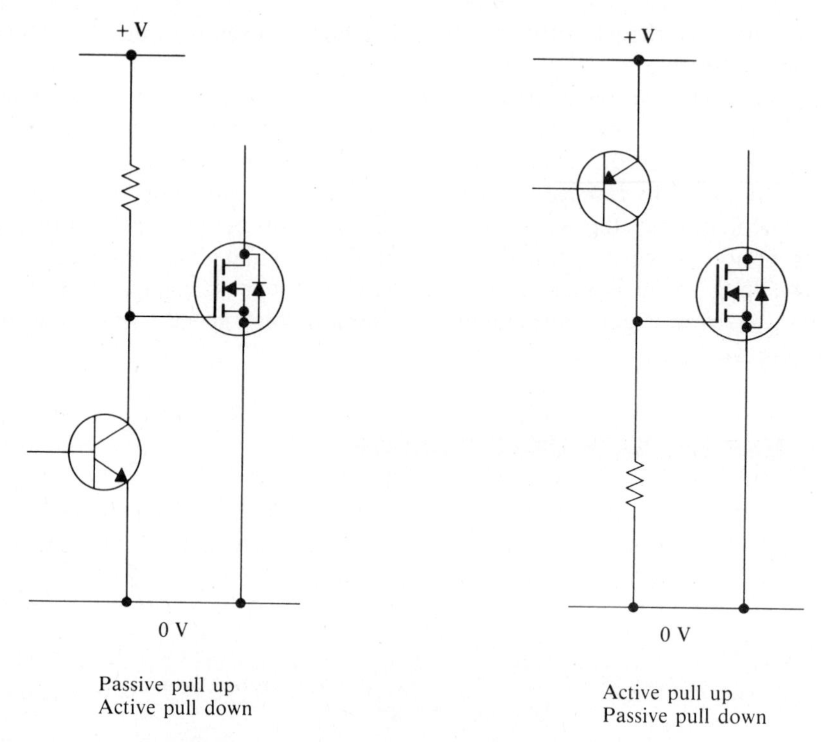

Figure 8.6. Gate drive circuits with inherent asymmetry.

gate driver circuit, as when the gate driver has an active pull down action but passive pull up or vice versa. Figure 8.6 shows two circuits with inherent asymmetry.

An asymmetric gate drive is sometimes used when a MOSFET is controlling a clamped inductive load. Turn-on is slowed in order to reduce the current surge that occurs in the MOSFET due to the reverse recovery charge of the free-wheeling diode. Turn-off is rapid in order to minimize the turn-off time and the switching losses.

8.7. COMPLEMENTARY EMITTER-FOLLOWER DRIVE CIRCUIT

To achieve switching speeds of the order of 100 ns or less requires a gate drive circuit with a low output impedance and the ability to sink and source relatively large currents. For example, a 100-V MOSFET with an $R_{DS(on)}$ of 0.55 Ω (e.g. an IRF 150) switching a load on a 50-V supply would typically require a gate charge transfer of 75 nC to raise the gate voltage from 0 to 10 V during turn-on of the MOSFET. If turn-on is completed within 100 ns, then the average gate current is 750 mA. A circuit capable of sourcing and sinking this level of current is shown in Figure 8.7. A pnp transistor and an npn transistor are connected in a totem-pole configuration so that they both act as emitter followers, thereby giving the circuit a low output impedance regardless of the direction of the load current. The

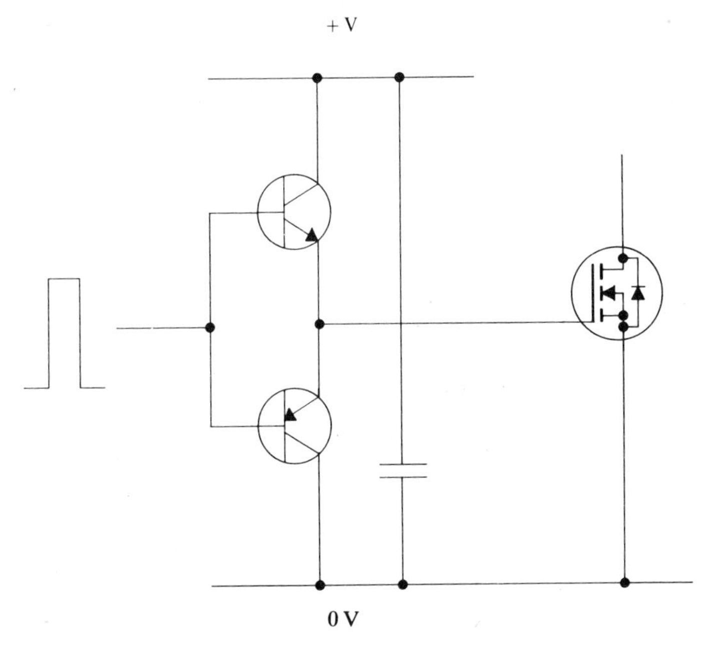

Figure 8.7. Double emitter-follower low-impedance gate drive circuit.

transistors operate in the linear regime, and the delay time associated with bipolar transistors operating in the saturated mode is thereby avoided. A low-value resistor may be connected in series with the output to adjust the switching performance of the MOSFET or to damp oscillations as the MOSFET passes through the linear region.

8.8. INTEGRATED-CIRCUIT GATE DRIVERS

There are a number of integrated circuits on the market which are designed to drive capacitive loads and are capable of sourcing or sinking large currents (several amperes) for a short time. Typical of these devices is the DS0026 clock driver and the TSC429 MOSFET driver. These devices incorporate level shifting as well as providing a buffered output. A decoupling capacitor must be connected across the power rails close to the driver IC to ensure low output impedance.

8.9. DRIVING MOSFETS FROM OPERATIONAL AMPLIFERS AND COMPARATORS

The gate signal for a power MOSFET may in some applications be generated by an operational amplifier or a comparator. For example, in pulse-width-modulated converters, the switching waveform may be generated by comparing the amplitude of two analogue signals. If the MOSFET gate is driven directly from an operational amplifier, the switching speed of the MOSFET will be limited by the

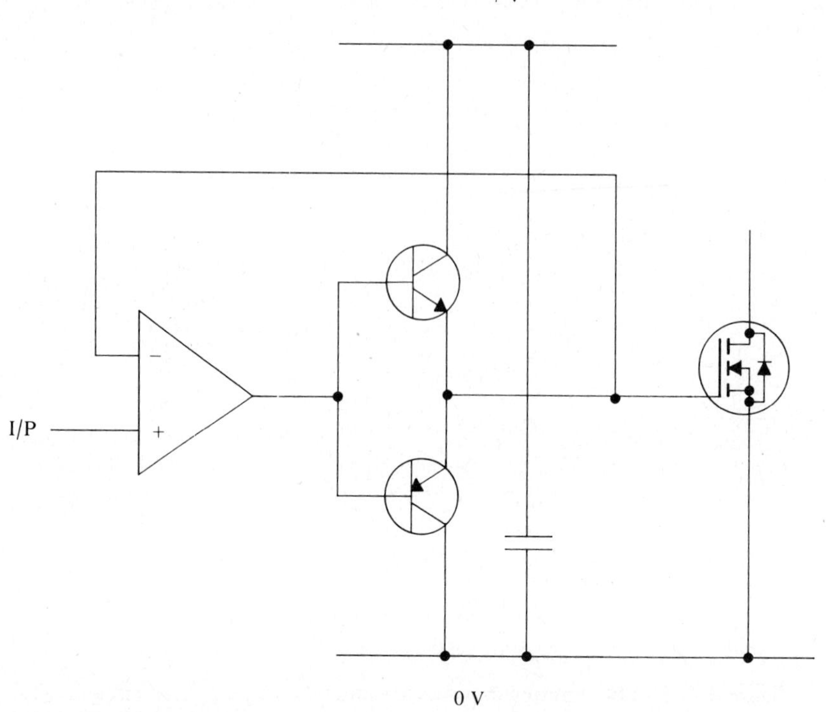

Figure 8.8. Driving a MOSFET from an operational amplifier.

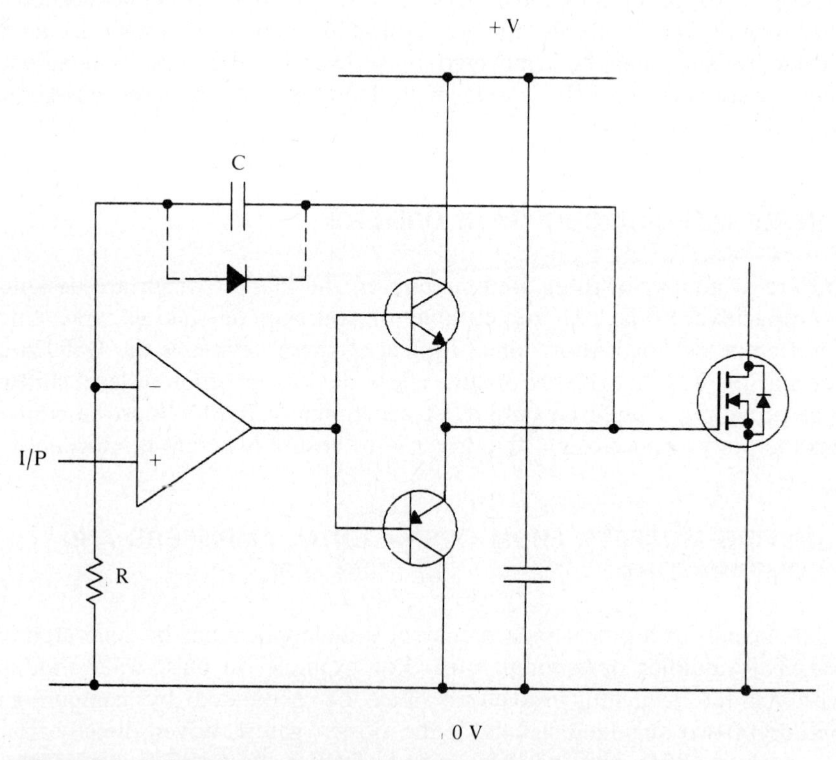

Figure 8.9. Driving from an operational amplifier with pulse edge shaping.

slew rate and by the current drive capability of the amplifier. Industry-standard operational amplifiers such as the 741S with an open-loop bandwidth of 1 MHz and a slew rate of 30 V/μs can be used to drive the MOSFET directly, but only where fast switching of the MOSFET is not required. For applications were fast switching is required, special-purpose current amplifiers may be used, such as the HA2630 with a bandwidth of 8 MHz, a slew rate of 500 V/μs, and an output current capability of 0.4 A. Rise and fall times of under 40 ns can be obtained with comparators at low output voltage. Level shifting may be required to attain the necessary gate voltage, and buffering may be necessary to provide the required value of gate current.

Figure 8.8 shows how a complementary emitter follower may be used to buffer the output of operational amplifiers and comparators. The shape of the output waveform may be modified by altering the feedback component as shown in Figure 8.9. Feedback via a capacitor allows the rate of rise and fall of the output voltage to be regulated, thereby controlling the rate of rise and fall of the MOSFET drain current. If a diode is connected in parallel with the capacitor, the output voltage will change rapidly in one direction but slowly in the other.

8.10. STORED-ENERGY GATE DRIVE CIRCUITS

The release of energy stored in an inductor may be used to drive the gate of a MOSFET as shown in Figure 8.10. Energy is stored in the inductance while Q1 is on and released into the zener diode and the gate of Q2 when Q1 turns off. The zener diode clamps the gate voltage at a level compatible with the maximum allowable value of the gate–source voltage of Q2.

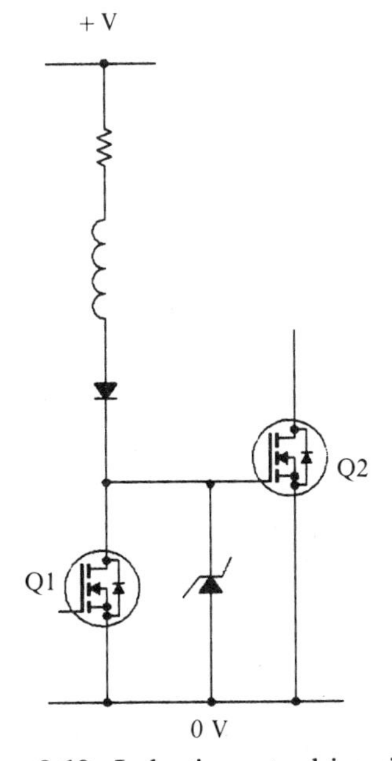

Figure 8.10. Inductive gate drive circuit.

The benefit of such a circuit is that a gate voltage can be generated which is higher than that of the supply rail. This permits full enhancement of a device where it might not otherwise be possible, for example when only a 5-V power source is available. The drawback to such circuits is that the current must be maintained in the zener clamp during the entire ON-period of Q2. This can result in high power loss in the drive circuit.

8.11. DRIVING MOSFETS FROM TTL

The high-state output voltage of TTL is barely sufficient to turn on a MOSFET of the industry-standard type with a threshold voltage between 2 and 4 V. As Figure 8.11 shows, in the basic TTL output stage, output load current from the positive rail has to pass through two forward-biased junctions, so that the maximum voltage obtainable from a TTL output in normal operation is 3.6 to 3.8 V. The minimum guaranteed high-level output when sourcing 0.4 mA is 2.4 V (2.7 V for 74LS and 74S TTL), inadequate for driving standard power MOSFETS.

The use of a pullup resistor, as illustrated in Figure 8.12, permits the gate voltage to rise to 5 V. The MOSFET will turn off more quickly than it turns on, since at turn-off the gate is pulled down by a transistor, whereas on turn-on the gate is charged at a rate determined by the pullup resistor. Even with a pullup

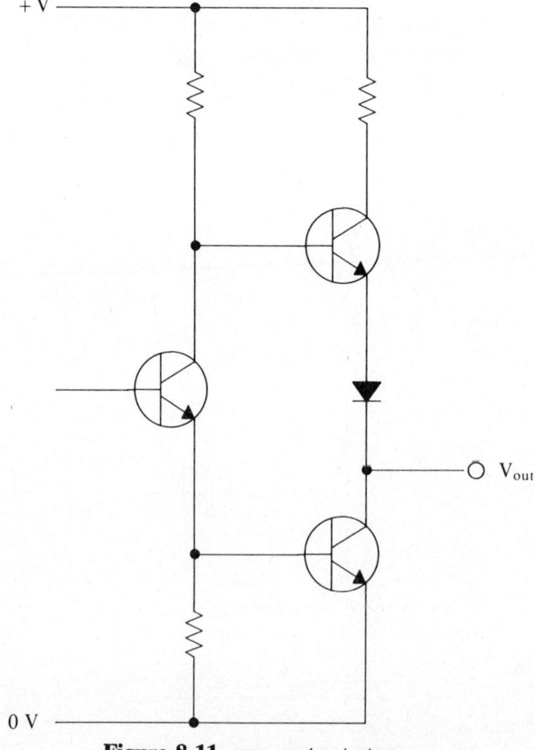

Figure 8.11. TTL output stage.

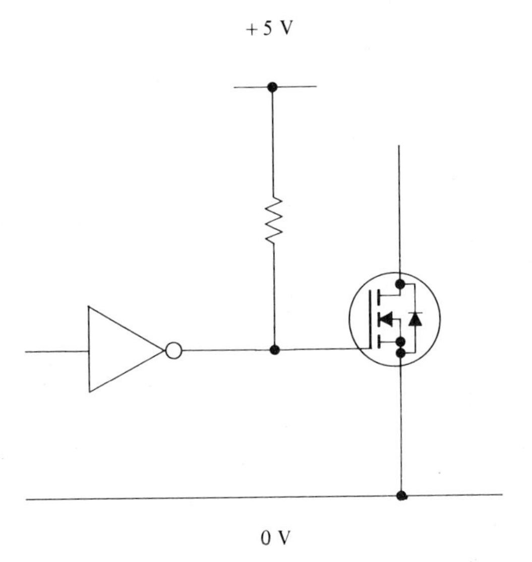

Figure 8.12. TTL gate drive with pullup resistor.

resistor the voltage obtainable from a TTL output stage is insufficient to enable standard MOSFETS to conduct rated drain current, as Figure 8.13 shows.

A higher gate drive voltage can be obtained by the use of open-collector TTL as shown in Figure 8.14. Switching of the MOSFET is asymmetric, since pullup is passive and pulldown is active. Active pullup can be achieved by use of a double

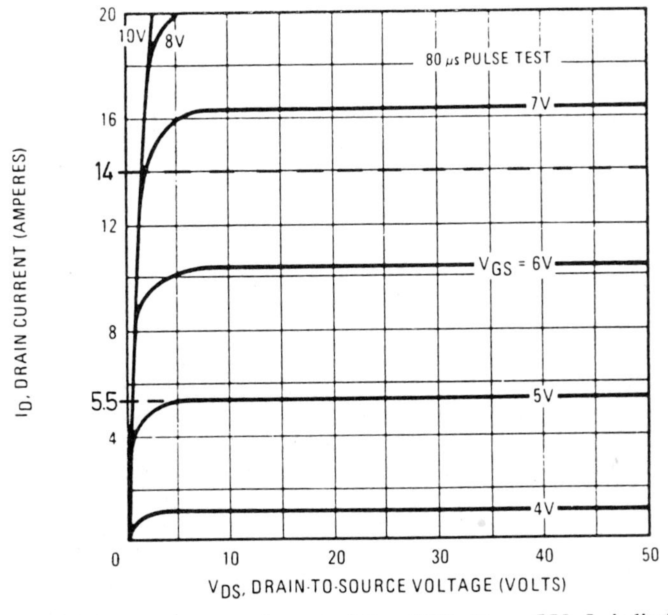

Figure 8.13. Typical MOSFET output characteristics. With $V_{GS} = 5$ V, I_D is limited to 5.5 A. The rated average value of I_D is 14 A.

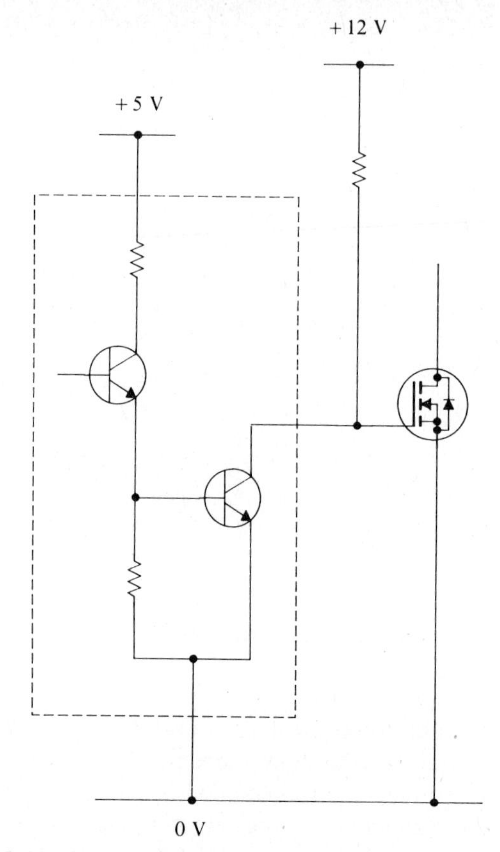

Figure 8.14. Open-collector TTL output with pullup resistor.

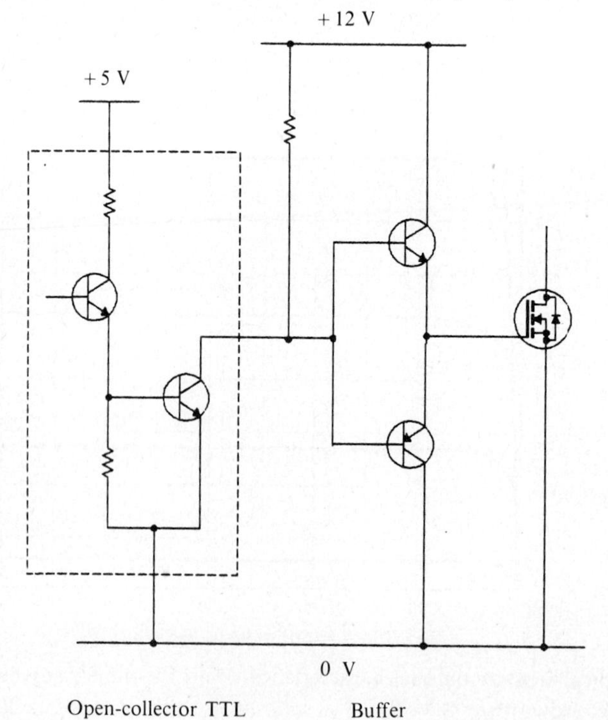

Figure 8.15. Open-collector TTL output with a buffer stage to provide active pullup.

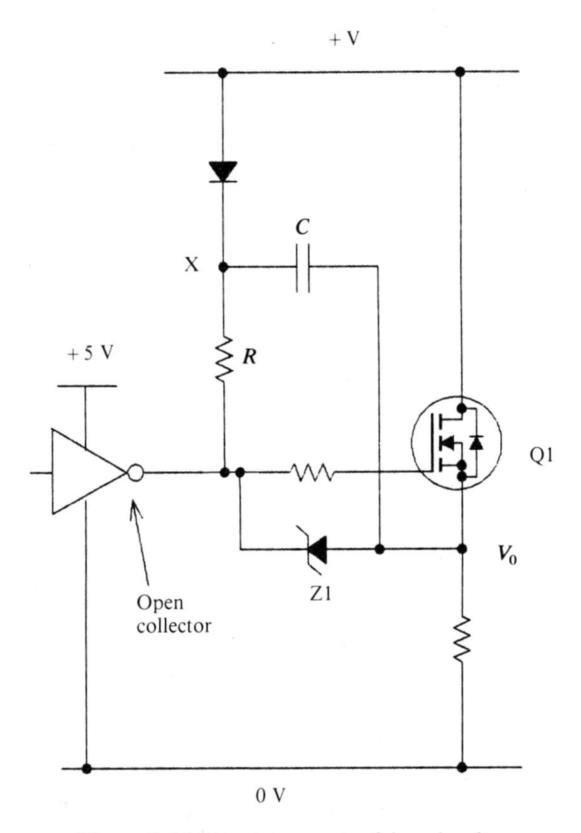

Figure 8.16. Bootstrap gate drive circuit.

emitter-follower bipolar transistor buffer, as shown in Figure 8.15, or by completing the TTL output stage using external components.

The bootstrap circuit shown in Figure 8.16 offers a method of driving a MOSFET from conventional TTL but with a gate voltage sufficient to ensure saturation at high levels of drain current. The load must be connected to the source terminal, as for instance in the case of the upper device in a bridge circuit. When Q1 is off, C is charged to approximately the supply voltage. When Q1 starts to turn on, V_0 rises. The voltage at node X also rises. Additional current is therefore supplied to the gate via R. The zener diode Z1 prevents the gate voltage rising to a level which could damage the gate oxide. A disadvantage of the bootstrap circuit is that the maximum pulse width is limited by the discharge time constant of R and C.

8.12. DRIVING MOSFETS FROM CMOS

Standard CMOS has a longer switching time than TTL, while high-speed CMOS has switching times of the same order as TTL. The output characteristics of CMOS are approximately the same whether sourcing or sinking current. The output impedance is about 500 Ω for supply voltages below 8 V, and about 1 kΩ for

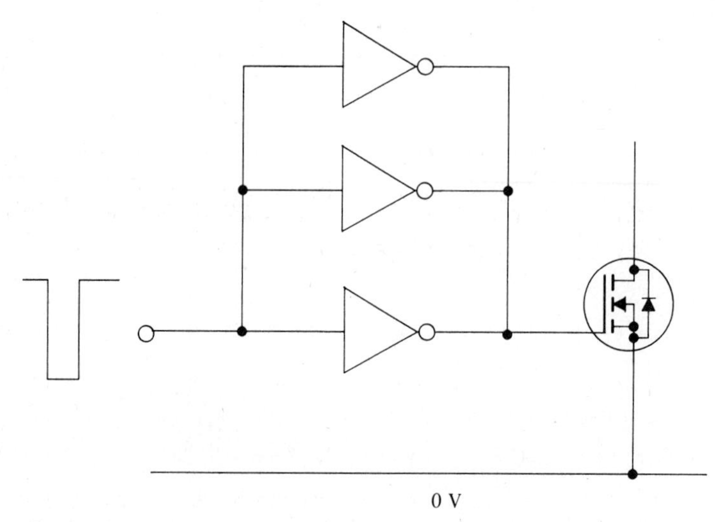

Figure 8.17. CMOS outputs paralleled to increase drive capability.

supply voltages over 8 V. The main advantage that CMOS has over TTL for driving MOSFETS is that CMOS can operate with a supply voltage of up to 15 V and can therefore supply sufficient voltage to ensure full turn-on of conventional MOSFETS at rated current. With a CMOS gate drive the MOSFET does not turn off as quickly as when driven by TTL, since the pulldown impedance of TTL is much lower than that of CMOS. The turn-on performance of the MOSFET when driven by CMOS operating from a 10-V supply is similar to that achieved with a TTL high-voltage open-collector buffer and a 680-Ω pullup resistor.

A lower output impedance can be achieved by connecting any number of CMOS gates in parallel (Figure 8.17). To achieve higher switching speeds than is possible with CMOS alone, the output of the CMOS may be buffered using a complementary emitter follower.

8.13. TRANSFORMER-ISOLATED GATE DRIVERS

Isolation between the power circuit and the control circuit may be achieved by the use of transformer coupling as shown in Figure 8.18. The output impedance of the transformer secondary will depend on the reflected impedance of the primary driver circuit and on the impedance of the transformer. Therefore for fast switching the transformer must have low leakage inductance and the primary driver should have low output impedance. A series capacitor is included to block the dc component in the primary-driver output waveform. The use of transformer coupling of the gate signal permits the optimum gate voltage to be obtained from a convenient supply voltage and may also facilitate impedance matching between the drive circuit and the MOSFET.

Due to leakage inductance, the impedance of the secondary winding of the transformer may not be low enough to prevent the gate voltage from rising above the threshold voltage under conditions of high drain voltage dv/dt. The

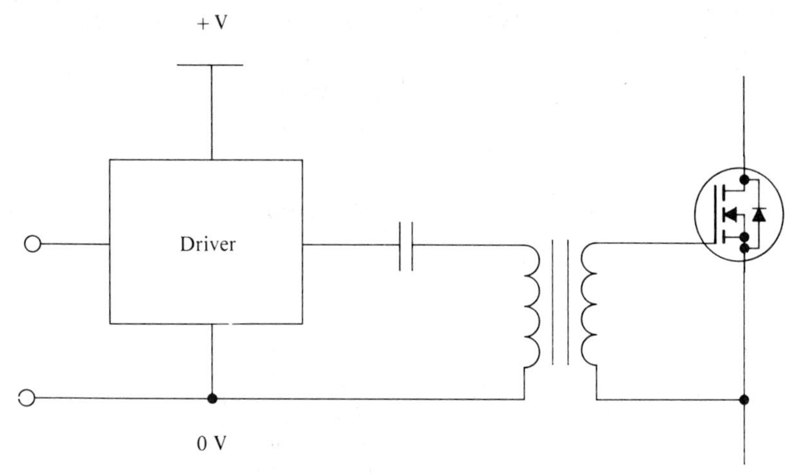

Figure 8.18. Transformer-isolated gate drive.

impedance presented to the gate during the time when the MOSFET is required to be off can be reduced by the use of a bipolar transistor as shown in Figure 8.19. Q1 provides a low impedance shunting of gate and source even when the transformer secondary voltage is zero.

A major restriction on the usefulness of a transformer-coupled gate drive is that the duty cycle over which it can operate is limited, as Figure 8.20 illustrates. The secondary voltage waveform cannot contain a dc component. Therefore the time integral of the negative part of the waveform must equal the time integral of the positive part of the waveform. If the duty cycle departs significantly from 50%, the excursions of the gate voltage will be much greater in one sense than in the other. Since the gate oxide breakdown voltage is the same for positive and negative gate voltage, excessive excursions of the transformer secondary voltage in either direction are unacceptable.

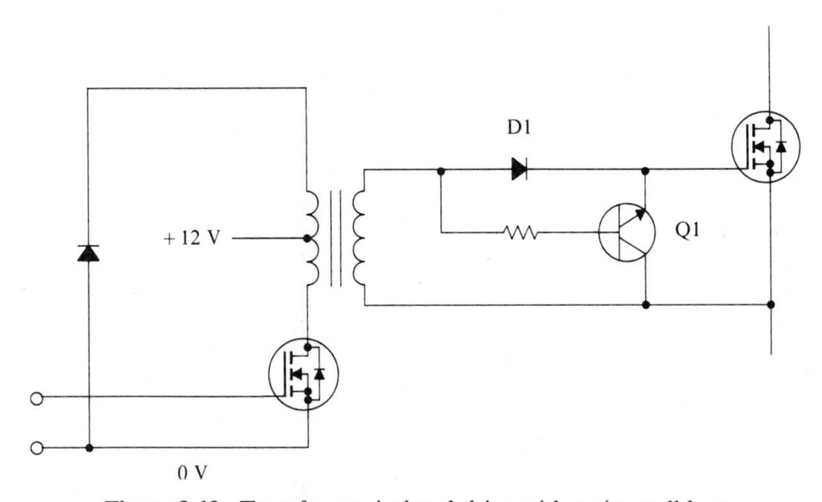

Figure 8.19. Transformer-isolated drive with active pulldown.

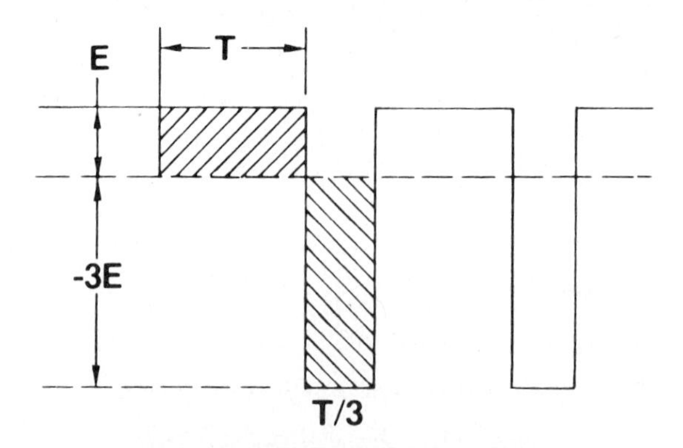

Figure 8.20. Constant volt–seconds characteristics of a transformer.

The circuit shown in Figure 8.21 achieves transformer isolation while permitting the duty cycle to vary between 1 and 99%. The circuit operates on the principle that since the leakage current between the gate and source of a MOSFET is very small, the charge on a MOSFET gate can be retained for some seconds or even minutes if the gate is suitably isolated. As the waveforms in Figure 8.22 show, the MOSFET is turned on when a positive-going pulse is applied to the transformer primary. The pulse appears across the secondary winding, and the gate of Q2 is charged positively by current flowing through the body–drain diode of Q1. At the end of the pulse, discharge of the gate is prevented by the blocking action of the MOSFET body-drain diode. Turn-off of Q2 is produced when a negative-going pulse is applied to the primary of the transformer. Appearing across the secondary winding, this pulse makes the gate of Q1 positive with respect to the source. Q1 turns on, permitting the gate of Q2 to acquire a negative charge, thereby turning off Q2. The gate of Q2 loses most of this charge as the transformer secondary voltage falls to zero at the end of the turn-off pulse,

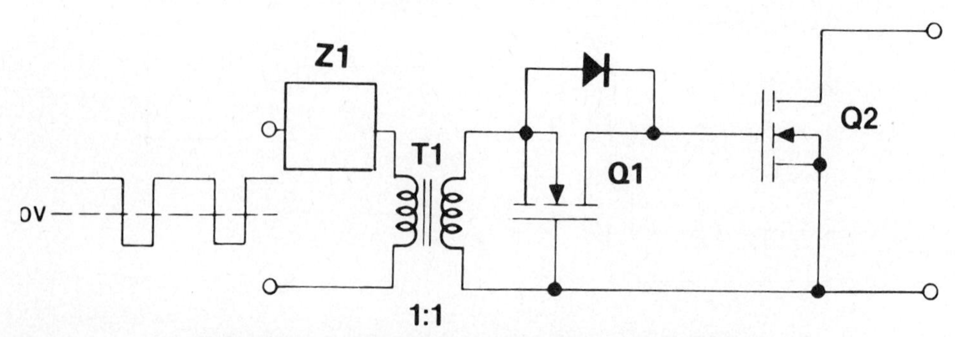

Figure 8.21. Wide-duty-cycle transformer drive circuit.

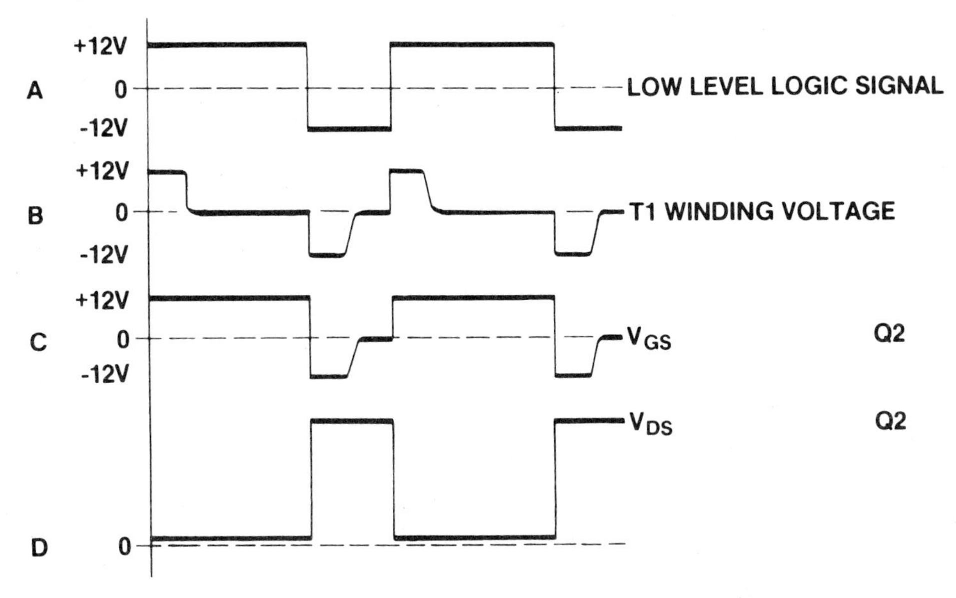

Figure 8.22. Transformer drive waveforms.

but Q2 nevertheless remains off until a positive pulse is again applied to the transformer. If the MOSFET is required to be on for a period long enough for there to be appreciable leakage of the gate charge, it may be necessary to increase the gate discharge time constant by connecting a capacitor between the gate and the source.

The circuit may be driven by a square wave rather than a pulse waveform provided some provision is made to limit the transformer primary current should the transformer core saturate. As Figure 8.22 shows, pulses are generated in the secondary winding whenever the primary voltage changes polarity. Thus Q2 is turned on by the secondary pulse generated by the positive-going transition of the primary waveform and turned off by the negative-going transition.

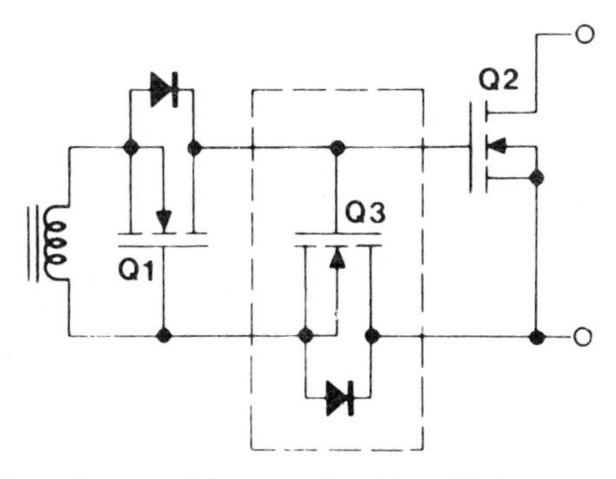

Figure 8.23. Transformer drive circuit with additional noise immunity.

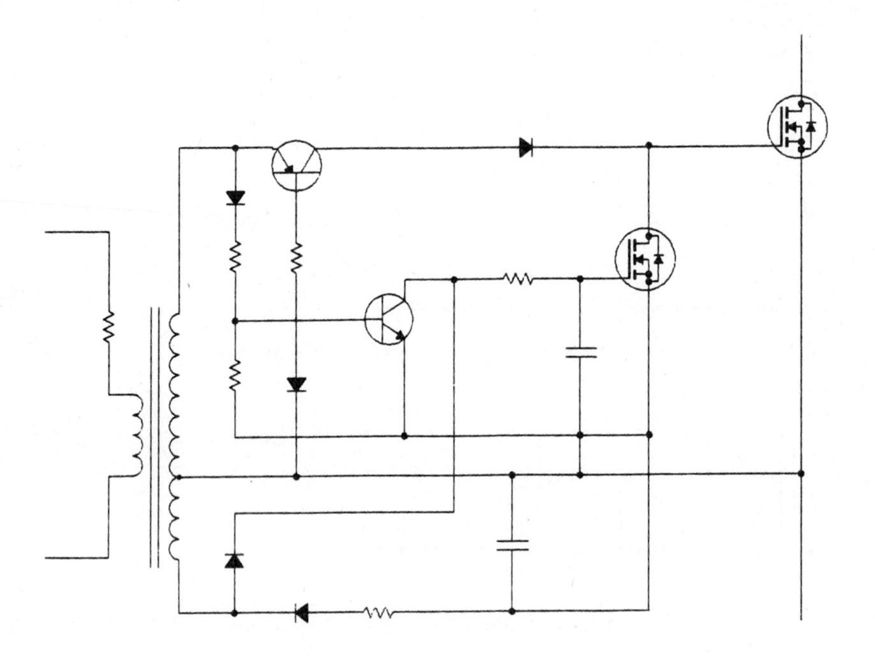

Figure 8.24. Transformer drive circuit with negative bias in the OFF state.

The noise immunity of the circuit shown in Figure 8.21 is approximately equal to the threshold voltage of the MOSFET. Greater noise immunity can be obtained from the circuit shown in Figure 8.23. In this circuit, the negative charge which the gate acquires when Q2 is turned off is retained due to the blocking action of Q3. The gate voltage of Q2 remains at some negative value while Q2 is off. This increases the margin between the OFF-state gate voltage and the threshold voltage of the MOSFET, thereby giving the circuit a noise immunity greater than the threshold voltage of the MOSFET.

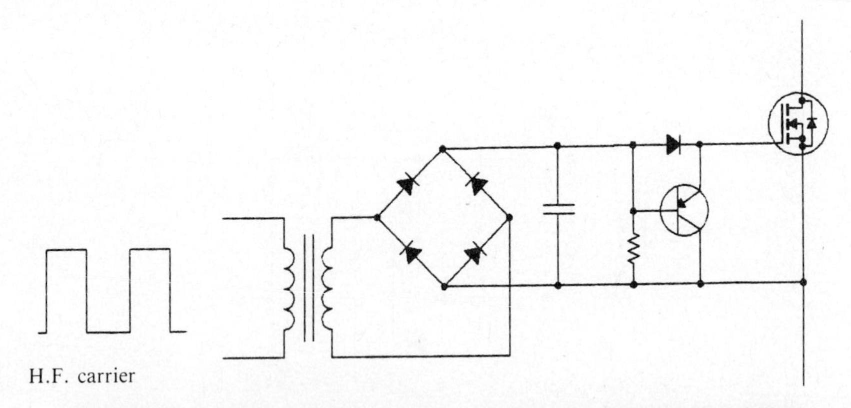

Figure 8.25. Transformer drive circuit with continuous-output capability.

A negative gate bias can also be provided by a transformer-isolated drive circuit in the manner shown in Figure 8.24 [1]. The transformer secondary winding is extended so that a negative voltage source is created to charge the reservoir during the turn-on period. The same secondary-winding extension provides the signal which controls the application of this negative voltage to the power-MOSFET gate. There are a number of possible circuit variations based on these principles [2].

Unlimited ON time can be obtained with the circuit shown in Figure 8.25 [3]. A 1-MHz carrier signal is rectified to provide a gate bias that turns on the FET. The

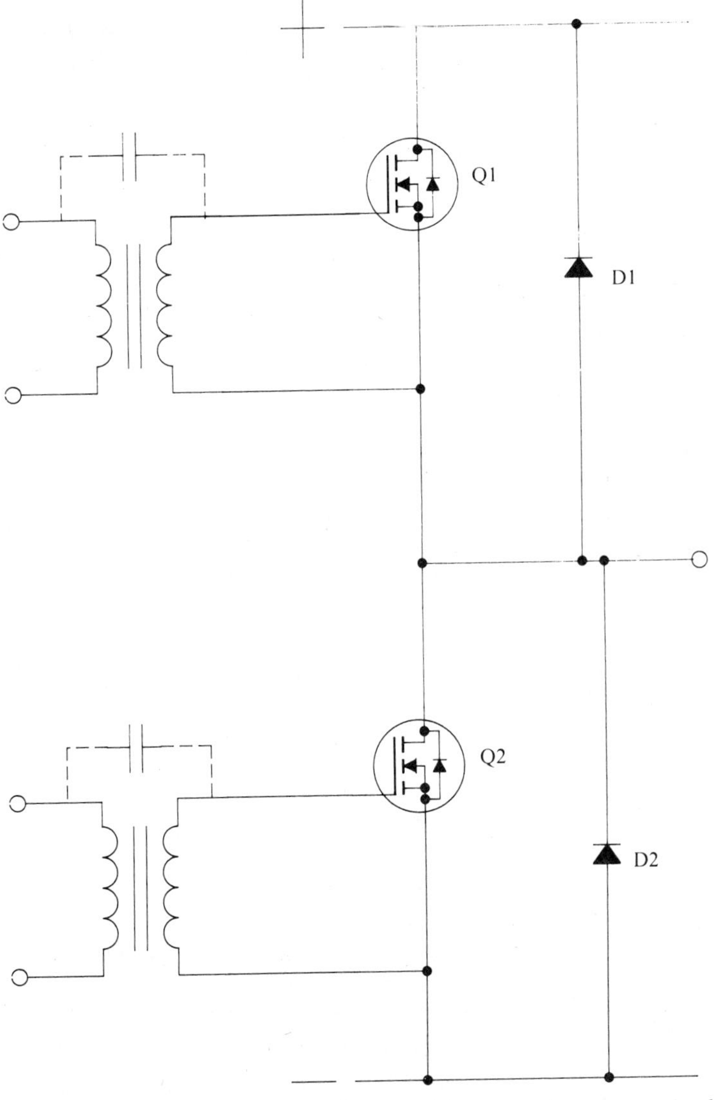

Figure 8.26. Interwinding capacitance in the gate-driver transformers can lead to spurious turn-on.

bipolar transistor ensures rapid discharge of the gate capacitance when the carrier is turned off. The bandwidth of this arrangement is limited by the use of a carrier signal to convey the switching signal.

Parasitic capacitance between the windings of a gate-driver transformer can cause the gate voltage to rise if there is a rapid rise of the potential of the primary winding with respect to the secondary winding. Such a situation commonly arises in the inverter circuit shown in Figure 8.26. If the free-wheeling diode D1 is conducting when Q2 turns on, the source voltage of Q1 will fall rapidly from $+V$ to $-V$ as Q2 takes over the load current from D1. Spurious turn-on of Q1 by currents flowing through the interwinding capacitance of the transformer can be prevented by the use of an electrostatic screen between the windings.

As advancing technology pushes operating frequencies higher and power-MOSFET die sizes increase, greater demands are placed on the driver transformer. The rise time must be fast to minimize crossover losses during switching, but at the same time voltage overshoot must be limited and ringing well damped. Also, the power requirements of the drive circuit must be kept within reasonable limits. Choosing the correct core material for the transformer becomes more critical as performance demands increase [4].

8.14. OPTICALLY ISOLATED GATE DRIVERS

The gate driver circuit may also be isolated from the power circuit by the use of an optical isolator. Figure 8.27 shows an example of an optically coupled drive circuit. Some optical-isolators incorporate a Schmitt trigger circuit to ensure that the output of the isolator changes state rapidly and cleanly, thereby allowing efficient operation of the power MOSFET [5]. These isolators usually operate from a 5-V supply. Level shifting as well as buffering of the output may therefore be required.

An optically coupled driver does not suffer from the duty-cycle limitations that are found in transformer-coupled drivers. The main disadvantage of an optically coupled gate drive is that it requires an isolated power supply. Figure 8.28 shows how the isolated power supply can be dispensed with by obtaining power for the gate drive from the power circuit. In applications in which the drain voltage only goes high for a short period each cycle or in which the operating frequency is low, the circuit shown in Figure 8.29 may be used.

The energy required to charge the gate of a MOSFET may be conveyed to the gate optically. If enough photovoltaic diodes are connected in series, sufficient voltage can be generated to turn on a MOSFET as shown in Figure 8.30. The gate-to-source leakage current in a MOSFET is so small that it can easily be supplied by the photovoltaic pile. The switching speed depends on the current that the pile can supply, and this depends on the conversion efficiency of the pile and the amount of light falling on it. Power MOSFETS may be packaged with a photovoltaic gate drive and a LED light source to provide a completely isolated gate drive (see Section 9.10).

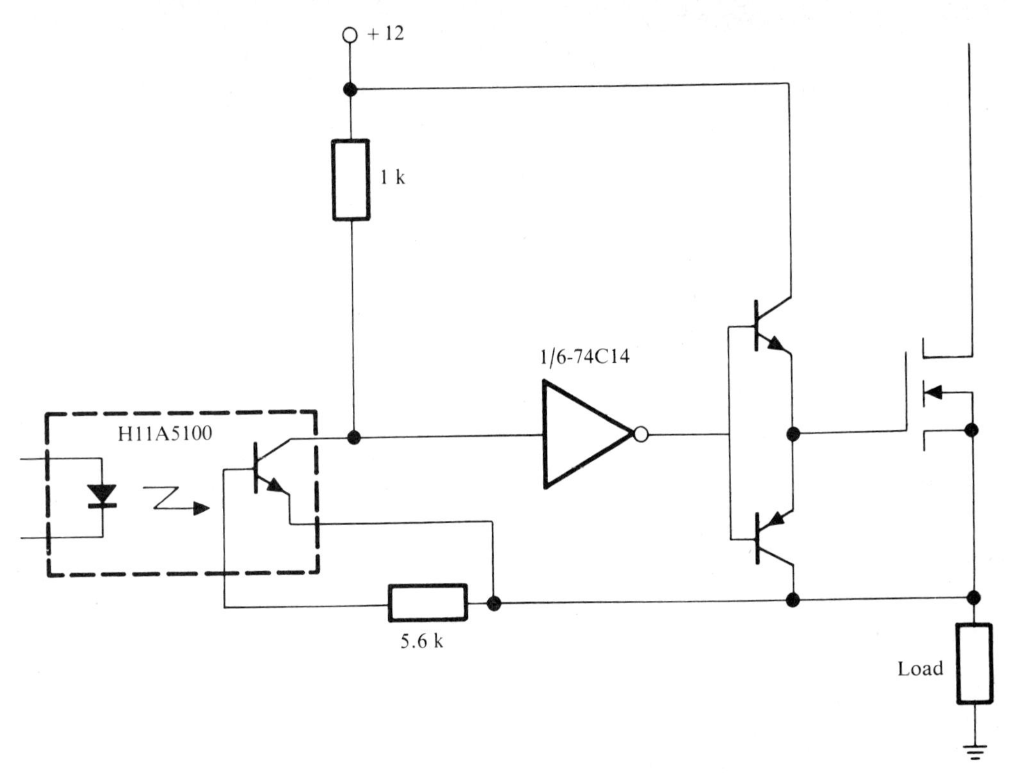

Figure 8.27. Optically isolated gate drive circuit.

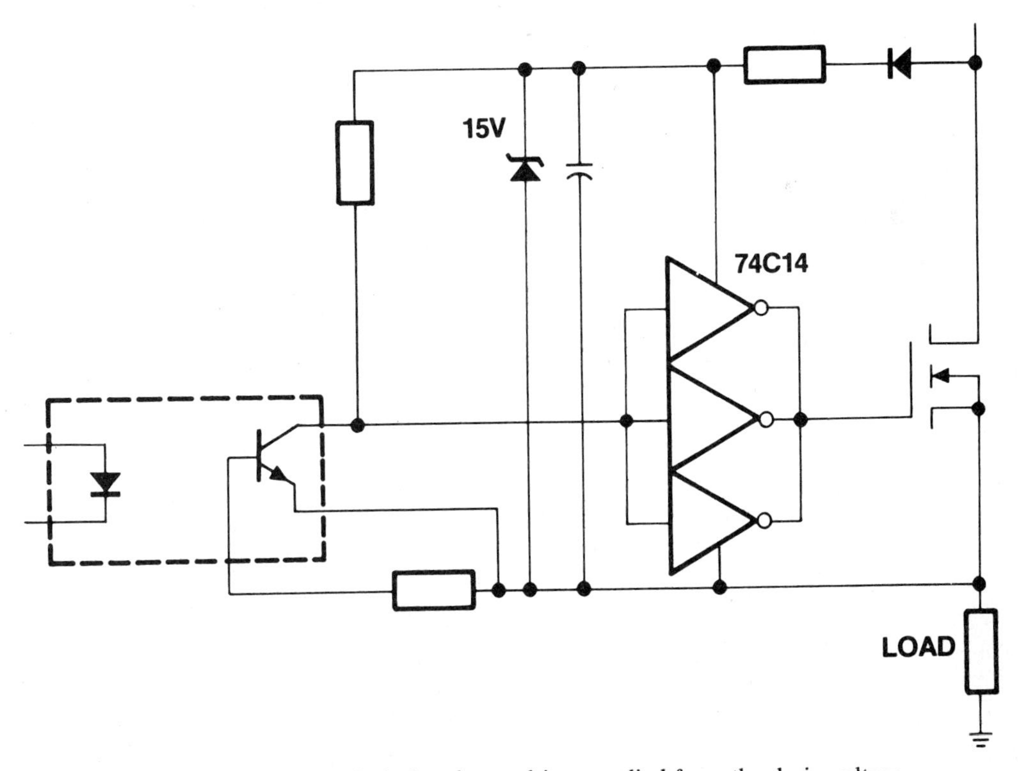

Figure 8.28. Optically isolated gate drive supplied from the drain voltage.

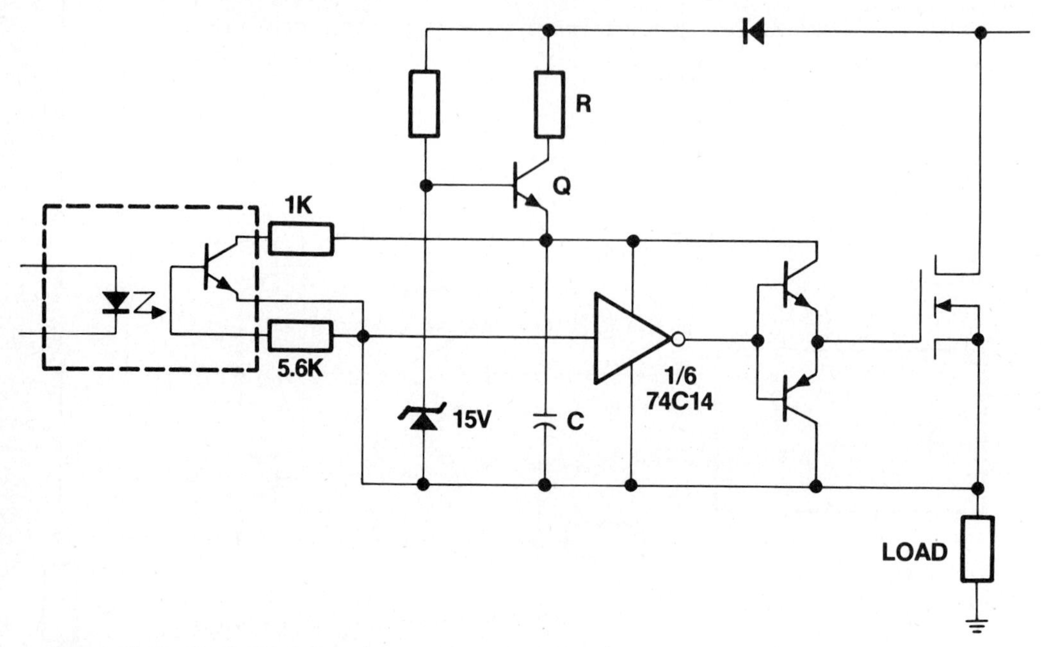

Figure 8.29. Optically isolated gate drive supplied from the drain voltage and capable of operating with small duty cycle at low frequency.

8.15. THE HIGH-SIDE SWITCH

In some switching applications it is preferable to use an n-channel MOSFET located between the positive rail and the load. When the MOSFET is on, its source will be almost at positive-rail potential, necessitating a gate-driver voltage that is greater than that of the positive rail. One means of providing this is a diode pump.

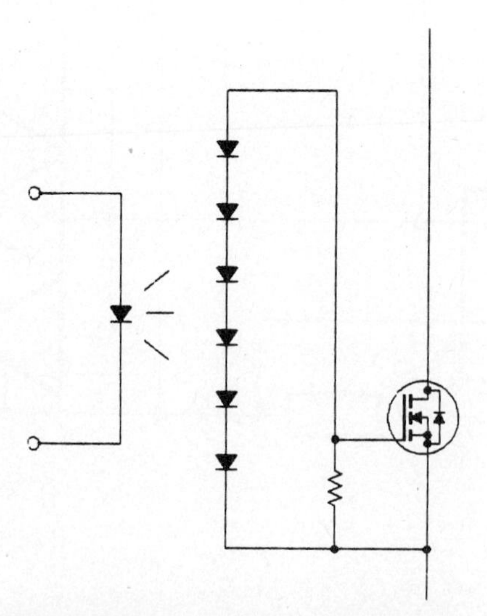

Figure 8.30. Photovoltaic gate drive.

High-side switches are used predominantly in the automotive industry, and a more extensive discussion of the subject will be found in Section 10.2.

8.16. H-BRIDGE DRIVER

The driving of the upper MOSFET in an H-bridge inverter circuit requires the provision of a gate drive supply of the order of 10 to 15 V above that of the dc supply to the bridge. The gate drive signal must also be conveyed from the control electronics, which are usually at ground potential, to the gate drive circuit. These requirements may be fulfilled by an isolated power supply and an optical isolator. An arrangement which achieves the same end but which is generally more economical is shown in Figure 8.31. Power for the gate drive is derived from the reservoir capacitor C_r. This capacitor is charged from the 15 V power supply via D1 whenever Q4 is on. Q1 and Q2 provide a low-impedance drive for Q3. Turn-off of Q3 is achieved by turning on current source Q5. Other configurations are possible, involving various combinations of bootstrap power sources, optical isolators, and level shifters [6]. Due to the need for level shifting and floating power supplies, the cost of driving an H-bridge is high. However, the advent of power integrated circuits which can perform this function [7] promises a considerable reduction in the cost of H-bridges.

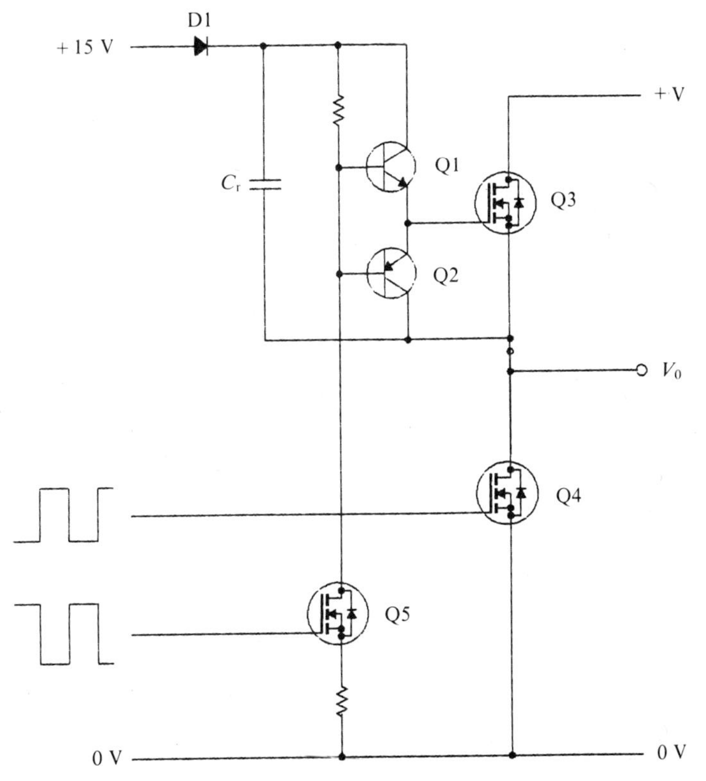

Figure 8.31. H-bridge gate drive circuit with diode-pumped power supply for upper gate drive.

REFERENCES

1. B. Taylor, "High frequency, high-power, off-line switching converters." *IEEE, Proc. Int. Telecomm. Eng. Conf. INTELEC* **86,** 623–626 (1986).
2. M. A. Preston, J. D. Edwards, and G. Williams, "A novel PWM inverter using transformer-isolated power MOSFETS." *Eur. Power Electron. Conf.* pp. 105–110 (1987).
3. N. Rasmussen, "MOSFET gate driver has unlimited on time." *Electron. Des. News* Nov. 28, p. 284 (1985).
4. V. J. Spataro, "The transformer isolated gate drive—analysis and design." *Proc. Power Convers. Int. Conf.* October, pp. 317–333 (1986).
5. W. H. Sahm, "Common drain power MOSFET gate drive solutions." *Proc Power Convers. Int. Conf.* October, pp. 203–209 (1986).
6. W. Schultz, "Drive techniques for high-side N-channel MOSFETS." *Power Convers. Intell. Motion (PCIM)* June, pp. 34–40 (1987).
7. S. Young, "High-speed., high-voltage IC driver for HEXFET or IGBT bridge circuits." *Int. Rect. Corp. Appl.* **AN-978,** (1988).

CHAPTER 9

The Power MOSFET as a Switch

9.1. INTRODUCTION

Like any other semiconductor switch, the power MOSFET is not ideal. It blocks voltage in one direction only. It passes a low but finite value of leakage current when off and has a nonzero value of resistance when on. The OFF-state blocking voltage and the ON-state current have limits, set respectively by the drain–source breakdown voltage and by the onset of active-region operation (equivalent to going out of saturation in the case of the bipolar transistor). The time taken to switch from one state to another is not zero.

Nevertheless, the power MOSFET is the device of choice for a wide range of switch applications. This chapter describes a number of such applications. Emphasis is laid on those applications in which the switch may be on or off for indeterminate periods. The role of the MOSFET switch in pulse-width-modulated and high-frequency power applications is discussed in other chapters.

9.2. BASIC SWITCHING CONFIGURATIONS

Figure 9.1 shows an n-channel power MOSFET in a simple switching circuit. An n-channel device is used where possible, since n-channel MOSFETS are generally cheaper than their p-channel equivalents. This is due to the lower mobility of holes than that of electrons, which requires the use of more silicon area in a p-channel device to achieve the same R_{DS}(on). Hence, n-channel MOSFETS tend to be more readily available and in a greater selection than p-channel ones.

The MOSFET is placed between the load and the power-source ground where possible, since the drive voltage may then be ground-referenced. Certain applications require that the switch be placed between the load and the positive rail—a technique commonly referred to as high-side switching. This is particularly prevalent in automotive applications, where the chassis of the vehicle is the negative ground. The load is connected to the positive supply rail by an n-channel MOSFET with its gate pulled high or by a p-channel MOSFET with its gate pulled low. Economics dictate which of these configurations is adopted (see Section 10.2).

Power MOSFETS incorporate an integral drain–source diode. Therefore they are capable of blocking voltage in one direction only. The ability of the power MOSFET to conduct in the reverse direction is put to use in ac switching applications. Two power MOSFETS can be configured to act as a single bidirectional ac switch (see Section 9.10).

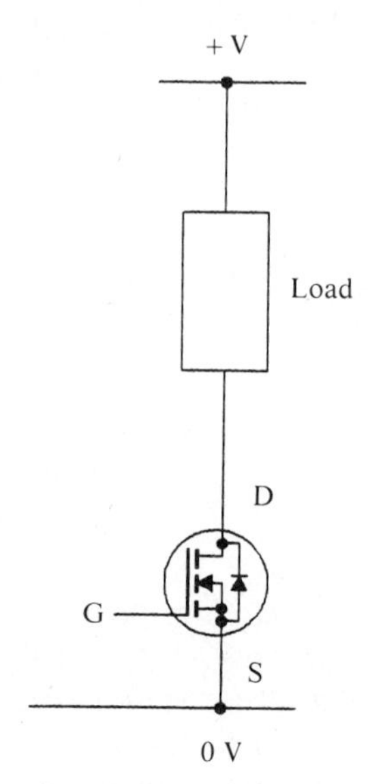

Figure 9.1. Basic switch configuration.

9.3. INDUCTIVE LOAD SWITCHING

During turn-off the collapsing magnetic field in an inductive load manifests itself as a voltage induced across the inductance, with a polarity such as to oppose the change in current. A free-wheeling diode is used to clamp the drain voltage and to carry the inductor current after turn-off (Figure 9.2*a*). The diode must become forward biased before the current in the MOSFET can start to reduce. Thus, when controlling a clamped inductive load, the device will experience simultaneously high values of current and voltage. This situation can lead to secondary breakdown in bipolar transistors, since the reverse-bias safe operating area does not usually include simultaneous application of rated voltage and rated current. Power MOSFETS do not suffer from this phenomenon and therefore can drive clamped inductive loads without the use of load-line shaping circuits.

On turn-off, overvoltage is still possible with a clamped inductive load if there is unclamped stray inductance in the circuit. This can be minimized through correct circuit layout, but is hard to avoid in some applications, such as the switching of a remote load, where significant lead inductances may be present. A snubber circuit may be placed across the MOSFET to limit peak voltage stress, as shown in Figure 9.2*a*. The unclamped inductance forms a damped resonant circuit with the snubber capacitance, resulting in a reduced voltage peak. However, the snubber dissipation is large and transition times are increased. The transient waveforms produced by power-MOSFET switching are studied in greater detail in Chapter 4. A zener (or avalanche) diode clamp ensures a better-defined maximum voltage (Figure 9.2*b*). If the MOSFET ratings permit (see Appendix 7, Notes 11 and 12), the drain voltage may be limited by avalanche breakdown of the MOSFET. The energy stored in the inductor is then dissipated in the MOSFET. Another means by which the energy stored in the inductor can be dissipated in the power MOSFET is

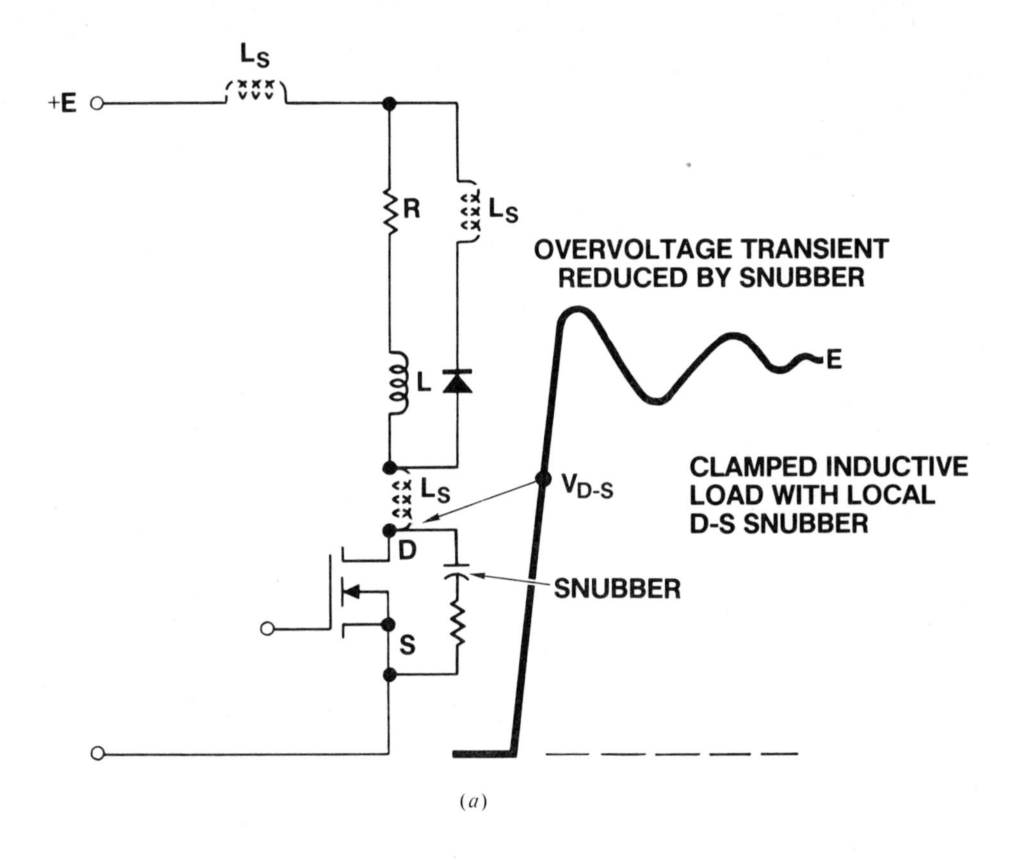

(a)

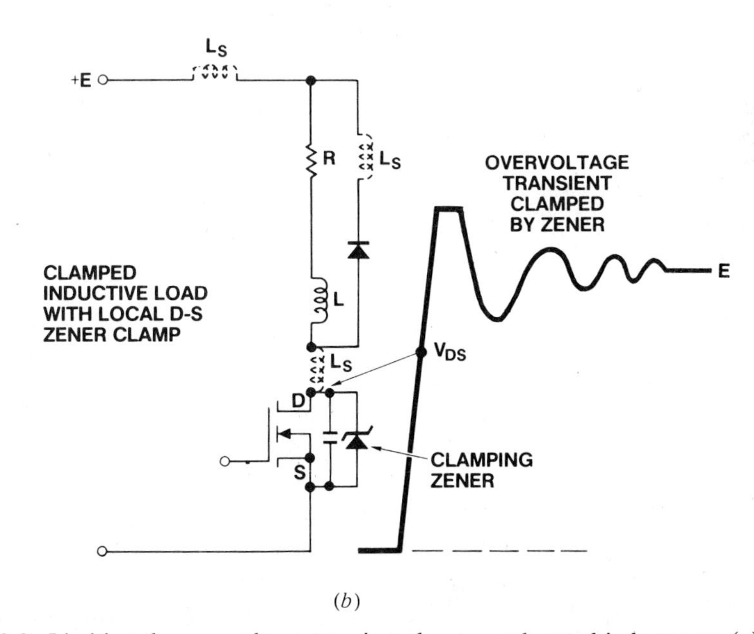

(b)

Figure 9.2. Limiting the overvoltage transient due to unclamped inductance: (*a*) Using a snubber; (*b*) using a zener (avalanche) diode.

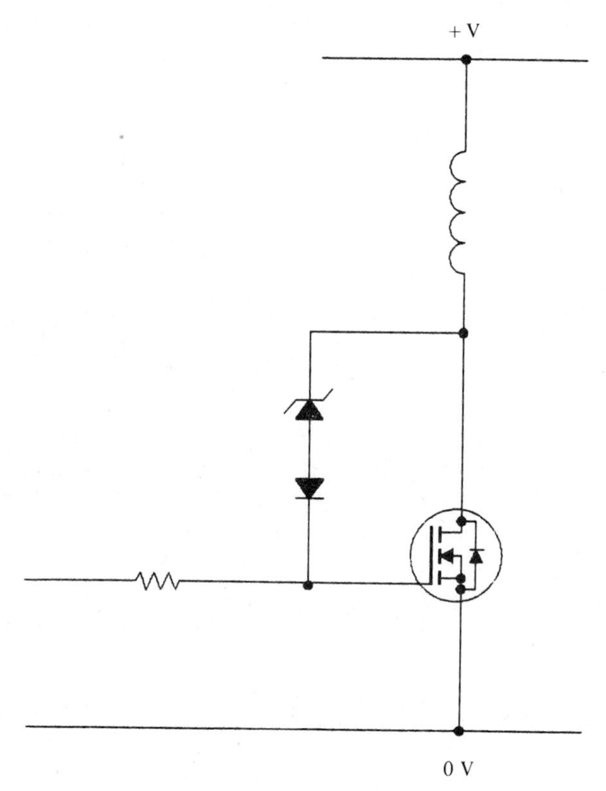

Figure 9.3. Active clamping of the drain–source voltage.

by use of the active clamp circuit shown in Figure 9.3. The drain voltage is clamped by turn-on of the power MOSFET when the breakdown voltage of the zener diode is reached. In this case the power MOSFET is not required to have any avalanche capability. A power integrated circuit has been proposed in which the control circuitry incorporated in the power device adjusts the clamping voltage and the current carried by the MOSFET, the object being the minimization of losses and a reduction in switching times [1].

9.4. HIGH INRUSH-CURRENT LOADS

Many switched loads draw high peak currents on turn-on. For example, a stationary dc motor has zero back emf, and therefore the starting current is limited only by the resistance of the windings. Incandescent lamps have a low resistance when the filament is cold, so that the current drawn at turn-on can be several times that drawn in steady operation (see Section 9.8). Power MOSFETS generally have a surge-current capability that is typically four times that of the average-I_D rating. The surge-current rating is determined principally by reliability considerations, in particular, stress on the wire bonds and the surface met-allization during high-current conditions. Since there is no second breakdown associated with power MOSFETS, it is usually permissible to switch an unsnubbered power MOSFET at the surge current rating with full rated voltage applied, although switching must be rapid and this condition is not specifically guaranteed on many data sheets.

Power MOSFETS have a further advantage over bipolar transistors in that they are voltage-controlled and consume no significant gate power in the conducting state, independent of drain current. In contrast, the current gain of a bipolar transistor falls as the collector current reaches high values, so that the cost of the base drive circuit increases rapidly with peak current capability of the switch.

It is necessary to ensure that the current inrush does not cause excess device dissipation. This is likely to occur if the gate drive voltage is insufficient to maintain a fully inverted channel region when carrying the inrush current, so that the device enters the linear region. If this happens, the device is called upon to support high values of current and voltage simultaneously, and the power dissipation will be high. The current is limited by the MOSFET when it enters the linear region, and this may be used to provide overcurrent protection for the load, but care must be taken that the MOSFET die temperature does not exceed the permitted maximum.

Another option is to sense the device current and turn off the MOSFET, diverting the current to a free-wheeling diode when the load current reaches a preset limit. This adds to the cost of the control circuit but may permit the use of a smaller, less expensive MOSFET.

9.5. LOGIC CONTROL OF LOADS

Since power MOSFETS have relatively simple drive requirements, they are often used as the interface between logic circuits and actuators or indicators. In many applications low switching speeds are acceptable and the current capability required of the gate driver is very low. The MOSFET can therefore be driven directly from the logic circuits in many cases. The surge-current capability of MOSFETS makes them suitable for driving loads that have a high ratio of peak to average current, such as print-hammer actuators.

A disadvantage of the MOSFET is that the standard threshold voltage is typically in the region of 3 V and the output voltage from 5-V logic is insufficient to allow the full current capability of the MOSFET to be used. The issue is essentially one of economics, since by using a larger MOSFET the transconductance can be raised to a level where the available gate drive voltage is sufficient to allow the MOSFET to carry the load current without entering the linear region.

Low-threshold MOSFETS (logic-level MOSFETS) have now become readily available. These typically have thresholds as low as 1.0 to 1.5 V (see Section 19.3). They also have higher transconductance and can carry the rated current of the device with a 5-V gate–source bias. These devices are directly compatible with TTL, CMOS, NMOS, and PMOS logic. However, since the threshold voltage has a negative temperature coefficient, standard threshold devices may have to be used in high-temperature applications to obtain the required noise immunity.

9.6. RELAY DRIVER

MOSFETS may be used to drive the coils that actuate relays, solenoids, and other electromechanical devices. A free-wheeling diode connected across the coil to clamp the coil voltage on turn-off will slow current decay, thus delaying the release of the mechanical element. This may not be acceptable, for instance when the coil controls the brake of an automated electric vehicle that is required to make accurate stops. A method is required for quickly reducing the current in the coil. Figure 9.4 shows a suitable circuit for this application. With the coil energized, the brake shoes are held out from the rotating axle. If power fails, or if the MOSFET is turned off, the brake is applied. A resistor is placed in series with the free-wheeling diode, reducing the decay constant of the inductive current. The required breakdown voltage rating of the MOSFET is increased by an amount equal to the product of the current at turn-off and the resistance value. A zener diode may be used in place of the resistor, allowing better control of the peak MOSFET voltage.

The coil current required to hold in solenoids or contactors is less than that required for pull-in. Initially current builds up in the coil, aiming at a maximum value given by the supply voltage divided by the load resistance. Current detection can be by means of either a series resistor or a current-sensing MOSFET (see Section 19.1). After a delay of 40 ms or so during which the solenoid pulls in, the MOSFET is pulsed at a frequency of 3 to 4 kHz with a duty cycle necessary to maintain the hold-in current. MOSFETS are particularly suitable in this application

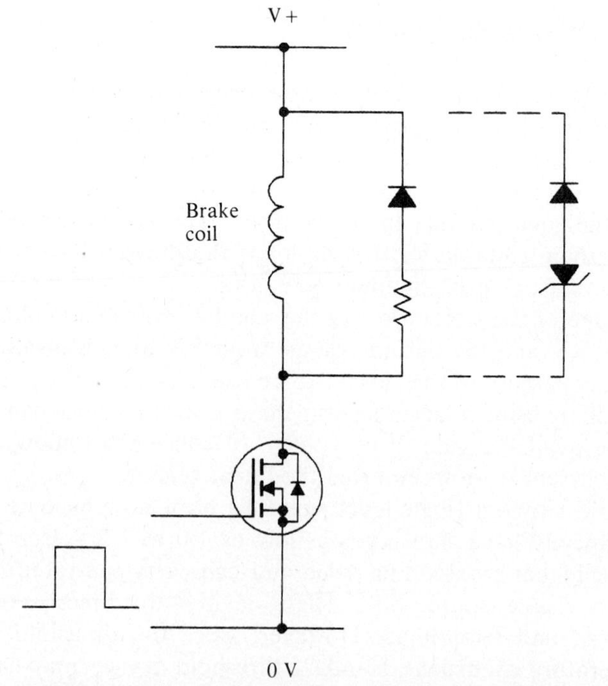

Figure 9.4. Switching an electromechanical brake.

because their peak current capability is greatly in excess of their average current capability and because their square SOA enables them to switch clamped inductive loads without the risk of second breakdown.

9.7. PRINTER HAMMER DRIVE USING MOSFETS

A dot-matrix printer system consists of a number of hammer printing heads, each hammer being operated by a solenoid or electromagnet. To obtain good print quality the coil current is profiled. Generally the coil current is required to rise from zero as rapidly as possible at turn-on, reach a set value, and remain there for a given period. It should then fall to zero as rapidly as possible.

A typical circuit used to obtain the desired current waveform is shown in Figure 9.5. At turn-on Q1 and Q2 are turned on applying the supply voltage across the drive coil. The supply voltage V_S is sufficiently high that the coil resistance is ineffective in limiting the coil current. Thus the current rises rapidly

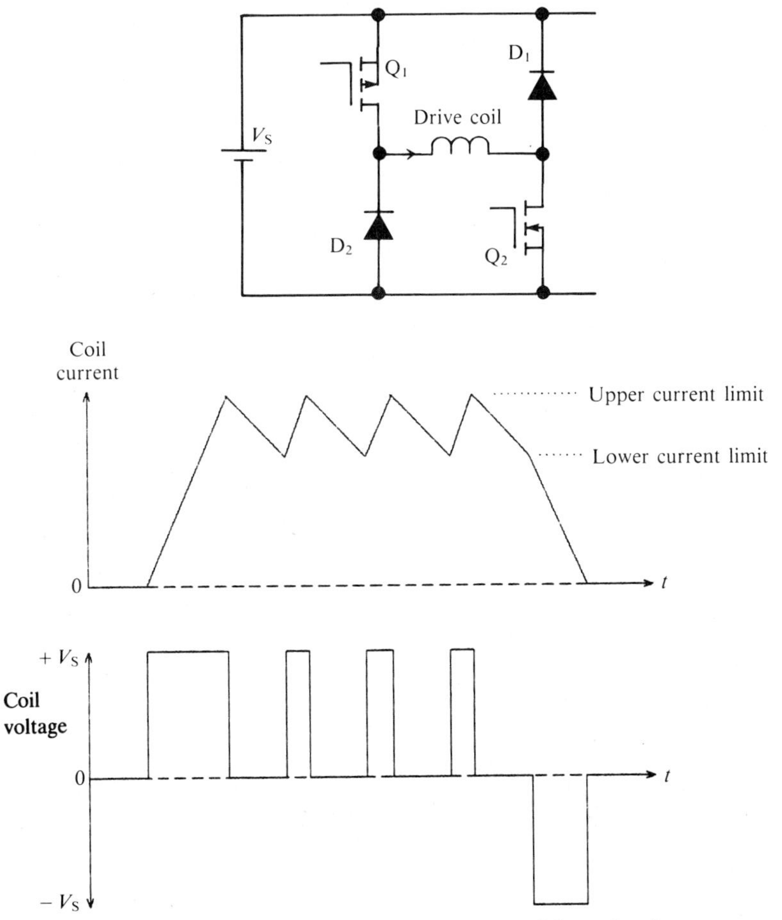

Figure 9.5. Hammer-printer zigzag driver circuit.

to the required value at a rate of rise determined by the coil inductance. Then Q1 is switched off and current freewheels in Q2 and D2. When the coil current has decayed to a predetermined value, Q1 is switched on, increasing the coil current again. This cycle repeats while the coil is required to be energized. When the coil is to be de-energized, Q1 and Q2 are both turned off, forcing the current to flow back to the supply via D1 and D2. Since the supply is applied in reverse across the coil, the current reduces rapidly to zero. Some form of current feedback is necessary to implement the current shaping. A sense resistor can be placed in series with the coil to detect current, or a current-sensing MOSFET can be used. A differential amplifier is used to sense load current, and a comparator, with added hysterisis, determines the maximum and minimum current levels.

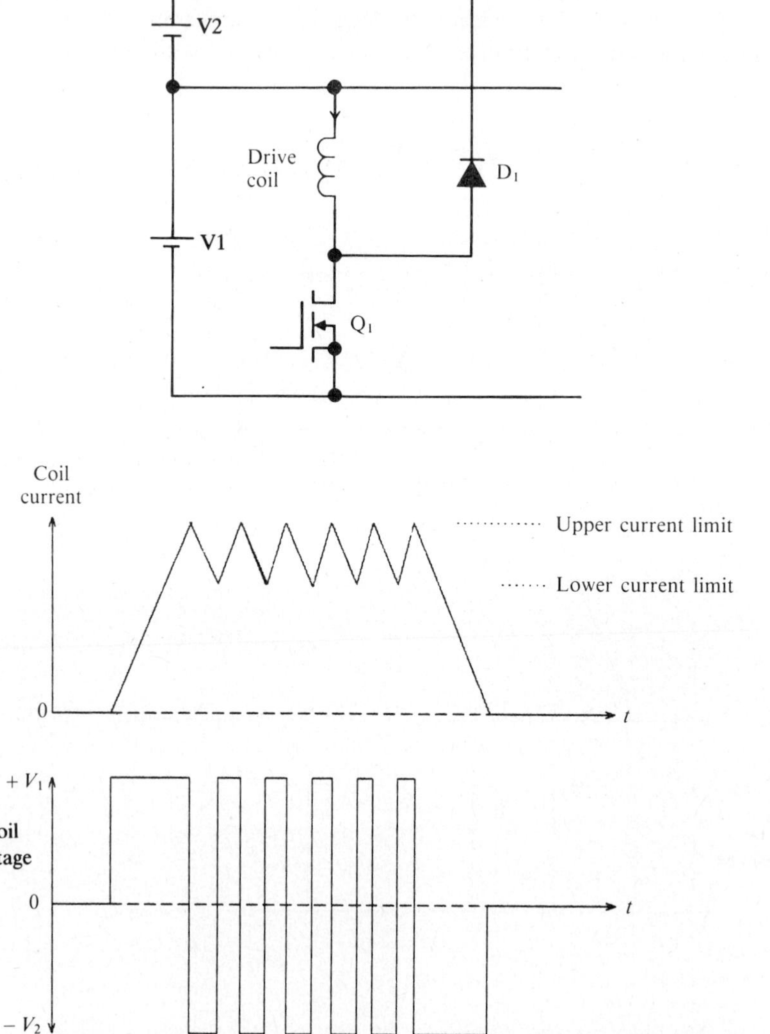

Figure 9.6. Hammer driver circuit with a single MOSFET.

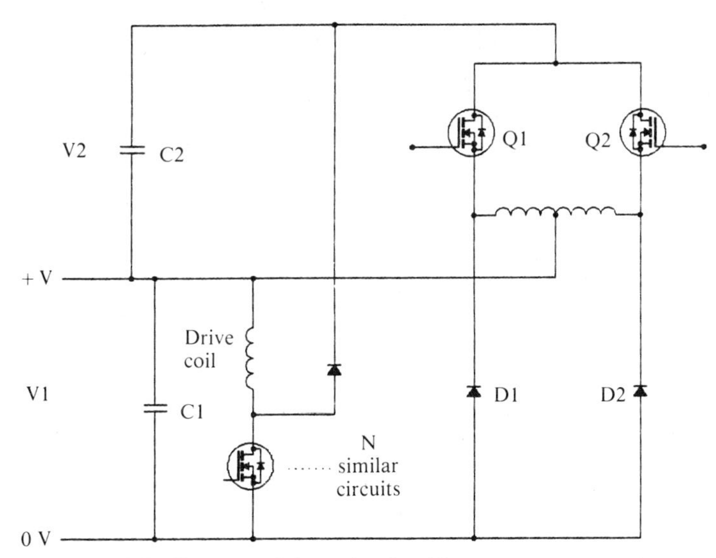

Figure 9.7. Hammer driver circuit with energy recuperation.

Figure 9.6 shows a print-hammer drive circuit that takes advantage of the fast switching capability of the power MOSFET [2]. Initially Q1 is switched on, and it remains on until the current reaches the desired value. Q1 is then switched on and off to control the coil current. While it is off, current flows through D1 into the low-impedance source V2. The amplitude of the current reduces rapidly as a reverse voltage is applied across the coil. When current reaches a set minimum, Q1 is switched on. When the print action is completed, Q1 is switched off and remains off.

This circuit has the advantage of only requiring one transistor and diode, as opposed to the pair of transistors and diodes required by the circuit shown in Figure 9.5. Better efficiency, higher reliability, and lower system cost are therefore possible. The switching frequency during the holding period will be much higher for a given value of current ripple than in the previous circuit, but due to the high switching speed and low switching losses associated with power MOSFETS, this is not a serious disadvantage.

This circuit requires a second power supply, V2, that is capable of sinking current. This supply is provided in the circuit shown in Figure 9.7 by the capacitor C2. An *energy recuperator* circuit is used to return energy from V2 back to V1. Q1 and Q2 are alternately switched on and off at about 40 kHz, with a duty cycle of 0.5. If V2 is less than V1, no current flows from C2 to V1. If V2 exceeds V1, due to energy being received from the print coils, the diodes D1 and D2 will conduct, returning energy to V1 and clamping the voltage V2. Only one such circuit is required for all the print-coil circuits.

9.8. LAMP APPLICATIONS

The main difficulty in switching an incandescent-lamp load is the inrush current drawn by this type of load. When cold, the resistance of an incandescent filament

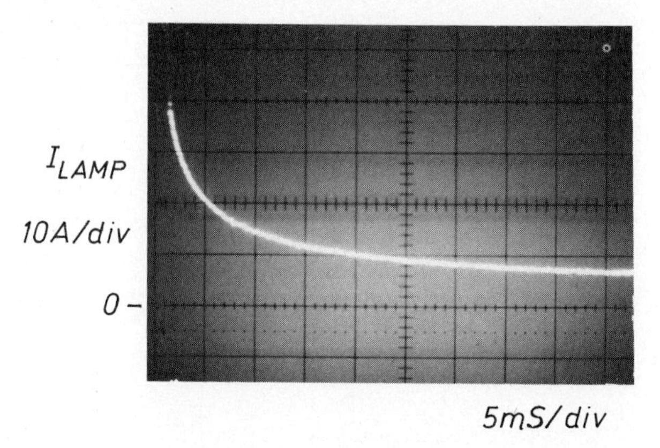

Figure 9.8. Turn-on current waveform for an incandescent lamp.

can be up to 10 to 15 times lower than its value at normal operating temperatures. Current in the lamp may still be twice the normal operating current 10 ms after turn-on (Figure 9.8). As well as imposing switch stresses, the current surge reduces the lamp lifetime considerably.

Furthermore, incandescent lamps are gas-filled to lengthen lifetime and improve efficiency. When such a lamp fails, flashover can occur, causing a large current surge to be drawn. This may be in order of 100 to 200 A, lasting 2 to 4 ms. Such surges make it difficult to achieve an economic design with bipolar transistors. The MOSFET, on the other hand, is better able to withstand the surge current and can also provide inherent current limiting if an appropriate gate voltage is used.

A simple circuit to extend lamp life is shown in Figure 9.9. This circuit eliminates current inrush at turn-on. Also, vibration stress due to ac current flow is eliminated. At turn-on, C1 and the input capacitance of the MOSFET are charged via R1. The charging time constant is such that the MOSFET reaches the fully ON state in about 0.5 s. The zener diode limits the gate voltage to 15 V. As the device passes through the active region, the current will be constrained by the gate–source bias and the transconductance. The *RC* time constant is matched to

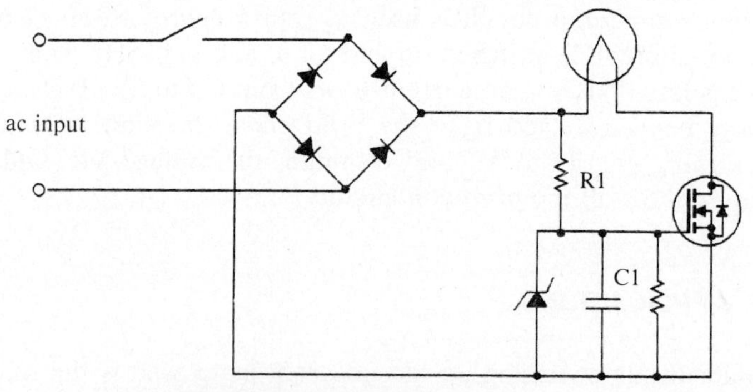

Figure 9.9. Lamp life extender.

the cold-start characteristic of the lamp, so that by the time the MOSFET is fully enhanced the lamp resistance is large enough to allow safe, reliable operation. Since the MOSFET requires only a tiny gate current, R1 can be very large, thereby ensuring that the drive power requirements are low. The MOSFET will normally operate in the fully enhanced region, so that efficiency will not be significantly impaired.

9.9. LOW-POWER LAMP DIMMER

Figure 9.10 shows a lamp dimmer circuit in which an oscillator generates a Pulse Width Modulated (PWM) drive signal for the MOSFET. The duty cycle varies the lamp output brightness. The supply for the control circuit is derived from the drain of the MOSFET thereby allowing a two wire connection to the control circuit. There is an upper limit on the drive duty cycle as the capacitor has to be recharged continuously. The use of PWM control and the low power requirements of the drive circuit ensure efficient operation making the circuit suitable for battery driver applications.

9.10. AC LOAD CONTROL

For the control of ac loads a bidirectional switch is required. This can be implemented with two power MOSFETS in series, so that one MOSFET controls the

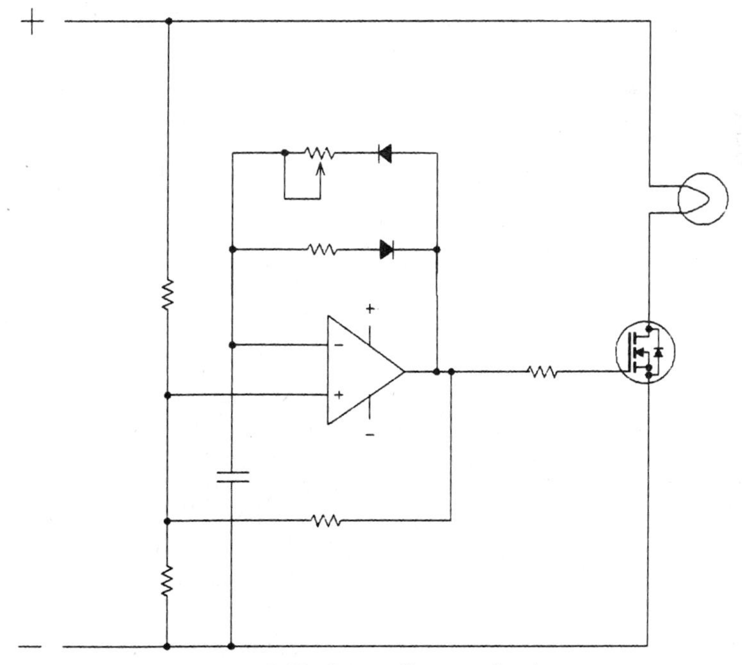

Figure 9.10. Lamp dimmer circuit.

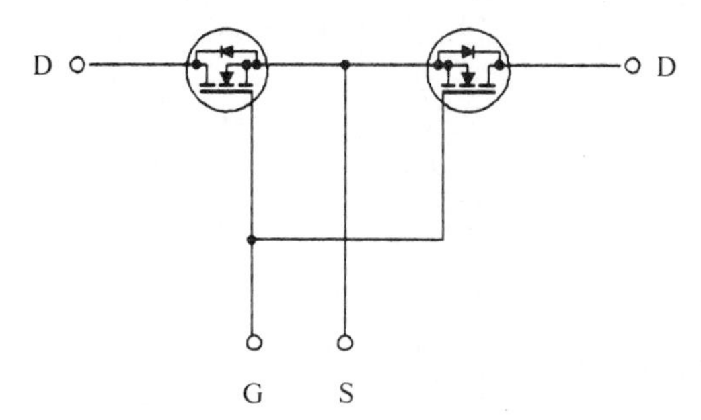

Figure 9.11. Birdirectional ac switch.

flow of current while the body–drain diode of the other completes the circuit (Figure 9.11) [3].

When the load is ground-referenced, a major problem with this arrangement is how to drive the switch, which is at line potential. When the switch is on, the gate voltage must be maintained sufficiently above the ac input to guarantee fully enhanced operation. Yet, when off or on, the gate–source voltage must not exceed the rated value, which is typically 20 V. One solution is to use optical isolators to couple a drive signal to the gate–source terminals of the ac switch. A power source is required for the drive circuit, and this is normally derived from the load power source. Gate-protection circuitry is generally necessary, and often zero-voltage turn-on and zero-current turn-off are desirable to reduce conducted and radiated electromagnetic interference. Up to twenty discrete

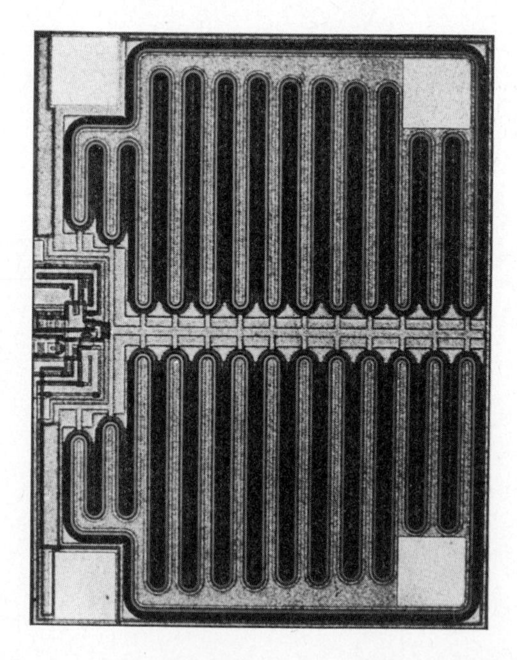

Figure 9.12. Monolithic bidirectional MOSFET switch (BOSFET). (Courtesy International Rectifier Corp.)

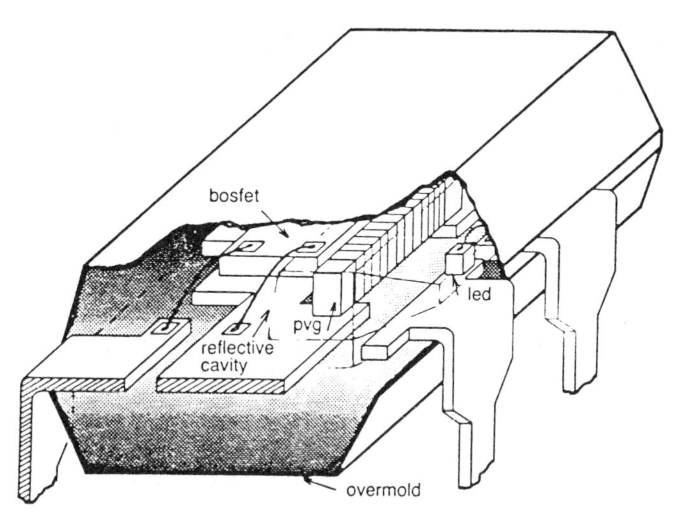

Figure 9.13. Photovoltaic relay using a bidirectional MOSFET and a photovoltaic gate drive. (Courtesy International Rectifier Corp.)

devices may be needed to implement the complete drive circuitry. It is not surprising therefore that power integrated circuits are being developed for this function.

Figure 9.12 shows a photograph of a bidirectional MOSFET variously called a bilateral MOSFET [4] or a BOSFET [5]. This device is a monolithic integration of two n-channel MOSFETS with a common source and a common gate. This effectively gives an ac switching device in which current flows in opposite directions each half cycle under control of a single gate.

The BOSFET also includes on-board low-voltage logic to ensure clean switching when driven from a photovoltaic pile. The use of an LED and photovoltaic pile to provide an isolated gate drive obviates the need for an isolated power supply [6]. This combination of bilateral MOSFET and photovoltaic gate drive, when encapsulated in a hollow package with the required separation of the LED and the photovoltaic pile as shown in Figure 9.13, form a photovoltaic relay. The conversion efficiency of photovoltaic generators is very low. Output is limited to a few microamperes, and cell sensitivity deteriorates when driving low-impedance loads. Therefore the MOSFET with its high impedance gate is the obvious device to use with a photovoltaic drive.

REFERENCES

1. M. I. Castro Simas and M. Simoes Piedade, "Active protection for power MOS transistors." *IEEE, Power Electron. Spec. Conf.* pp. 69–75 (1987).
2. B. Pelly, "New circuit for printer hammer drive using HEXFETs." *Int. Rectifier Appl.* (1982).
3. W. Fragale, B. Pelly, and B. Smith, "Using the power MOSFETS integral reverse rectifier." *Proc. Powercon* **7,** J.2.1–J.2.11 (1980).

4. "The bilateral MOSFET: A new device sees new applications." *Electron. Des.* Sept. 6 (1984).

5. International Rectifier, "The BOSFET photovoltaic relay." *Int. Rectifier HEXFET Data Book,* pp. F.40–F.42 (1985).

6. W. Collins, "Power interfaces: The missing links evolve." *Powertechnics* November, pp. 26–28 (1985).

CHAPTER 10

Automotive Applications

10.1. INTRODUCTION

The automotive market is predicted to be one of the largest for power MOSFETS. The reason is that it is essentially a low-voltage application and it is at low voltages that MOSFETS show their best cost advantage over bipolar transistors. Since $R_{DS(on)}$ per unit area of silicon is approximately proportional to $V^{2.6}$ (see Section 3.5), low-voltage devices offer the lowest $R_{DS(on)}$ for a given size of die and therefore the best value in terms of amperes per unit cost. Forward voltage drop is also an important consideration. Safety regulations dictate an upper limit on the voltage drop that can be allowed across a switch in an automobile. Typically this is of the order of 0.5 V. It may be difficult for a bipolar transistor to meet this specification, while a power MOSFET can always be chosen with sufficiently low $R_{DS(on)}$ to meet the forward-drop requirements. MOSFETS may be paralleled to reduce the total $R_{DS(on)}$ of the switch.

The avalanche rating, which is likely to become a standard feature of power-MOSFET data sheets, is of major benefit in the electrically noisy automotive environment. Line-voltage surges are common due to the opening and closing of switches, the connection and disconnection of inductive loads during maintenance, loose connections, and pickup from adjacent circuits.

Amongst other features which make the power MOSFET particularly suited to automotive applications are its ability to interface directly with logic circuits, for example in a multiplexed wiring (MUX) system, and its square safe operating area, which gives it good reliability when switching clamped inductive loads.

10.2. HIGH-SIDE SWITCHING

Most vehicles operate with a negative vehicle ground and a positive supply, and most loads have one side connected to ground while the other is connected to the positive battery rail through a switch. Thus the load is switched on the "high side". One reason for this is that corrosion due to galvanic action between the ground and electrically live components within the load would result if the load were not disconnected from the supply when the vehicle was switched off. Another reason is that if the internal components are electrically live with respect to the chassis of the vehicle, even when the vehicle is switched off, a short circuit of the supply is likely should anyone attempt maintenance without disconnecting the battery.

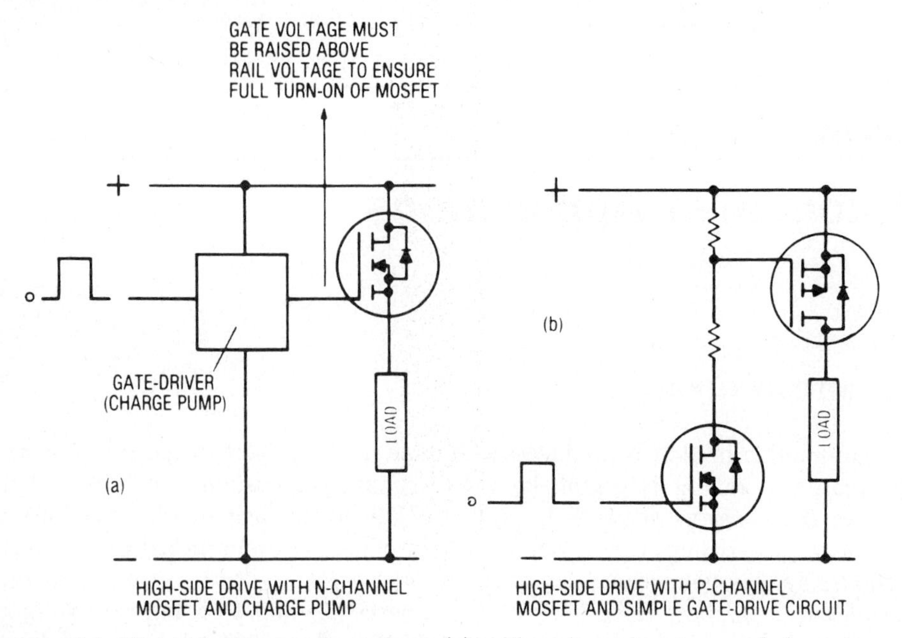

Figure 10.1. High-side drive configurations: (*a*) with n-channel MOSFET and charge pump; (*b*) with p-channel MOSFET and simple gate drive.

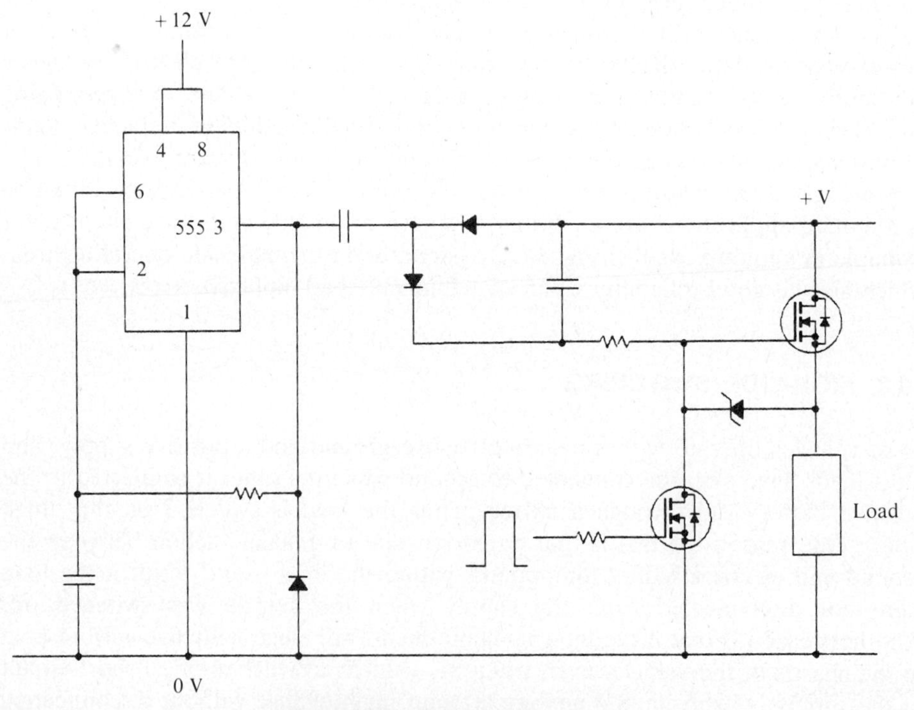

Figure 10.2. Charge-pump circuit.

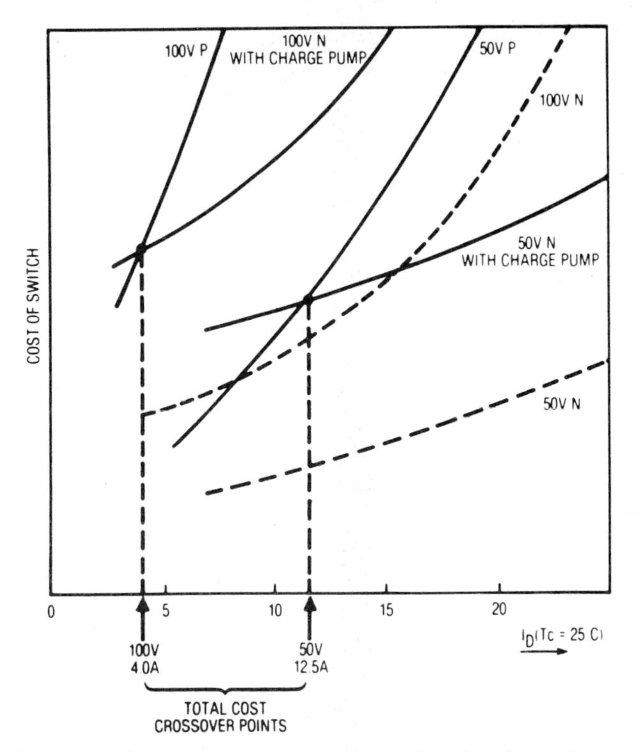

Figure 10.3. Comparison of the cost of n-channel and p-channel high-side switch.

It matters little with a mechanical switch whether the switch is in the high side or the low side. However, the dominance of high-side switching has important consequences for the use of power MOSFETS in automotive applications. As Figure 10.1 shows either a p-channel or an n-channel MOSFET may be used, but while the p-channel MOSFET can obtain its gate voltage from the negative supply rail, the n-channel MOSFET requires a separate supply, more positive than the positive battery rail. This supply is commonly generated by a diode charge pump (Figure 10.2), adding to the cost of the switch. Because electrons have a higher mobility than holes, n-channel MOSFETS have a lower $R_{DS(on)}$ per unit area than p-channel MOSFETS of the same voltage rating. n-channel MOSFETS are therefore usually cheaper than p-channel devices of the same current rating. The choice is thus between an n-channel MOSFET plus charge pump and a more expensive p-channel MOSFET without a charge pump. The charge-pump cost is fixed, while the MOSFET cost is a function of $R_{DS(on)}$ or the current rating. There is therefore a value of load current below which a p-channel MOSFET is the more economical choice, while above that value an n-channel MOSFET and a charge pump is the cheaper solution (Figure 10.3).

10.3. VOLTAGE-RATING REQUIREMENTS

Power-MOSFET voltage ratings in automotive applications are primarily determined by the *load-dump* requirement. This is the voltage generated by the alternator if it suddenly sheds a heavy load as well as losing the clamping action of the battery, for example as the result of a broken cable. Due to the time constant of the alternator field winding, the field current cannot be reduced rapidly enough to match the decrease in load current. The excessive excitation produced by the field without armature reaction to offset it, causes the alternator to produce a high output voltage. The two most common load-dump voltage requirements for 12-V systems are 50 V and 60 V. Higher requirements prevail in 24-V systems. Both 50-V and 60-V power MOSFETS have appeared on the market to meet these requirements, since optimization of the performance–cost ratio is essential in such a price-conscious market. There is a move amongst vehicle and alternator manufacturers to attach a clamp capable of absorbing the load-dump energy across the alternator terminals, so that the load-dump voltage can be limited to a value closer to the nominal battery voltage. This would allow the use of lower-voltage MOSFETS, thereby reducing the cost of solid-state switches in automobile applications. However, this is not yet general practice. While power MOSFETS with an avalanche capability can be used to suppress rail transients, the load-dump energy would be more than a single power MOSFET could absorb. Equal sharing of the avalanche current between many MOSFETS could not be guaranteed, due to variation in their breakdown voltages.

10.4. THRESHOLD VOLTAGE

An important requirement of MOSFETS in automotive applications is that they should be able to remain fully on while carrying full load current under cranking conditions. This is a low-battery-voltage condition encountered when the starter motor is cranking the engine under the worst conditions of temperature and battery state. Vital systems such as the ignition are required to function at the minimum specified cranking voltage, which may be as low as 4.5 V. This is barely above the threshold voltage of most standard types of power MOSFET.

The problem of cranking voltage may be overcome either by using an outsize device, so that it has sufficient transconductance to permit it to carry full load current with only the cranking voltage on the gate, or by providing a charge pump or some other form of auxiliary supply to generate adequate gate voltage. Where the MOSFET is driven from logic circuits, it is usually possible to make provision in the logic power supply for an adequate source of gate voltage. An alternative solution is offered by low-theshold devices, which typically have a threshold voltage half that of standard power MOSFETS. Therefore, under cranking conditions a low-threshold device can carry significantly more current than a standard MOSFET of the same die area. The gate oxide of low-threshold MOSFETS is thinner than that of standard MOSFETS (typically half as thick), so that its puncture voltage is correspondingly lower. Use of low threshold MOSFETS in the electrically noisy environment of an automobile therefore requires more careful attention to the prevention of gate overvoltage.

10.5. REVERSE BATTERY VOLTAGE

Automobile electronic equipment is required to survive the accidental reversal of the battery terminals. A bipolar transistor can withstand this condition if the

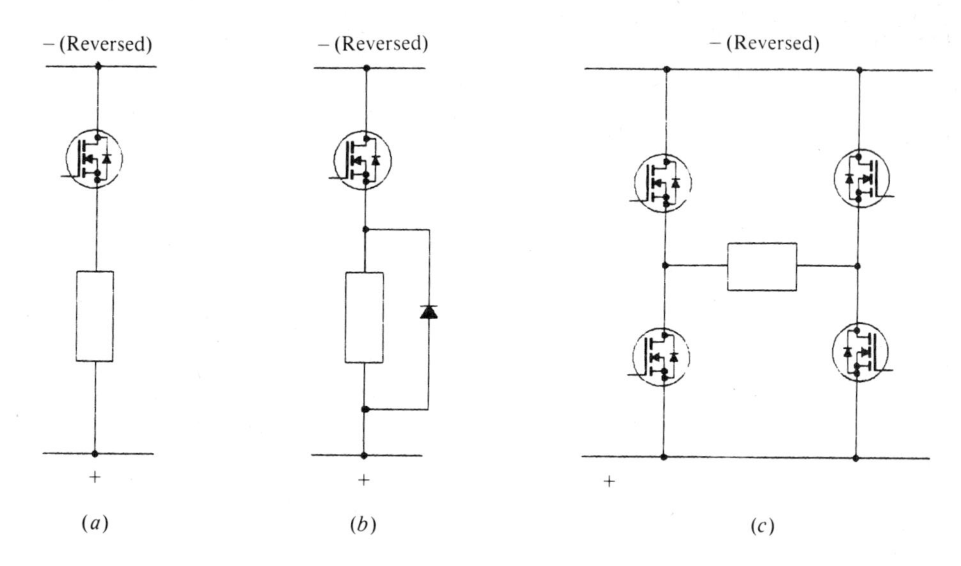

Figure 10.4. Effect of reversed battery on typical circuit configurations. (*a*) Reverse current limited by the load; (*b*) short circuit between the power rails; (*c*) short circuit between the power rails.

reverse breakdown voltage of its base–emitter junction is greater than the battery voltage. A power MOSFET, on the other hand, with its integral antiparallel diode, will conduct current in the reverse direction. Figure 10.4 shows three common circuit configurations. In the first (Figure 10.4*a*) the current that flows under reversed-battery conditions is limited by the load, and no damage will result provided the load can accept reverse current flow. In the other two circuits (Figure 10.4*b* and *c*) the integral diodes of the MOSFETs form a short circuit between the battery rails, so that either a series diode must be included to prevent reverse current flow, or fuse blowing must be accepted as a consequence of battery reversal.

10.6. OPERATING TEMPERATURE

The two main operating environments for automobile equipment are *engine-compartment* applications and *body* applications. For body applications a maximum ambient temperature of 85°C is normally specified, while for engine-compartment applications the ambient temperature in which the MOSFET has to operate can be as high as 125°C. Since the maximum allowable junction temperature for power semiconductors is of the order of 150°C, their power dissipation capability in engine-compartment applications is severely limited.

The power-handling capability of a power MOSFET at the high ambient temperature of the engine compartment can be increased significantly by raising the maximum allowable junction temperature of the device. The usual maximum allowable junction temperature for power MOSFETS is 150°C, but at least one manufacturer has introduced devices in which this has been increased to 175°C [1]. This permits the MOSFET to dissipate more power and to carry a greater current. The power dissipation capability of the transistor is given by

$$P = \frac{T_{J(max)} - T_C}{R_{thJC}} \tag{10.1}$$

where $T_{J(max)}$ is the maximum allowable junction temperature, T_C is the case temperature, and R_{thJC} is the junction-to-case thermal resistance.

Assuming an ambient temperature of 125°C and a case temperature of 135°C, the power-dissipation capability of the device with a $T_{J(max)}$ of 150°C is given by

$$P = \frac{150 - 135}{R_{thJC}} = \frac{15}{R_{thJC}} \tag{10.2}$$

For the device with a $T_{J(max)}$ of 175°C the power dissipation capability is given by

$$P = \frac{175 - 135}{R_{thJC}} = \frac{40}{R_{thJC}} \tag{10.3}$$

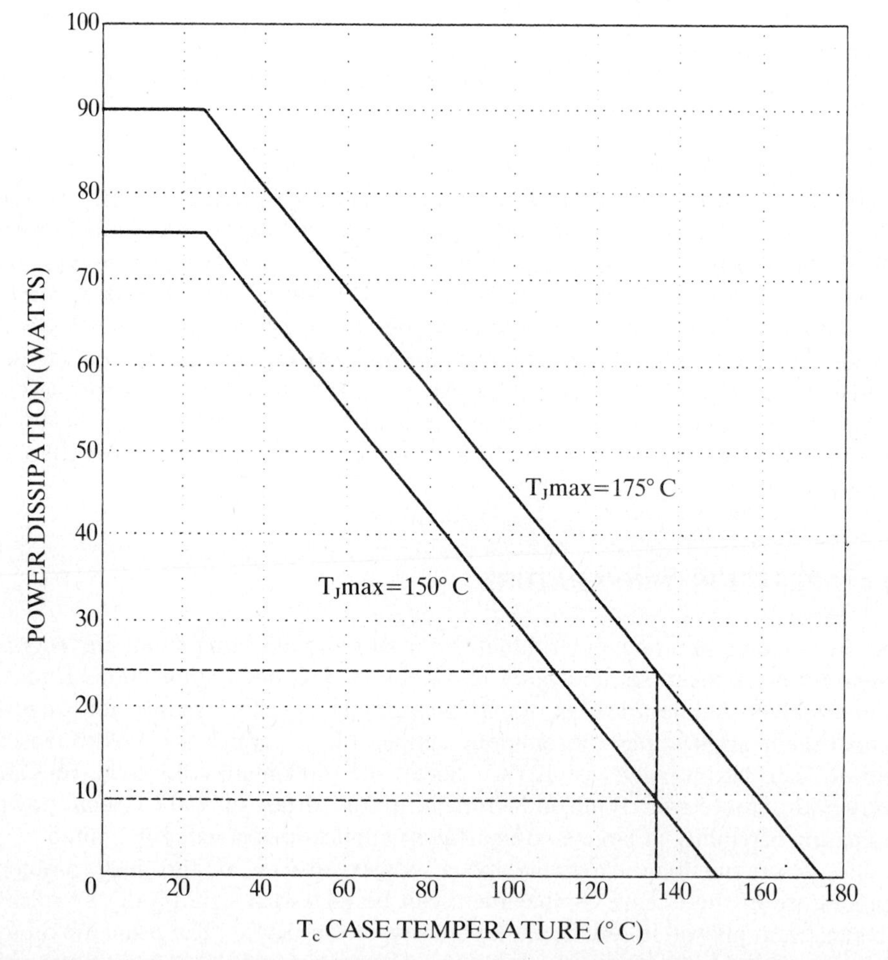

Figure 10.5. Power-dissipation capability of MOSFETS with different maximum allowable junction temperatures.

The result is a 166% increase in the power that can be dissipated, as Figure 10.5 illustrates. Since current and power are related by

$$I_D = \sqrt{P/R_{DS(on)}} \tag{10.4}$$

if we assume that $R_{DS(on)}$ increases by 20% as the temperature of the junction is increased from 150°C to 175°C, the ratio of the current handling capabilities of the two devices is given by

$$\begin{aligned} \frac{I_{D(175)}}{I_{D(150)}} &= \sqrt{\frac{40R_{DS(on)}}{15R_{DS(on)} \times 1.2}} \\ &= 1.49 \end{aligned} \tag{10.5}$$

However, to hold the case temperature of the high-temperature device at 135°C would require an increase in heatsink size because of the increased power dissipation. In practice economics might well dictate a slightly higher case temperature, so that the increase in current-handling capability resulting from the use of the higher-temperature device would be somewhat less than indicated by Equation (10.5).

10.7. APPLICATIONS

As cars incorporate more electronic equipment, the applications for a semiconductor switch multiply. Many of these applications involve motor control—for example, for electronic power steering, window motion, door locking, mirror positioning, cooling fans, seat movement, and dynamic suspension control. Current limiting is a common requirement with motor control, and this is therefore an important area of applicaion for current-sensing MOSFETS. Speed control of windshield wipers and air-conditioning fans can be achieved efficiently and cost-effectively using pulse-width modulation voltage control, although with such systems careful control of electromagnetic interference (EMI) is required to prevent interference with the car radio.

Many mechanical relays are still used in cars, despite the problems of contact welding and contact wearout. However, semiconductor switches may be expected eventually to replace these. The power MOSFET is particularly suited to replacing relays which control loads with a high inrush current, such as a car headlamp (see Figure 9.8). The high peak currents can cause relay-contact sticking, whereas a power MOSFET can tolerate high current overloads. An additional benefit of power-MOSFET switching is that the life of incandescent lamps can be prolonged by operating the MOSFET in the constant-current mode at turn-on, thereby relieving the mechanical stress on the lamp filament.

The absence of second breakdown in power MOSFETS makes them suitable for clamped inductive-load applications such as the control of fuel injectors. In this application it is important to provide a high-voltage dump for the current in the injector coil so that the current rapidly reduces at turnoff. Due to the square safe operating area of the MOSFET, active clamping of the drain voltage (Figure 9.3) can be used without fear of SOA failure. When the drain voltage exceeds the zener breakdown voltage, the gate voltage is raised above the threshold value, thereby turning on the power MOSFET and limiting the drain voltage. Alternatively, if the MOSFET carries a repetitive avalanche rating, its own clamping capability may be employed.

10.8. AUTOMOBILE IGNITION SYSTEMS

Another major application in which SOA is critical is the electronic ignition system. An important consideration for any automobile ignition system is that the aiming voltage across the spark plug electrodes should be sufficient to enable firing. Higher aiming voltages will be required to guarantee firing with fouled plugs and low battery voltages. Also, the time duration of the spark should be such as to provide good combustion and reliable cold start.

In an inductive discharge ignition system energy is stored in the primary magnetizing inductance of the transformer while the switch is closed. When it is opened, a high voltage is induced in the high-tension (HT) secondary. A capacitor is connected across the contact breaker to suppress sparking. The capacitor will limit the rate of fall of the primary current, thus limiting the secondary aiming voltage. The suppressor capacitor discharges via the contact breaker points with current limited only by parasitic elements. This results in contact erosion, affecting the ignition timing and the need for regular replacement of the points.

Performance can be improved and maintenance eliminated if the contact breaker points are replaced with a solid-state switch. However, bipolar transistors used in this application require a load-line shaping network to prevent second breakdown during turnoff. Snubber circuits are not required with power MOSFETS, and high aiming voltages are obtainable. In practice the MOSFET circuit can produce values of aiming voltage of sufficient magnitude to cause breakdown of

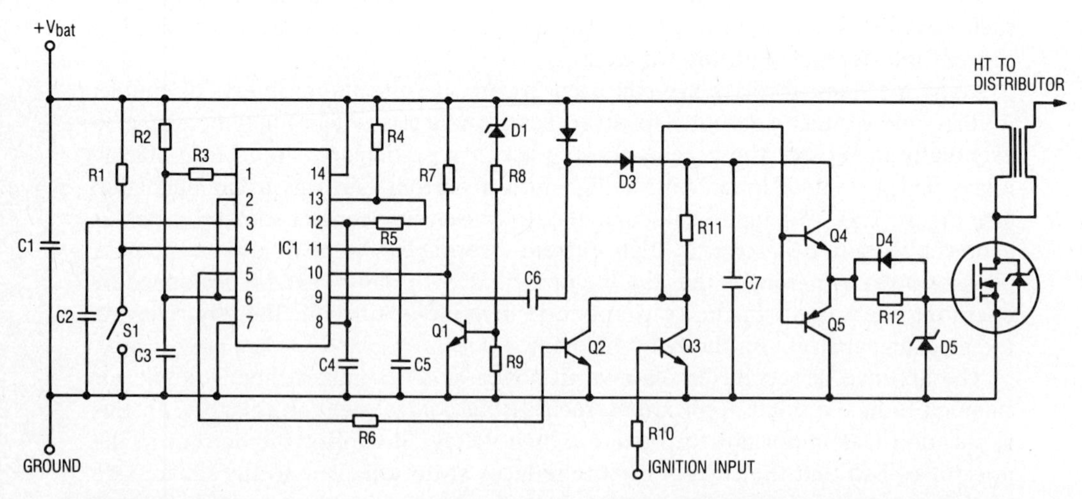

Figure 10.6. Electronic ignition circuit. IC1 is a 556 dual multivibrator which drives the output-transistor charge pump and also provides a test signal for the unit.

the induction coil should the HT lead be disconnected from the coil. To prevent this it is necessary either to clamp the primary voltage with an avalanche diode or to use the MOSFET's own avalanche capability. With the secondary of the coil open, the energy stored in each cycle in the coil is dumped into the MOSFET, so it is necessary to ensure that the resultant junction heating is within allowable limits.

Figure 10.6 shows an inductive discharge ignition system based on a power MOSFET. The MOSFET's own avalanche capability is used to clamp the primary voltage. However, if a MOSFET without adequate avalanche capability is used, a snubber or zener clamp will be required between drain and source [2].

Another MOS-gated device which appears suited to electronic ignition systems is the IGBT (see Chapter 20). The IGBT is a conductivity-modulated device and therefore for high-voltage devices offers the advantagc of a lower forward drop for a given forward current and a given silicon area. Early versions of IGBTS were prone to latching as a result of the high values of dv/dt experienced at turnoff. The wide safe-operating area required is also an obstacle to the use of a minority-carrier device in this type of electronic ignition. However, IGBTS now commercially available are very much less prone to latching and have improved SOA characteristics, so that the IGBT must also be considered a suitable device for this application.

REFERENCES

1. D. Grant, "HEXFET III: A new generation of power MOSFETS." *Int. Rectifier Appl. Note* **AN-966** (1987).
2. B. E. Taylor, "Economic, high performance, high efficiency electronic ignition with avalanche-rated HEXFETs." *Int. Rectifier Appl. Note* **AN-969** (1986).

CHAPTER 11

Power Supplies

11.1. INTRODUCTION

It is in power supplies that the power MOSFET has had the most impact to date, because of the economic advantages resulting from the higher operating frequencies which power MOSFETS make possible. The main advantages are a reduction in weight and volume of the power supplies, mainly due to a reduction in the size of magnetic components and filter capacitors. Generally the design effort required with power MOSFETS is less than that required when bipolar transistors are used, so that design costs are reduced. Also, a MOSFET-based power supply can readily be adapted to a different power requirement without the radical redesign of the base drive and snubber circuits required with bipolar-based power supplies.

Power supplies may be divided into two main classes—switching and linear. This chapter deals only with power supplies in which the power MOSFET is used in the switching mode. Linear power supplies are dealt with in Chapter 15. Switching power supplies may be divided into two basic categories—*switched-mode power supplies* (SMPS) and *resonant power supplies*. These two categories may be further subdivided according to circuit configuration and control strategy. This chapter describes where power MOSFETS have proved most useful and gives the principles governing the use of power MOSFETS in power supplies.

11.2. OPERATING FREQUENCY

Power MOSFETS, with their high switching speed and low switching losses, quickly gained acceptance in SMPS applications because they allowed operating frequencies to be raised from the 20 to 40 kHz common with bipolar transistor models to values of 100 to 150 kHz. Magnetic components and capacitors account for much of the bulk and weight of power supplies. By elevating the operating frequency of an SMPS the size of these components can be reduced, thereby reducing the size and weight of the power supply, and therefore its cost.

Power-MOSFET-based switched-mode power supplies typically have an operating frequency of the order of 100 to 200 kHz. Above this frequency the benefit of the higher operating frequency is diminished by influence of parasitic circuit elements. Figure 11.1 shows how the impedance of a typical output filter capacitor varies with frequency, showing that at 100 kHz it is beginning to lose its effectiveness as the parasitic inductance begins to dominate the impedance.

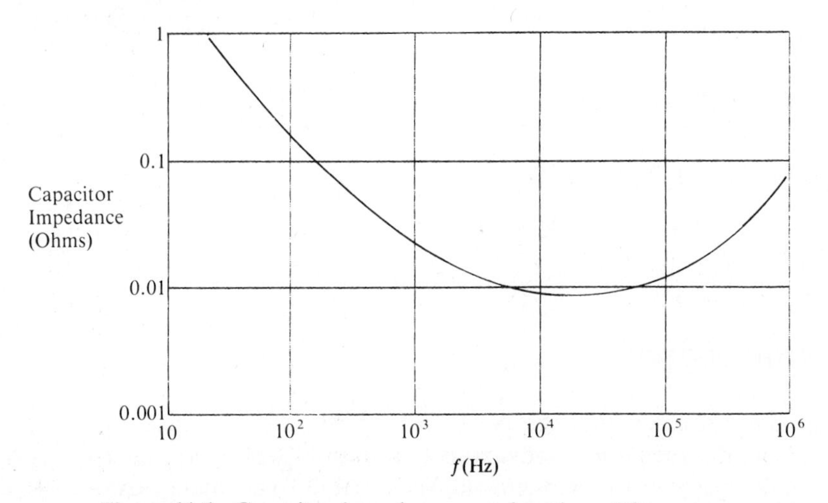

Figure 11.1. Capacitor impedance as a function of frequency.

Similarly, an increase in the core losses of transformers and inductors, as well as skin effect in the windings and the requirements of meeting international safety standards, militate against a further reduction in the size of these components. Furthermore, high operating frequencies imply high switching speeds. In single-ended power supplies the voltage overshoot caused by transformer leakage may be difficult to control when very high switching speeds are used [1]. There are therefore a number of factors which have tended to produce a plateau in SMPS operating frequencies in the region of 150 kHz. Notwithstanding this, the benefits of operating rectangular waveform PWM circuits at higher frequencies remain attractive, and operating frequencies continue to edge higher as solutions are found to the difficulties. It has been shown that this type of supply can be made to operate at frequencies as high as 1 MHz [2, 3]. Careful layout and selection of components, particularly in the gate circuit [4], are required to achieve efficient operation at these switching frequencies. As the switching frequencies approach the megahertz range, losses associated with charging and discharging of the gate become significant. New MOSFET designs therefore seek to minimize this capacitance [5].

The frequency of single-ended switched-mode power supplies operating from 240 V ac is also limited by the energy lost as a result of charging and discharging the parasitic drain–source capacitance. The voltage to which this capacitance is charged is twice the dc voltage produced by rectification of the ac supply, so that with a 240-V supply $V_{DS(max)}$ is approximately 679 V. Taking a typical maximum value for the output capacitance of 200 pF (IRF 830), the energy stored in the output capacitance of the MOSFET is given by

$$E = \tfrac{1}{2}CV^2$$
$$= \tfrac{1}{2} \times 200 \times 10^{-12} \times (679)^2 = 0.046 \text{ mJ} \qquad (11.1)$$

At a switching frequency of 100 kHz this can result in a power loss that is given by

$$P = Ef$$
$$= 0.046 \times 10^{-3} \times 10 \times 10^{3} = 4.6\ \text{W} \tag{11.2}$$

This represents a barely tolerable loss of efficiency in some designs and a barrier to further increase in the operating frequency. The problem becomes less acute if a double-ended circuit is used, since the maximum drain–source voltage is then limited to the dc rail voltage.

11.3. SMPS CIRCUITS

A representative selection of common switched-mode power-supply configurations is shown in Figure 11.2 together with waveforms indicating typical conditions of current and voltage experienced by the power MOSFET. In single-ended designs the power MOSFET experiences approximately twice the dc rail voltage plus the voltage spike generated at turnoff by falling current in the transformer leakage inductance. The amplitude of this spike is generally limited by a snubber circuit, by a zener clamp, or by avalanching of the power MOSFET where the MOSFET characteristics permit this. For single-ended supplies operating from 240 V ac the minimum acceptable voltage rating for the MOSFET is generally

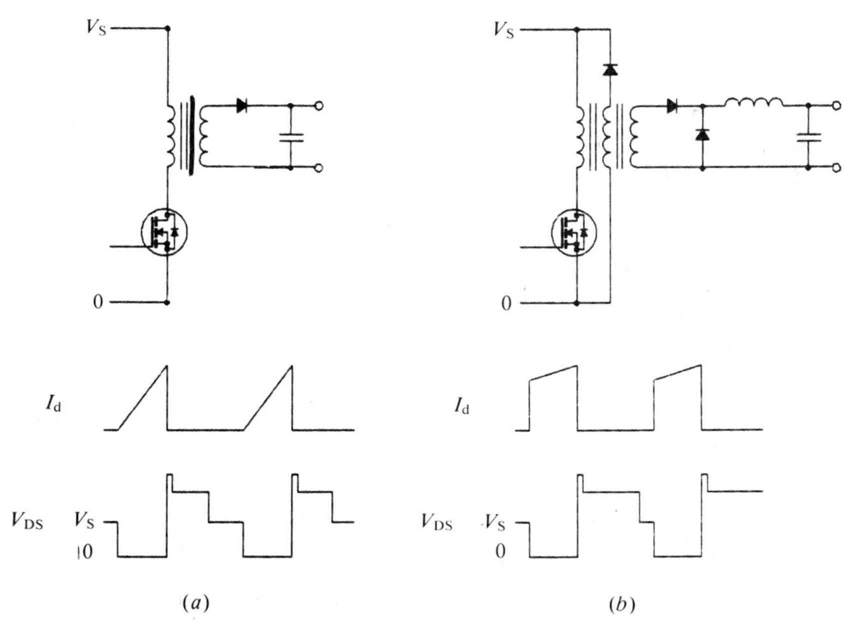

Figure 11.2. Common switched-mode power-supply circuits. (*a*) Flyback; (*b*) forward; (*c*) push–pull; (*d*) half-bridge; (*e*) full-bridge; (*f*) half-bridge diagonal.

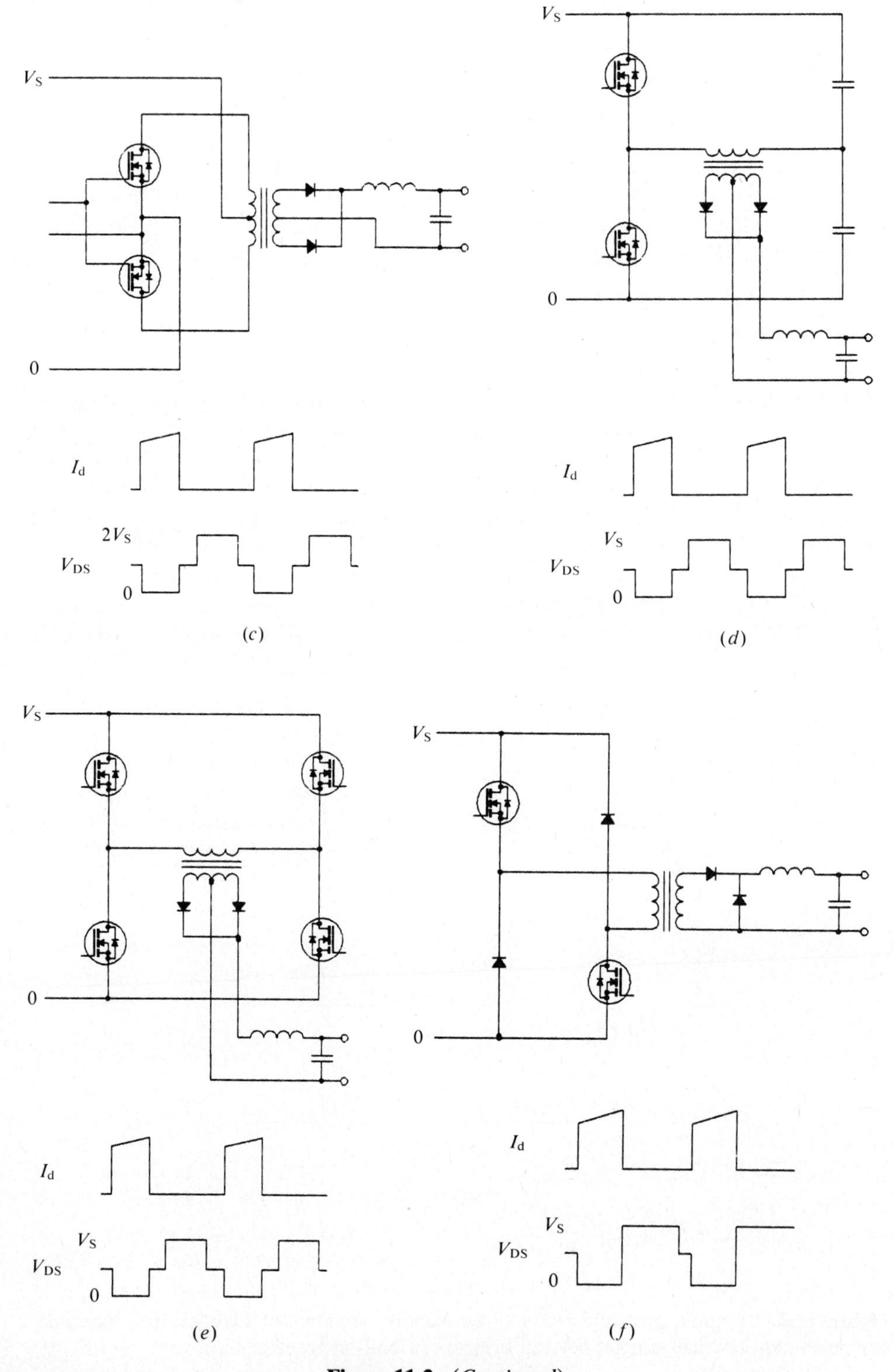

Figure 11.2. (*Continued*)

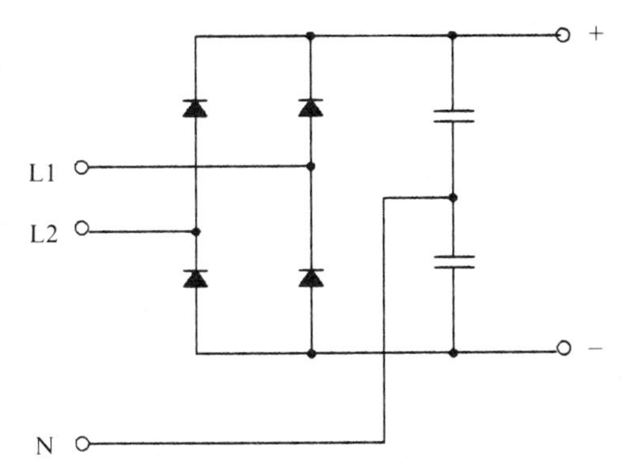

Figure 11.3. Dual-voltage input circuit.

800 V, though 900 V or occasionally 1000 V is used despite the $R_{DS(on)}$ penalty. The advent of power MOSFETS with a repetitive avalanche capability [6] may be expected to result in a reduction in the voltage rating of MOSFETS used in this application as either the device's own clamping capability is employed or designers decide that they can reduce safety margins.

In half-bridge and full-bridge SMPS circuits, theoretically the MOSFETS are not exposed to voltages greater than the dc rail voltage. Given the $V^{2.6}$ relationship between voltage rating and $R_{DS(on)}$, bridge circuits are often the most suitable arrangement for 240-V ac off-line switchers. For operation from 110 V ac, single-ended designs may be suitable. However, manufacturers looking to make their products suitable for the international market favor the dual-input-voltage front-end arrangement shown in Figure 11.3, which allows operation from 120 or 240 V ac but requires the converter stage to operate from a nominal 339-V dc rail. For operation from 240 V ac the supply is applied across terminals L1 and L2; for operation from 110 V ac the supply is connected to terminal N and either L1 or L2.

The zigzag circuit of Figure 11.2*f* permits operation in the forward and flyback converter modes while limiting the MOSFET voltage to the dc rail voltage, but at the expense of the use of two switching devices instead of one.

11.4. CURRENT-MODE CONTROL WITH CURRENT-SENSING MOSFETS

The two principal SMPS control strategies are voltage control and current-mode control. In voltage control the voltage feedback signal is compared with a reference and the error signal used to control the duty cycle of the switching of the power MOSFET. In current-mode control (Figure 11.4) the voltage error signal is used as a current-demand signal which forms the input to a current-control loop. This current control loop utilizes a current feedback signal from the primary side of the transformer to regulate the MOSFET duty cycle. Due to the presence of transformer magnetizing current and the nature of the secondary-filter inductor current, the primary current waveform includes a ramp. This ramp is compared with the current demand signal to determine the instant of turn-off of the MOSFET.

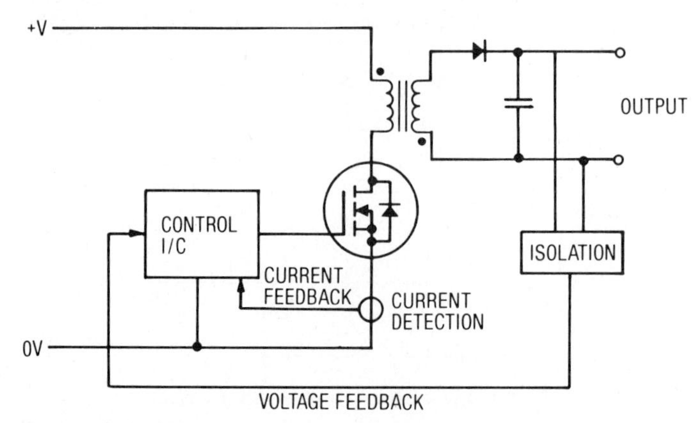

Figure 11.4. Flyback power supply with current-mode control.

The advantages of current-mode control are improved stability (a pole is removed from the main control loop), automatic feed-forward compensation, pulse-by-pulse current limiting, and easy paralleling of supplies. The major disadvantage of current-mode control is the need to sense the current in the switching device. Traditionally this sensing function was performed by either a series resistor or a current transformer with a resistor load. However, there are major disadvantages with both methods. The series resistor gets hot, wastes energy, and reduces reliability. Choosing a resistor involves a difficult compromise between keeping the dissipation low and generating a signal that is large enough to swamp any electrical noise. Finally, there remains the problem of locating a low-value, noninductive, high-current resistor that is readily available. The disadvantages of the current transformer mainly derive from the fact that it is a magnetic component not readily compatible with automatic assembly.

The current-sensing MOSFET (see Chapter 19) offers a third method of deriving a current sense signal. The use of a current-sensing MOSFET rather than an ordinary MOSFET causes negligible additional MOSFET losses and provides a signal that can be used directly by many of the popular integrated circuits used for current-mode control [7, 8]. Figure 11.5 shows a typical power supply circuit (non-isolated) with current feedback from a current-sensing MOSFET.

The size of the feedback signal will depend on the value of the sensing resistor (see Section 19.1.8). A high-value resistor will be needed if a large-amplitude current signal is required. However, the higher the value of the sensing resistor, the greater the change in current-sensing ratio resulting from a change in MOSFET temperature. A change in current-sensing ratio will have little effect on the normal operation of the current-mode control system. A change in current sense ratio can be compensated for by the voltage feedback loop, since the rate of change of current-sensing ratio will be very small compared with the response time of the voltage feedback loop. The change in ratio and consequent change in the gain of the current feedback loop is unlikely to be large enough to affect the stability of the system. However, a change in sensing ratio is significant when output-current limiting is a required function. The accuracy of the current limit will be directly affected by any change in the current-sensing accuracy. To

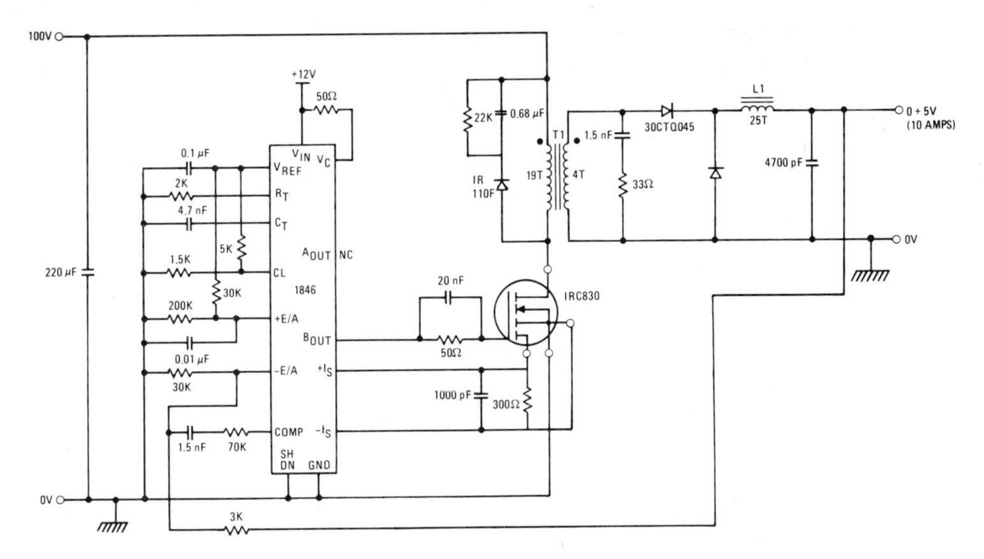

Figure 11.5. Forward converter using current-sensing MOSFET for current-mode control. (Courtesy of International Rectifier Corp.)

maintain the required accuracy over the operating temperature range it may be necessary to use a low value of the sense resistor and to amplify the current sense signal. Fortunately, some integrated circuits for current-mode control incorporate an operational amplifier capable of amplifying the current signal.

The virtual-earth method of deriving the current sense signal (see Section 19.1.7) avoids a variation in the current-sensing ratio due to a change in the temperature of the power MOSFET. Figure 11.6 shows such a circuit. A second operational amplifier is required to give a positive going signal. The additional complexity of this circuit may be acceptable if operational amplifiers are already

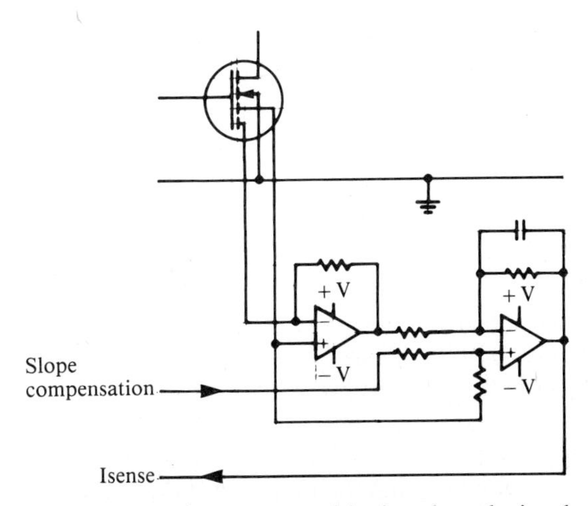

Figure 11.6. Current-sensing MOSFET with virtual-earth signal generation.

employed in the control circuit. A negative supply for the operational amplifiers may be obtained from an auxiliary winding on the power transformer. Slope compensation for current mode control may be added via the operational amplifiers [9]. Note that in both circuits (Figure 11.5 and Figure 11.6) filtering of the current sense signal is necessary to control the amplitude of spikes produced at turn-on and turnoff due to capacitive and mutual inductance effects in the MOSFET. The filter rolloff point is chosen so as to limit the amplitude of the spikes while having negligible effect on the performance of the current-control loop.

11.5. RESONANT POWER SUPPLIES

As the frequency of operation of an SMPS is increased, there comes a point where the MOSFET switching time becomes a significant proportion of each cycle. Losses due to switching and the charge and discharge of the MOSFET parasitic capacitance become significant and effectively impose a limit on the switching frequency. To achieve further increases in operating frequency requires the use of resonant circuits in which switching of the MOSFET occurs at either zero voltage or zero current and the MOSFET output capacitance is incorporated into the resonant circuit.

Resonant circuits have become popular for use in power supplies where a high power density is required, principally in dc–dc converters for military applications. Power densities of up to 25 W/in.^{3} (without integral heat sinks) have been developed, and there is speculation that densities up to 100 W/in.^{3} are achievable. Resonant circuits offer less advantage in off-line power supplies because the size and weight of the "front end"—the input rectifiers, filter capacitors, and filter inductors—are not reduced by the adoption of resonant technology in the converter circuit. Nevertheless, some reduction in the overall bulk of the supply can be achieved, and other advantages accrue, such as improved efficiency, improved response time, and reduced EMI. Therefore resonant circuits can also be expected to become more prevalent in off-line power supplies.

The main disadvantage of resonant circuits is the high peak values of current which the semiconductors must carry. The control circuit is also more complex than that for square-wave PWM power supplies, although this problem will be dispelled as integrated circuits for the control of resonant power supplies become more sophisticated and more plentiful.

11.6. RANGE OF RESONANT FREQUENCIES

The frequencies used in resonant supplies tend to be located around the decades 100 kHz, 1 MHz, and 10 MHz. There may be some rationale for this in that each of these frequency bands requires the adoption of different construction techniques [10].

The 100 kHz range is occupied by high-power converters in which resonance is used as a means of achieving higher efficiency [11]. Figure 11.7 shows such a converter. The advantages of sine-wave operation over square-wave switching are

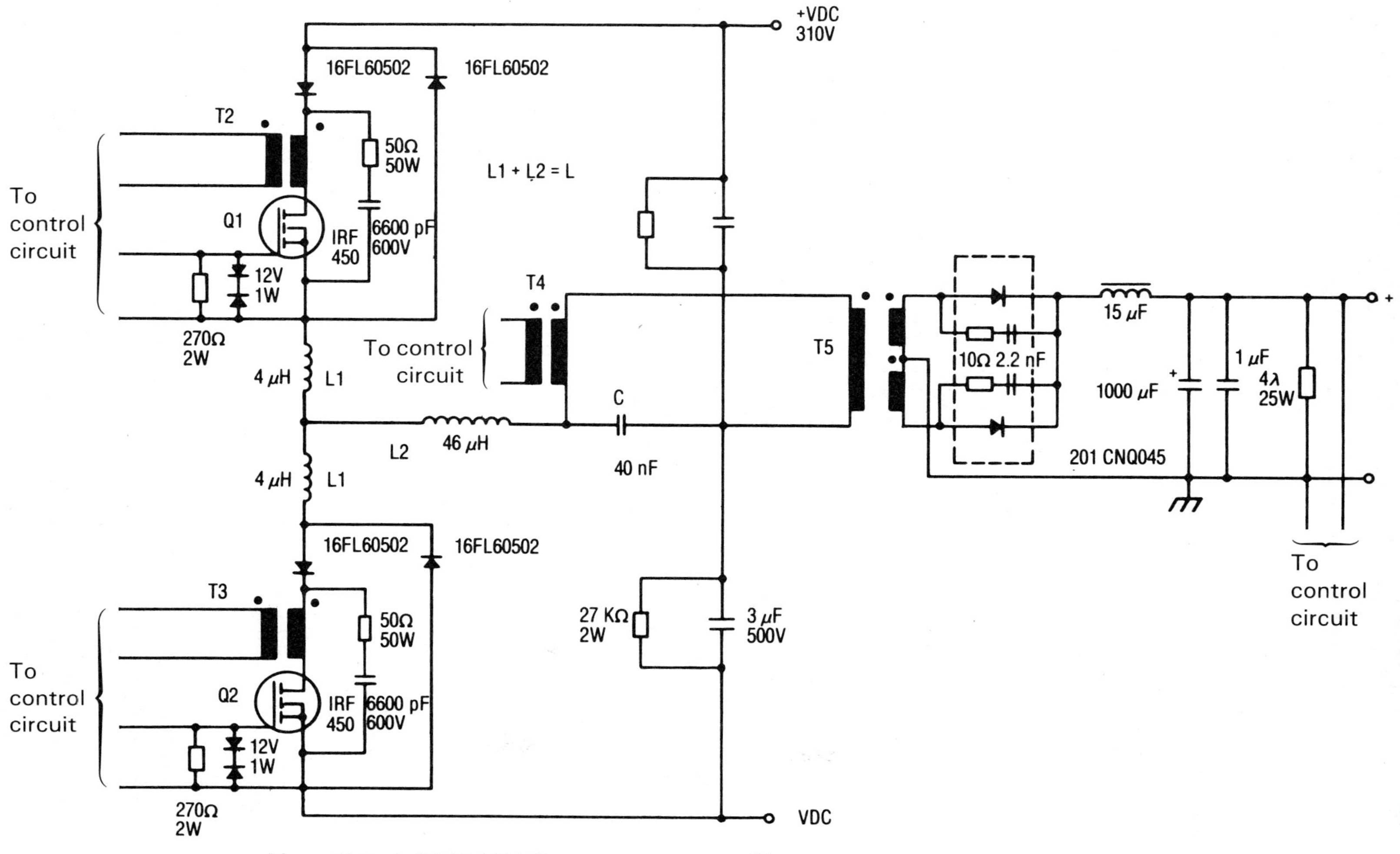

Figure 11.7. A 500-W 100-kHz resonant converter. (Courtesy of International Rectifier Corp.)

the absence of high values of di/dt and dv/dt, which make filtering and EMI control easier; that switching losses are lower, since the current is zero at the moment of switching; that rectifier losses are low, since di/dt is low; and that the bandwidth of the transformer is lower, since it does not have to handle harmonics. Despite higher peak current values in the primary circuit, the efficiency of such a converter can be higher than that obtainable with a square-wave converter.

The 1-MHz range arises from the switching speed of the power MOSFET when driven in the conventional manner and when contained in a conventional package. Under these circumstances the switching speed of the MOSFET is of the order of 50 ns, so that turn-on and turnoff occupy 10% of a cycle at an operating frequency of 1 MHz. The operating frequency is therefore constrained by the proportion of time that it is acceptable for the MOSFET to be in the switching regime. The difficulty in maintaining rectifier efficiency above 1 MHz and the increasing difficulty in obtaining magnetic material with acceptable performance also form barriers to pushing resonant frequencies beyond the 1-MHz range.

The 10-MHz range is reached by the adoption of a range of new techniques such as integrated magnetics, and the use of special high-frequency power MOSFETS. Conventional MOSFET performance starts to be degraded to a noticeable degree by polysilicon gate resistance at about 5 MHz. Losses in the internal gate resistance become significant due to the rate at which charge flows into and out of the gate (see Section 4.5). Furthermore, since some areas of the die are further from the gate pad than others, the internal gate resistance between the pad and these areas varies. This gate propagation delay results in uneven distribution of drain current and uneven heating [12,13]. By providing sufficient gate vias or by the use of metal gates and an open cell structure, the operating range can be extended to at least 25 MHz. Clearly, special low-inductance packaging is required at this frequency, or, more likely, the MOSFET die will be used unpackaged in a hybrid circuit. Fast switching also requires specialized gate drive circuits and even the integration of the gate driver onto the power MOSFET die [14]. Traditionally power-MOSFET design has sought as its primary aim the minimization of $R_{DS(on)}$. For very high-frequency operation minimization of parasitic capacitances may become the dominant requirement.

11.7. RESONANT SCHEMES

An almost infinite variety of resonant converter circuits have been proposed. Amongst the menu of options which the designer has to choose from are the following:

- Series or parallel resonant
- Fully resonant or quasi-resonant
- Full-wave or half-wave switches
- Zero-voltage or zero-current switching
- Continuous resonance or discontinuous resonance
- Fixed frequency or variable frequency
- Tuned or untuned secondary

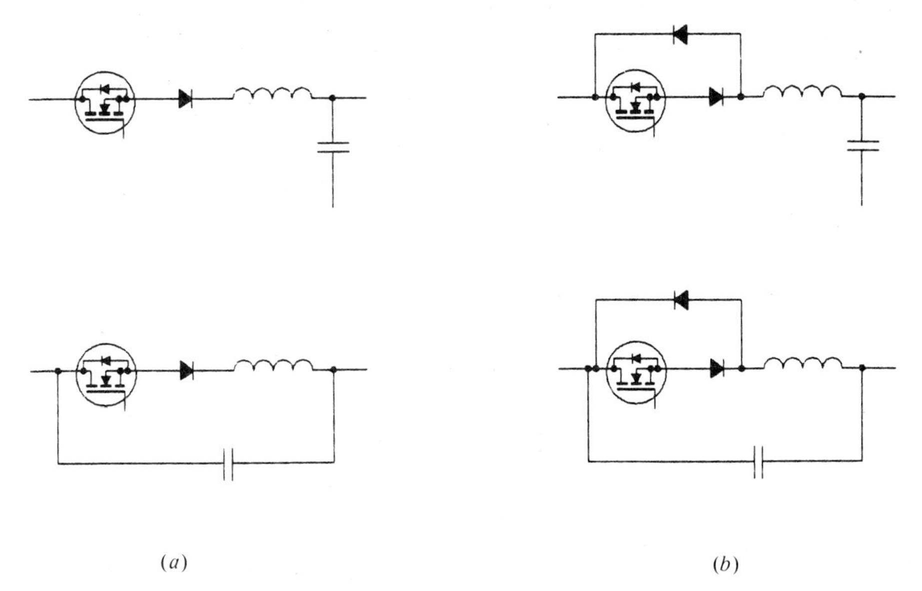

Figure 11.8. Resonant switch configurations.

Add to this the choices of one or two switching devices, isolated or nonisolated secondary, forward or flyback, etc., and it is clear that the number of possible distinct systems is very large. The choice is dictated in part by the need to accommodate parasitic circuit elements. Parasitic elements often form part of the resonant circuit, and layout materially affects circuit operation. In some resonant circuits in which a transformer is used, the tuned circuit may consist almost entirely of the transformer leakage reactance and the MOSFET output capacitance. Where Schottky diodes are used in the output rectifier, the capacitances of these devices may also be incorporated in a secondary side tuned circuit.

Given the wide range of configurations and operating modes that are possible when resonance is employed in a power-supply circuit, an adequate treatment of the subject is necessarily lengthy and is not therefore included in this work. Rather the reader is referred to the references, particularly Refs. 13 and 15–18, which incorporate overviews of the subject. We concentrate on the manner in which the power MOSFET operates in the environment of a resonant circuit.

As Figure 11.8 shows the semiconductor switch can be either a half-wave or a full-wave device [15]. Resonance involves current flow into and out of the tuned circuit. In a single-switch resonant circuit, current will flow in one direction through the MOSFET channel and a current path must be provided for flow in the opposite direction during the other half cycle. If the power MOSFET is used as a full-wave switch, it is usually necessary to isolate the integral diode of the power MOSFET by use of a series diode, since the integral MOSFET diode is likely to be too slow for operation at typical resonant circuit frequencies.

Voltage control may be achieved either by controlling the switching frequency, so that the circuit operates closer to or further away from resonance [16], or the frequency may be kept constant and the amplitude of the single or half cycles of resonance controlled by modulating the ON time of the switching device. Yet another option is to use discontinuous resonance with variation of the repetition

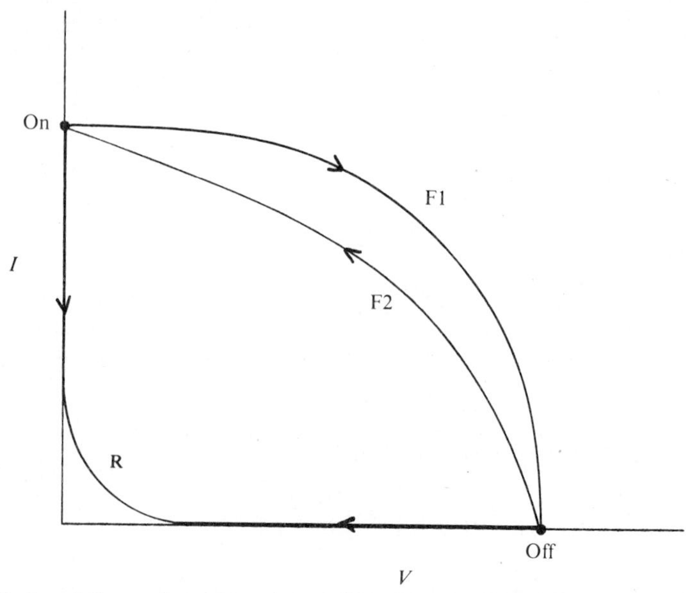

Figure 11.9. Load-line trajectories. F1 and F2: forced switching; R: resonant switching.

frequency. Patent applications have apparently been made for some combinations [17].

In quasi-resonant converters resonance is used to aid switching in what are basically square-wave converter circuits operating in the discontinuous mode [18, 19]. Alternatively, the MOSFET can be used to control the flow of current through a resonant circuit, with switching occurring at times when the voltage across the MOSFET is zero [20, 21]. In this case the drain–source capacitance of the MOSFET often forms part of the resonant circuit. Both methods achieve one of the aims of resonant operation, which is to reduce the switching losses. Figure 11.9 shows a comparison of the switching trajectories for square-wave and resonant systems. Zero-current switching avoids losses associated with simultaneous high values of current and voltage, while zero-voltage switching avoids the losses associated with discharge of the MOSFET output capacitance. However, resonant circuits can expose the power MOSFET to peak values of current or voltage that are several times greater than the values they would experience in PWM-controlled square-wave circuits.

11.8. SYNCHRONOUS RECTIFICATION

The principle of synchronous rectification is illustrated in Figure 11.10. Power MOSFETS rather than diodes are used as the rectifying elements in the output stage of the power supply. These MOSFETS are switched in synchronism with those on the primary side of the transformer. The secondary-side MOSFETS are connected so that their body–drain diodes take the place of the normal rectifier diodes. The gates of the MOSFETS are driven by extensions of the secondary winding.

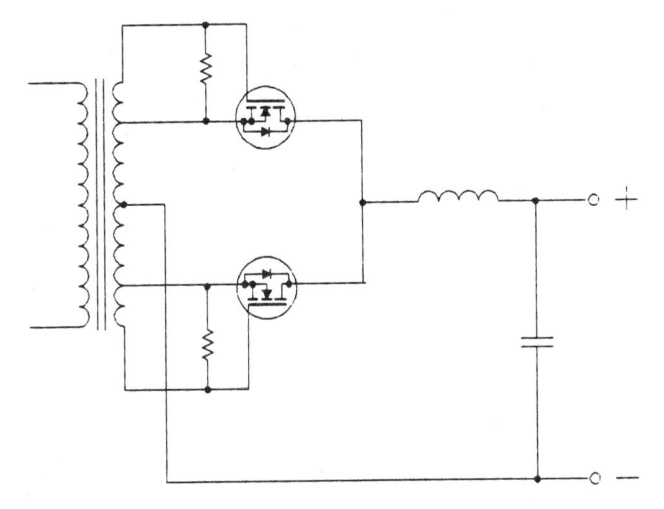

Figure 11.10. Synchronous rectification.

The forward blocking direction of the MOSFETS is the reverse blocking direction of their body–drain diodes, so that a MOSFET is maintained in the OFF state while the diode is reverse biased. When the body–drain diode is forward biased, however, the MOSFET is turned on by making the gate positive with respect to the source in the normal manner (n-channel device). As Figure 11.11 shows, with the gate positive and the channel region inverted, the MOSFET can conduct current through the channel in either direction. If the $R_{DS(on)}$ of the MOSFET is low enough, the voltage drop across the device will be insufficient to forward-bias the

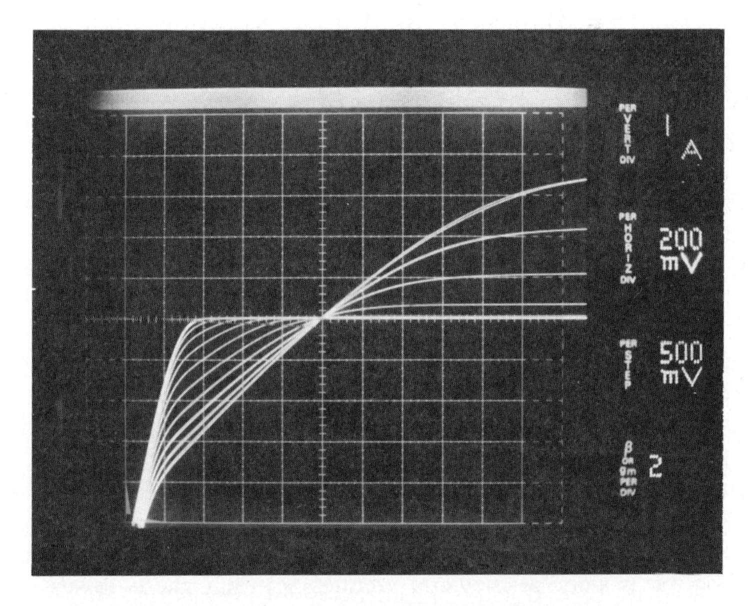

Figure 11.11. Forward and reverse conduction characteristics of a power MOSFET (IRF530).

diode and all the current passing through the device will flow through the channel. The voltage drop across the MOSFET can be reduced to any level desired by use of a large enough MOSFET die or by paralleling MOSFETS.

The power loss in a synchronous rectifier can be compared with that in a conventional rectifier [22] as follows:

The power lost in a Schottky diode is given by

$$P_s = I^2_{L(rms)} R_B + I_L V_{OS} + P_1 \tag{11.3}$$

where I_L is the average output current, $I_{L(rms)}$ is the rms diode current, R_B is the diode bulk resistance, V_{OS} is the diode offset voltage, and P_1 is the loss due to leakage.

The power lost in a MOSFET rectifier is given by

$$P_M = I^2_{L(rms)} R_{DS(on)} \tag{11.4}$$

The power saved by the use of a synchronous rectifier is then

$$P_{saved} = I^2_{L(rms)}(R_B - R_{DS(on)}) + I_L V_{OS} + P_1 \tag{11.5}$$

As an example consider a 72-W, 3-V power supply employing full-wave rectification in the output stage. The diodes conduct with a 50% duty cycle. The Schottky diodes have a voltage rating of 20 V, an offset voltage of 0.35 V, a bulk resistance of 7.5 mΩ, and a leakage current of 10 mA at 100°C when blocking 6 V. The power saved in each phase of the rectifier when these diodes are replaced by synchronous rectifiers is given by

$$\begin{aligned} P_{saved} &= \tfrac{1}{2}(24)^2(0.0075 - R_{DS(on)}) + 12 \times 0.35 + 6 \times 0.010 \\ &= 288(0.0075 - R_{DS(on)}) + 4.2 + 0.06 \end{aligned} \tag{11.6}$$

The power saved will thus be 4.26 W per phase if $R_{DS(on)}$ can be made equal to the bulk resistance of the Schottky diode. A 30-V MOSFET with an $R_{DS(on)}$ of 15 mΩ is a practical proposition. Two such devices in parallel would therefore have a combined resistance of 7.5 mΩ.

The foregoing example ignores the losses associated with the charging and discharging of the gate. If the rectifier were required to operate at high frequency—for example, in the 1-MHz region—these losses would be significant and would have to be included in the calculations. One proposed variation of the conventional MOSFET structure reduces the input capacitance by the elimination of the middle section of the polysilicon gate where it overlaps the epitaxial drain region between cells [23].

A reduction in forward conduction losses in low-voltage rectifiers is not the only motive for employing synchronous rectification. It may also be used advantageously to reduce diode reverse-recovery losses in higher-voltage applications. Again, it is only at elevated frequencies that these losses become of sufficient magnitude to justify the additional cost of synchronous rectification.

In a synchronous rectifier it is important that the diode should never conduct, or minority-carrier storage will cause turnoff delays, thereby producing a momentary short circuit of the secondary winding. This means that there must be no dead time when both MOSFETS are off, since this will allow load current to flow through the body–drain diodes of the MOSFETS rather than the channels. Since it is difficult to arrange for one MOSFET to turn on at the exact instant the other turns off, some slight overlap of the MOSFET conduction periods is advisable. In a voltage-fed converter the momentary short circuit of the secondary winding which this produces will cause losses. These losses are avoided when the converter is current-fed. Conduction times greater than 50% per cycle are possible with transformer-coupled gate drives if depletion-mode MOSFETS are used in the rectifier stage [24]. This keeps the transformer duty cycle below 50% and permits a volt-second balance to be maintained.

Figure 11.12 shows a current-controlled SMPS employing synchronous rectification. A preregulator forms a current source for the push–pull transformer converter stage. The use of a preregulator permits the dc link voltage to be reduced to a value which allows the use of 400-V MOSFETS when operating from a 240-V ac supply. Eight power MOSFETS form the synchronous rectifier, resulting in rectifier losses that are 40% lower than would be expected if Schottky diodes were used for rectification. Two parallel MOSFETS are used in the preregulator stage to further reduce losses (a single device of half the $R_{DS(on)}$ would do as well). The gates of all the power MOSFETS are transformer-driven. The conduction period of Q10 to Q13 slightly overlaps that of Q14 to Q17. This prevents the body–drain diodes of these transistors from ever conducting. The momentary short circuit produced by this overlap does no harm, since the converter is current-fed.

The primary objective of the design is the minimization of losses rather than equipment cost. The cost of synchronous rectification is higher than when simple diodes are used as the rectifying elements. Therefore synchronous rectifiers are

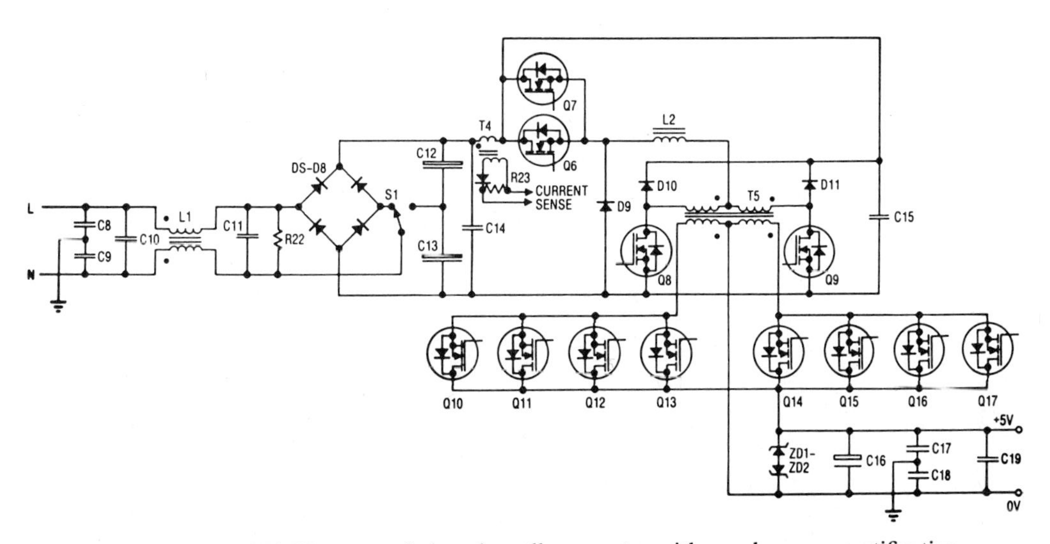

Figure 11.12. A 250-W current-fed push–pull converter with synchronous rectification.

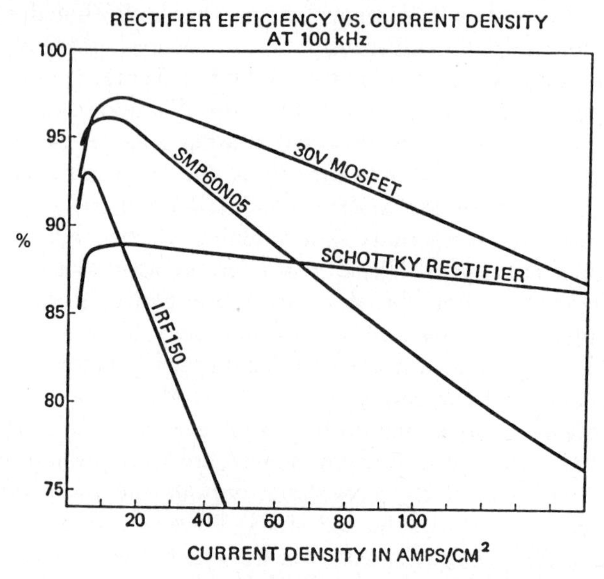

Figure 11.13. Comparison of rectifier efficiencies versus current density at 100 kHz. From Blanc et al. [25]. By kind permission of The Power Electronics Conference 1986.

used when there are other overriding requirements, such as the need to achieve high efficiency in low-output-voltage supplies. While a diode drop of 0.5 V would be acceptable in the rectifier stage of a 5-V dc power supply, 0.5 V would represent a high proportion of the output voltage of a 1.3-V dc supply and would cause a severe loss in efficiency. Synchronous rectifiers also find application in very high-frequency power supplies, although the levels of parasitic capacitance

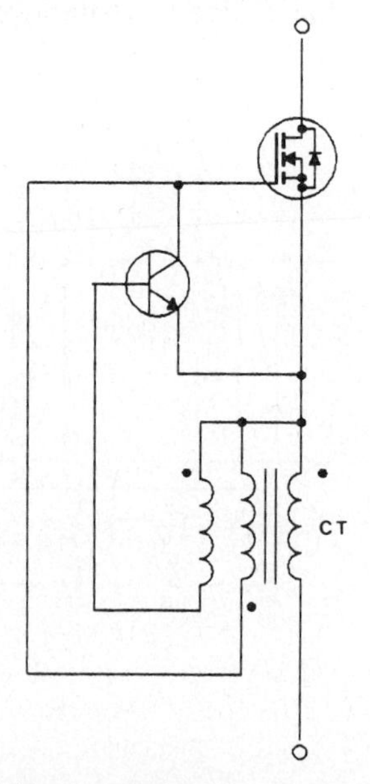

Figure 11.14. A two-terminal synchronous rectifier.

found in present technology devices would seem to preclude their use at frequencies much above 250 kHz [25].

An analysis of the relative efficiencies of a MOSFET synchronous rectifier and a Schottky rectifier for a 3-V output are made in Ref. 25. Figure 11.13 shows the results for 100-kHz operation. SMP60N05 and IRF150 are commercially available devices. The 30-V MOSFET is a hypothetical device with a specific $R_{DS(on)}$ of $3\ m\Omega\ cm^2$, a specific gate charge of $500\ nC/cm^2$, and a specific output capacitance of $8\ nF/cm^2$. The Schottky diode was modeled as a bulk silicon resistance of $5\ m\Omega$ in series with an offset voltage of 0.35 V. The capacitance was 4000 pF, and the die area $0.165\ cm^2$. As well as the losses associated with device resistances and capacitances, eddy-current losses in the packages were taken into account. Since the normal operating current density for the Schottky diode was of the order of $300\ A/cm^2$, it is clear that to obtain increased efficiency, a MOSFET synchronous rectifier has to operate well below the current density of the Schottky diode. This result suggests that synchronous rectifiers will only be economic at this output voltage in situations where efficiency is paramount or cooling costs are high, such as in aerospace applications. Synchronous rectifiers offer particular advantages in cryogenically cooled applications, since cooling under these conditions is expensive and $R_{DS(on)}$ falls to a low value at low temperatures, thereby reducing dissipation [26].

In the applications described so far the MOSFET is not able to function as an autonomous rectifier but requires the provision of a synchronized gate drive signal. One method [27] of producing a two-terminal rectifier which does not require any external gate signal source is shown in Figure 11.14. A current transformer detects incipient conduction of the body–drain diode and, before saturating, charges the gate of the MOSFET. The charge remains on the gate, keeping the channel in conduction as long as the source current continues to flow. When the source current attempts to reverse, the current transformer supplies base drive to the bipolar transistor, which discharges the gate and turns the channel off.

REFERENCES

1. W. Schultz, "Cellular anode zeners clamp high speed power MOSFETS." *Power Convers. Intell. Motion* (*PCIM*) July, pp. 60–62 (1986).
2. R. P. Severns, "High frequency switching regulator techniques." *IEEE, Power Electron. Spec. Conf.* pp. 290–298 (1978).
3. H. Martin, "Design of an efficient high density 1 MHz switching converter." *Proc. Powercon* **11,** B.1.1–B.1.6 (1984).
4. R. Vinsant, "Driving MOSFETs at megahertz speed." *Powertechnics,* pp. 18–22 (1987).
5. T. Daly, "New technology makes power MOSFETS faster, more efficient." *Power Convers. Intell. Motion* (*PCIM*) January, pp. 14–18 (1988).
6. D. Grant, "HEXFET III: A new generation of power MOSFETS." *Int. Rectifier Appl. Note* **AN-966,** 1–15 (1987).
7. H. Ishi, S. Young, R. Pearce, and D. Grant, "Using HEXSense current sense

HEXFETs in current mode control power supplies." *Int. Rectifier Appl. Note* **AN-961,** 1–5 (1986).

8. J. Alberkrack, "A new high performance current-mode controller teams up with current sensing power MOSFETS." *Proc. Power Convers. Int. Conf.* October, pp. 303–315 (1986).
9. R. Pearce and D. Grant, "A 70 W Boost-Buck converter using HEXSense Current mode control." *Int. Rectifier Appl. Note* **AN-962** (1986).
10. R. Redl and N. O. Sokal, "High-frequency switching-mode power converters: general considerations, and design examples at 0.6, 1, and 14 MHz." *Int. High-Freq. Power Convers. Conf., 1st,* Virginia Beach pp. 265–296 (1986).
11. S. Young and G. Castino, "High frequency converter using power MOSFETS." *High Freq. Power Convers. Conf.* pp. 21–35 (1986).
12. H. C. Lin, Y. F. Arzoumanian, J. L. Halsor, M. N. Giuliano, and H. F. Benz, "Effect of silicon-gate resistance on the frequency response of MOS transistors." *IEEE Trans. Electron Devices* **ED-22**(5), pp. 255–264 (1975).
13. G. M. Dolny, C. F. Wheatley, and H. R. Ronan, "Computer-aided analysis of gate-voltage propagation effects in power MOSFETS." *Proc. High Freq. Power Convers. Conf.* pp. 149–154 (1986).
14. A. F. Goldberg and J. G. Kassakian, "The application of power MOSFETS at 10 MHz." *IEEE, Power Electron. Spec. Conf.* pp. 91–100 (1985).
15. K. Liu, R. Oruganti, and F. Lee, "Resonant switches—topologies and characteristics." *IEEE, Power Electron. Spec. Conf.* pp. 106–116 (1985).
16. W. C. Kilbourne, "Using frequency modulation to optimize switching converter performance at 1 MHz and beyond." *Proc. Powercon* **11,** B.2.1–B.2.10 (1984).
17. P. C. Todd and R. W. Lutz, "Practical resonant power converters—theory and application." *Powertechnics* April, pp. 30–34 (1986).
18. S. Freeland and R. D. Middlebrook, "A unified analysis of converters with resonant switches." *IEEE, Power Electron. Spec. Conf.* pp. 20–30 (1987).
19. M. F. Schlect and L. F. Casey, "Comparison of the square-wave and quasi-resonant topologies." *IEEE, 2nd Annu. Appl. Power Electron. Conf.,* San Diego pp. 124–135 (1987).
20. K.-H. Liu and F. C. Lee, "Zero voltage switching technique in dc/dc converters." *IEEE, Power Electron. Spec. Conf.* pp. 58–70 (1986).
21. W. A. Tabisz, P. Gradzki, and F. C. Lee, "Zero-voltage-switched quasi-resonant buck and flyback converters—experimental results at 10 MHz." *IEEE, Power Electron. Spec. Conf.* pp. 404–413 (1987).
22. R. Blanchard and R. Severns, "MOSFETS move in on low voltage rectification." *Proc. Power Convers. Int. Conf.* October, pp. 213–220 (1984).
23. Y. Shimada, K. Kato, S. Ikeda, and H. Yoshida, "Low input capacitance and low loss VD-MOSFET rectifier element." *IEEE Trans. Electron Devices* **ED-29**(8), 1332–1334 (1982).
24. R. A. Blanchard and P. E. Thibodeau, "Use of depletion mode MOSFET devices in synchronous rectification." *IEEE, Power Electron. Spec. Conf.* pp. 81–86 (1986).
25. J. Blanc, R. Blanchard, and P. Thibodeau, "Use of enhancement and depletion mode MOSFETS in synchronous rectification." *Power Electron. Conf.,* San Jose pp. 1–8 (1986).
26. R. Blanchard and R. Severns, "Designing switched-mode power converters for very low temperature operation." *Proc. Powercon* **10,** D.2.1–D.2.11 (1983).
27. W. R. Archer, "Current drives synchronous rectifier." *Electron. Des. News* Nov. 28, p. 279 (1985).

CHAPTER 12

Motor Drives

12.1. INTRODUCTION

This chapter reviews the use of power MOSFETS in the control of the speed and torque of electric motors and includes descriptions of the schemes applicable to each class of motor. In these applications the power MOSFET will be noted for its capability as a high speed switch with full safe-operating-area rating, rather than its use as a linear series controller—although, as will be seen in the application to small brushless dc motors, both techniques can be used within the same system. With the availability of a wide range of power MOSFETS both in discrete form and paralleled in modules, the drive power can usefully span a few watts for small stepper motors up to tens of kilowatts for large dc and ac motors.

Power MOSFETS are attractive in this application because the low gate currents needed can be obtained directly from 15-V CMOS integrated circuits. A further advantage of the power MOSFET is its ability to chop at high frequency. For low-speed high-torque operation, pulse-width modulation is a valuable technique for obtaining a good form factor and for minimizing audible noise. For mobile applications, high-frequency switching offers a useful decrease in the size and weight of filter components.

12.2. DC MOTOR DRIVES

The general requirement in series, shunt, and brushless motor drives is to control torque and speed and hence the power from the motor. Voltage control can be used to control the speed of both shunt and series motors, although in the case of the series motor, feedback is required for accurate speed control. Although the shunt or series motor can be driven directly from a switching power supply, it is more common to consider the motor inductance as part of the filter and to consider the switching unit as part of a speed, position, and current limit control loop. The torque ripple and noise typical of phase-controlled drives is avoided by the use of high switching frequencies. The response of the speed control loop is similarly improved by raising the switching frequency.

The motor torque is proportional to the armature current in shunt motors and to the square of the current in series motors, while the conduction loss in both motors and power MOSFETS is proportional to I^2. Therefore, for best efficiency, as well as improved brush gear life and lower probability of field permanent-magnet demagnetization, a low form factor (I_{rms}/I_{avg}) is desirable. This is achieved by raising the switching frequency to reduce ripple amplitude.

12.3. DC MOTOR DRIVES—CHOPPER CONTROL

Chopper drives may be classified as unidirectional or bidirectional, with or without dynamic braking. The basic unidirectional chopper circuit without braking is shown in Figure 12.1. The average voltage supplied to the motor is controlled by varying the duty cycle of the switching waveform. The two common methods of achieving this are:

1. Fixed Pulse Width, Variable Repetition Rate. This method is appropriate for some semiconductor switches, such as GTOs, in which it is essential to maintain a minimum ON period of the switch. An undesirable feature of this arrangement is a poor current waveform when starting the motor from rest. During the ON time of the switch the current rises to a high value because of the absence of back emf. A long OFF time is then required to restore the average value of the current. This results in undesirable torque pulsations and mechanical noise. Also, the load current reduces to zero for a proportion of the OFF time of the switch. Although in an open-loop system this is not significant, in closed-loop control discontinuous current can cause instability, due to a change in gain and phase characteristics of the amplifier-and-load combination.
2. Pulse-Width-Modulated Chopper Running at a Fixed Frequency. The power MOSFET has no minimum ON-time limitation, so that with this scheme current peaking and torque ripple can be minimized.

The current waveforms are obtained by considering the motor as an inductor,

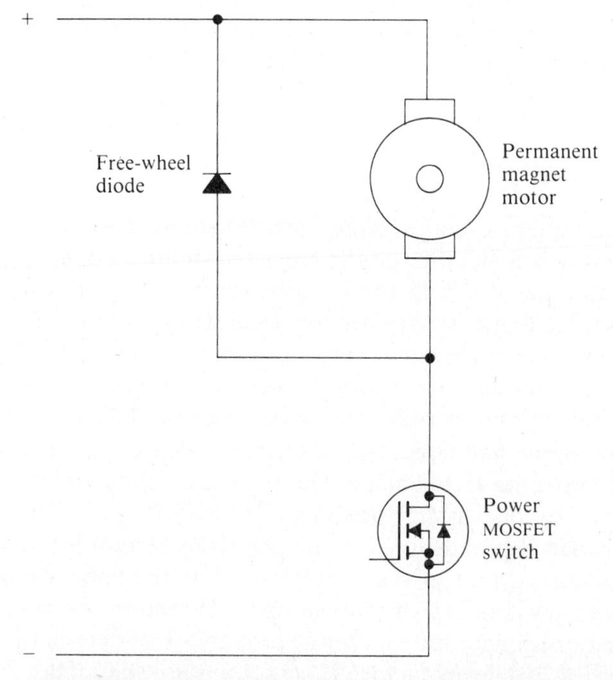

Figure 12.1. The basic single-MOSFET chopper circuit.

increased. The optimum di/dt, which can be controlled by dv_G/dt, occurs where the summation of these losses is a minimum. During the latter half of the diode reverse-recovery period the diode also experiences high dissipation due to rising voltage across it. The discussion of diode recovery losses in PWM inverters (Section 13.8) is also relevant to diode losses in choppers.

As the MOSFET turns off, the drain–source voltage rises rapidly, forward-biasing the free-wheeling diode. During turn–on the diode forward voltage remains high for a brief time before falling to its normal value. This forward recovery voltage is due to turn-on delay in the diode and to the package inductance. The voltage drop can be of the order of 40 V, depending on the applied di/dt. The power MOSFET sees this voltage in addition to the supply voltage, and must be rated to handle it.

The choice of MOSFET rating must take into account the maximum allowable die temperature under peak load conditions such as may be encountered during motor starting. For switching frequencies above 500 Hz the cyclic variation in MOSFET die temperature is not a significant constraint. Figure 12.4 shows the MOSFET drain current. The conduction loss for the single-ended chopper MOSFET is given by

$$P_C = \frac{1}{3}\frac{t}{T} R_{DS(on)}(I_a^2 + I_b^2 + I_a I_b) \tag{12.3}$$

To this must be added the switching loss, which may be calculated in the manner described in Section 13.8.

From the total power dissipation and the power-versus-temperature derating curve of the MOSFET data sheet, the maximum allowable case temperature can be found. If this temperature is unreasonably low, a smaller die size can be used; if it is too high, a larger die size must be chosen. For high-current choppers, multiple-die devices may be required (see Section 17.1.2). Various strategies of current control can be applied to the chopper system. Using low-value sensing resistors or current-sensing MOSFETS (see Section 19.1), the peak value of current can be monitored on a pulse-by-pulse basis to curtail the pulse width during starting or in the event of a load short circuit. A further average-current sensing

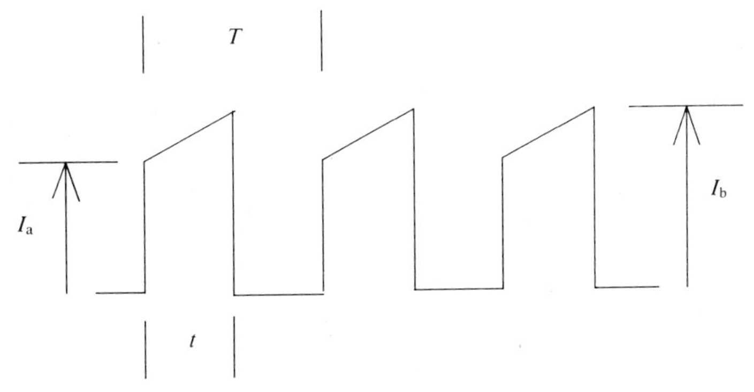

Figure 12.4. Idealized MOSFET current waveform.

circuit can be added to the control loop to protect the motor and the power MOSFET against sustained high current.

12.4. BRAKING

In the case of a series motor, controlled braking of the single-ended chopper may be achieved by field reversal with a plugging diode (Figure 12.5). When the field contacts of the motor change over for braking, the armature potential reverses to forward-bias the plugging diode. The MOSFET, in turning on, applies the full supply voltage across the field, the magnetizing current of which circulates via the free-wheeling diode during the OFF period of the MOSFET. In this situation the power MOSFET is controlling the field current, not the armature current Therefore, if braking torque control is required, current sensing must be located in the armature circuit rather than in the source lead of the MOSFET.

In the case of a permanent-magnet motor, braking is provided by the addition of a second MOSFET switch across the motor as shown in Figure 12.6*a* [1]. This arrangement allows the motor energy to be returned to the supply. During braking MOSFET A will be off. PWM switching is applied to MOSFET B causing armature current (I_1) to ramp up and then circulate via the supply and the body–drain diode of MOSFET A (I_2). Figure 12.6*b* shows the waveforms associated

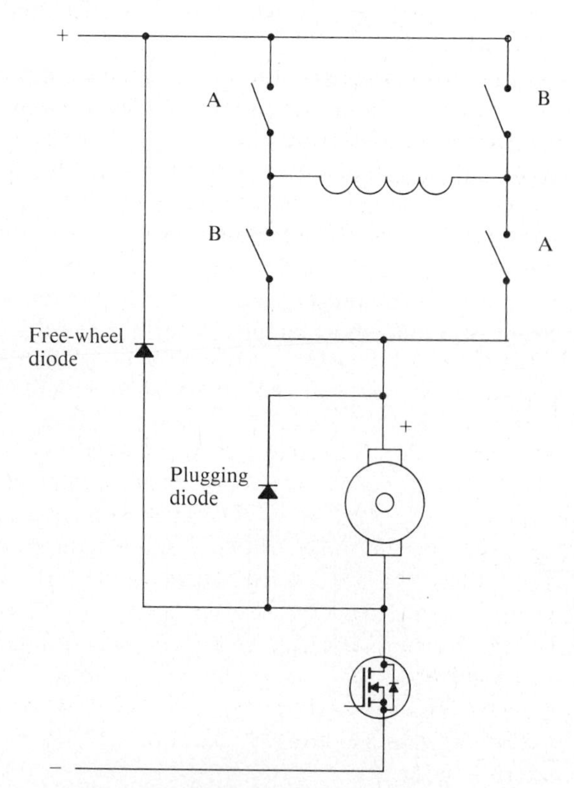

Figure 12.5. Braking by field reversal.

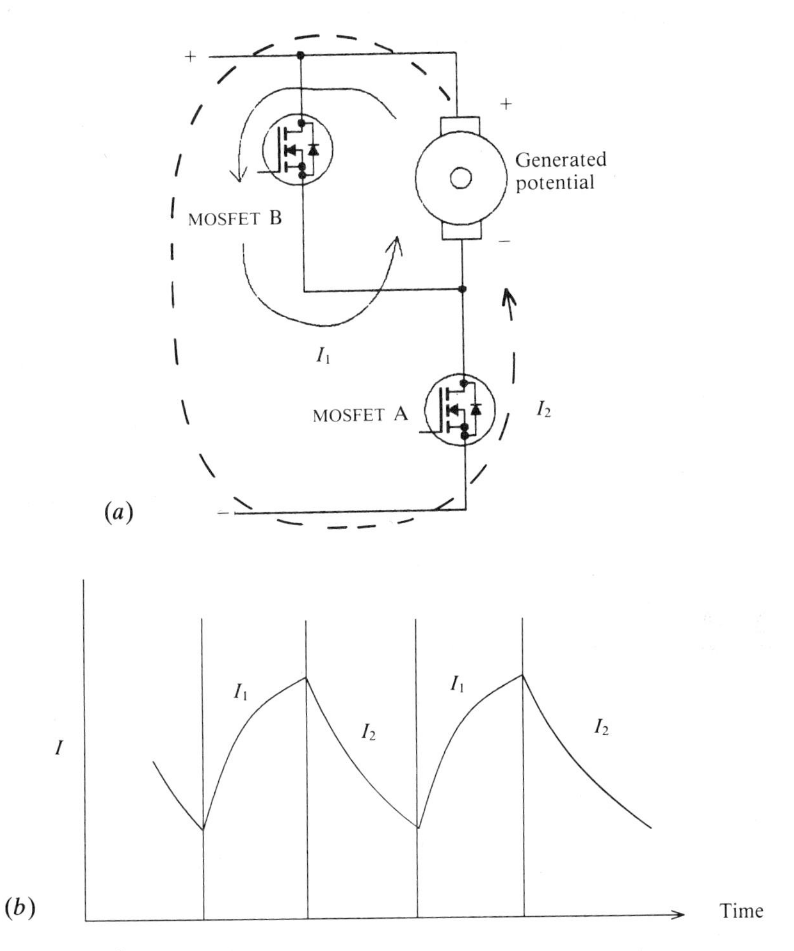

Figure 12.6. Braking of a permanent magnet motor by addition of a second MOSFET. (*a*) Current paths; (*b*) armature current waveform.

with this circuit. Since MOSFET B is associated only with braking, where high efficiency is not essential, this switch could be replaced by a p-channel device, enabling a simplification of the gate drive.

The situation again arises of reverse recovery currents adding to the motor current. The body–drain diode of MOSFET B becomes the free-wheeling diode in the driving or motoring mode, while the body–drain diode of MOSFET A carries the free-wheeling current during braking. This approach is satisfactory at low switching frequencies, where the rate of change of current can be reduced sufficiently to limit the reverse recovery current while still maintaining a low average value of switching losses. It may also be necessary to limit dv/dt during switching in order to protect MOSFETs with low dv/dt capability, for example by increasing the Miller capacitance. Alternatively, a separate fast free-wheeling diode may be used with a diode in series with the MOSFET to force the current to flow through the external diode.

A third method of preventing body–drain diode conduction by use of a

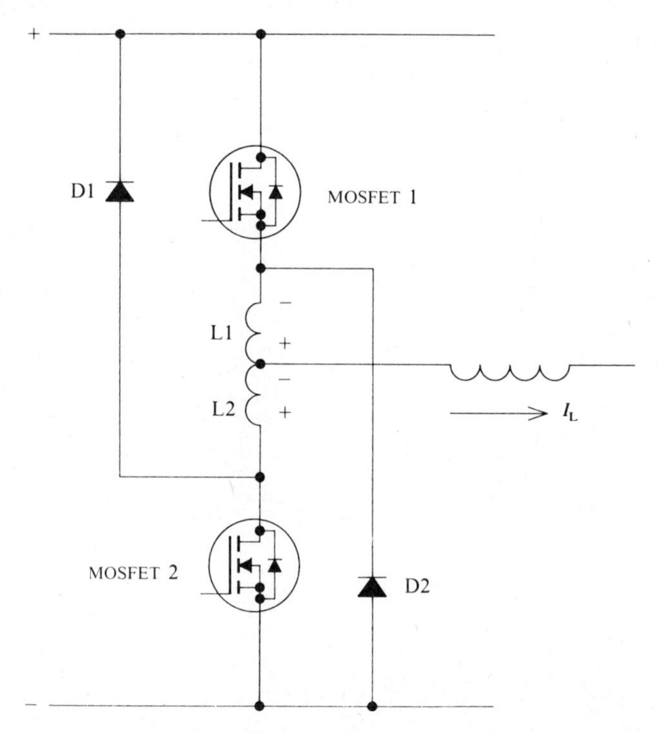

Figure 12.7. Use of center-tapped inductor to prevent body–drain diode conduction.

center-tapped inductor is shown in Figure 12.7. During braking, as MOSFET B turns off, the current in the motor reduces at a rate determined by the components L, R, V_{gen}, and V_S defined in Figure 12.2a. Consider the effect of this reducing current in inductance L1, which is one half of a center-tapped inductor. Potentials are set up in L1 and L2 in the directions shown, levels of 0.5 V each being reasonable. These potentials bias diode D1 on and the body–drain diode of MOSFET 1 off, so that the free-wheeling current during braking flows via the fast diode D1. A similar effect occurs during motoring, where diode D2 conducts in preference to the body–drain diode of MOSFET 2. This method minimizes switching losses, permitting ultrasonic switching frequencies to be achieved.

12.5. REVERSIBLE FULL-BRIDGE CHOPPER DRIVE

The basic full-bridge drive circuit is shown in Figure 12.8. To drive in the forward direction, MOSFETS A and D are off, MOSFET C is on, and MOSFET B is switched with a PWM signal. This gives similar performance to that obtained with the single-ended chopper, with the MOSFET-A body-drain diode acting as the free-wheeling diode. An alternative arrangement is to switch MOSFETS B and C together so that free-wheeling current now flows via the supply and the

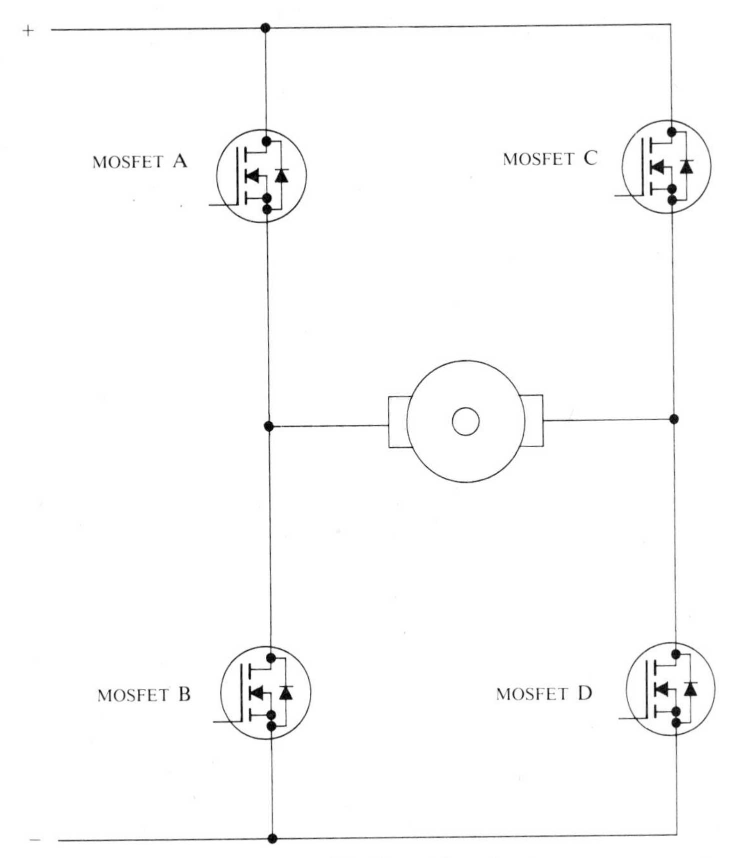

Figure 12.8. Full-bridge drive circuit.

body–drain diodes of MOSFET A and MOSFET D. Here the current decay during the OFF period of the switch is greater because the decay is due to the generated voltage plus the supply voltage, instead of the generated voltage alone. To limit the level of diode reverse-recovery-current losses, an external choke may be added in series with the supply with its own dissipative freewheel path, as shown in Figure 12.9 [2].

Another method of slowing turn-on, thereby reducing the amplitude of the diode recovery current spike, is to add common source inductance as shown in Figure 12.10. During turn-on, the rise of current in L creates a voltage in opposition to the gate drive voltage. This negative feedback controls the gate-to-source voltage, thereby limiting $\mathrm{d}i/\mathrm{d}t$. To prevent ringing between the inductance and the output capacitance of the MOSFET, a low-value damping resistor R may be included in parallel with the inductor. If it is necessary to control $\mathrm{d}v/\mathrm{d}t$ as the free-wheeling diode recovers, this can be achieved by the use of capacitive feedback as shown in Figure 12.11.

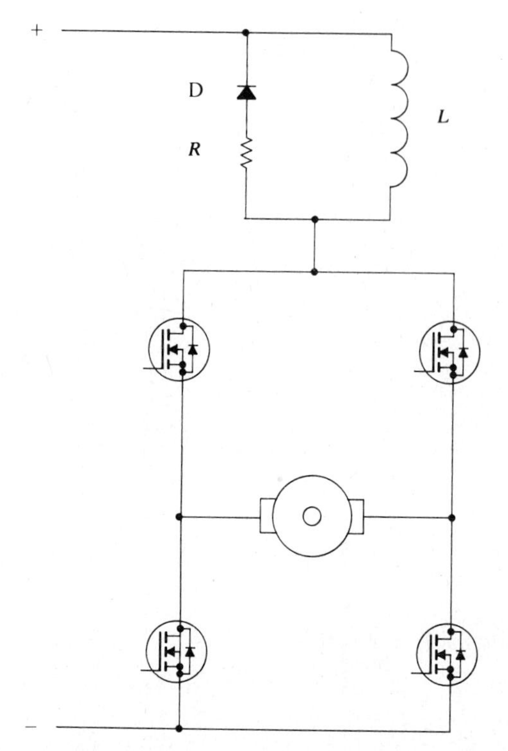

Figure 12.9. Addition of series inductance to reduce diode recovery losses.

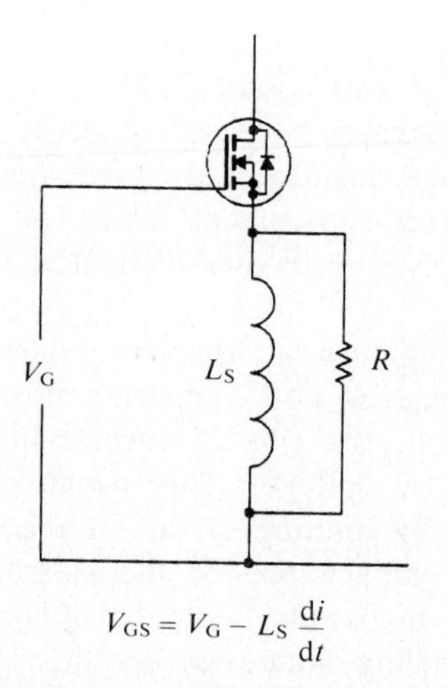

$$V_{GS} = V_G - L_S \frac{di}{dt}$$

Figure 12.10. Use of source inductance to slow MOSFET switching.

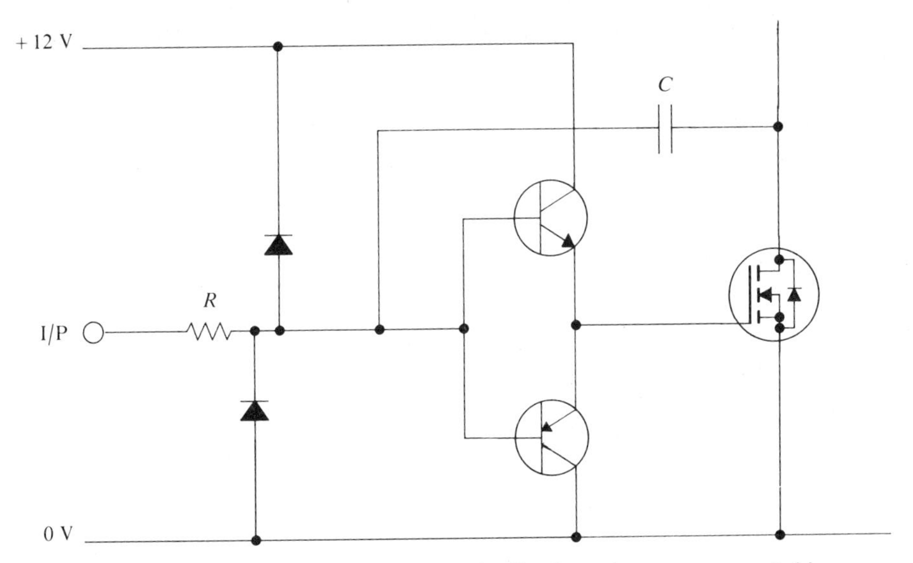

Figure 12.11. Use of capacitive feedback to slow MOSFET switching.

12.6. BRUSHLESS MOTOR DRIVE

The brushless dc motor is essentially a synchronous motor with rotor position feedback, contructed so that the motor windings are stationary while the magnets rotate. Switching of the supply to the phases of the stator in the correct sequence is arranged by a control circuit which receives absolute rotor-position information from a Hall-effect or some other form of position sensor. Power MOSFETS are widely used in this application due to their good overload capability and high reliability.

The simplest brushless system commonly used is a star-connected, two-phase, four-pole motor with single-ended drive (Figure 12.12). Such an arrangement would be employed typically in small, constant-speed hard-disc drives; the

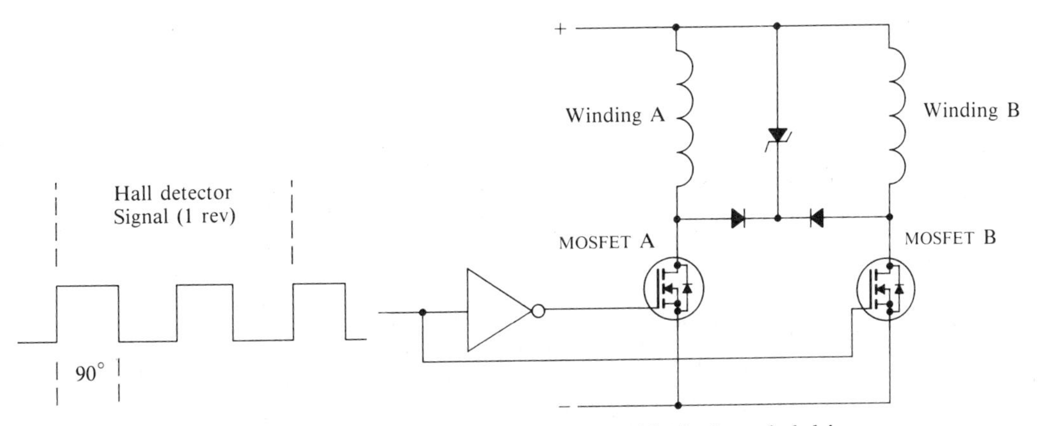

Figure 12.12. Two-phase, four-pole motor with single-ended drive.

high-torque ripple produced by this kind of drive can be smoothed by the inertia of the rotor. The peak current is defined by the winding resistance and is typically three times the rated average current. A zener diode is used to clamp the motor drain voltage on turnoff.

The most popular brushless configuration is probably the three-phase, four-pole motor with trapezoidal waveform switching (Figure 12.13*a*). Half a turn of the rotor is equivalent to one complete electrical cycle of the motor. This period is divided into six intervals controlled by Hall-effect or optical switches, as shown in Figure 12.13*b*.

Figure 12.14*a* shows a brushless dc motor drive scheme in which Q1, Q3, and Q5 provide voltage control as well as switching current between the windings.

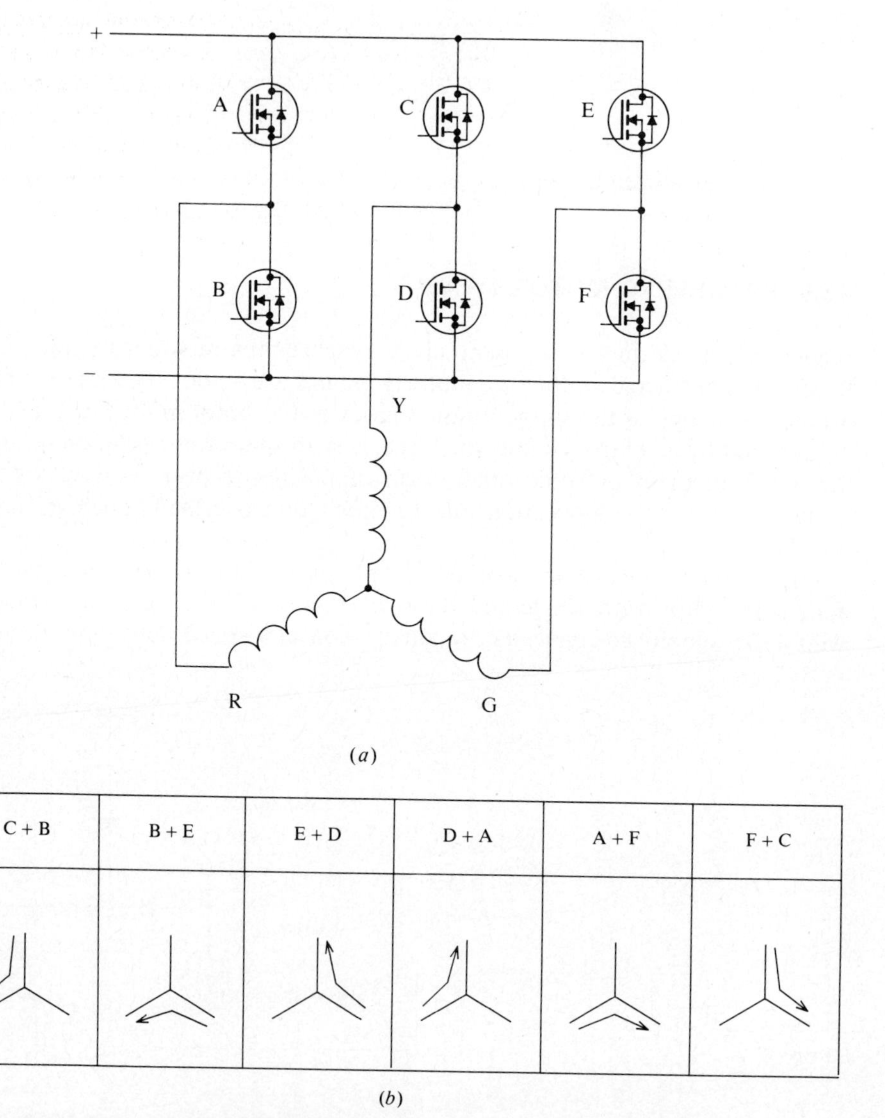

Figure 12.13. Three-phase motor drive. (*a*) Power circuit; (*b*) switching sequence.

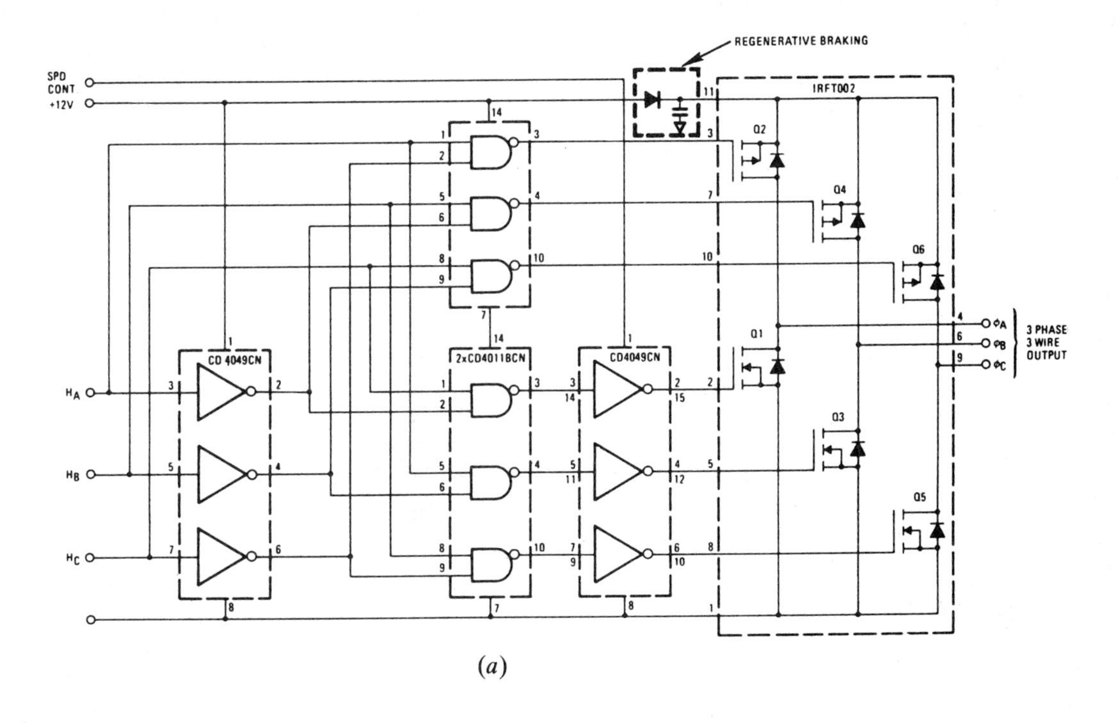

(*a*)

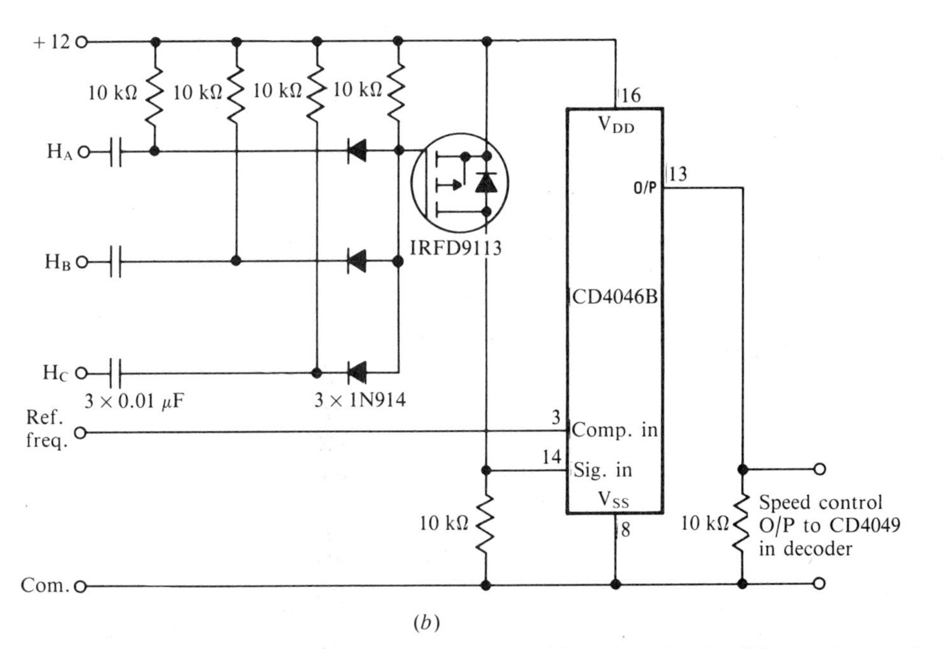

(*b*)

Figure 12.14. Three-phase brushless motor drive. (*a*) Drive circuit; (*b*) speed-control signal decoder. (Courtesy of International Rectifier Corp.)

The amplitude of their gate voltages is varied in relation to the motor speed. This method of voltage control is dissipative and therefore only suitable for small motors such as those used in computer disc drives. Figure 12.14*b* shows a voltage control circuit which generates from the MOSFET switching signals a voltage control signal which is linearly proportional to the motor speed.

From the switching sequence it can be seen that each MOSFET conducts for 120° of the output cycle. Figure 12.15 shows the phase-voltage, line-voltage, and typical line-current waveforms resulting from the six steps. The partially inductive nature of the load maintains a phase current for longer than 120°, while the MOSFET conducts for only 120°. The body–drain diode of the other MOSFET in the leg of the same phase provides a path for the free-wheeling current.

The shape and amplitude of the current waveform change with speed and supply voltage. An estimation of power dissipated in the MOSFET can be made by determining the rms current during MOSFET conduction and hence arriving at the $I^2R_{DS(on)}$ loss, to which must be added the dissipation in the body–drain diode determined from the current waveform and the forward characteristics of the diode. At low frequency and during starting, the value of the load current used to determine I^2R, and thus the MOSFET die temperature, must be the peak current of the waveform, since the cycle will be too long to permit averaging of the temperature. Also it should be borne in mind that the current waveform shape will be altered by the use of load-angle control in which the "neutral" or commutation position of the Hall devices is electrically altered.

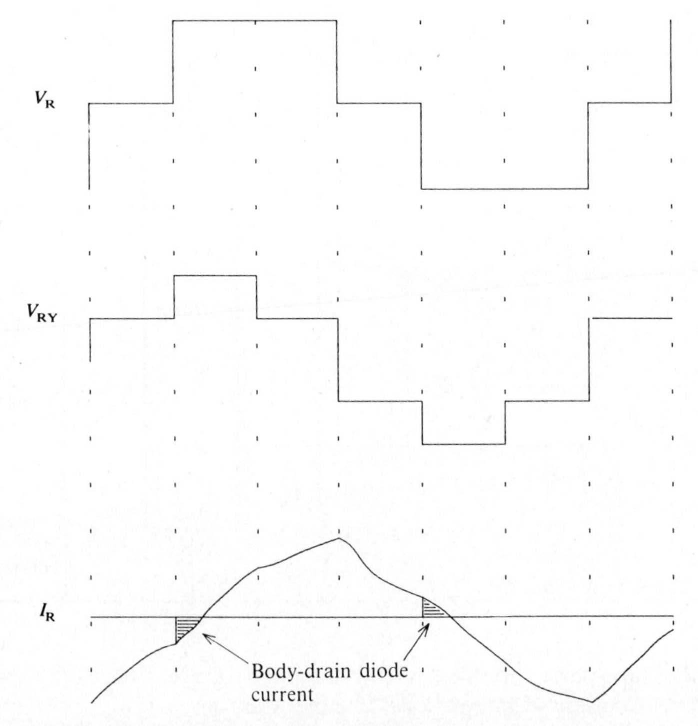

Figure 12.15. Voltage and current waveforms in the brushless motor drive.

To protect the MOSFETS and the motor from excessive dissipation during low-speed operation, some form of current limiting is usually necessary. The following methods are used:

1. Reducing the supply voltage with open-loop control linked to the drive frequency.
2. Linearly controlling the conductivity of the MOSFETS by reducing the gate drive voltage.
3. High-frequency chopping current control.
4. Variable-pulse-width sine-weighted switching.

Method 3 gives the highest efficiency. When the load current reaches its limiting value, the upper MOSFET is turned off. The current decays by free-wheeling around the circuit as shown in Figure 12.16 until the current reduces to a predetermined value, whereupon the upper switch turns back on. This repeats at several kilohertz, effectively providing a switch-mode current source. This chopping sequence, applied to each of the upper three MOSFETS, is superimposed on the basic switching cycle. Method 4 uses a more complex control scheme to control the chopping of the upper MOSFETS so that the three phase currents are sinusoidal rather than trapezoidal. This scheme requires a continual update of motor position but results in lower torque ripple, particularly at low speed. Methods 3 and 4 involve reverse recovery of the body–drain diodes because of the use of multipulse pulse-width modulation. Steering diodes or a center-tapped choke may therefore be needed to isolate the body–drain diode and to reduce diode recovery losses.

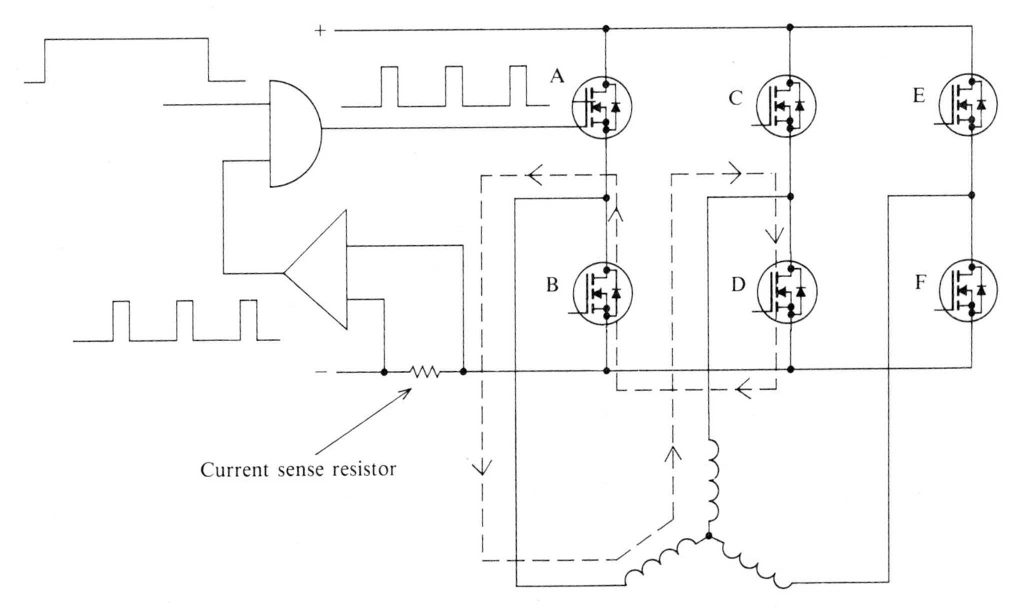

Figure 12.16. Current control by high-frequency chopping.

12.7. STEPPING-MOTOR DRIVES

The concept behind the stepping motor is that a rotor containing permanent magnets aligns itself with the field generated by current applied to the stator windings. On switching current to the various stator windings in a regular cyclic manner, the rotor will follow by continually seeking alignment with the energized poles. Beyond this simple concept lies a multitude of variations both in motor design and drive design. The permanent-magnet rotor design, while certainly the most common among small motors, is not the only form in which stepping motors are to be found. The variable-reluctance motor uses a rotor of ferromagnetic material which is not permanently magnetized. The hybrid stepping motor has a rotor formed by combining pairs of ferromagnetic cores sandwiched with permanent magnets. The variations in the drive systems for stepping motors arise from the tradeoff between economy and efficiency with regard to limiting the stator current, and also from the requirements for subdividing the natural step angle.

A basic two-phase unipolar drive with resistive current limit is shown in Figure 12.17. Only one pole pair is shown in the rotor, although in practice as many as twelve pairs may be used. Two excitation modes are shown in Figure 12.18, where either a single phase is energized, causing alignment with that phase, or two phases are excited simultaneously, causing alignment between these two phases. This second approach achieves a higher working torque and positional accuracy, but at the expense of higher input current and MOSFET duty cycle.

At high speed the available torque drops off due to the winding inductance

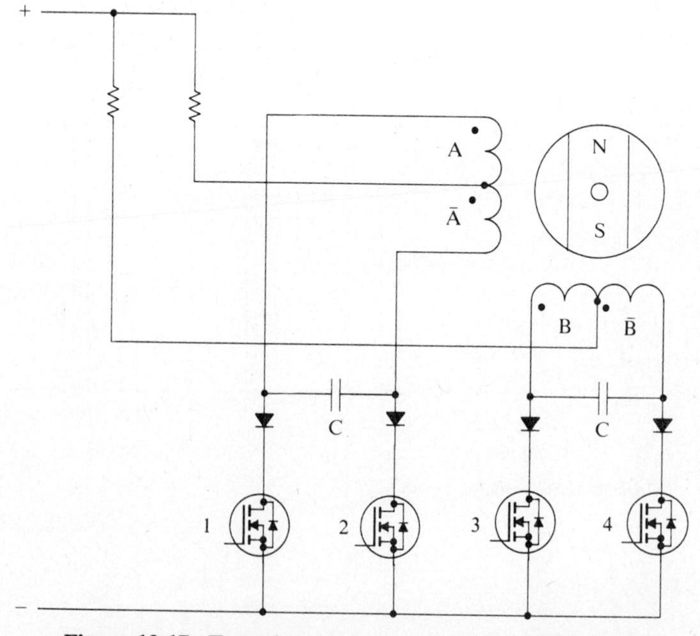

Figure 12.17. Two-phase unipolar stepping-motor drive.

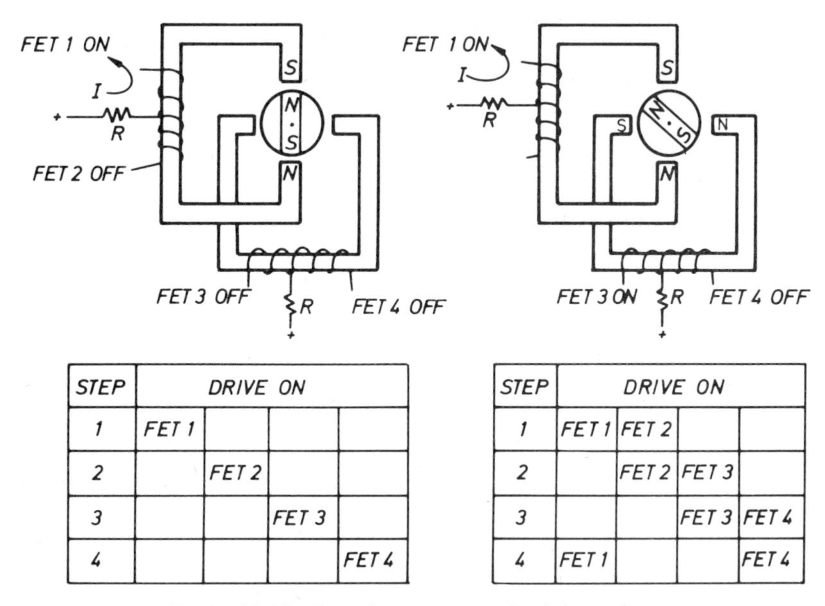

STEP	DRIVE ON			
1	FET 1			
2		FET 2		
3			FET 3	
4				FET 4

STEP	DRIVE ON			
1	FET 1	FET 2		
2		FET 2	FET 3	
3			FET 3	FET 4
4	FET 1			FET 4

Figure 12.18. Stepping-motor excitation modes.

limiting the rate of change of current, which in turn limits the peak current at each step. To overcome this, a high supply voltage may be used with current-limiting resistors in series with the windings to protect the motor and MOSFETS. A variety of other schemes for protecting power MOSFETS in this application have been developed [3].

In the circuit shown in Figure 12.17 diodes are located in series with each MOSFET to prevent the body–drain diode of that MOSFET acting as a clamp when the opposite MOSFET turns off. A voltage greater than twice the supply voltage can therefore be developed across each winding. This allows the current in the winding to fall at a greater rate than would otherwise be the case. The capacitor across the winding pair, together with the inductance and resistance of the coils, forms a damped resonant circuit which determines the peak voltage appearing at the drains of the MOSFETS.

Alternatively the drains may be zener-clamped, as shown in Figure 12.19, so that $V_S + V_Z < BV_{DSS}$ (where V_S is the supply voltage and V_Z the zener diode breakdown voltage). A resistor may be used instead of the zener diode, with the resistor value chosen so that the drain voltage remains below BV_{DSS} when carrying peak current.

Bipolar drive motors are also available without a center-tap connection for use with a full bridge as shown in Figure 12.20. A similar bridge is required for other phases. Automatic clamping of drain–source voltage occurs via the body–drain diodes and the supply. To simplify the gate-drive requirements, p-channel MOSFETS may be used in the upper positions, as Figure 12.20 illustrates.

To improve both the high-speed torque and the efficiency, a switching current regulator can be built around the motor. A low-value resistor R is added to the negative supply of the bridge and is used to sense the motor current, as shown in

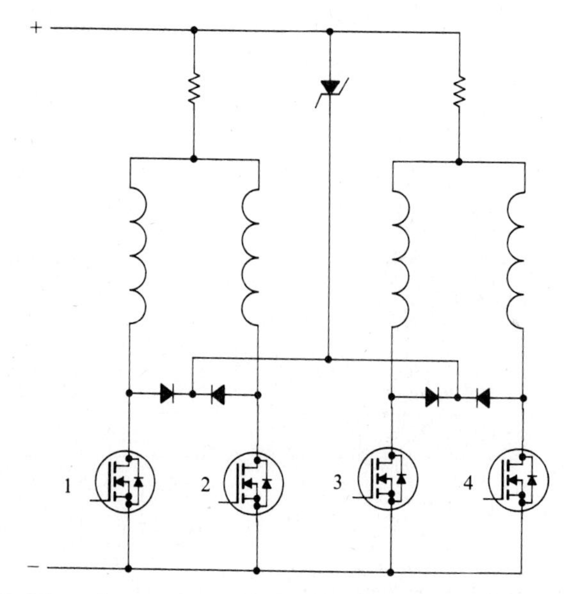

Figure 12.19. Two-phase unipolar stepping-motor drive with zener clamping.

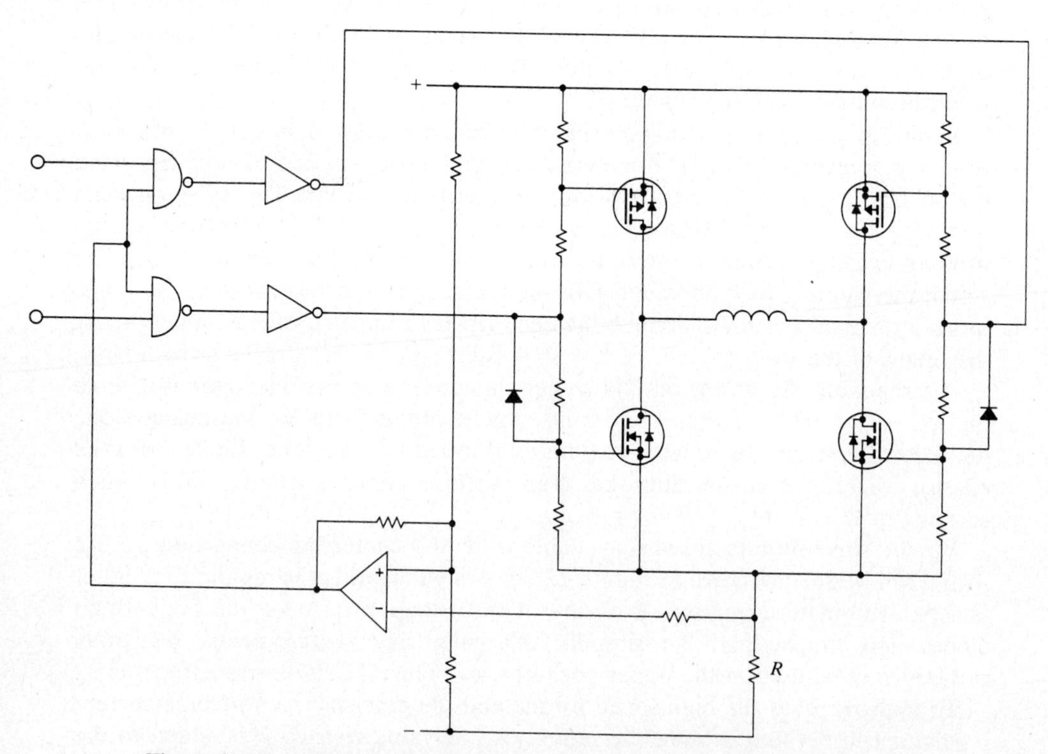

Figure 12.20. Full-bridge bipolar stepping-motor drive with current control.

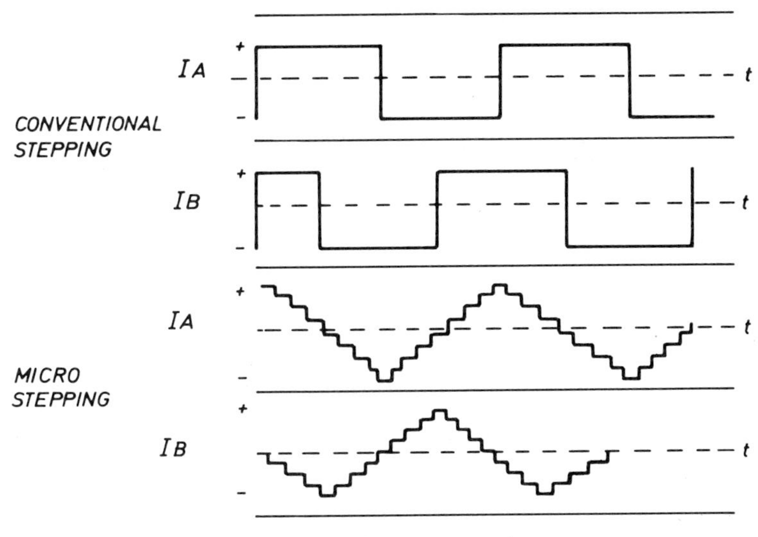

Figure 12.21. Microstepping.

Figure 12.20. No series current-limiting resistor is used in this circuit, allowing the current to ramp up at a rate dependent on the supply voltage and the winding inductance. When the voltage across R exceeds the comparator threshold, both p-channel MOSFETS turn on and both n-channel MOSFETS turn off. The input filter of the comparator discharges to the reset level while the coil current decays around the p-channel MOSFETS. The bridge then reverts to its initial state of one n-channel and one p-channel device conducting. This sequence repeats at a high frequency to control the average current in the coil. This approach uses multipulse PWM, and therefore the MOSFETS should have an adequate diode-recovery dv/dt rating.

Microstepping provides fine control over the rotor position. By using a current control loop to control the pulse-width modulation of the drive bridge, the current levels may be proportioned between the two phases. The rotor may thus be drawn to any position between two basic steps. In this way the basic step may be subdivided into a number of microsteps as shown in Figure 12.21. One way of obtaining this current waveform from a full bridge is always to have two of the MOSFETS conducting, either MOSFET 1 and MOSFET 4, or MOSFET 2 and MOSFET 3 (see Figure 12.22). This has the benefit that the channels conduct the free-wheeling current. If the current levels are below $(0.6\,V)/R_{DS(on)}$, then the body–drain diodes will not conduct, thereby avoiding diode-recovery problems.

12.8. BATTERY-DRIVEN MOTORS

Power MOSFETS have proved advantageous in the control of motors driven from batteries in such applications as hand-held tools [4]. With the reduction of power-MOSFET prices to commodity levels, toy manufacturers have also discovered the benefits of power MOSFETS in the control of motors and other battery-driven equipment.

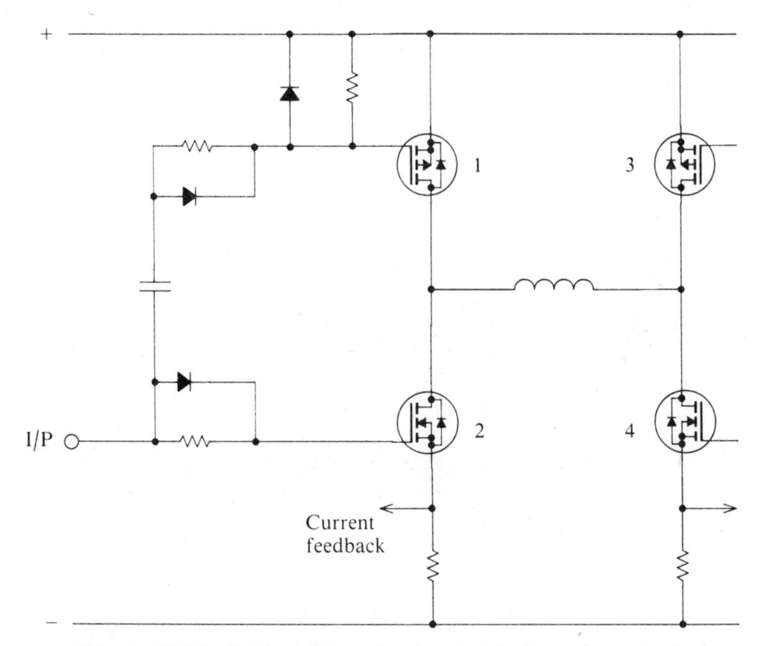

Figure 12.22. Bridge drive circuit suitable for microstepping.

The motor used is often a dc motor with mechanical commutator and permanent-magnet field. Speed control is obtained by chopper control, or control may be limited to a simple on-off function. Where chopper control is used, a power MOSFET is a very appropriate device because of its ability to switch efficiently at frequencies above the audible range with little power loss. Gate-drive power requirements are negligible. The average current drawn from the battery for the charging and discharging of the gate of a 15-A power MOSFET operating at 20 kHz is typically less than 1 mA.

The principal advantage that the power MOSFET offers in battery applications is the improvement in efficiency over that obtainable when using a simple bipolar transistor or a Darlington bipolar transistor. The forward drop of a power MOSFET can be reduced to any desired level by using sufficient silicon area. Since most battery applications are low-voltage and since power-MOSFET resistance falls steeply with the voltage rating, the use of a power MOSFET is economic. If a bipolar transistor is used, the designer is faced with the need of supplying a large base current, with the attendant power loss, in order to ensure saturation of the transistor and reasonable conduction losses. The alternative of using a Darlington bipolar transistor to minimize base-drive requirement brings with it increased conduction losses due to the higher forward drop of the Darlington transistor.

Another major factor in favor of the power MOSFET in this application is the high surge-current capability of the power MOSFET compared with that of the bipolar transistor. The power MOSFET, with its high overload capability, is better suited to handling stall currents, starting currents, and other overload situations commonly associated with motor operation. The square safe operating area also tends to give the power MOSFET a higher reliability in such applications, due to its lack of second breakdown, an important freature when controlling clamped inductive loads.

12.9. SIX-STEP THREE-PHASE DRIVE

A three-phase bridge circuit (Figure 12.23) is commonly used to drive three-phase ac motors—either induction motors, synchronous motors with permanent-magnet excitation, or reluctance motors. The motor windings may be connected in either delta or star, star connection being more often used for high-voltage motors in order to minimize the voltage across each phase of the winding.

The current waveform shown in Figure 12.23*b* is typical for the motor running at the rated speed. At reduced speeds the current waveform becomes more "peaky", resulting in increased torque ripple. As, for example, MOSFET A turns off, due to the lagging nature of the load, the current free-wheels through the body–drain diode of MOSFET B. MOSFET B is gated on at this time. The current reverses and commutates from the diode of B to the channel of B during the period that B is on and no stress is applied to the body–drain diode. However, below base frequency, where constant-power gives way to constant-torque operation, some means of reducing the effective motor voltage is required. A straightforward approach is to derive a variable dc supply by phase control of the

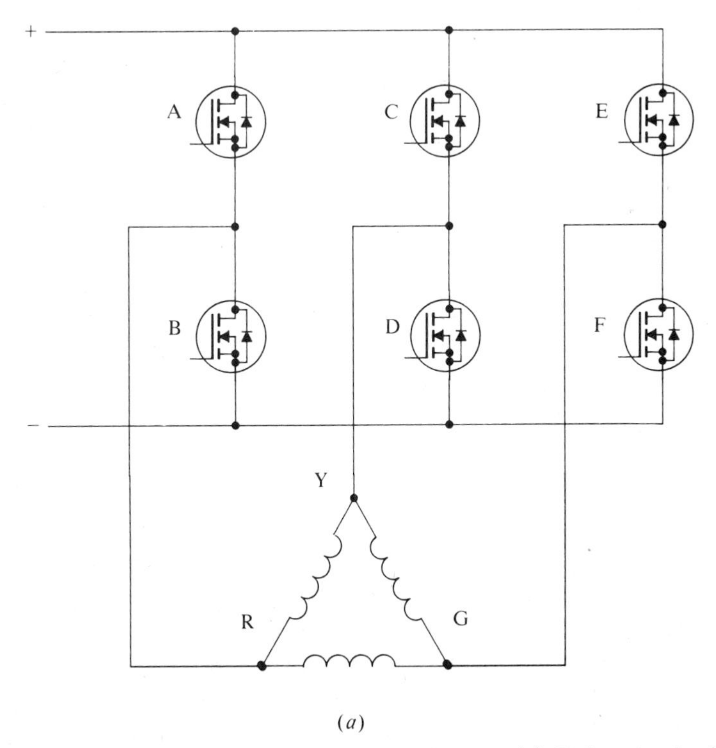

Figure 12.23. Six-step, three-phase drive for ac motors. (*a*) Drive circuit; (*b*) switching sequence and winding-current waveform.

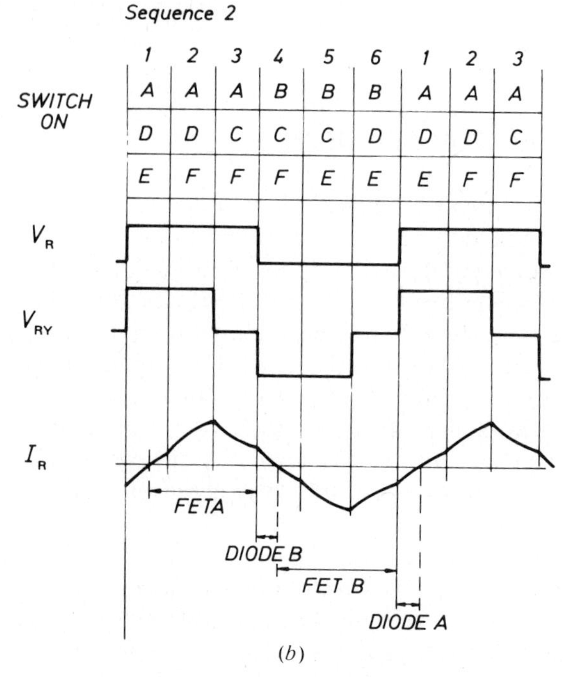

(*b*)

Figure 12.23. (*Continued*)

ac mains. However, it is usually more economical to achieve voltage control by chopping the conduction periods of the upper or lower MOSFETS at several kilohertz. During this chopping action the free-wheeling current from the body–drain diode of one MOSFET commutates into the channel of the opposite MOSFET in the same limb, so that the problems of diode recovery current and diode recovery dv/dt must be taken into consideration.

12.10. SIX-STEP THREE-PHASE DRIVES WITH HARMONIC ELIMINATION

In the previous section, constant-duty-cycle PWM was considered as a method of reducing the effective drive voltage of six-step inverters operating below base frequency. Improved performance can be obtained by employing a more sophisticated form of PWM.

At low drive frequency the current and torque ripple occurring in six-step drives is not effectively smoothed out by the mechanical inertia. In this situation it is beneficial to eliminate the low-order harmonics, in particular the 5th and 7th. Since 3rd and 9th harmonics are absent from the line-to-line waveforms in a three-phase system, the 11th harmonic is then the first to cause torque ripple. By inserting four notches into each of the three switching waveforms of the basic six-step drive, the 5th and 7th harmonics are canceled (Figure 12.24). The notch angles are 16.25° and 22.06°. At a maximum frequency of 120 Hz, typical of inverters driving motors designed for 50 or 60 Hz, this notch period is 159 μs. Therefore, timing errors or delays in MOSFET switching of up to a microsecond or

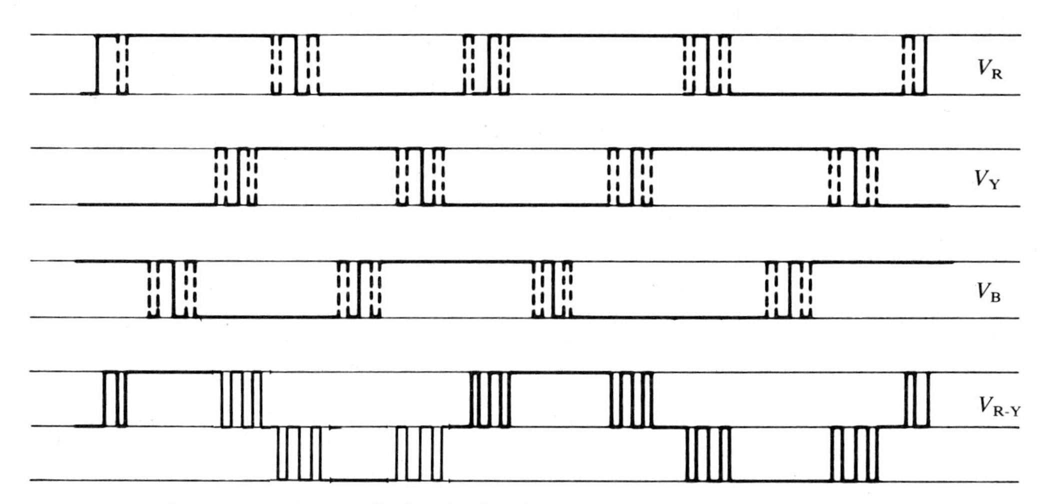

Figure 12.24. Harmonic cancellation in the six-step waveform by the addition of notches.

more can be tolerated. By reducing di/dt and dv/dt, as described in the previous section, the body–drain diode recovery current can be reduced to acceptable levels.

To synthesize a reduced voltage for low-frequency operation would require a much higher superimposed chopping frequency to preserve the harmonic elimination, and it then becomes advantageous to use sine-weighted modulation. The current waveform depends on the drive frequency and loading on the motor. Figure 12.25 gives a typical example for half a cycle of one phase. From this waveform, conduction losses in the MOSFETS can be found by summing the I^2R losses and the diode forward conduction losses at each switching interval over the whole output cycle.

The usual method of braking is to ramp the drive frequency down at a controlled rate. If the energy recovered from the rotor and its load is greater than the losses in the stator and inverter, then the dc supply voltage rises due to regeneration. An additional MOSFET may be employed to switch on and off a shunt load resistor to dissipate this energy as shown in Figure 12.26.

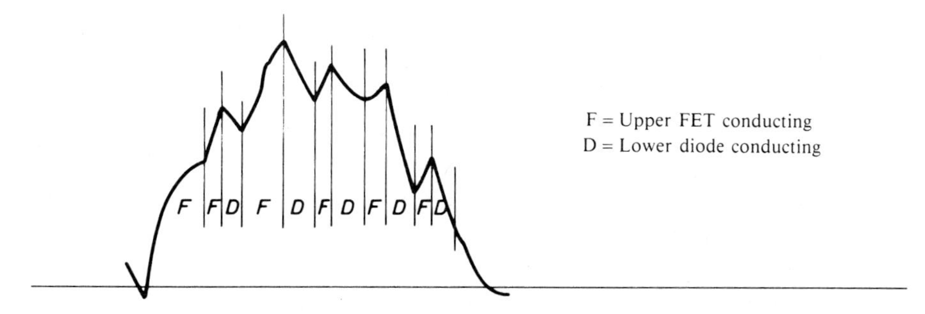

Figure 12.25. Waveform obtained with sine-weighted switching.

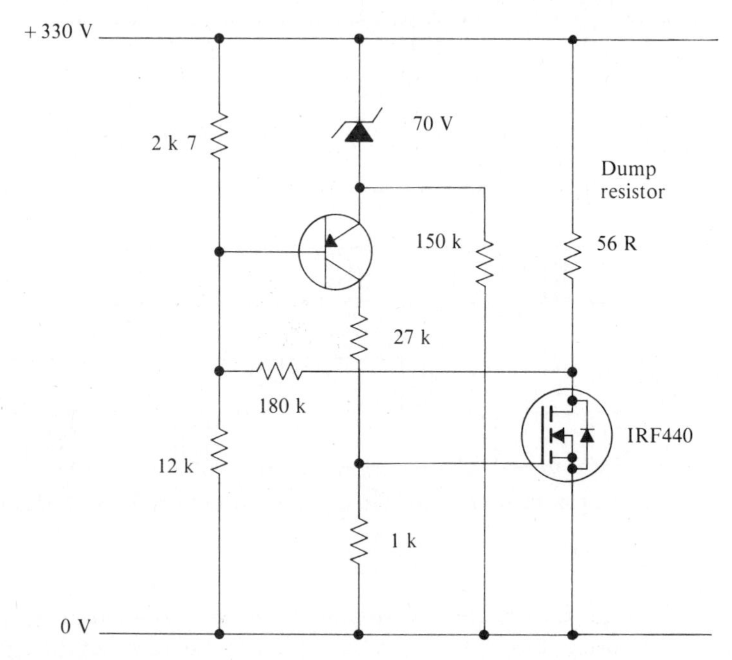

Figure 12.26. Energy-dump circuit for controlling the dc rail voltage during regeneration.

12.11. SILENT THREE-PHASE PWM DRIVES

Although basic six-step drives, and to some extent harmonic-canceled six-step drives, are attractive because of their simplicity, there are a number of applications where these methods are inadequate. The current waveforms are distorted and have high peak values, so that there may be high motor losses due to the poor current form factor and considerable vibration due to torque ripple. By chopping the output of each phase with a mark-to-space ratio which varies sinusoidally throughout the output cycle, the effective output voltage from each phase is made to vary sinusoidally. Due to the motor inductance, or by adding output filters, the resulting current waveforms will be approximately sinusoidal. If the chopping frequency is raised above 18.5 kHz, the noise produced by the switching becomes inaudible. This type of drive is discussed more fully in Chapter 13.

REFERENCES

1. S. Clemente and B. Pelly, "A chopper for motor speed control using parallel connected power MOSFETS." *Int. Rectifier Appl. Note* **AN941A** (1983).
2. E. Hebenstreit, "Overcoming the dv/dt problem in power MOSFET switching stages

during commutation." *Proc. Power Convers. Int. Conf.* September, pp. 147–153 (1982).
3. C. D. Moyer, "Using power MOSFETS in stepping motor control." *Proc. Powercon* **9,** C1.4.1–C1.4.9 (1982).
4. P. W. Neumann, "FET dc speed control switch investigation in cordless portable tools." *IEEE, Power Electron. Spec. Conf.* June, pp. 157–161 (1984).

CHAPTER 13

PWM Inverters

13.1. INTRODUCTION

The technique of pulse-width modulation allows an inverter operating from a fixed-voltage dc supply to generate an ac supply with variable frequency and variable output voltage. The principal applications for this kind of inverter are ac motor drives and uninterruptible power supplies. Figure 13.1 shows the basic power circuit for operation from a single-phase ac supply. The load is switched alternately between the positive and negative rails of the dc supply. By appropriate control of the switching instants of the power devices, an output voltage waveform can be produced whose fundamental is of the required frequency and amplitude (Figure 13.2).

Power-MOSFET switching speeds permit the use of a switching frequency outside of the audible frequency range. Since in a typical application the switching frequency is at least an order of magnitude higher than the output frequency, output filtering is practical. This makes it possible to economically generate variable-frequency ac supplies that are virtually free of harmonics and RFI problems. Audible acoustic noise is also eliminated, making such systems acceptable for use in the home and in the office. Applications include variable-frequency ac motor drives and uninterruptible power supplies for computer equipment [1]. Figure 13.3 shows typical output waveforms from a power-MOSFET inverter switching at 20 kHz.

13.2. CURRENT WAVEFORMS IN PWM INVERTERS

The load connected to the output of each phase of the inverter will generally be either a filter inductor or an inductive load, such as one phase of the stator winding of a motor. Due to the wide separation of the switching frequency and the output frequency, the load will appear highly inductive relative to the switching frequency, and therefore the load current will be approximately constant and continuous during each switching cycle. During a switching cycle the load current commutates between one of the switching devices and the diode of the other. The output current and output voltage waveforms for one leg of such an inverter at various points during the output cycle are shown in Figure 13.4. When load current is flowing out of the center point of the bridge, the top MOSFET and bottom diode conduct alternately. When load current is flowing into the center point, the bottom MOSFET and top diode conduct.

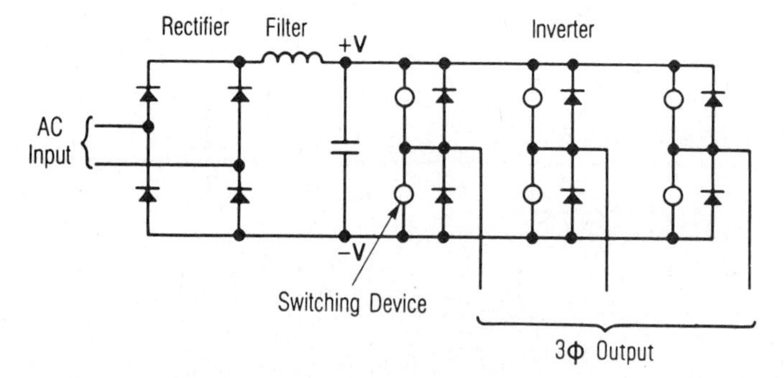

Figure 13.1. PWM inverter circuit.

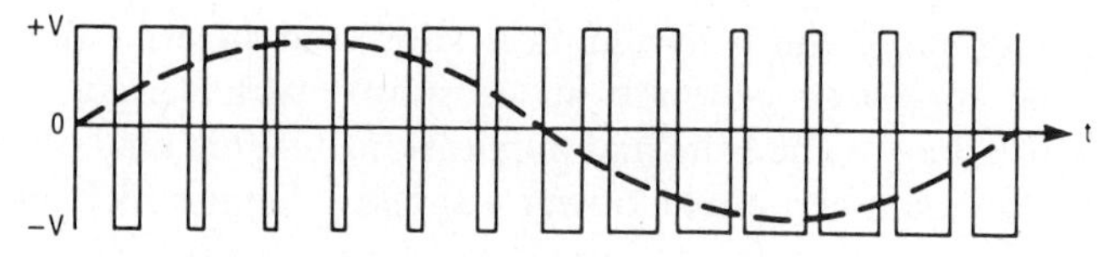

Figure 13.2. PWM waveform.

13.3. USING THE BODY–DRAIN DIODE

The body–drain diode of a power MOSFET can be used as the free-wheeling diode in PWM circuits provided the device carries a rating which enables the designer to know for certain that the device is operating within its capabilities. The relevant rating is known by several terms—diode-recovery dv/dt, commutating dv/dt, or dynamic dv/dt. We have chosen diode-recovery dv/dt, since it most clearly

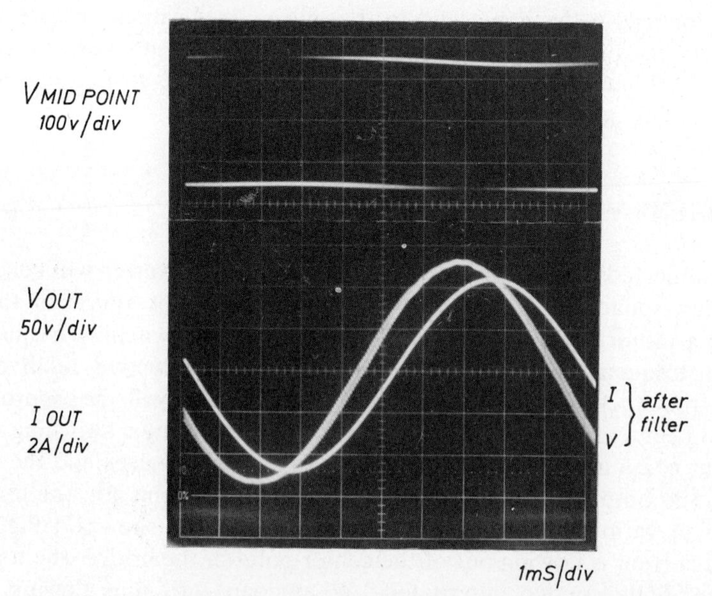

Figure 13.3. Output waveforms of a three-phase 1-kVA inverter.

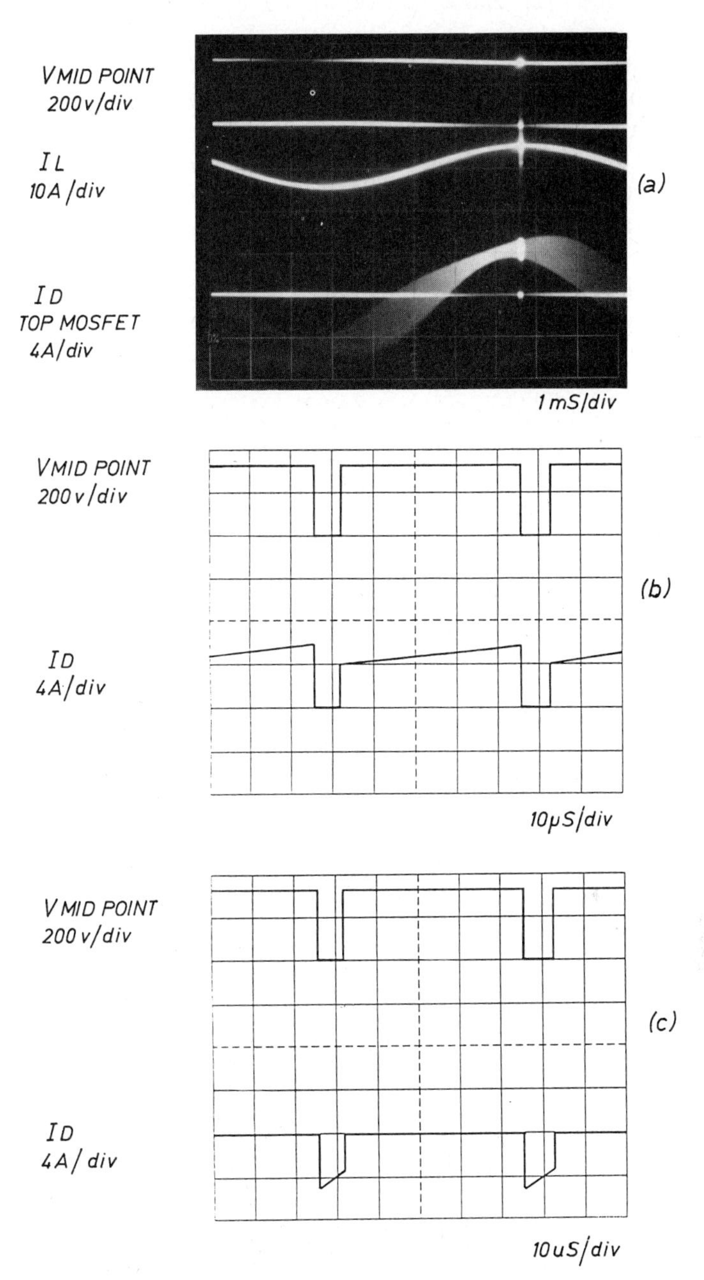

Figure 13.4. Current and voltage waveforms in the MOSFETS of one phase of a three-phase 1-kVA inverter. (a) Waveforms for one output cycle with short period intensified. (b) Expanded waveforms from intensified period showing V_{MIDPOINT} and I_D for the top MOSFET. (c) Expanded waveforms showing I_D for the lower MOSFET.

identifies the phenomenon with which it is associated—namely the recovery of the body–drain diode [2]. Diode-recovery dv/dt capability is an important rating in PWM inverter applications and other applications, such as dc servo motor drives [3], in which pulse-width modulation is employed.

It has long been recognized that power MOSFETS can fail in certain situations when their integral diode is used [4]. The relationship between device design and susceptibility to this type of failure is discussed in Section 3.6. Failure occurs as the drain–source voltage rises rapidly upon recovery of the body–drain diode—a situation encountered in PWM inverters. The critical moment for the device comes as the diode is recovering and the MOSFET (and its internal parasitic bipolar transistor) is called upon to block forward voltage. The more rapid the rise of drain voltage, the more likely is activation of the parasitic bipolar transistor, since

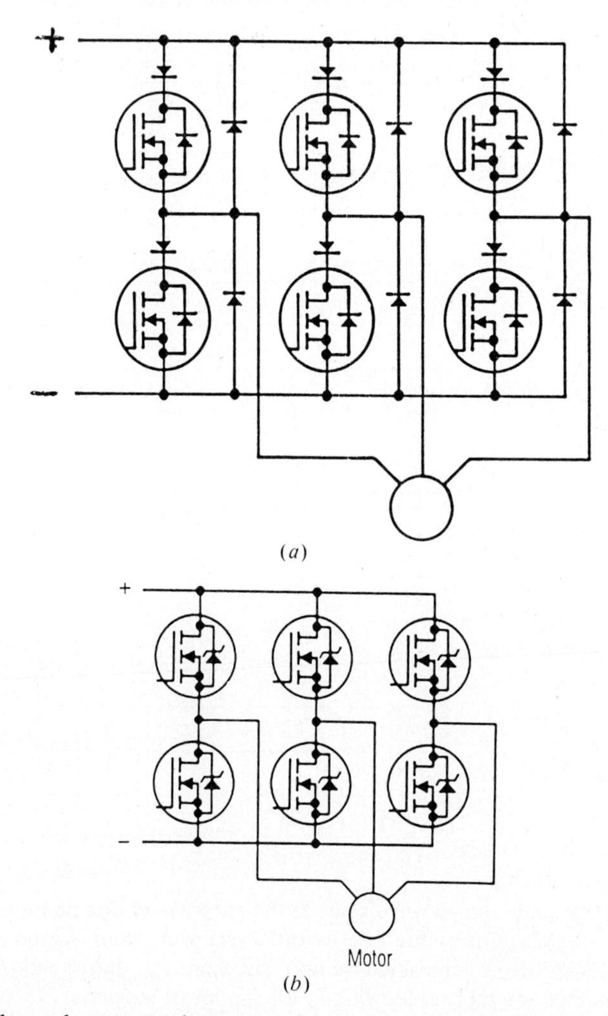

Figure 13.5. Three-phase PWM inverters. (*a*) External diodes are used to isolate the MOSFET body–drain diodes. (*b*) If the MOSFETS possess an adequate diode-recovery dv/dt rating, the body–drain diode may be used as the free-wheeling diode.

a higher dv/dt is associated with higher values of diode recovery current at the critical period. Prior to the introduction of diode-recovery dv/dt ratings, designers had to establish the allowable limits of dv/dt by painful experience with no firm guarantee that the devices were being operated within their capabilities. This led to the artifice of putting a diode in series with the power MOSFET to prevent conduction of the body–drain diode, with an external antiparallel diode providing a free-wheeling path for the load current (Figure 13.5*a*). MOSFET manufacturers are now providing the necessary information on the data sheets (see Appendix 7), so that the use of external diodes may no longer be necessary (Figure 13.5*b*), except where the diode recovery current of the intrinsic MOSFET diode is unacceptably large.

The rate of rise and fall of the drain voltage can be restrained by the use of a snubber circuit [5]. This has the benefit of reducing the dv/dt applied to the MOSFET, but at the expense of significant losses in the snubber and the extra burden on the MOSFET of carrying the snubber discharge current.

The use of a series diode and an external fast free-wheeling diode reduces the magnitude of the diode recovery current, but at the expense of increased forward conduction drop. If a Schottky diode is used as the series diode, it should be capable of withstanding avalanching. This requirement arises from charge stored on the output capacitance of the MOSFET. In the circuit shown in Figure 13.6 almost the whole of the dc supply voltage appears across C2, the output capacitance of the lower MOSFET. The Schottky capacitance, C1, will be roughly of the same order as C2. When the upper MOSFET turns off and D2 carries the

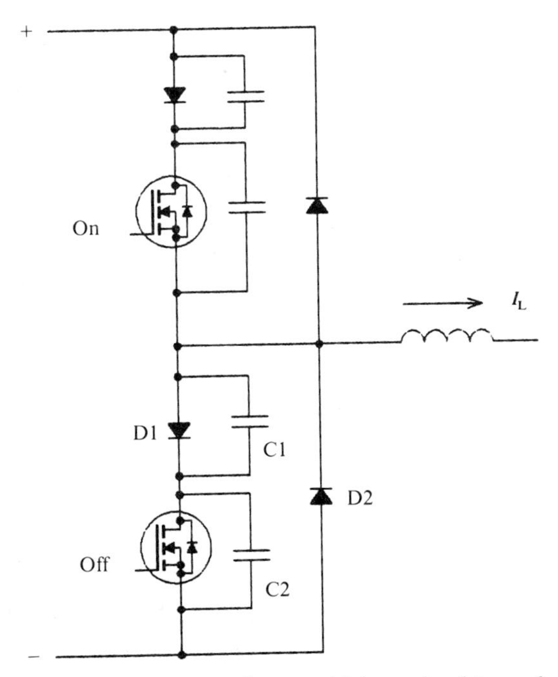

Figure 13.6. Inverter leg, showing capacitance which can lead to avalanching of the series diode.

free-wheeling current, the total voltage across C1 and C2 is close to zero. Charge therefore shifts from C2 to C1 to create a potential balance by settling with equal voltages across C1 and C2. This will normally cause the Schottky-diode to avalanche.

13.4. DIODE RECOVERY CURRENT

Whichever direction the load current is flowing, each switching cycle sees the commutation of the load current from the body–drain diode of one MOSFET to the channel of the other. This occurs when the nonconducting MOSFET turns on and results in a pulse of diode recovery current flowing between the rails of the dc power supply. The reverse recovery time of the diode, t_{rr}, increases with voltage rating typically as shown in Figure 13.7. For a constant value of di/dt, data-sheet values of t_{rr} and Q_{RR} are closely related, and Q_{RR} bears a similar relationship to the voltage rating. As well as voltage rating, t_{rr} is also a function of forward current, di/dt, reapplied reverse voltage, and junction temperature. These conditions can have a significant effect on the measured value of t_{rr}. Therefore, when comparing devices on a basis of t_{rr} it is important to take note of the conditions under which it is measured.

Diode recovery losses can be reduced by slightly overlapping the conduction periods of the upper and lower MOSFETS [6]. This ensures that the channel carries some of the load current when the diode conducts, thereby reducing Q_{RR} during recovery of the diode. However, the MOSFET voltage drop in high-voltage, high-$R_{DS(on)}$ MOSFETS is likely to be considerably greater than the forward drop of the diode, so that the diode continues to carry most of the current. Figure 13.8 shows the reverse conduction characteristics for a 100-V and a 500-V MOSFET of

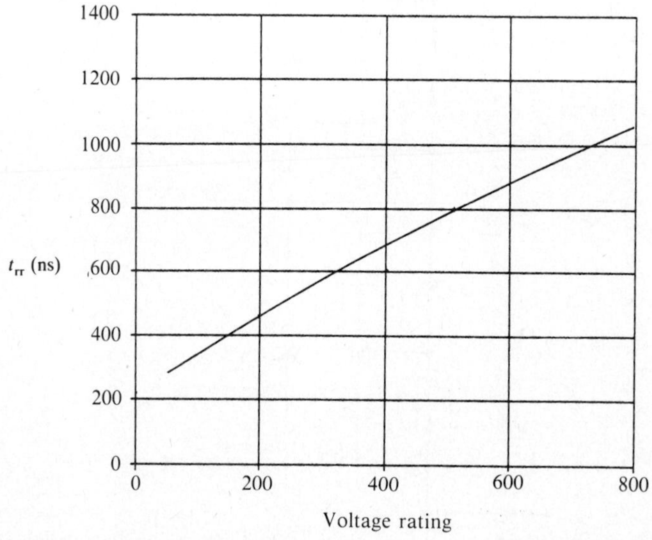

Figure 13.7. Typical reverse recovery time of a MOSFET body–drain diode as a function of voltage rating. (Die size 4.45 mm by 2.92 mm).

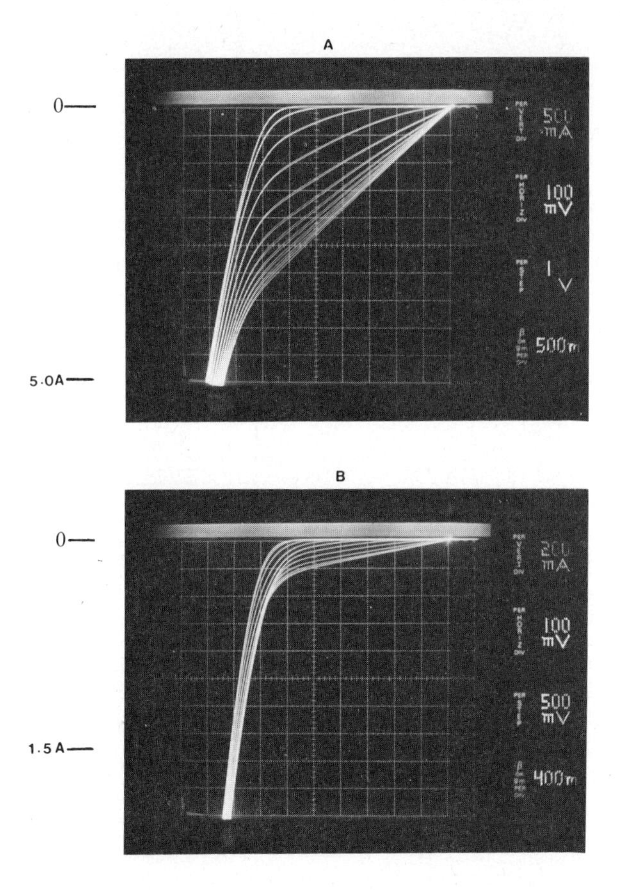

Figure 13.8. Reverse-conduction characteristics of MOSFETS compared with I_D rating. (*a*) 100 V MOSFET. I_D rating = 5.0 A (*b*) 500 V MOSFET. I_D rating = 1.5 A.

the same die size. In both cases diode conduction commences at values of the source current below the I_D rating of the devices, but especially in the case of the 500-V device.

13.5. MOSFET TURN-ON

During turn-on of one of the MOSFETS, for example Q2, in one arm of an inverter, the dc rail voltage will be supported by the circuit inductance, the body–drain diode of Q1, and the drain–source voltage of Q2 (Figure 13.9). During the period when the body–drain diode of Q1 is still saturated with carriers and is unable to support reverse voltage, the dc rail voltage is approximately equal to the voltage drop across the circuit inductance plus the drain–source voltage of Q2.

The rate of rise of current in Q2 is controlled by the rate of rise of its gate voltage, since drain current and gate voltage are linked by transconductance. The rate of rise of current determines the voltage drop across the circuit inductance (since $V = L\ di/dt$). The rest of the dc rail voltage appears across Q2. Therefore,

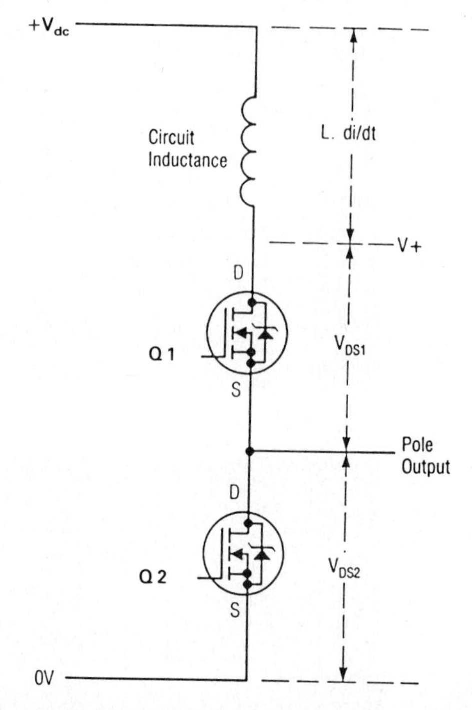

Figure 13.9. Equivalent circuit of inverter during commutation.

as long as some of the dc rail voltage is dropped across Q2, the value of the circuit inductance (either parasitic inductance or an inductor deliberately located in the dc rail) is not the principal factor in determining di/dt during turn-on; but only determines the extent to which the dc rail voltage is divided between the inductance and the MOSFET. Neither does it directly affect the magnitude of the diode recovery charge. However, the inductor does reduce the voltage across the MOSFET during the time that it is supporting the supply voltage as well as carrying the load current and the diode recovery current of the other MOSFET. This reduces the losses in the MOSFET, although energy is stored in the inductor which must later be dissipated. If the gate voltage is raised sufficiently rapidly, Q2 will turn fully on, and the rate of rise of diode recovery current is then determined by the inductance.

13.6. CONTROLLING THE RATE OF TURN-ON

The rate of turn-on is essentially determined by the rate of rise of gate voltage (see Section 4.2). There are a number of ways in which this can be controlled.

A MOSFET gate driver in its simplest form consists of a voltage source and a series resistor (Figure 13.10*a*). The voltage provided by the source should be adequate to ensure that the MOSFET remains fully on when carrying peak drain current. A voltage somewhat higher than the bare minimum is advisable to ensure that the channel is fully enhanced and the forward drop across the MOSFET

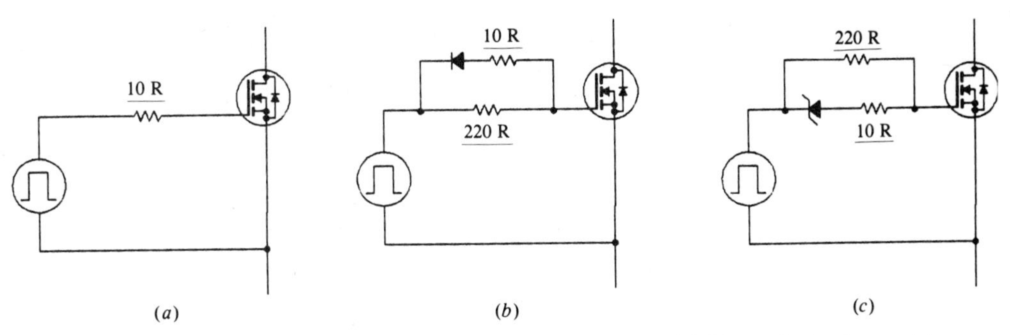

Figure 13.10. Gate drive circuits for switching speed control. (*a*) Nonpolarized gate drive; (*b*) polarized gate drive; (*c*) polarized gate drive with transconductance-limited drain current.

is minimized. Figure 13.10*b* shows a modification of the simple driver circuit which permits independent adjustment of turn-on and turnoff times. This circuit is applicable to bridge inverter circuits, since turnoff can normally proceed as rapidly as possible, while turn-on must be slowed to mitigate the effects of the shoot-through produced by the diode recovery process.

The series-resistor method has the disadvantage that in a typical PWM application, in which it is desired to hold the diode recovery current well below the I_{DM} rating of the MOSFET, the value of the gate resistor will be of the order of several hundred ohms. This means that the delay time between the application of the gate pulse and the gate voltage reaching the threshold value will be unacceptably long. Furthermore, with such a high value of the series gate resistance, the MOSFET will be prone to oscillation due to feedback. A further disadvantage of a high-impedance gate drive is that it causes turn-on tailing. This can be seen in Figure 13.11, which shows the voltage and current waveforms during turn-on of one MOSFET in a bridge circuit. The current waveform includes the diode recovery current of the other device. Tailing occurs at turn-on after the

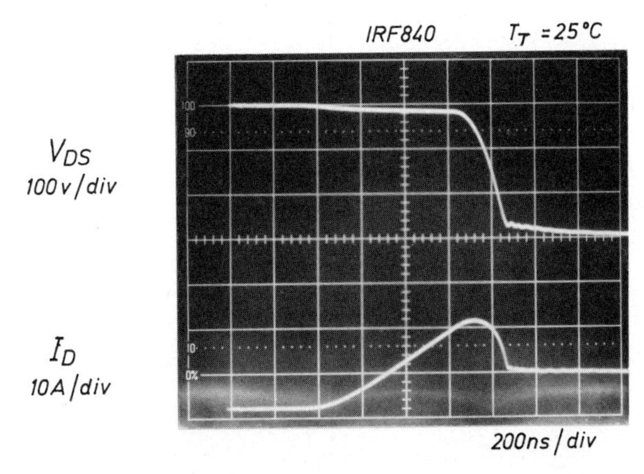

Figure 13.11. MOSFET turn-on with voltage tailing.

drain voltage has fallen to a value approximately equal to the gate voltage. At this point there is a sharp increase in gate drain capacitance, so that the Miller effect becomes very powerful. The strong Miller feedback action slows down the turn-on, so that there is a delay in fully turning on the MOSFET. During this time the MOSFET is carrying the full load current, so that this tail of drain voltage represents significant losses within the device. A low-impedance gate drive minimizes the length of the tail period.

In a PWM inverter switching at 20 kHz, waveform quality requires that the commutation process should be accomplished as quickly as possible but without infringing any of the device limits. In particular, it is necessary to respect the limit on peak drain current. The peak diode recovery current can be controlled by adjustment of the gate voltage of the incoming MOSFET during the period of diode recovery of the other device. A circuit which accomplishes this is shown in Figure 13.10*c*. The series resistance at turn-on and turnoff is only 10 Ω, but the available voltage during the initial stage of turn-on is reduced by the zener diode to a value that ensures that constant-current operation occurs at a suitable value of drain current.

The problem of high gate-driver impedance can be avoided by use of a low-value gate series resistor and ramping gate drive voltage. The required rate of rise of gate voltage is obtained by adjusting the rate of rise of the leading edge of the gate drive pulse (Figure 13.12). The series resistance then plays a less important role in determining the rate of rise of drain current, and a lower value of the resistor can be used.

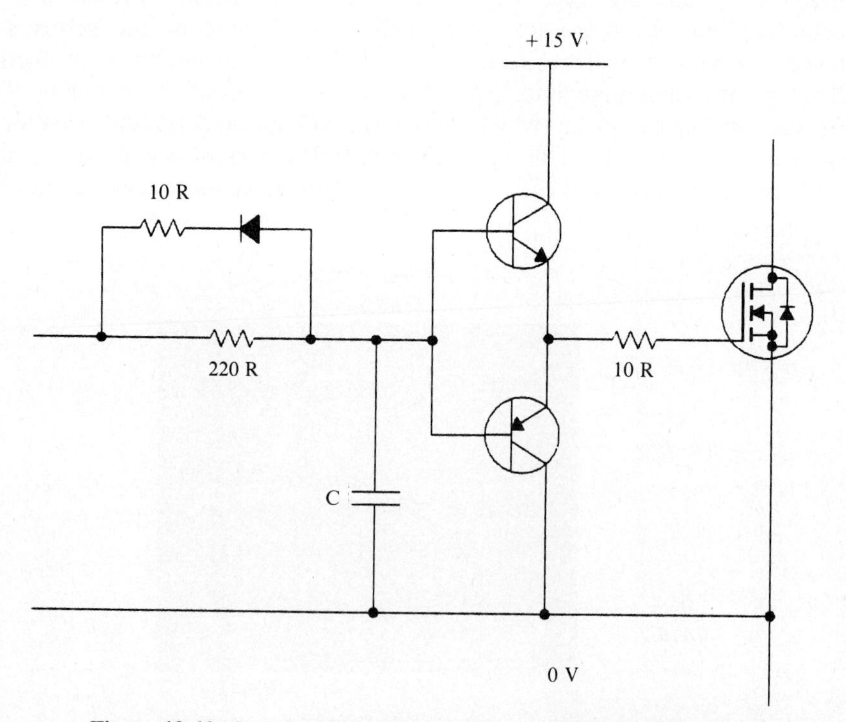

Figure 13.12. Low-impedance gate drive circuit with slow turn-on.

13.7. USING INDUCTANCE TO CONTROL DIODE RECOVERY CURRENT

The diode recovery current and the losses associated with it can be controlled by inserting inductance in the circuit at various locations, such as in the source leads of the MOSFETS, at the center point of the bridge, and in the dc supply. Inductance may also be used to relieve the stress on power MOSFETS with an inadequate diode-recovery dv/dt rating. These methods were outlined in Chapter 12.

The di/dt during turn-on of a power MOSFET may be controlled by locating an inductor in the source lead, common to both the load and the gate drive (Figure 12.10). The voltage drop across the inductor detracts from the gate drive voltage V_G, so that di/dt will stabilize at a value given by

$$L_S \frac{di}{dt} = V_G - V_{GS} \tag{13.1}$$

where V_{GS} is the gate–source voltage required to maintain the drain current at any given time. Since V_{GS} will vary as the current rises, di/dt is only approximately constant. To damp the resonance of this choke with the MOSFET capacitance, a low-value resistor should be placed across the choke.

An alternative to using Schottky diodes to steer the load current away from the body–drain diodes of the MOSFETS (Figure 13.5) is to locate a center-tapped choke at the midpoint of the bridge (see Section 12.4 and Figure 12.7). The slopes of each segment of current I_L need to be found either by calculation or by measurement. These values of di/dt are a function of the motor equivalent-circuit components, loading, and supply voltage. With MOSFET 1 of Figure 12.7 chopping and supplying I_L, we want D1 to free-wheel the current in preference to the body–drain diode of MOSFET 2. Suppose during a 5-μs OFF period of MOSFET 1 the current in the load falls by 0.5 A. This slope of current induces a voltage across L1 given by

$$V = -L1 \frac{di}{dt} \tag{13.2}$$

L1 can therefore be selected to produce a voltage of say 1 V. If L1 and L2 are wound on the same core, then an additional 1 V is induced in L2 by transformer action. There then exists a 2-V bias difference across D2 relative to the body–drain diode of MOSFET 2, such that only D2 will conduct. The choke must of course be designed so as not to saturate at the highest value of load current.

MOSFET losses may be reduced by including a series inductor in the dc supply (Section 12.5 and Figure 12.9). As the diode recovery current increases, di/dt produces a voltage drop across the inductor. This reduces the voltage across the MOSFETS and reduces dissipation in them. However, energy is trapped in the inductor during diode recovery, and this has to be dissipated in the discharge resistor in parallel with the inductor. Furthermore, the presence of the inductor results in both dips and peaks being impressed on the dc voltage supplied to other phases.

13.8. DIODE RECOVERY LOSSES AND SWITCHING LOSSES

At the beginning of every switching cycle the load current is circulating through the body–drain diode of one MOSFET prior to the turn-on of the other MOSFET. Before the diode can sustain a reverse voltage, a quantity of charge must be extracted from it. Further charge is extracted as the voltage of the center point of the bridge swings between power rails with a corresponding rise in the reverse voltage of the diode. The total charge associated with diode recovery flows between the rails of the dc supply without doing useful work. The diode recovery current therefore leads to a loss of energy, which appears as heat in the MOSFETS.

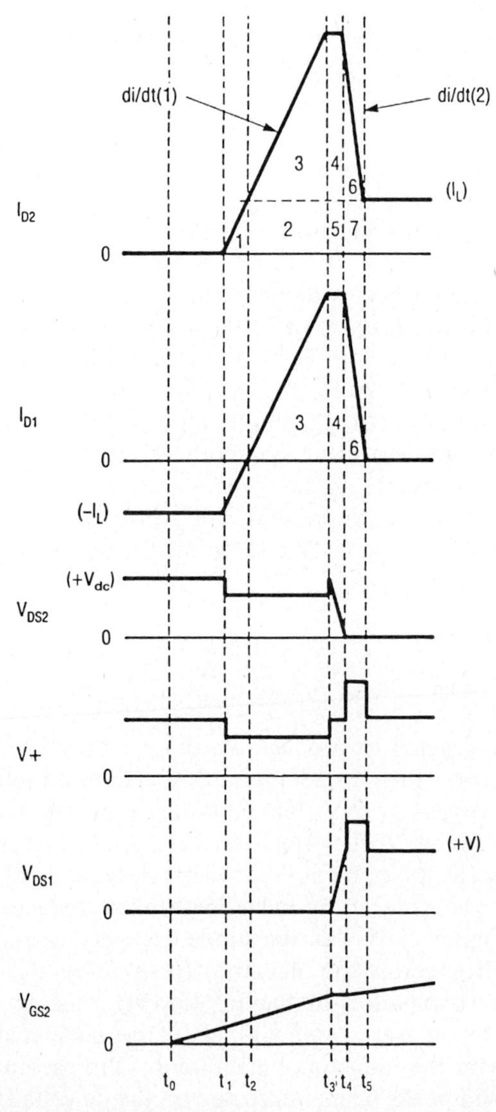

Figure 13.13. Indealized inverter commutation waveforms.

Figure 13.13 shows an idealized version of the waveforms associated with the turn-on of the lower MOSFET in one leg of an inverter (see Figure 13.9). Figure 13.14 shows photographs of actual waveforms corresponding to these. The origin of these waveforms is as follows:

t_0–t_1. Delay time. The gate capacitance is charging to the threshold value, which it attains at time t_1.

t_1–t_2. MOSFET Q2 starts to conduct with a di/dt determined by the rate of rise of the gate voltage. The di/dt causes a voltage drop in the circuit inductance. The rest of the dc rail voltage is supported by the MOSFET. By t_2, MOSFET Q2 has taken over the load current from the diode of MOSFET Q1 and the diode current has fallen to zero.

t_2–t_3. Negative diode current in MOSFET Q1 (shown as positive drain current) removes charge from the diode of MOSFET Q1. At time t_3 sufficient charge has been removed to enable MOSFET Q1 to block voltage.

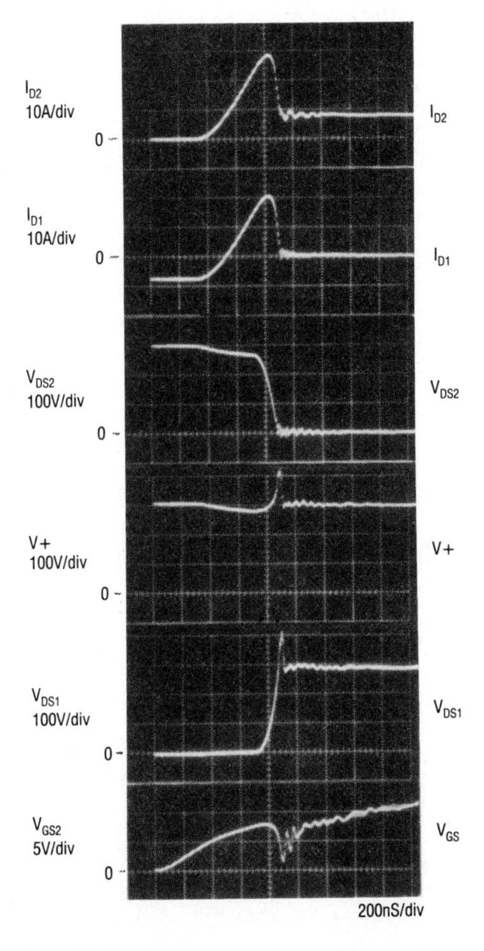

Figure 13.14. Inverter commutation waveforms.

t_3–t_4. The drain voltage of MOSFET Q2 falls as the voltage across MOSFET Q1 rises. Miller feedback now influences the switching of MOSFET Q2. The diagram shows a constant current during this period. However, the current waveform during this period is determined both by the Miller feedback and the relationship between rate of change of voltage and diode current during this stage of the recovery.

t_4–t_5. The dc rail voltage, aided by the voltage across the circuit inductance produced by the changing current, now appears across MOSFET Q1, and the current in its diode decays to zero.

In practice the periods t_3–t_4 and t_4–t_5 are very short and merge together. The fall in diode current is in fact very rapid at this time, and in practice the gate voltage falls to the threshold level as the drain current falls to zero. The gate voltage then rises as the gate capacitance is charged through the series gate resistor.

13.9. EFFECT OF d*i*/d*t* AND TEMPERATURE ON Q_{RR}

Figure 13.15 illustrates the effect of switching speed on the diode reverse recovery current. The switching speed is determined by the value of the series gate resistor. The diode recovery charge Q_{RR} falls as the recovery time is extended, due to the increased opportunity for minority-carrier recombination. Figure 13.16 shows an example of how Q_{RR} varies with di/dt.

For very low values of the series gate resistance a large dip occurs in the dc supply rail voltage, due to the parasitic inductance of the reservoir capacitors and the system wiring. While this may not be too detrimental to waveform quality, it is likely to contribute to the level of RFI generated by the circuit.

Figure 13.17 illustrates the effect of increasing temperature on the diode recovery current (and hence on Q_{RR}) for a typical high-voltage device. The relationship between Q_{RR} and temperature is shown in Figure 13.18.

13.10. CALCULATION OF INVERTER SWITCHING LOSSES

The losses incurred during turn-on may be predicted from Figure 13.13. Normal conduction losses are not included, and only those conduction losses produced when devices are simultaneously carrying current and blocking voltage are considered. Turnoff losses have also been neglected, as these are small compared with the turn-on losses. The losses in MOSFET Q2 (Figure 13.9) occur during the following intervals:

t_1–t_2. The drain current builds up to the value of the load current while supporting the available voltage (dc rail voltage minus the drop in the circuit inductance).

t_2–t_3. Both the load current and the rising diode recovery current are carried while blocking the available voltage.

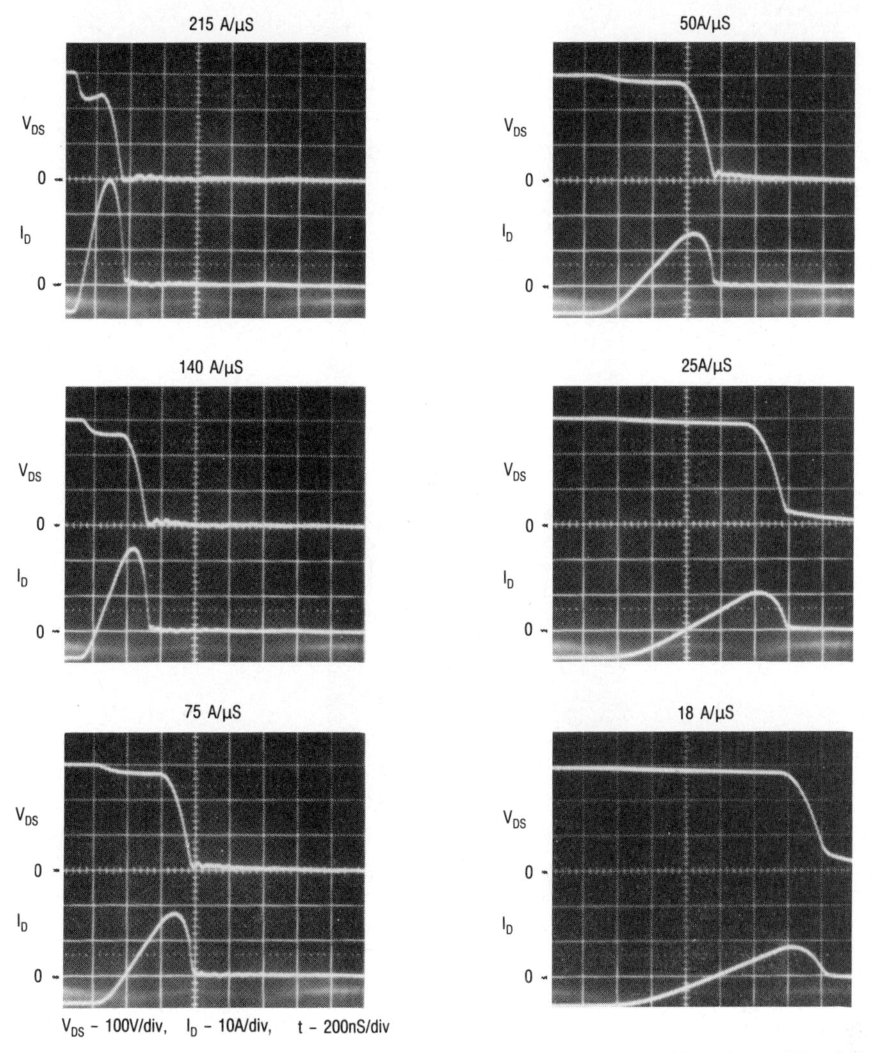

Figure 13.15. Diode recovery waveforms for a range of di/dt values. ($T_J = 25°C$)

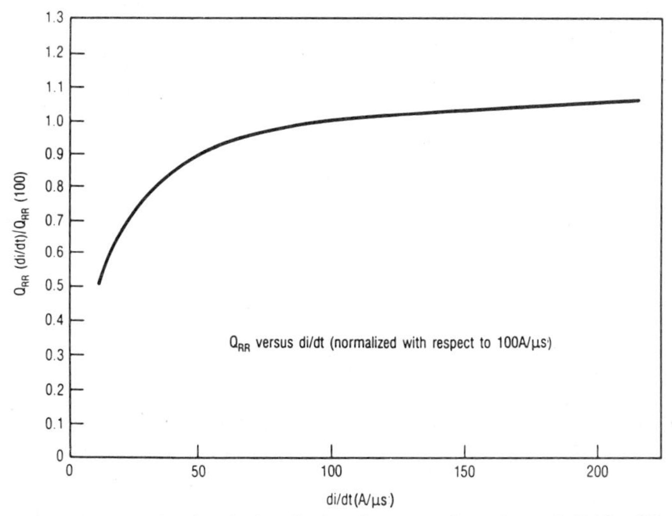

Figure 13.16. MOSFET body–drain diode Q_{RR} as a function of di/dt. ($T_J = 25°C$)

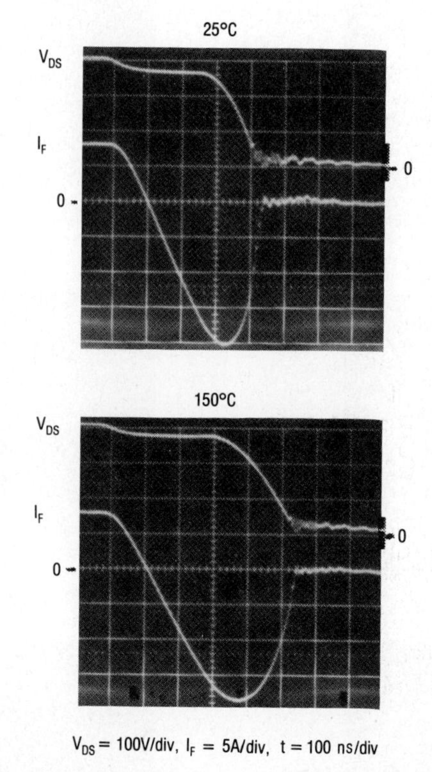

V_{DS} = 100V/div, I_F = 5A/div, t = 100 ns/div

Figure 13.17. Effect of temperature on Q_{RR}.

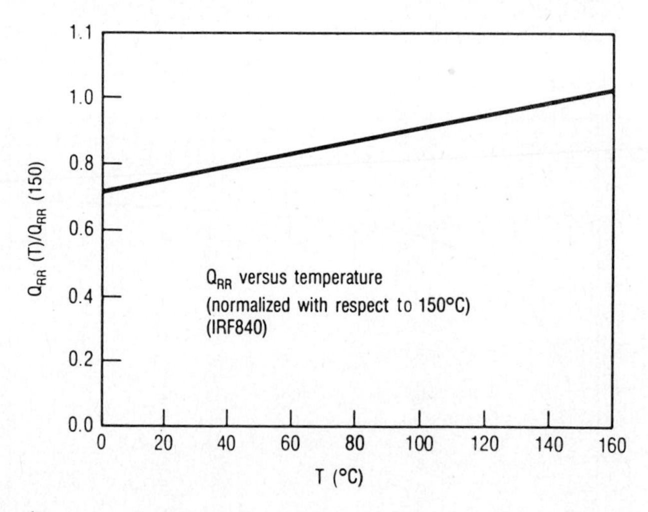

Figure 13.18. Q_{RR} as a function of junction temperature (normalized with respect to 150°C)

t_3–t_4. Some losses are incurred during this period, but not a significant amount, since the drain voltage is falling rapidly.

As the losses in MOSFET Q2 decline, the losses in MOSFET Q1 increase:

t_3–t_4. Some losses are incurred as the voltage across the diode increases.

t_4–t_5. This is the final phase of the diode recovery current, resulting in losses, since the diode now blocks the available voltage.

The losses in MOSFET Q1 and MOSFET Q2 have two basic origins: the conduction of the load current by MOSFET Q2 during the period t_0–t_3, and the passage of diode recovery current through both devices during the period t_1–t_5.

Estimation of the total losses is simplified if it is assumed that the effect of the circuit inductance is neutral in that though it reduces the available voltage during one part of the switching cycle, it increases it during another. (If the breakdown voltage of the MOSFET is exceeded, then this energy will be dissipated in avalanche breakdown—permitted with MOSFETS having a repetitive avalanche rating.) Therefore, if it is assumed that all charge transfer takes place between supply rails whose voltage remains constant at the nominal value, the total losses during commutation of the inverter leg may be calculated approximately by multiplying the dc rail voltage by the various transfers of charge that occur. If a series inductor with its own clamping circuit is located in the dc rail, then energy stored in this inductor is not dissipated in the MOSFET, but in the clamp circuit.

Thus the losses in MOSFET Q2 are given by the rail voltage times the charge represented by areas 1, 2, and 3 in Figure 13.13. The losses in MOSFET Q1 are given by the rail voltage times the charge represented by area 6. Losses associated with areas 4 and 5 are distributed between the two MOSFETS. The total losses are therefore approximately equal to the rail voltage multiplied by the charge represented by areas 1, 2, 3, 4, 5, and 6. Areas 3, 4, and 6 represent the Q_{RR} of the diode. Areas 1, 2, and 5 are a function of the load current and the rate of rise of diode recovery current. Thus the total commutation energy loss may be expressed as

$$\begin{aligned} E &= V_{dc}[(\text{area }3 + \text{area }4 + \text{area }5) + (\text{area }2 + \text{area }5) + (\text{area }1)] \\ &= V_{dc}\left[Q_{RR} + I_L(t_4 - t_2) + (t_2 - t_1)^2 \frac{di/dt}{2}\right] \\ &= V_{dc}\left(Q_{RR} + I_L\sqrt{\frac{2Q_{RR}}{di/dt}} + \frac{I_L^2}{2\,di/dt}\right) \end{aligned} \tag{13.3}$$

where I_L is the load current at the time of commutation.

This gives the energy loss for one pulse at one value of drain current. From Figure 13.19 it can be seen that to a first approximation Q_{RR} is linearly related to the drain current and therefore, assuming a sinusoidal load current, can be expressed as

$$Q_{RR}(t) = Ki(t) = K\hat{I}_L \sin \omega t \tag{13.4}$$

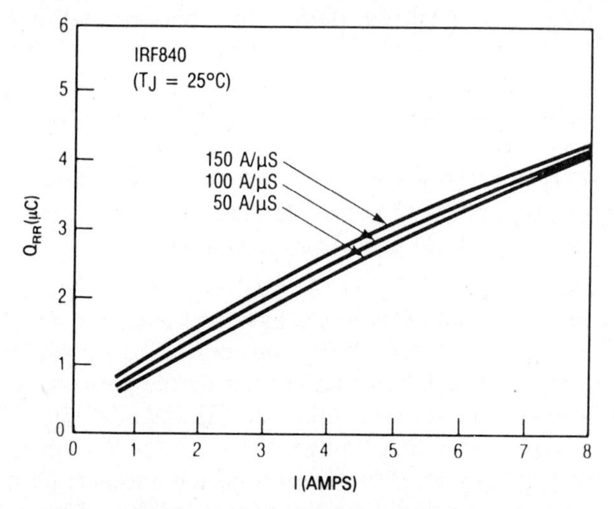

Figure 13.19. Q_{RR} as a function of source current.

and

$$Q_{RR}(\theta) = K\hat{I}_L \sin\theta \tag{13.5}$$

where $\theta = \omega t$. Substituting into (13.3) for Q_{RR} gives the energy dissipated when a commutation occurs at a current value of $\hat{I}_L \sin\theta$ $(0 < \theta < \pi)$:

$$E(\theta) = V_{dc}\left[K\hat{I}_L \sin\theta + \hat{I}_L \sin\theta \sqrt{\frac{2K\hat{I}_L \sin\theta}{di/dt}} + \frac{\hat{I}_L^2 \sin^2\theta}{2\, di/dt}\right] \tag{13.6}$$

The average power loss due to commutation during that switching cycle is given by

$$P(\theta) = V_{dc}f_s\left[K\hat{I}_L \sin\theta + \hat{I}_L \sin\theta \sqrt{\frac{2K\hat{I}_L \sin\theta}{di/dt}} + \frac{\hat{I}_L^2 \sin^2\theta}{2\, di/dt}\right] \tag{13.7}$$

where f_s is the switching frequency. Since the switching frequency is approximately two orders of magnitude greater than the output frequency, Equation (13.7) can be taken as representing the power loss due to commutation expressed as a continuous function of θ. The average power loss due to commutation during one half cycle of output current is therefore given by

$$\begin{aligned} P_{ave} &= \frac{V_{dc}f_s}{\pi}\left[K\hat{I}_L \int_0^\pi \sin\theta\, d\theta + \hat{I}_L \sqrt{\frac{2K\hat{I}_L}{di/dt}} \int_0^\pi \sin^{3/2}\theta\, d\theta + \frac{\hat{I}_L^2}{2\, di/dt}\int_0^\pi \sin^2\theta\, d\theta\right] \\ &= \frac{V_{dc}f_s\hat{I}_L}{\pi}\left[2K + 2.47\sqrt{\frac{K\hat{I}_L}{di/dt}} + 0.785\frac{\hat{I}_L}{di/dt}\right]. \end{aligned} \tag{13.8}$$

This represents the sum of the commutation power losses in both MOSFETS in the inverter phase. The process for determining the commutation losses is therefore:

1. Read the typical value for Q_{RR} from the data sheet.
2. Adjust Q_{RR} for di/dt and temperature, using the graphs of the type shown in Figure 13.16 and Figure 13.18.
3. Using this value of Q_{RR}, obtain K from the equation

$$K = Q_{RR}/I_F \tag{13.9}$$

   where I_F is the test current specified in the typical Q_{RR} rating in the data sheet.
4. Substitute K along with di/dt and $\hat{I}_L$ in Equation (13.8) to obtain the commutation power loss.

This must be an approximate result in view of the assumptions made.

13.11. CONDUCTION LOSSES

The load current is carried either by the channel of one MOSFET or by the body–drain diode of the other. If carried by the channel of a MOSFET, the power dissipation in the MOSFET channel during that period is given by

$$P_D = I_D^2 R_{DS(on)} \tag{13.10}$$

$R_{DS(on)}$ is a function of the die temperature and to a lesser extent of the instantaneous value of the drain current. Data sheets should provide information relating $R_{DS(on)}$ to I_D and temperature. During the period when the load current is being carried by the body–drain diode of one of the MOSFETS, the power dissipation is given by

$$P_D = I_D V_{SD} \tag{13.11}$$

V_{SD}, the forward drop of the body–drain diode, is a function of the load current. MOSFET data sheets usually include a graph showing V_{SD} versus diode current.

Since the direction of current flow at any time is not predetermined, but rather is dictated by load conditions, the upper MOSFET is maintained on during positive segments of a switching cycle, and the lower MOSFET is maintained on during the negative segments of a switching cycle, except for short periods when both devices are off to prevent overlapping conduction and shoot-through. Therefore, when load current is flowing through a body–drain diode, the channel associated with that diode will be in the conducting state. A proportion of the load current will flow through the channel, since current can flow through the channel of a MOSFET in either direction. This lowers the forward drop across the MOSFET. In high-voltage devices with high values of $R_{DS(on)}$ this effect is negligible, but in low-voltage inverters diode conduction losses can be mitigated by channel conduction. Diode recovery losses, however, are not affected, since the channel is turned off prior to commutation, to avoid shoot-through, and the free-wheeling current reverts to the diode path. However, if the deadband is very short the charge which is established in the diode may be less than that normally associated with the given level of forward diode current.

As Figure 13.20 shows, during each switching cycle the proportion of time spent in each mode (channel conduction or diode conduction) depends on the modulation duty cycle δ prevailing during that switching cycle. The modulation duty cycle is directly related to the required magnitude of fundamental output voltage at that time. Therefore, for a sinusoidal output waveform, the division of losses between channel conduction and diode conduction varies sinusoidally throughout the output cycle. At the same time the magnitude of the load current

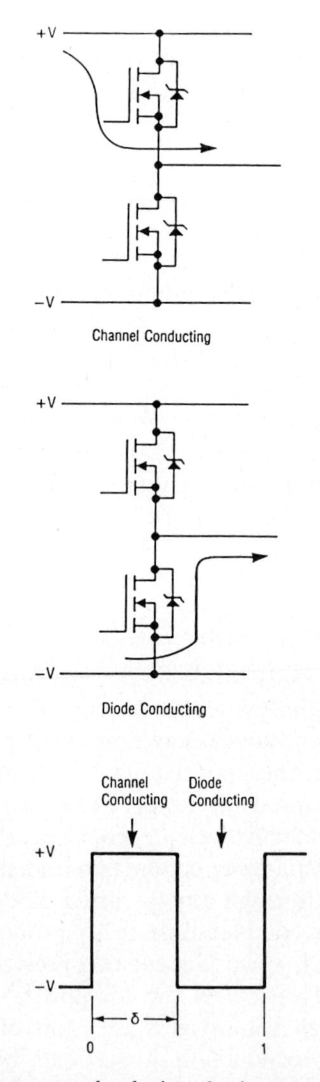

Figure 13.20. Current paths during the inverter commutation cycle.

varies with a phase relationship to the fundamental voltage output waveform that is determined by the load power factor.

Heatsinking must be sized for worst-case dissipation, which occurs when the load power factor is unity and the longest MOSFET conduction periods coincide with the peaks of the load current. In high-voltage inverters diode conduction losses will only be a small fraction of the MOSFET forward-conduction losses, so that although diode losses increase as the load phase angle moves away from 0°, the total losses are reduced. Similarly, total losses are greatest when the modulation depth is unity (assuming sinusoidal modulation).

To calculate MOSFET losses under these conditions, assume a load current waveform of the form $\hat{I}_L \sin\theta$. During a switching cycle occurring at a time corresponding to the angle θ, the MOSFET will conduct in the forward direction for a proportion of the cycle given by

$$\delta_{\text{MOSFET}} = \tfrac{1}{2}(1 + \sin\theta) \tag{13.12}$$

The proportion of the cycle for which the diode conducts is given by

$$\delta_{\text{DIODE}} = \tfrac{1}{2}(1 - \sin\theta) \tag{13.13}$$

Therefore the MOSFET power loss during a pulse occurring at angle θ is

$$\begin{aligned} P(\theta) &= I^2 R \\ &= \hat{I}_L^2 \sin^2\theta \cdot \tfrac{1}{2}(1+\sin\theta) R_{\text{DS(on)}} \\ &= \tfrac{1}{2}\hat{I}_L^2 R_{\text{DS(on)}}(\sin^2\theta + \sin^3\theta) \end{aligned} \tag{13.14}$$

The average power loss due to MOSFET conduction over one output half cycle is therefore given by

$$\begin{aligned} P_{\text{ave}} &= \frac{\hat{I}_L^2 R_{\text{DS(on)}}}{2\pi} \int_0^\pi (\sin^2\theta + \sin^3\theta)\, d\theta \\ &= 0.462 \hat{I}_L^2 R_{\text{DS(on)}} \end{aligned} \tag{13.15}$$

The diode conduction losses for a switching cycle at angle θ are given by

$$\begin{aligned} P(\theta) &= \hat{I}_L V_{\text{SD}} \\ &= \hat{I}_L \sin\theta \cdot \tfrac{1}{2}(1 - \sin\theta) V_{\text{SD}} \end{aligned} \tag{13.16}$$

where V_{SD} is the forward diode drop. To simplify calculations this is assumed to be constant for all values of current. The average power loss due to diode conduction over one output half cycle is therefore given by

$$\begin{aligned} P_{\text{ave}} &= \frac{\hat{I}_L V_{\text{SD}}}{2\pi} \int_0^\pi (\sin\theta - \sin^2\theta)\, d\theta \\ &= 0.068 \hat{I}_L V_{\text{SD}} \end{aligned} \tag{13.17}$$

Figure 13.21 shows how these losses vary with load phase angle.

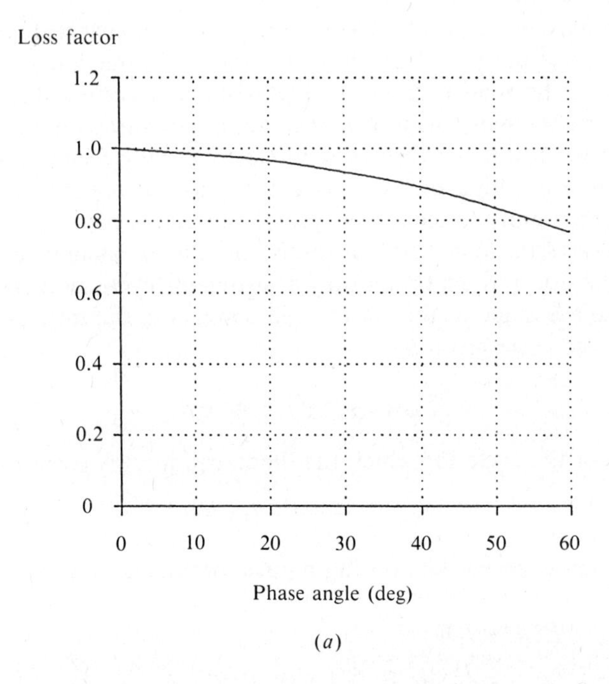

(*a*)

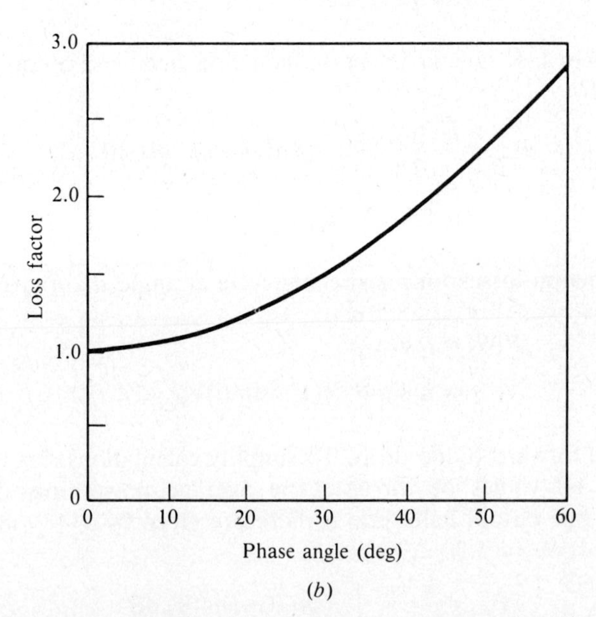

(*b*)

Figure 13.21. (*a*) MOSFET conduction losses versus load phase angle. (*b*) Diode conduction losses versus load phase angle.

13.12. FAST-RECOVERY DIODE MOSFETs

During the manufacture of standard power MOSFETS no special attempt is made to control the recovery time of the body–drain diode. In MOSFET design and fabrication heavy emphasis is placed on minimizing $R_{DS(on)}$, and from this point of view minority-carrier lifetime killing is undesirable due to the attendant rise in silicon resistivity. However, a long minority-carrier lifetime results in a long diode recovery time.

Power-MOSFET diodes, with a recovery time typically of the order of 100 ns for 50-V devices and 600 ns for 500-V devices, might be classified as fast, but in comparison with the switching speed of a power MOSFET they are slow. Furthermore, the diode recovery charge Q_{RR} tends to be large, due to the large area which the diode occupies. Power-MOSFET die designs seek to optimize the performance of the MOSFET, and the area of silicon occupied by the body–drain diode is usually a secondary consideration. Most MOSFET designs result in a body–drain diode whose current-carrying capability is considerably in excess of the forward current capability of the MOSFET, although this is not necessarily reflected in the current rating given to the diode. The diode is commonly rated at the same values of average and surge current as the MOSFET.

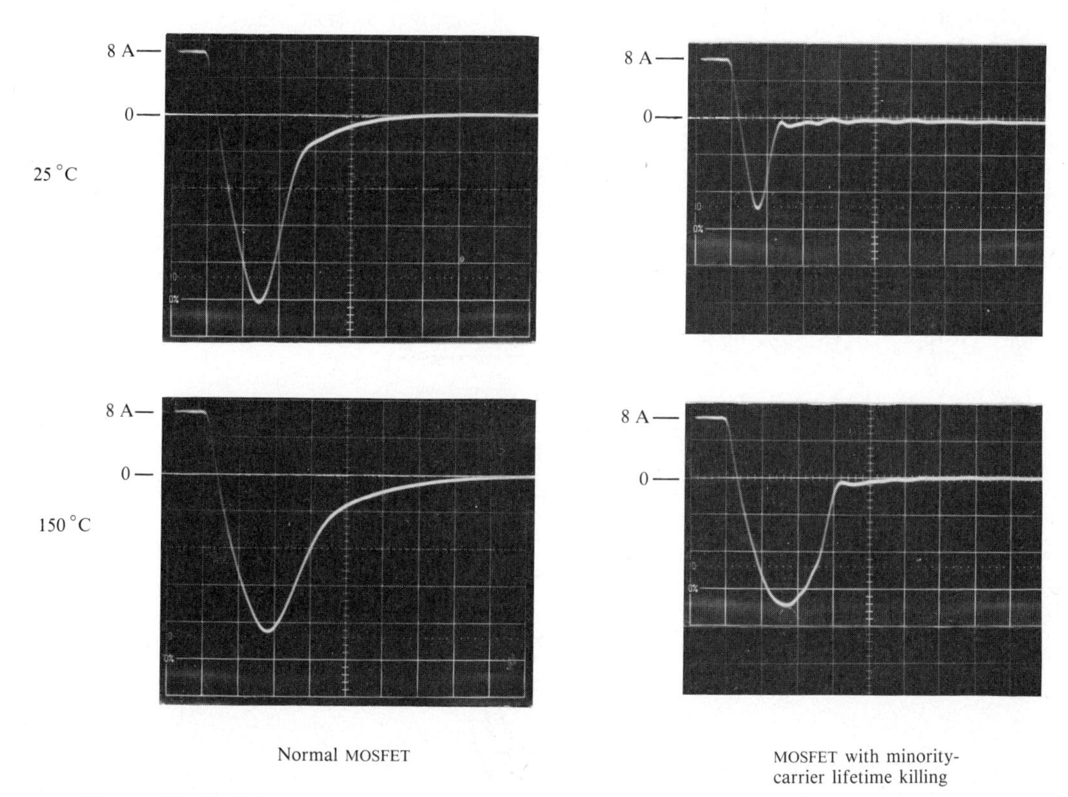

Figure 13.22. Effect of temperature on the reverse recovery current in the body–drain diode of similar MOSFETS, with and without minority-carrier lifetime killing. (Low reverse voltage.)

The body-drain diode may be made faster and its Q_{RR} lowered by reducing the minority-carrier lifetime in the epitaxial region. This may be achieved by doping with a heavy metal, such as gold or platinum, or by irradiating the die with electrons, fast protons, or other particles which create recombination centers in the crystal lattice. Both heavy-metal doping and irradiation cause difficulties in manufacture. The heavy metals may contaminate the gate oxide, and irradiation causes gate oxide damage which cannot be totally annealed out. Nevertheless, power MOSFETS with fast diodes have been made successfully [7, 8].

The effectiveness of minority-carrier lifetime killing is reduced as the temperature of the silicon increases. Recombination centers become less effective as thermal energy increases. Therefore it is important to consider the diode characteristics at the maximum junction temperature at which the device will operate. At a typical maximum allowable junction temperature of 150°C some types of minority-carrier lifetime killing will be ineffective and the difference in the speed of a doped and an undoped device will be reduced. Figure 13.22 shows an example of the effect of temperature on devices with and without lifetime killing.

The price which has to be paid for a fast body–drain diode is an increase in $R_{DS(on)}$ resulting from the heavy-metal doping or irradiation of the epitaxial region. Increased conduction loss must be taken into consideration when assessing whether total losses are reduced by the use of a power MOSFET whose diode has been made faster by minority-lifetime killing.

13.13. AVALANCHE REQUIREMENTS

Avalanche capability adds to the suitability of the MOSFET for inverter applications. As well as being able to tolerate voltage spikes produced by such events as load connection and disconnection and contactor operation, an avalanche rating

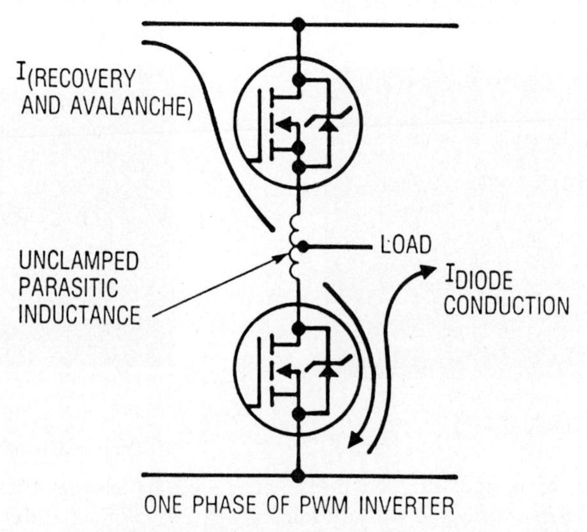

Figure 13.23. Unclamped parasitic inductance in inverter bridge circuit.

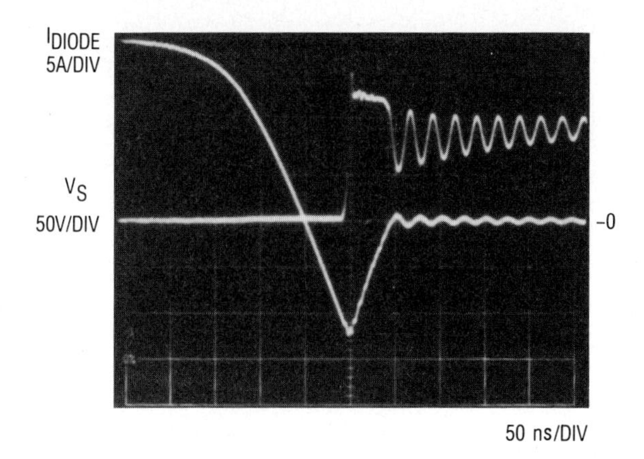

Figure 13.24. A MOSFET (IRF 530) subject to diode-recovery dv/dt and avalanche.

enables MOSFETS to tolerate overvoltage transients produced during normal operation. For example, if a large amount of parasitic inductance is included at the midpoint of the inverter, as shown in Figure 13.23, energy will be stored in this inductance during diode recovery. As the waveforms in Figure 13.24 demonstrate, this energy results in avalanche breakdown of the MOSFET. The device used in Figure 13.24 is experiencing first diode-recovery dv/dt followed immediately by avalanche breakdown.

Another important benefit of avalanche capability is that if a power MOSFET can tolerate avalanche breakdown, the safety margins traditionally applied when selecting the voltage rating of the switching devices can be relaxed. This is particularly beneficial in the case of power MOSFETS, since their $R_{DS(on)}$ increases rapidly with voltage rating. Where it is not possible to use the MOSFET's own avalanche capability, then some external form of voltage clamping must be used. A low-loss clamping circuit is given in reference [9].

13.14. GATE-DRIVER DESIGN

The negative power rail of a bridge circuit can usually be linked to the ground of the control electronics as shown in Figure 13.25. This allows the bottom devices of the bridge to be driven directly from the control circuit without the need for isolation. However, the potentials of the sources of the top devices swing between the positive and negative rails of the dc supply, making it impossible for the sources of the upper devices to be linked directly to the control electronics. This difficulty may be resolved in two basic ways. The first is to employ level-shifting circuitry in which the gate of the upper device is pulled above the positive rail voltage by the generation of a gate supply voltage source using a charge-pump arrangement (see Figure 8.31). Some degree of sophistication is required in the level-shifting circuitry to limit dissipation.

The second basic method of driving the upper power devices is to provide an isolated power supply for the top gate driver and to control the application of

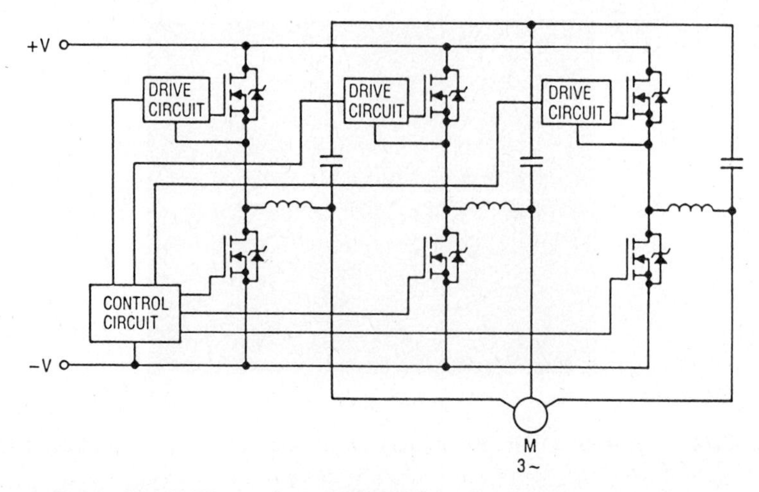

Figure 13.25. Three-phase MOSFET inverter with output filtering.

the gate voltage through some isolated signaling arrangement. The gate driver must meet the following requirements:

- Short delay time.
- Low impedance output
- Capability of accepting duty cycles between 0% and 100%
- Insusceptibility to dv/dt triggering during switching
- Low capacitance between the two sides of the power supply.

A suitable arrangement is shown in Figure 13.26. Isolation is provided by the 2601 optical coupler. This has a Faraday screen between the LED and the phototransistor to prevent spurious triggering of the phototransistor by capacitive

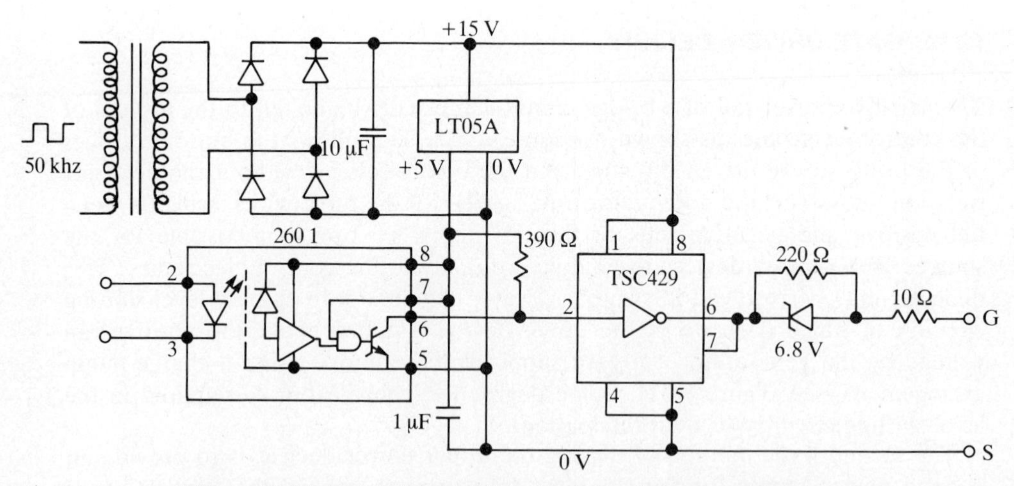

Figure 13.26. Isolated gate drive circuit for inverter applications.

currents during switching of the power circuit. The gate drive is provided by a MOSFET driver IC which is capable of providing 6 A of charging current. The output of the driver is moderated by the gate drive circuit described earlier. Power for both integrated circuits is obtained from a rectified high-frequency ac supply. The primary and secondary windings of the ferrite transformer are wound on independent bobbins with air-gap separation of the two windings to reduce capacitive coupling between them. This minimizes the flow of capacitive current produced by the voltage swings of the midpoint of the power circuit.

13.15. PWM WAVEFORM GENERATION

The traditional method of PWM waveform generation has been to input to a comparator a triangular timing wave and a sinusoidal reference waveform. This method has to a large extent been replaced by digital waveform generation. The digital method has the advantages of freedom from drift, absence of dc components in the output, perfect phase balance, etc. [10–12]. Overmodulation may be used to obtain maximum output from the inverter. While overmodulation increases the amplitude of the fundamental of the output waveform, it also introduces distortion, as Figure 13.27 illustrates. A preferable method of increasing the amplitude of the output is the addition to the reference sine wave of a measure of third harmonic as shown in Figure 13.28. Since the third harmonic is a triplen harmonic, the third harmonics of all three phases will be cophasal. Therefore the third harmonic will be eliminated from the line-to-line voltage waveform. By adding a third harmonic of one-sixth amplitude of the fundamental, the output voltage of the inverter and its kVA rating can be increased by 15% without distortion of the output. The use of power MOSFETS and an ultrasonic switching frequency gives the power stage a bandwidth which enables it to satisfactorily reproduce third-harmonic waveforms when the output frequency is as high as 400 Hz [13].

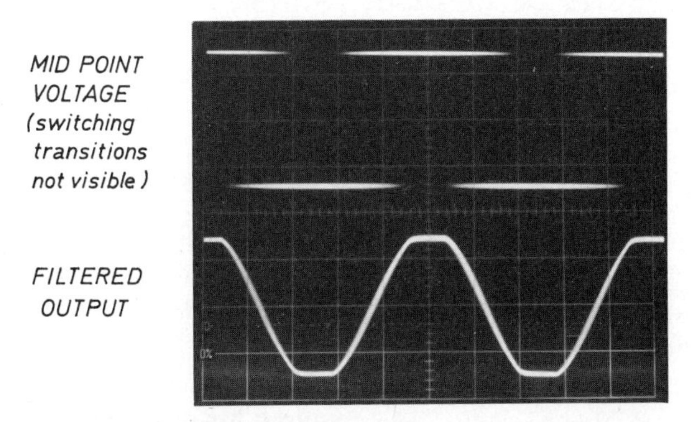

Figure 13.27. Distortion introduced by overmodulation.

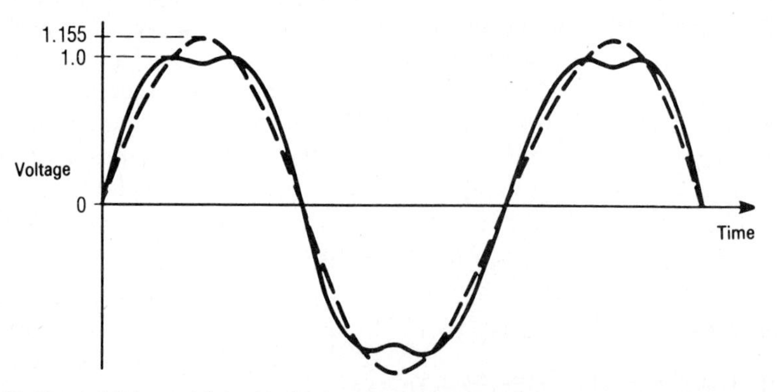

Figure 13.28. Addition of $\frac{1}{6}$th third harmonic to the phase voltage (solid line) permits the line-to-line voltage (dashed line) to be increased by 15.5%.

13.16. INVERTER RATINGS VERSUS MOSFET TYPE

The power rating obtainable from a three-phase inverter using a particular MOSFET type depends on two considerations—efficiency and equipment size. The governing factor is the size of the heatsink that the designer considers acceptable. A small heatsink will, for a given power output, give a high junction temperature, high $R_{DS(on)}$, and therefore higher conduction losses. The efficiency will be reduced, but the equipment will be small. A large heatsink, on the other hand, will give a lower junction temperature and a lower $R_{DS(on)}$, so that although the equipment will be large, the efficiency will be high.

Conduction losses can be reduced by using MOSFETS with a larger die and a lower $R_{DS(on)}$, although commutation losses will increase due to the higher Q_{RR} associated with the larger die. Q_{RR} losses are most important in high-voltage designs because of the large amount of charge stored in the broad epitaxial layer required to block high voltage. Conduction losses are most important in low-voltage, high-current designs. Therefore most benefit is obtained from an investment in extra silicon area in low-voltage designs.

13.17. COMPARISONS OF BIPOLAR TRANSISTORS AND MOSFETS

The following table compares the suitability of power MOSFETS and bipolar transistors for use in inverters:

MOSFET	Bipolar Transistor
Integral free-wheeling diode.	Only some Darlingtons have diode.
No snubber required	Snubber required.
Low switching losses.	High switching losses.
Can be driven direct from logic.	Base drive circuits required.
Same design for different power ratings.	New base drive and snubber design required for different power rating.
High overload capability.	Limited overload capability.
Avalanche capability.	Usually no avalanche capability.

Because of the lower switching losses associated with power MOSFETS, the MOSFET is likely to be the device of choice for 20-kHz PWM inverters at all voltages. However, for six-step and low-frequency inverters, the choice is not so clear. An important consideration not addressed in the above table is forward voltage drop. At low voltages, where $R_{DS(on)}$ for a given area of silicon is low, again power MOSFETS are likely to be the optimum device. However, at high voltages the higher $R_{DS(on)}$ of the power MOSFET militates against its use, especially since at low switching frequency there is no significant advantage in the MOSFET's low switching losses. The power MOSFET still has a number of system advantages, such as ease of drive, snubberless operation, square SOA, and high overload capability, although the degree to which these can be traded against forward drop losses will depend on the application.

In the case of power MOSFETS the question of forward voltage drop is largely one of price, since the voltage drop in a power MOSFET can be reduced to any desired value by using sufficient silicon area. In the case of bipolar transistors the question is more complicated, since forward voltage drop can be traded for other features. By using a Darlington transistor the base current requirement is reduced at the expense of increased voltage drop. The delay time associated with a bipolar transistor can be reduced by operating in quasi-saturation, again at the expense of increased voltage drop.

The move towards power MOSFETS is likely to continue, since the price of the older, bipolar transistor technology is well established, while low-cost power MOSFETS are a recent phenomenon. The extent to which this trend will be affected by the emergence of the insulated-gate bipolar transistor (IGBT) and the MOS-controlled thyristor (MCT) (see Chapter 20), with their mixtures of MOSFET and bipolar-transistor characteristics, has yet to become clear. IGBTS are now available from a number of sources, and low-frequency, high-voltage inverters are seen as one of the main areas of application for these devices.

REFERENCES

1. G. Capoling, "Survey of PWM techniques for single phase transistor inverters." *Eur. Power Electron. Conf.* pp. 2.93–2.98 (1985).
2. D. A. Grant, "Avalanche and dv/dt ratings for power MOSFETS—measures of ruggedness." *Proc. Power Convers. Int. Conf.* May pp. 148–158 (1987).
3. W. Taylor, "Designing a high power MOSFET-based servo drive." *Proc. Powercon* **10D1-1,** pp. 1–13 (1983).
4. L. Lorenz and H. Amann, "Commutation behaviour in dc/ac converters with power MOSFETS." *Proc. Power Convers. Int. Conf.* June, pp. 316–329 (1986).
5. L. Lorenz, "Power MOSFET using for high power switching applications." *Proc. Power Convers. Int. Conf.* May, pp. 369–387 (1987).
6. P. Freundel, "Power MOSFETS or bipolar power transistors for converter circuits?" *IEEE Int. Semiconductor Power converter Conf.,* Orlando, FL, May 24–27, pp. 38–44 (1982).
7. A. P. Connolly, "The FREDFET—a new MOSFET for motor speed controls." *Proc. Power Convers. Int. Conf.* April, pp. 105–109 (1984).
8. B. J. Baliga, "Improving the reverse recovery of power MOSFET integral diodes by electron irradiation." *Solid-State Electron.* **26**(12), 1133–1141 (1983).

9. B. R. Nair and P. C. Sen, "Voltage clamp circuits for a power MOSFET PWM inverter." *IEEE Ind. App. Soc. Ann. Meet.* pp. 797–806 (1984).
10. D. A. Grant, "The use of ratio-changing in pulse width modulated inverters." *Proc. Eur. Conf. Electrotech., 4th,* pp. 420–424 (1980).
11. D. A. Grant, M. Stevens, and J. A. Houldsworth, "The effect of word length on the harmonic content of microprocessor-based PWM waveform generators." *IEEE Trans. Ind. Appl. Soc.*, **IAS-21**(1), 218–225 (1985).
12. Y. Y. Tzou, R. W. Wang, and Y. C. Wu, "Design and implementation of a multiprocessor-based digital PWM inverter controller." *IEEE Power Electron. Spec. Conf.* June, pp. 136–145 (1987).
13. D. A. Grant, J. A. Houldsworth, and K. N. Lower, "A new high-quality PWM AC drive." *IEEE Ind. App. Soc. Ann. Meet.*, pp. 530–535 (1982).

CHAPTER 14

High-Frequency Applications

14.1. SOLID-STATE BALLASTS

Electronic ballasts for fluorescent tubes, mercury arc lamps, and sodium vapor lamps have not replaced conventional chokes to the degree once expected. The reason for this is that it is difficult for a system that involves two conversion stages (ac to dc followed by dc to ac) to compete economically with a choke and a starter switch. Solid-state ballasts therefore have tended to be used where they have an advantage which compensates for the higher initial cost. The advent of power MOSFETS has moved the balance in favor of the solid-state ballast because of the low switching loss of MOSFETS, which enables them to operate efficiently at high frequencies, and because of their ease of drive, which reduces equipment cost. The major disadvantage associated with power MOSFETS in this application is that the solid-state ballast is generally a high-voltage application, and high-voltage MOSFETS have a high $R_{DS(on)}$ for a given area of die. However, this disadvantage has largely disappeared as the cost of power MOSFETS has fallen, since $R_{DS(on)}$ can be reduced to any level desired by using a die of sufficient size.

If the use of a solid-state ballast cannot be justified by lower purchase cost, other considerations may prevail. In terms of light output per watt, solid-state ballasts can have a higher overall efficiency than conventional inductor-ballast systems due to the greater efficiency of fluorescent lamps when operating from a high-frequency supply. This can result in a fairly short pay-back time for lamps which operate continuously. Higher switching frequencies result in smaller magnetic components, resulting in units that are smaller and lighter [1]. Power MOSFETS enable very high operating frequencies to be achieved efficiently. The general arguments in favor of solid-state ballasts, such as flicker-free light, silent operation, and ability to operate from a dc supply, apply to power-MOSFET-based ballasts as much as to any other type.

The high-frequency capability of power MOSFETS has made the use of resonant circuits economic. The advantages of resonant circuits include the low level of RFI which resonant circuits generate and the possibility of employing resonance to generate a high voltage for striking the tube.

14.2. CLASS-D AMPLIFICATION

14.2.1. Introduction

Class-D amplifiers employ pulse-width modulation (PWM) of a high-frequency rectangular wave to produce an output whose average value tracks that of the input signal. For minimum distortion, the switching frequency should be at least five times that of the highest frequency component in the audio signal, so that if an audio frequency of 25 kHz is to be accommodated, the switching frequency will need to be at least 125 kHz. To maximize the amplifier output it is necessary to be able to modulate the pulse widths to the point where extremely short pulses are required. Thus for maximum output with minimum distortion the output devices of a class-D amplifier must be capable of very fast switching (e.g. 10 ns). Furthermore, for good efficiency they must be capable of switching at these speeds without incurring significant losses. The characteristics ideally required of the power switches in a PWM amplifier are rapid low-loss switching, large SOA, ability to conduct current in both directions (for driving a reactive load, i.e. the filter) and good dv/dt and di/dt capability. All these requirements can be fulfilled by the power MOSFET.

14.2.2. Pulse-Width Modulation (PWM)

The principle of pulse-width modulation as employed in a class-D amplifier is illustrated in Figure 14.1. For a given modulation duty cycle δ, the average load voltage is given by

$$V_L = V_S(2\delta - 1) \tag{14.1}$$

A PWM waveform is traditionally obtained by comparing a triangular wave of fixed frequency with the modulating signal (in this case the audio waveform) as shown in Figure 14.2. The comparator output is used to control the switching of the upper and lower switching transistors in the output stage, thereby varying the average output voltage in accord with the audio signal. The output waveform will contain high-frequency components due to the switching process, and these must be removed by filtering.

14.2.3. Design Example

The circuit diagram of the amplifier is shown in Figure 14.3. There are a number of features which are critical to obtaining good reproduction of the audio input.

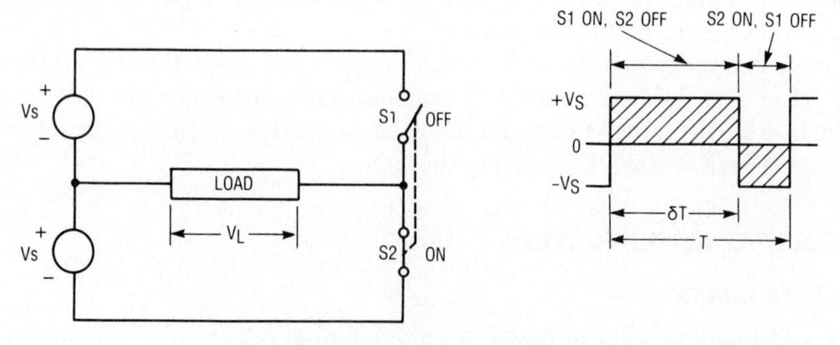

Figure 14.1. Pulse-width modulation.

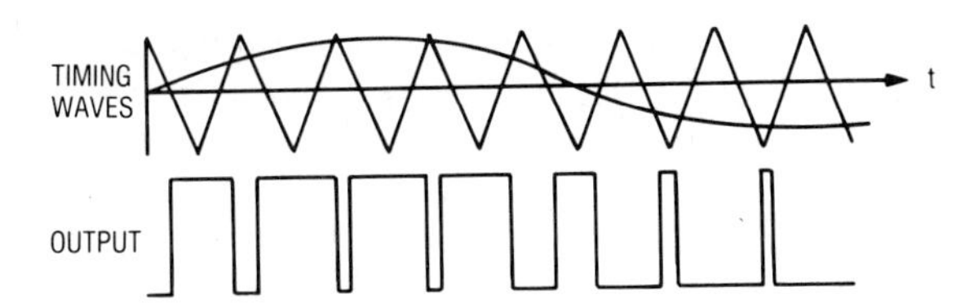

Figure 14.2. Generating PWM waveforms.

The input stage consists of a high-speed comparator. Since fast switching is essential in the output stage if distortion is to be minimized, fast switching is required in all stages in which the signal is manipulated. The amplitude of the triangular waveform will need to match the range of the input signal. If at any time the amplitude of the input signal exceeds that of the triangular waveform, the output waveform will be clipped, resulting in distortion. The level of output distortion will also depend on the linearity of the triangular waveform.

Complementary n-channel and p-channel MOSFETS are employed in the output stage. The use of an n-channel MOSFET for the upper switch would require the use of an isolated gate driver or some other arrangement such as a bootstrap gate drive arrangement. Transformer or optical-isolator coupling is not feasible because of the brevity of the pulse length required under conditions of maximum modulation, while the bootstrap method incurs excessive losses. The use of complementary MOSFETS with direct gate drive therefore offers the best prospect of obtaining fast symmetrical switching of the output devices.

Transistors TR1 and TR2 buffer the output of IC1 and provide a low-impedance gate drive for TR3 and TR4. TR3 and TR4 act as signal splitters and

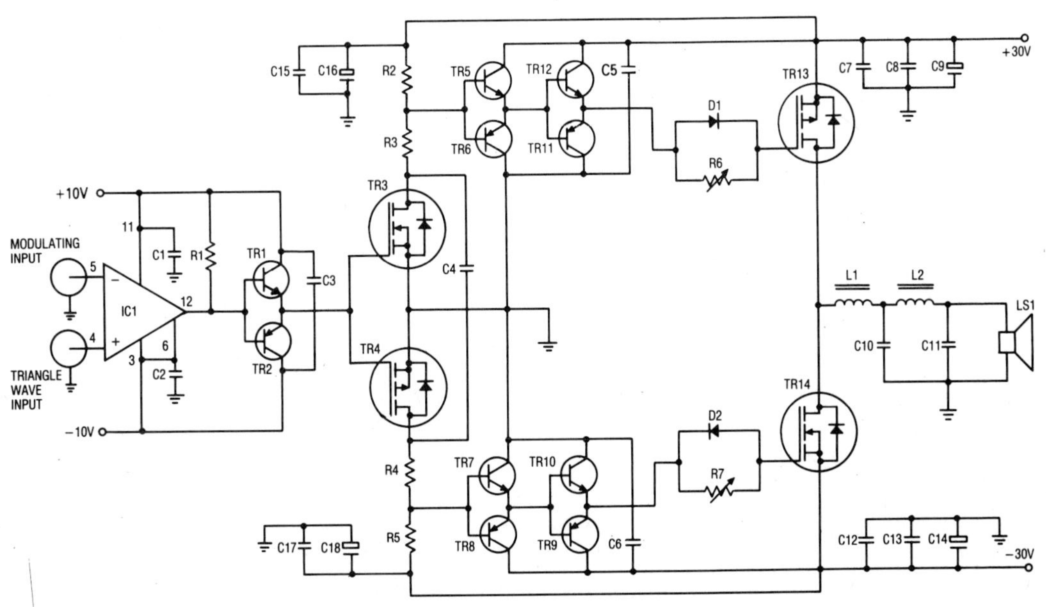

Figure 14.3. Class-D amplifier.

level shifters for driving the output MOSFETS TR13 and TR14. TR5, TR6, TR11, and TR12 provide a low-impedance gate drive for TR13. TR7, TR8, TR9, and TR10 provide the same function for TR14. D1 with R6 and D2 with R7 shape the switching performance of TR13 and TR14 respectively. Some adjustment of the rate of switching of the output MOSFETS is required to obtain the waveform symmetry needed for minimum output distortion. The switching times of the complementary n-channel and p-channel MOSFETS will not be exactly the same, due to differences in their parasitic capacitances. Furthermore there may be sufficient variation in the parasitic capacitances of devices of the same type to make adjustment necessary. Diodes D1 and D2 ensure that TR13 and TR14 are turned off as rapidly as possible, but turn-on is slightly delayed by R6 and R7 to ensure that TR13 and TR14 do not conduct simultaneously and create a short-circuit across the supply rails. Furthermore, since the body–drain diode of the outgoing MOSFET may have been in conduction prior to turn-on of the incoming device, a slight slowing of the turn-on of the incoming MOSFET gives the diode longer to recover and reduces the amplitude of the diode recovery current which flows through the incoming MOSFET. The MOSFET integral diodes can serve as free-wheeling diodes if their dv/dt ratings are adequate. Output filtering is provided by L1, L2, C10, and C11.

14.2.4. Avoiding Distortion

In order to avoid distortion, the MOSFETS must switch quickly. This results in a current of several amperes being switched in about 10 ns. Because of the high values of di/dt which this implies, it is essential to pay careful attention to track routing on the PCB. The use of a ground plane is beneficial. Ringing due to airborne interference is likely. Decoupling should be used extensively, particularly at the output stage, because of the high values of di/dt present. Decoupling is also required in the driver stage, which draws sharp peaks of current.

The output transistors are matched on the basis of ON-resistance. Any great disparity between their resistances would result in distortion of the output waveform when the output current is large.

14.2.5. The Output Filter

The output filter, consisting of L1, L2, C10, and C11, is required to remove switching-frequency components and their sidebands from the output voltage. A switching frequency of 150 kHz minimizes the total harmonic distortion. A higher switching frequency would, in theory, improve the quality of the output waveform, since the higher the ratio of switching frequency to modulating frequency, the smaller are the amplitudes of the unwanted modulation products in the output waveform. However, a higher switching frequency implies shorter minimum pulse widths. This results in increased distortion due to switching rise times and fall times, as well as delay times in the power circuit. A 150-kHz switching frequency is optimum in this design, since it yields the best compromise between these two conflicting requirements. However, switching frequencies up to 0.5 MHz have been achieved [2]. If the switching frequency is crystal-

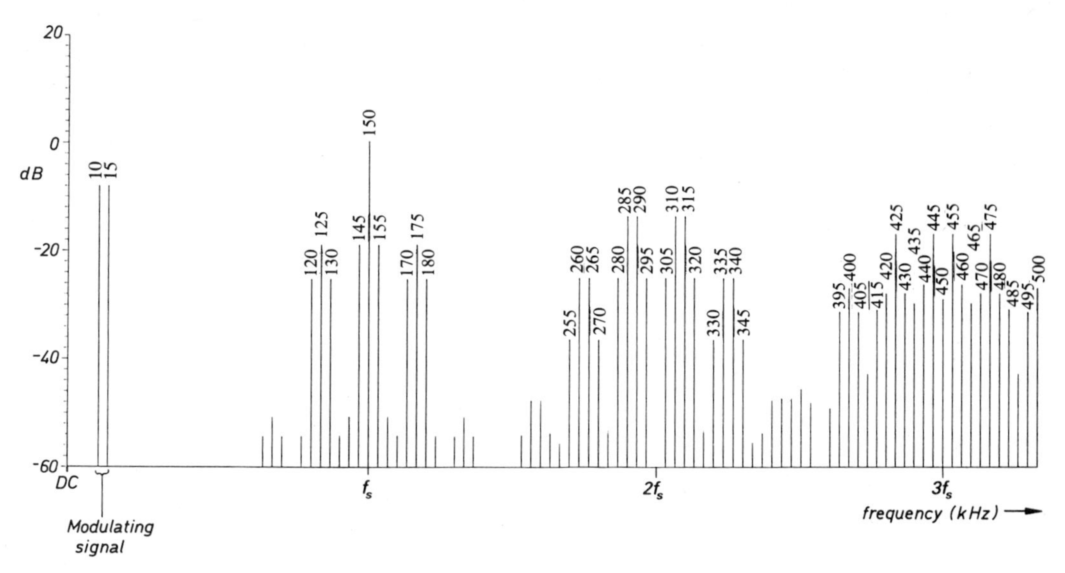

Figure 14.4. PWM waveform spectrum (half bridge). f_s is the switching frequency.

controlled, tuned filter sections can be used in the output filter to give good attenuation of the switching frequency components and their harmonics [3].

A typical frequency spectrum for the output waveform prior to the filter is shown in Figure 14.4. The input signal consists of two equal-amplitude components at 10 and 15 kHz. The modulation depth for each component is 0.4 (−8 dB). 0 dB corresponds to the amplitude obtained with a single-frequency signal and unity modulation (the maximum output obtainable without clipping). It can be seen that there are many unwanted components in the output voltage waveform of the switching stage. The inductance of the loudspeaker coil is insufficient to limit in a significant way the currents resulting from these voltage components. It is therefore necessary to employ a filter which has minimum effect on components in the audio frequency range but produces maximum attenuation of the switching frequency component and its sidebands. The second-order *LC* filter used in this design represents a compromise between the need to minimize losses, the desire to eliminate as many unwanted components as possible from the output waveform, and cost.

14.2.6. Efficiency

Figure 14.5 shows the efficiency versus output power. The input power is determined from a measurement of power-supply voltages and currents. The output power is obtained by measuring the amplitude of the output (at 1 kHz) across a 5-Ω resistor. The resistance of the load is measured immediately at the end of each test to avoid inaccuracy due to temperature drift. The harmonic content of the waveform is small enough (less than 1%) to have little influence on the efficiency measurement.

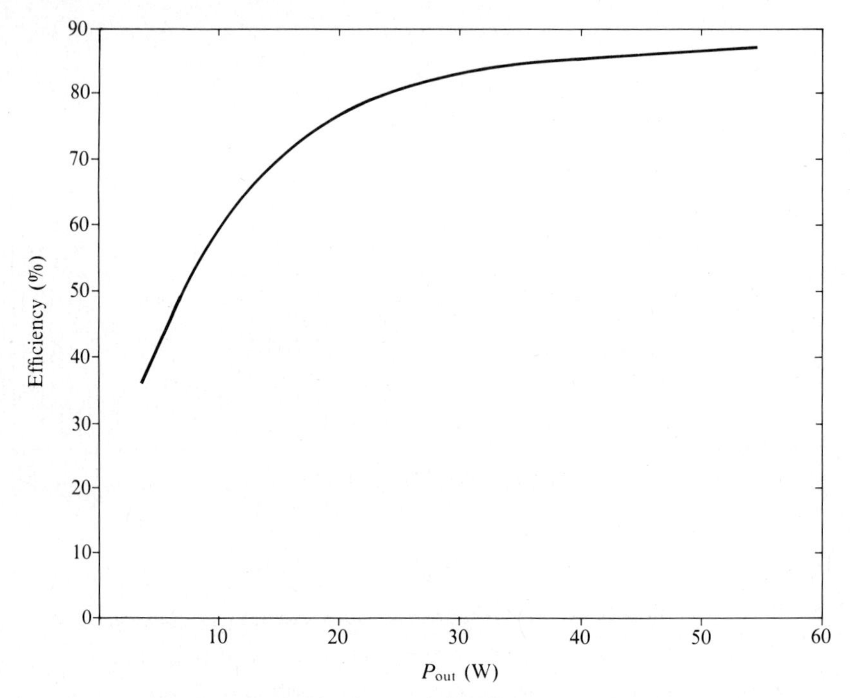

Figure 14.5. Efficiency as a function of power output.

14.2.7. Other Circuit Topologies

More sophisticated power-circuit arrangements than the one described here are possible, with a corresponding improvement in output power and waveform quality. Figure 14.6 shows a full-bridge arrangement. This can be operated in two ways. If Q1 and Q4 are switched alternately with Q2 and Q3, the output spectrum prior to the filter is the same as that obtained with the half bridge (Figure 14.4) except that the amplitudes of all components are doubled. Therefore, for a given load impedance the output power is quadrupled. In automobile applications, for example, it would be possible to put nearly 36 W into a 2-Ω loudspeaker from a 12-V supply.

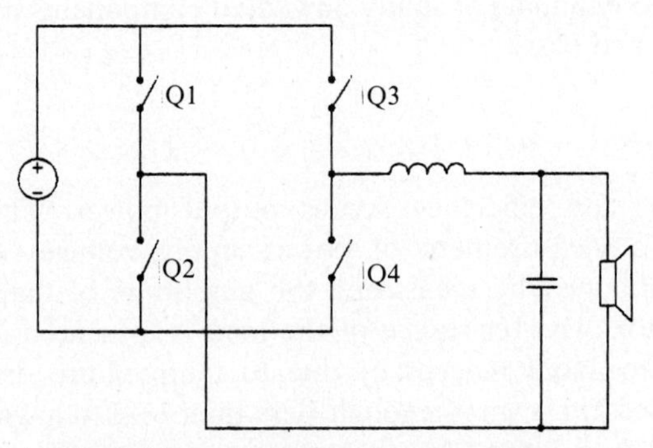

Figure 14.6. Full-bridge PWM amplifier.

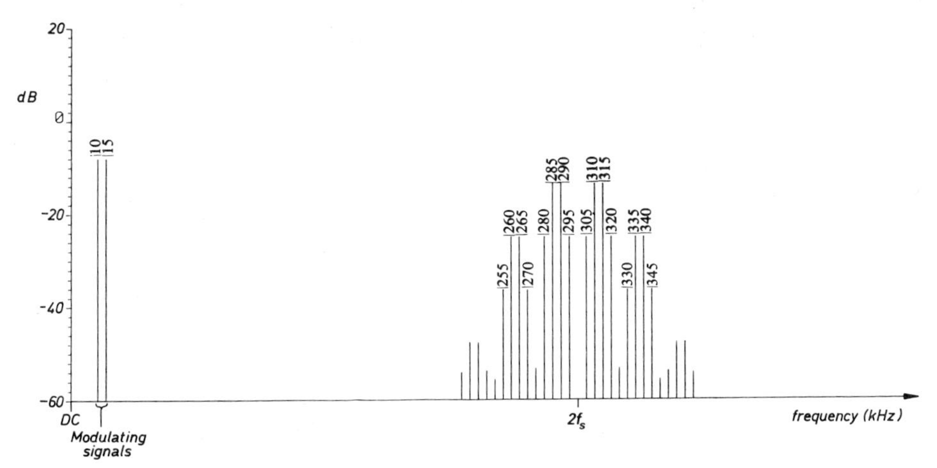

Figure 14.7. PWM waveform spectrum for full-bridge amplifier. f_s is the switching frequency.

Alternatively, the two legs of the bridge may be driven as though they were two separate half bridges. If the modulator of one half bridge is supplied with an inverted version of the modulating signal applied to the other, but the same triangular wave is used for both modulators, then the amplitude of the audio signal in the output will be doubled as before, but the harmonic content of the PWM waveform will be much reduced, as Figure 14.7 shows. This eases the filter requirements, thereby reducing the size of the filter components.

14.3. INDUCTION HEATING

Radio-frequency power for induction heating has in the past been generated by vacuum-tube oscillators. The high-frequency capability of the vacuum tube has allowed operation at several hundred kilohertz. This is the frequency range required for surface heating and for nonmagnetic parts. Thyristors and bipolar transistors have been unable to operate efficiently at these frequencies. The advent of the power MOSFET has made it possible for solid-state induction heaters to operate in the 100- to 200-kHz range both for industrial and domestic applications [4]. The power range for which the power MOSFET is appropriate is determined only by the number of MOSFETS that can realistically be operated in parallel. At high powers the static induction transistor is perhaps a more suitable device than the power MOSFET, although—being essentially a junction field-effect transistor—the SIT too suffers from the phenomenon of rapidly rising ON-state resistance as the voltage rating increases.

The most common configuration for a MOSFET induction heating inverter is the H-bridge as shown in Figure 14.8. This arrangement has the advantage that the maximum voltage experienced by the MOSFETS is equal to the power supply voltage. The load can be either series resonant, as shown in Figure 14.8, or parallel resonant (current-fed). The parallel resonant scheme has the advantage that it does not require the body-drain diodes of the power MOSFETS to conduct. These diodes are relatively slow and if allowed to conduct may seriously impair

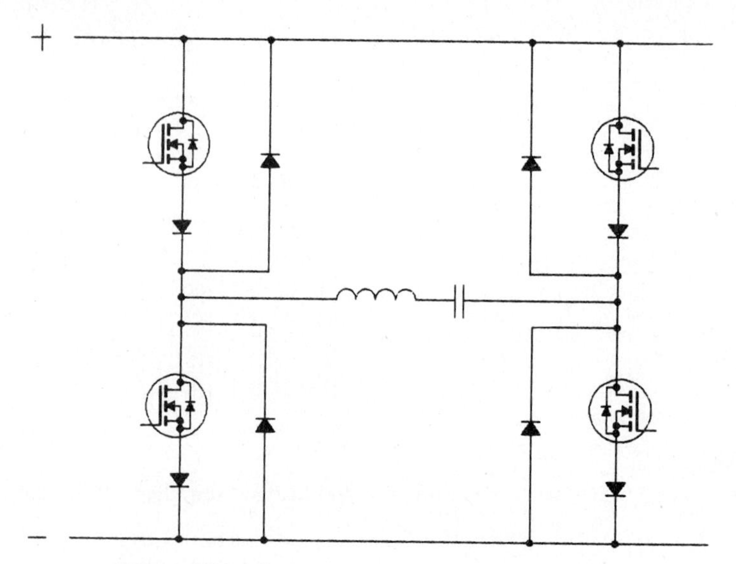

Figure 14.8. Series-resonant induction heating.

the efficiency of the inverter due to shoot-through of the diode recovery current. These diode recovery losses can also be avoided with a series resonant load if the MOSFETS are switched above resonant frequency so that the load current is always lagging with respect to the applied voltage [5]. Then, the load current commutates from the body-drain diode of a MOSFET to the channel of the same MOSFET, rather than from the body-drain diode of one MOSFET to the channel of the other. If the inverter is operated in a mode which requires diode conduction with commutation of the load current from one MOSFET to the other, the integral diode of the MOSFET must be isolated by a series diode and a fast external anti-parallel diode [6].

A further advantage of the parallel-resonant load is that the output voltage is sinusoidal, so that the inverter may be commutated near the voltage zero. This means a minimum rate of rise and fall of drain voltage, thereby minimizing the current flowing in the drain–gate capacitance and thus reducing the gate drive power requirements [7]. On the other hand, the series-resonant circuit requires a higher-voltage capacitor in the tuned circuit. Both series and parallel resonant circuits are widely used [8, 9]. Using power MOSFETS, resonant converters can attain efficiencies as high as 95% while operating at several megahertz [10].

REFERENCES

1. R. J. Haver, "Electronic ballasts." *Power Convers. Intell. Motion* (*PCIM*) April, pp. 52–58 (1987).
2. B. E. Attwood, "Design parameters important for optimization of very high fidelity (class D) audio amplifiers." *Proc. Conv. Audio Eng. Soc., 71st* (1982).
3. B. E. Attwood, "A 0.5 MHz switching dc–ac inverter topology provides low EMI environment for critical spacecraft applications." *Proc. Powercon* **11,** B.3.1–B.3.16 (1984).

4. L. Hobson, D. W. Tebb, and D. Turnbull, "Dual-element induction cooking unit using power MOSFET." *Int. J. Electron.* **59**(6), 747–757 (1985).
5. W. E. Frank and C. F. Der, "Solid state RF generators for induction heating applications." *IEEE, Ind. Appl. Soc. Annu. Meet.* pp. 939–944 (1982).
6. Z. D. Fang, D. Y. Chen, and F. C. Lee, "Designing a high frequency snubberless FET power inverter." *Proc. Powercon* **11,** D1.4.1–D1.4.10 (1984).
7. S. Bottari, L. Malesani, and P. Tenti, "High-efficiency 200 kHz inverter for induction heating applications." *IEEE, Power Electron. Spec. Conf.* pp. 308–316 (1985).
8. D. Tebb, "An induction heating power supply using high voltage MOSFETS." *Proc. Power Convers. Int. Conf.* May, pp. 236–244 (1987).
9. L. Piglione and A. Vagati, "Power M.O.S. in induction heating applications." *Proc. Power Convers. Int. Conf.* October, pp. 150–159 (1984).
10. E. Hebenstreit, "New type of class D sinusoidal converter constructed with FETS." *Eur. Power Electron. Conf.* pp. 163–167 (1987).

CHAPTER 15

Linear Applications

15.1. LINEAR POSTREGULATORS

In multiple-output switched-mode power supplies it is common for voltage feedback to the control circuit to be obtained from one of the outputs, for example the 5-V output, so that this output alone benefits from precise voltage control. If precise control of the other outputs is required, they must be subject to additional regulation. This is typically achieved with a linear regulator, or postregulator, in the output.

A critical factor in the efficiency of linear regulators is the headroom, or the difference between input and output voltage. Regulators designed to operate with low headroom are sometimes referred to as low-dropout regulators. Regulators based on bipolar transistors have difficulty in achieving low headroom voltages, since to obtain low values of $V_{CE(Sat)}$ high values of base current are required. Furthermore, it may not be acceptable to operate the bipolar transistor in saturation, since the rate at which it can respond to changing conditions will be limited by the long storage time resulting from operation in the saturated mode. Power MOSFETS, on the other hand, are voltage-controlled, and the forward drop across the MOSFET during conduction can be reduced to any value required by use of an appropriately sized device. Figure 15.1 shows a comparison of the voltage drop associated with MOSFET and bipolar regulators.

The presence of an integral antiparallel diode is a further advantage of power MOSFETS in this application. The diode protects the MOSFET against damage by reverse voltage should the input to the regulator be crowbarred, whereas a bipolar transistor would require an extra device to protect it. The surge current capability of a power MOSFET, which is typically four times its average current rating, also gives the MOSFET regulator better ability to withstand overloads.

When an n-channel device is used as a linear regulator of a positive supply rail, a gate voltage greater than the positive rail is required. Fortunately the current drawn from this supply is very small, and the cost of providing the supply is therefore low. Figure 15.2 shows three possible means of providing the gate voltage in n-channel linear regulators.

Figure 15.3 shows a negative voltage regulator using a p-channel MOSFET. Figure 15.4 shows how simple current limiting may be achieved with few additional components at the expense of increased voltage drop.

Low-dropout regulation is especially important in battery-operated equipment in prolonging useful battery life. As Figure 15.5 shows, a battery used to power an instrument, tool, or toy will have a longer useful life when regulated by a

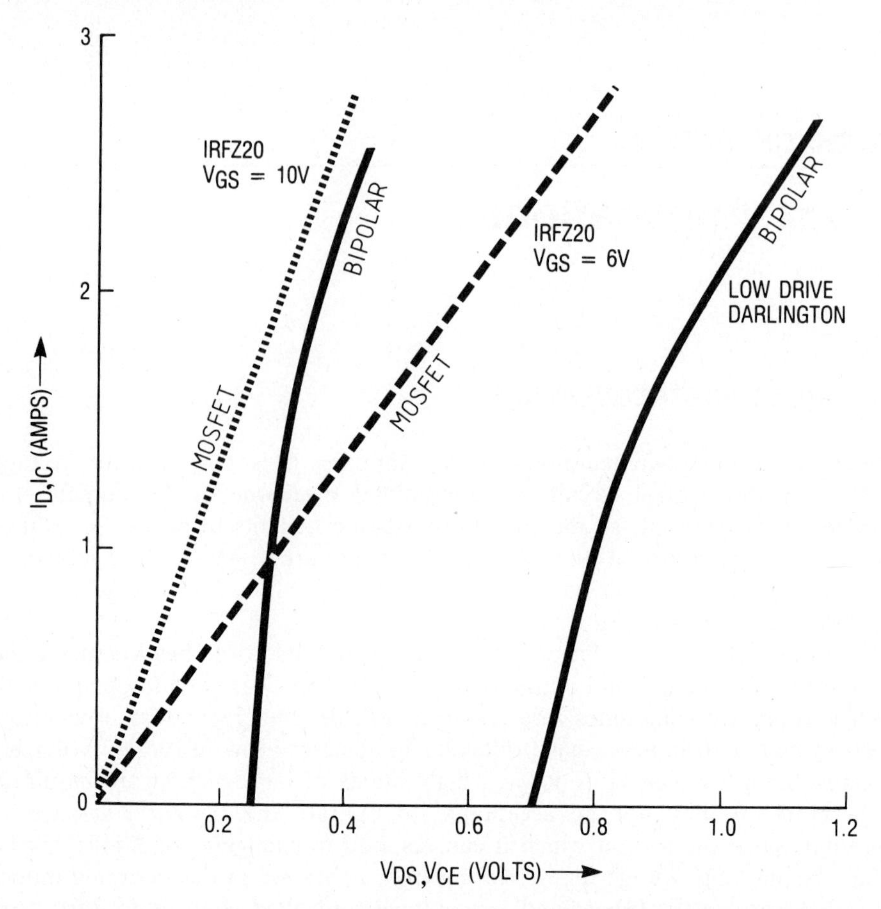

Figure 15.1. Forward voltage drop of various power semiconductor devices.

low-dropout MOSFET regulator than by a bipolar regulator with a greater headroom requirement.

15.2. AUDIO AMPLIFIERS

15.2.1. Introduction

Power MOSFETS, when used in audio amplifiers [1–3] and other forms of linear amplifier [4, 5], avoid some of the problems encountered when using bipolar transistors. Two important advantages of the power MOSFET are its lack of second breakdown and its negative transconductance coefficient, although some form of bias-point stabilization is often required.

Other advantages include the high gate input impedance, although the gate driver impedance should be low enough to prevent loss of bandwidth due to the gate capacitance. Fortunately, in the commonly used source–follower configuration the gate–source capacitance is "bootstrapped", reducing its effect typically

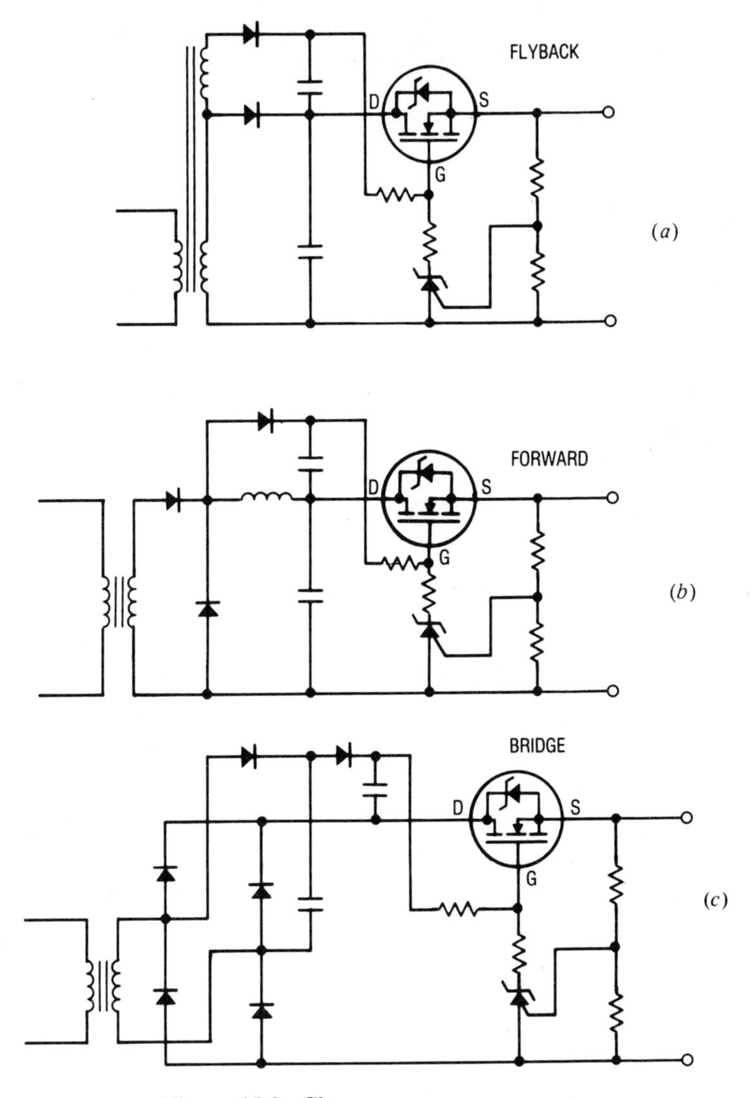

Figure 15.2. Charge-pump arrangements.

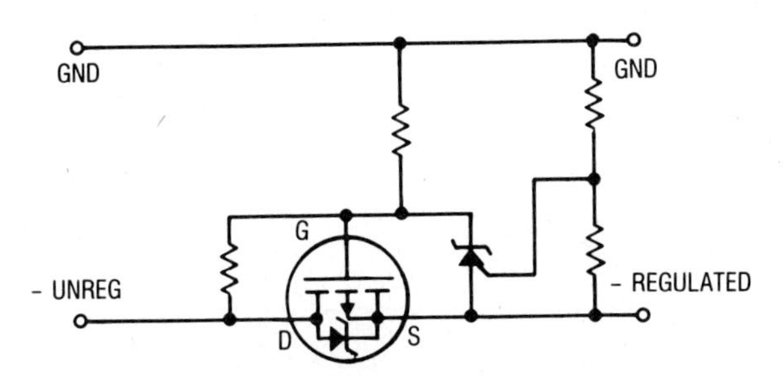

Figure 15.3. Negative voltage regulator.

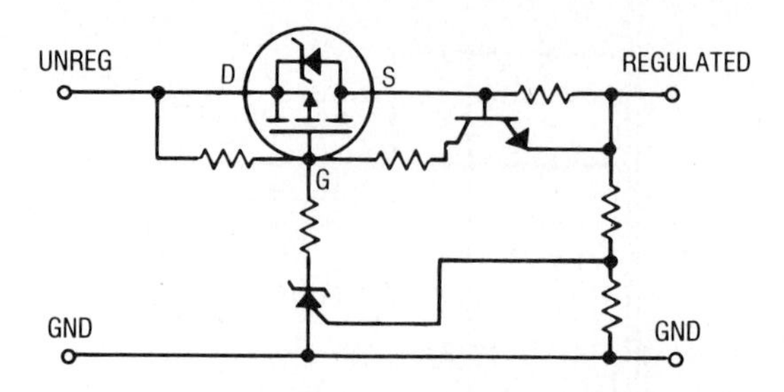

Figure 15.4. Positive voltage regulator with current limit.

by an order of magnitude [6]. The power MOSFET is not subject to minority-carrier storage-time delay, which can cause distortion and reduce bandwidth in bipolar-transistor amplifiers. Power MOSFETS can be paralleled for high power output, although device selection for matching threshold voltage and transconductance is often necessary to ensure similar bias points and good current sharing. Current sharing can be promoted at the expense of bandwidth by the use of source ballast resistors [7]. The MOSFET is usually considered to be a more linear device than the bipolar transistor in that the relationship between the gate voltage and drain current in a MOSFET is more linear than that between base current and collector in a bipolar transistor. This is truest at higher current values, where the gain of the bipolar transistor falls off rapidly while the transconductance of the power MOSFET tends to a constant value due to velocity saturation (see Figure 15.6). This greater linearity reduces the level of feedback required, allowing operation with a higher closed-loop gain.

Bipolar transistors have their safe operating area restricted by second breakdown. This necessitates load-line limiting circuitry to prevent the operating point entering the region of high voltage and high current where second breakdown occurs. The power MOSFET's square SOA characteristic makes such measures unnecessary.

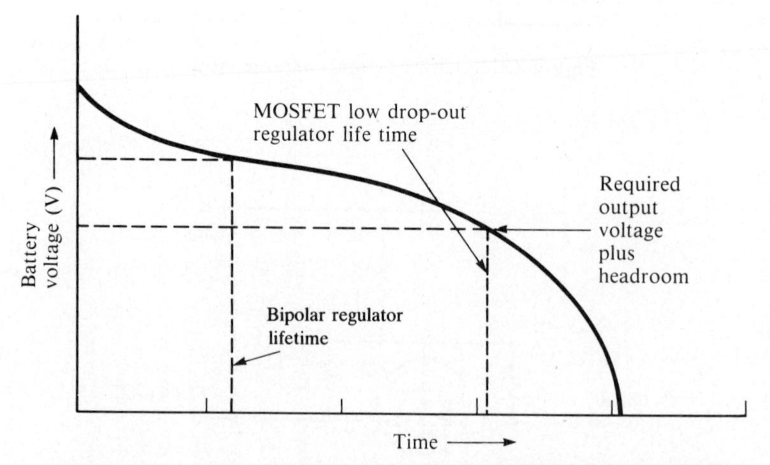

Figure 15.5. Battery lifetime extended by low-dropout regulator.

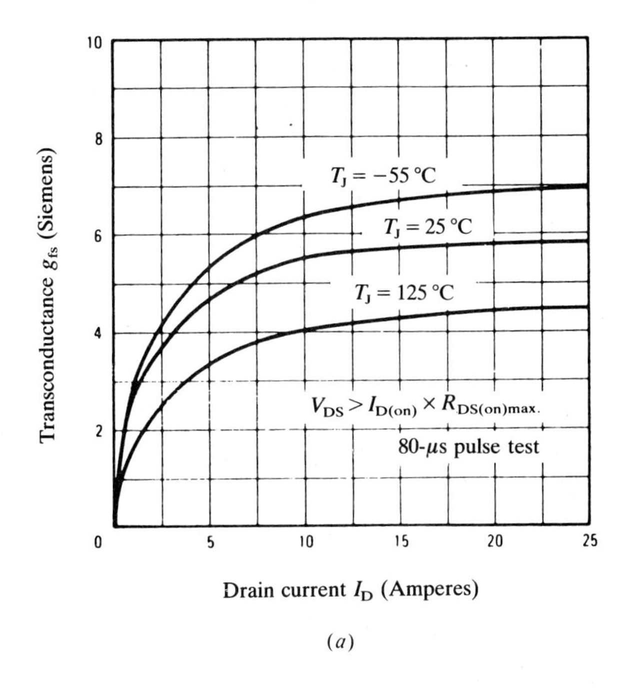

(*a*)

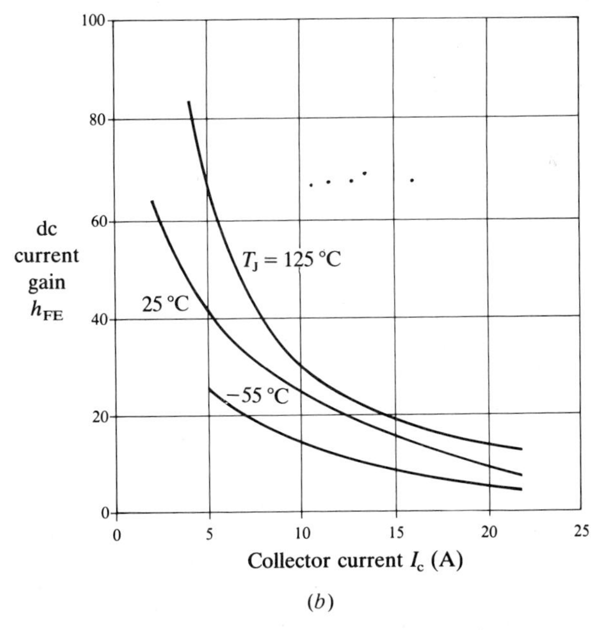

(*b*)

Figure 15.6. Gain characteristics of a power MOSFET and a similarly rated bipolar transistor. (*a*) 14-A, 100-V MOSFET; (*b*) Bipolar transistor with similar current and voltage ratings.

The advantage of the simple gate-drive requirement of the power MOSFET is diminished somewhat by the higher gate-drive voltage requirements, although this is not generally a major drawback. The power MOSFET has the disadvantage of a smaller transconductance at low current levels. This can result in distortion in the crossover region unless an adequate bias current is maintained. The drain–gate capacitance in a power MOSFET can interact with inductance in the gate circuit to form a Colpitts oscillator, so that good PCB layout practice is required to avoid the possibility of parasitic oscillation. A small resistance in the gate circuit may be required to provide damping.

15.2.2. Bias-Point Stabilization

The negative temperature coefficient of the transconductance of the power MOSFET is often cited as an advantage in audio amplifiers, but this requires qualification. In a bipolar-transistor amplifier it is common practice to sense the temperature of the transistors and to alter the bias conditions to prevent thermal runaway, since at constant base-emitter voltage the collector current has a positive temperature coefficient. With a power MOSFET, a temperature increase results in a decrease in transconductance, a decrease in current, a decrease in dissipation, and stabilization at a new current level. However, power MOSFETS vary in their suitability for this application. Figure 15.7 shows the transfer characteristics of two power MOSFETS. Particularly in Figure 15.7*a,* it can be seen that for high current values the drain current decreases as temperature increases with the gate voltage held constant. For low values of drain current the drain current increases with temperature, and for one value of drain current—the zero-temperature-coefficient point—there is no change. At lower currents, the decrease in threshold voltage with increasing temperature is the dominant effect.

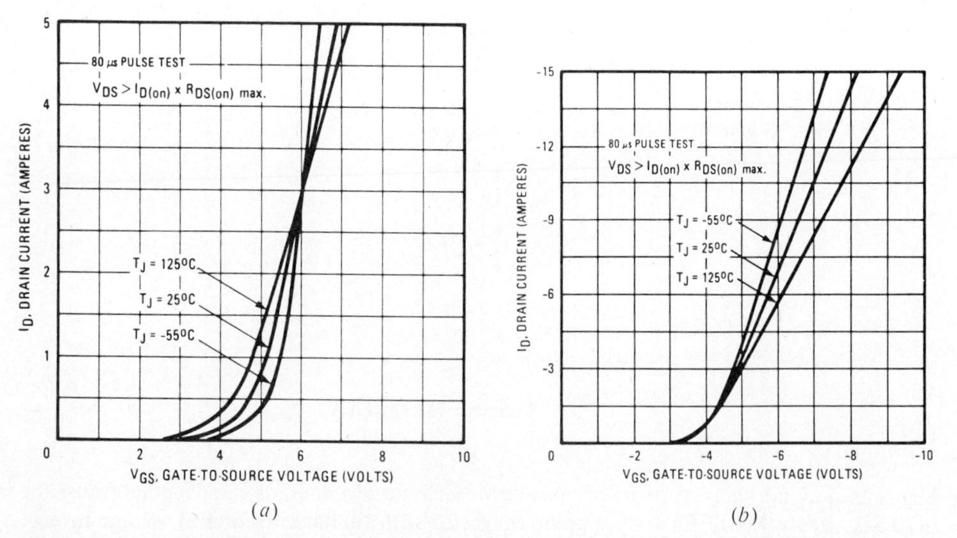

Figure 15.7. Typical transfer characteristics of power MOSFETS, with the zero-temperature-coefficient point at (*a*) a high value and (*b*) a low value of I_D.

At high current values the fall in transconductance with increasing temperature dominates. A stable bias point is only obtained if the bias current is above the drain current corresponding to the zero-temperature-coefficient point. Unfortunately, as Figure 15.7*a* shows, for some devices this current can be many amperes, whereas the quiescent current at the bias point in an audio amplifier is typically of the order of 100 mA. More appropriate power-MOSFET characteristics for audio-amplifier use are shown in Figure 15.7*b*. Devices having this characteristic have tended to be lateral MOSFETS, although vertical devices with the required characteristics for this application have begun to appear. The theoretical basis for the variation in MOSFET characteristics with temperature change is discussed in Section 3.7.

The typical audio amplifier operates in the class-AB, mode in which the class-B push–pull configuration is operated with a degree of class-A bias to minimize crossover distortion. This results in a continuous quiescent current flowing through the two devices. If stabilization of the quiescent current level is necessary, it can be achieved by using either another device to sense the temperature and exert negative feedback, or a specifically designed feedback system such as that shown in Figure 15.8 [8]. In this system the level of bias current is extracted from the large half cycles of load current by detecting the asymmetry in the zero crossings of the supply current waveform.

The class-AB amplifier shown in Figure 15.9 uses a complementary pair of power MOSFETS in the output stage [9]. Split power rails are used to give improved rejection of power-supply ripple and to allow the load to be directly coupled. The output devices Q5 and Q6 operate in the source–follower configuration. The use

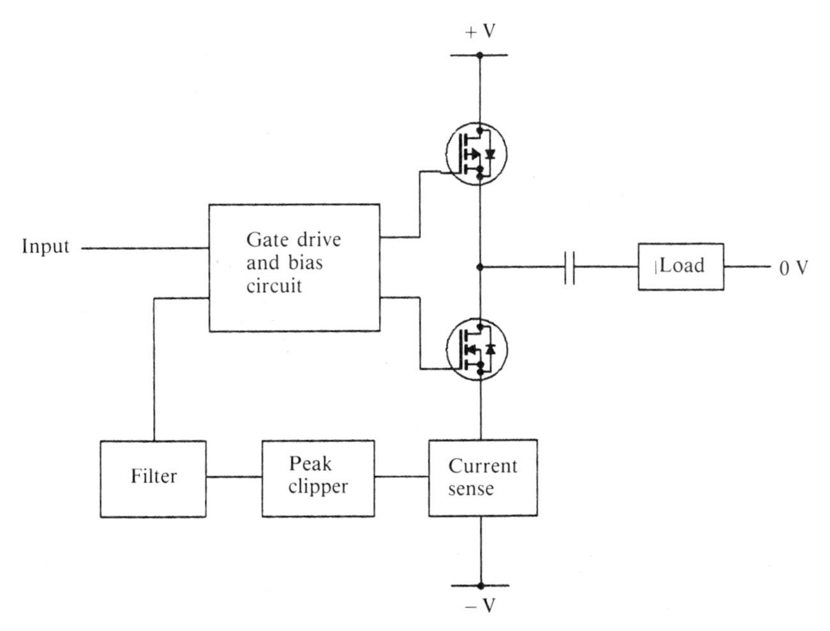

Figure 15.8. Feedback bias arrangement for a class-AB amplifier. The zero crossings of the current waveform are detected to form a trapezoidal waveform whose average value is a measure of the bias current flowing through the power MOSFETS.

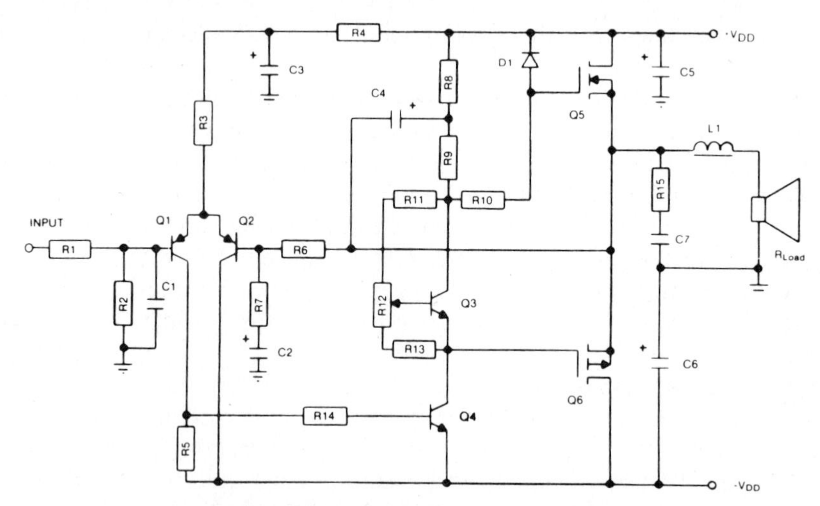

Figure 15.9. Class-AB amplifier circuit.

of this configuration is common in audio amplifiers; it offers two main advantages: (a) the possibility of oscillation in the power stage is reduced as the voltage gain is less than unity, and (b) signal feedback through the heatsink on which the devices are mounted is eliminated, since the drain, which is electrically connected to the metal tab of each of the TO-220 transistors (or can in the case of TO-3 transistors), is connected to one of the dc supply rails.

A symmetrical output is achieved by providing a "bootstrapped" drive to the gate of the n-channel device, Q5, from the output. The use of the bootstrap circuit C4, R8, R9 also allows the driver transistor Q4 to operate at near-constant current, which improves the linearity of the driver stage. The diode D1 acts as a clamp for the bootstrap circuit, restricting the positive voltage at the gate of Q5 to $+V_{DD}$. This allows symmetry to be maintained under overload conditions.

Transistor Q3 and resistors R11, R12, R13 provide gate–source offset voltage for the output devices. R12 is used as a potentiometer, allowing adjustment of the output quiescent current for variation in MOSFET threshold voltage. A degree of temperature compensation is built into the circuit, as both the emitter–base voltage of the bipolar transistor Q3 and the combined threshold voltages of the MOSFETS Q5, Q6 have a temperature coefficient of $-0.3\%/°C$.

REFERENCES

1. J. L. Linsley Hood, "80–100 watt MOSFET audio amplifier." *Wireless World* June, pp. 40–42 (1982).
2. T. Sampei, S. Ohashi, and S. Ochi "100 watt super audio amplifier using new MOS devices." *IEEE Trans. Consum. Electron.* **CE-23**(3), 409–416 (1977).
3. E. Borbely, "High power high quality amplifier using MOSFETS." *Wireless World* March, pp. 69–75 (1983).
4. M. Alexander, "Voice coil drivers using complementary MOSFETS." *Motor-Con* April, pp. 223–228 (1984).

5. M. Alexander, "Boost OP-AMP output power with complementary power MOSFETS." *Siliconix Appl. Note* **AN83–5** (1983).
6. R. R. Cordell, "A MOSFET power amplifier with error correction." *Proc. Conv. Audio Eng. Soc., 72nd* pp. D9, 1–12 (1982).
7. K. Gauen, "Paralleling power MOSFETS in linear applications." *Proc. Power Convers. Intell. Conf.* April, pp. 144–151 (1984).
8. B. Roehr, "The autobias amplifier." *Siliconix Tech. Artic.* **TA82-1** (1982).
9. P. Wilson, "Linear power amplifier using complementary HEXFETs." *Int. Rectifier Appl. Note* **AN-948A** (1983).

CHAPTER 16

MOSFETS and Bipolar Transistors

16.1. COMPARISON OF MOSFETs AND BIPOLAR TRANSISTORS

16.1.1. Introduction

The important differences between a power MOSFET and a bipolar transistor originate in the conduction mechanisms at work in each device. In the power MOSFET majority carriers alone are responsible for current flow, while in the bipolar transitor the flow of current is regulated by minority-carrier injection. As Figure 16.1 shows, both the MOSFET and the bipolar are three-layer devices which normally block the passage of current due to the presence of a reverse-biased pn junction in the current path. In the case of the n-channel MOSFET shown, conduction is initiated when the gate is made positive with respect to the source. The surface of the p region is inverted, thereby producing a continuous channel of n-type material between the drain and source terminals through which majority carriers—electrons in this case—can flow. In the case of the NPN bipolar transistor, conduction is initiated by forward-biasing the base–emitter junction with current flowing from base to emitter. Due to the relative doping densities of the base and emitter region, this results principally in the injection of electrons into the base region. Since the electrons are traversing a p-type region, they are minority carriers. If the base is narrow relative to the diffusion length of the electrons in the p-type base region, most of the electrons will diffuse into the depletion region formed by the reverse bias applied to the collecter–base junction. Here they encounter an electric field which sweeps them into the collector region.

16.1.2. The MOSFET and the Bipolar Transistor as Charge-Controlled Devices

It is possible to view both the MOSFET and the bipolar transistor as charge-controlled devices. In the case of the MOSFET, conduction is initiated when the input capacitance of the MOSFET has received sufficient charge to elevate the gate-to-source voltage to the level necessary to achieve the required degree of inversion of the channel region (see Figure 16.2*a*).

In a bipolar transistor, the injection of holes from the base terminal creates a positive charge in the p base region, which, following the principle of charge neutrality, is balanced by an inflow of electrons. The base region may therefore be considered to have been enhanced with respect to the electron concentration,

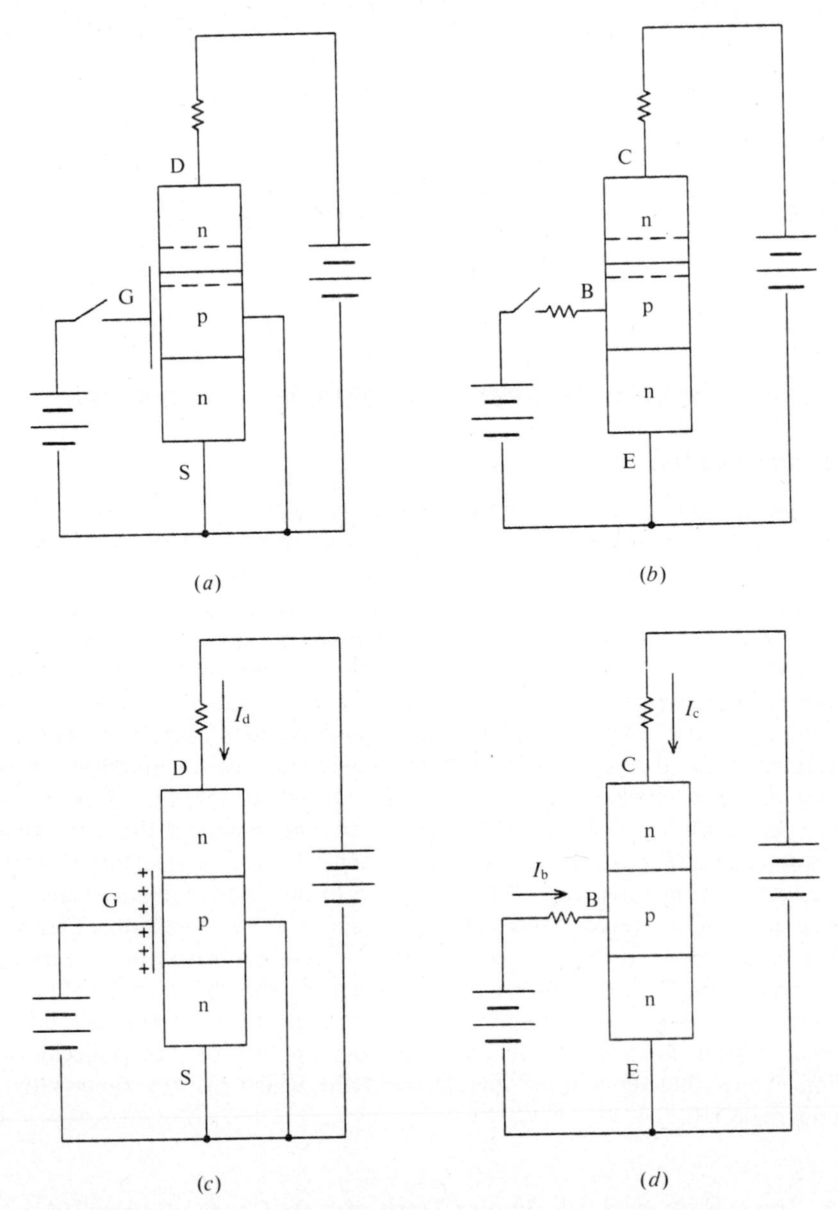

Figure 16.1. Operation of MOSFET and bipolar transistors. (*a*) MOSFET in the OFF state; (*b*) bipolar in the OFF state; (*c*) MOSFET conducting; (*d*) bipolar conducting.

thereby permitting electrons to flow from the emitter to the collector (see Figure 16.2*b*).

The significant difference between the two devices is that in the case of the bipolar transistor, recombination will take place between the electrons and the holes, and it is therefore necessary to inject base current continuously in order to maintain the required hole concentration, as Figure 16.3 illustrates. In the

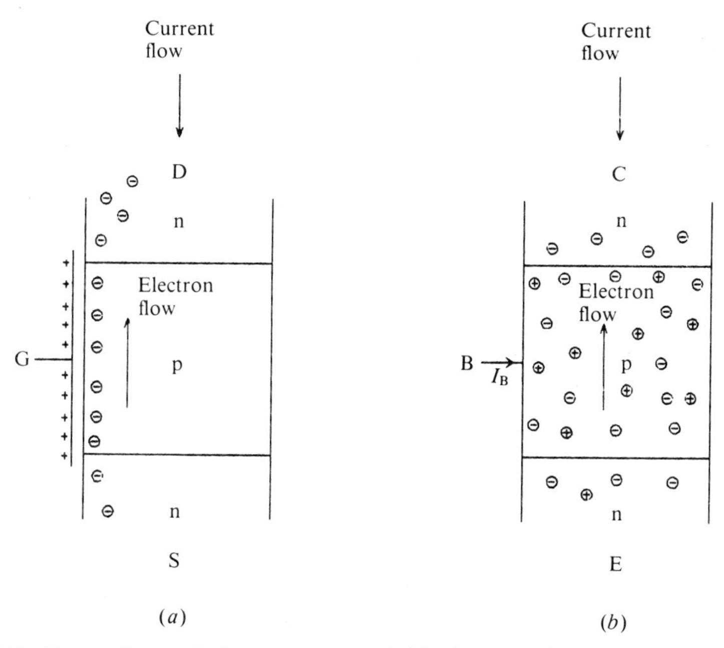

Figure 16.2. Comparison of the MOSFET and bipolar transistors as charge-controlled devices. (*a*) MOSFET; (*b*) bipolar.

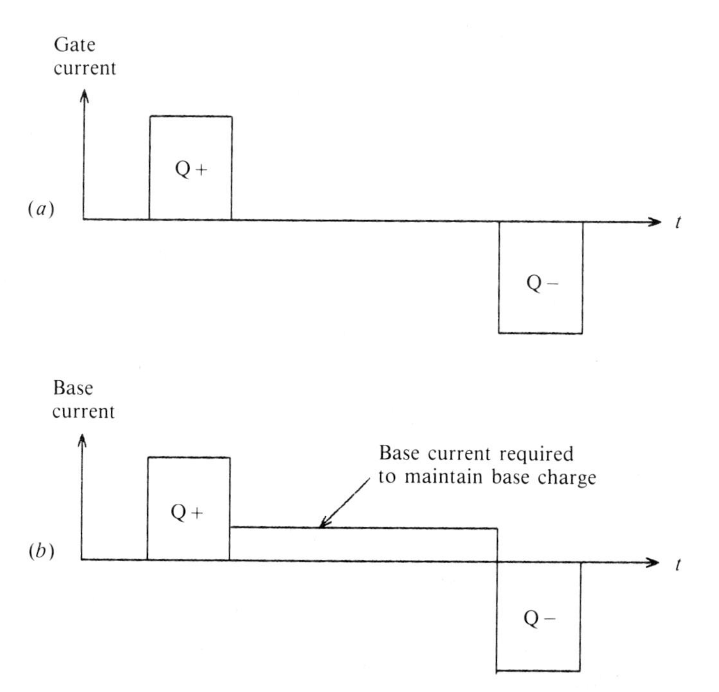

Figure 16.3. Charge requirements for controlling MOSFETS and bipolar transistors. (*a*) MOSFET; (*b*) bipolar.

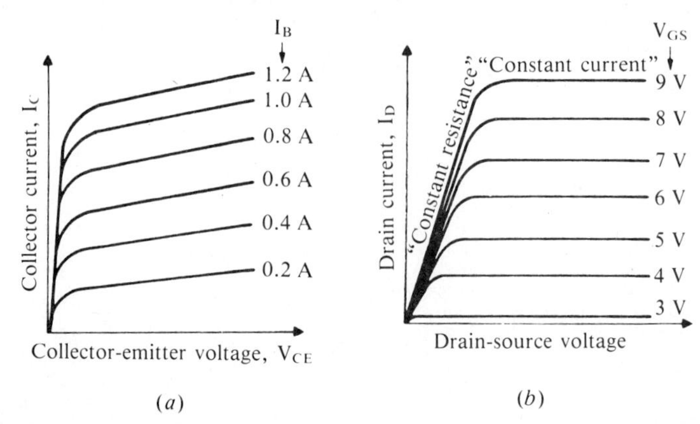

Figure 16.4. Comparison of the transfer characteristics of MOSFETS and bipolar transistors. (*a*) Bipolar; (*b*) MOSFET.

MOSFET, electrons and holes are separated by a dielectric, so that no recombination is possible and gate current only flows when charge is being established or removed. Generally, the amount of charge required to turn on a MOSFET is small, being that required to establish inversion and accumulation layers, and charge the gate–drain and gate–source capacitances.

The amount of charge stored in a bipolar transistor may be relatively large, particularly if the transistor is driven into saturation with excess base current. The result of overdriving the base is to cause the base region to expand beyond its metallurgical boundary into the collector region. The minority carriers diffusing across the base–collector boundary produce conductivity modulation of the collector region, thereby lowering the collector–emitter voltage. Bipolar power transistors usually exhibit the type of characteristics shown in Figure 16.4*a*. The quasi-saturation region which precedes saturation is caused by the base region expanding into the collector region (see Figure 16.5). When the whole of the

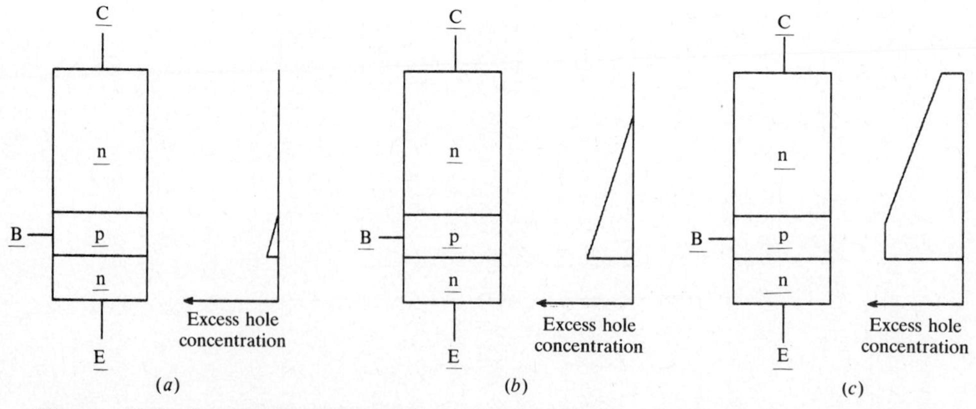

Figure 16.5. Conductivity modulation in a bipolar transistor. (*a*) Nonsaturated condition—normal base width; (*b*) Quasisaturation—base region exceeds width of metallurgical base; (*c*) Oversaturation—base region completely occupies collector region.

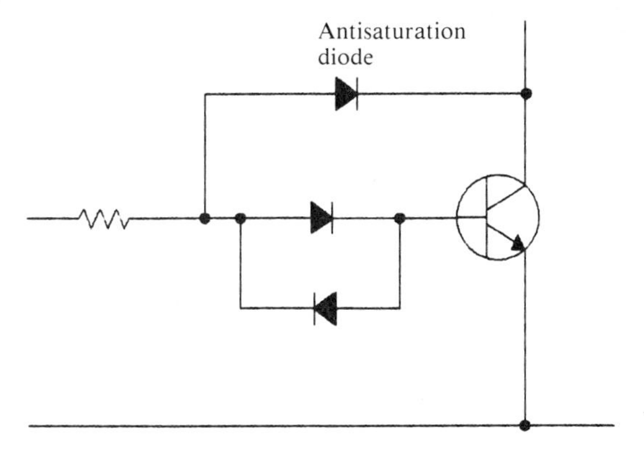

Figure 16.6. Antisaturation clamp.

collector region is conductivity-modulated, no further reduction in collector–emitter voltage is possible and the transistor is in saturation. Although in a power circuit it is usually advantageous to reduce the forward voltage drop across the transistor, the turnoff delay time resulting from saturation and the difficulties of extracting the base charge can be unacceptable. If this is the case, saturation of the bipolar transistor can be prevented by the use of an antisaturation diode as shown in Figure 16.6. The positive slope of the curves in the linear region (Figure 16.4*a*) is caused by a narrowing of the base width as the depletion region expands into the base. A similar shortening of the channel occurs in power MOSFETS, but the doping levels used for the channel and drain regions produce characteristic curves with almost zero slope in the linear region. (See Appendix 1).

The characteristics of a power MOSFET are shown for comparison in Figure 16.4*b*. In the power MOSFET, conduction through the drain region, which corresponds to the collector region of a bipolar transistor, is by the drift of majority carriers under the influence of the applied drain–source voltage. There is no mechanism for producing conductivity modulation of the drain region. The factors determining the shape of these characteristics are discussed in Chapter 3.

16.1.3. Delay Times and Switching Times

Delay times and switching times in both the MOSFET and the bipolar transistor are determined by the need for charge to be established or removed. The switching performance of both devices depends to some extent on the nature of the load. However, the underlying principles are adequately illustrated by comparing the operation of the devices when controlling a resistive load. Figure 16.7*a* and *b* show the idealized switching waveforms associated with each device. The switching times of both the bipolar transistor and the power MOSFET are dependent on the manner in which the devices are driven and are only valid for the conditions used in the manufactuter's test. Nevertheless, they do provide a useful yardstick for comparing the switching speed of the devices of a single manufacturer and, if the test conditions are not too much at variance, may also be useful for comparing devices made by different manufacturers.

One essential difference between the power MOSFET and the bipolar transistor

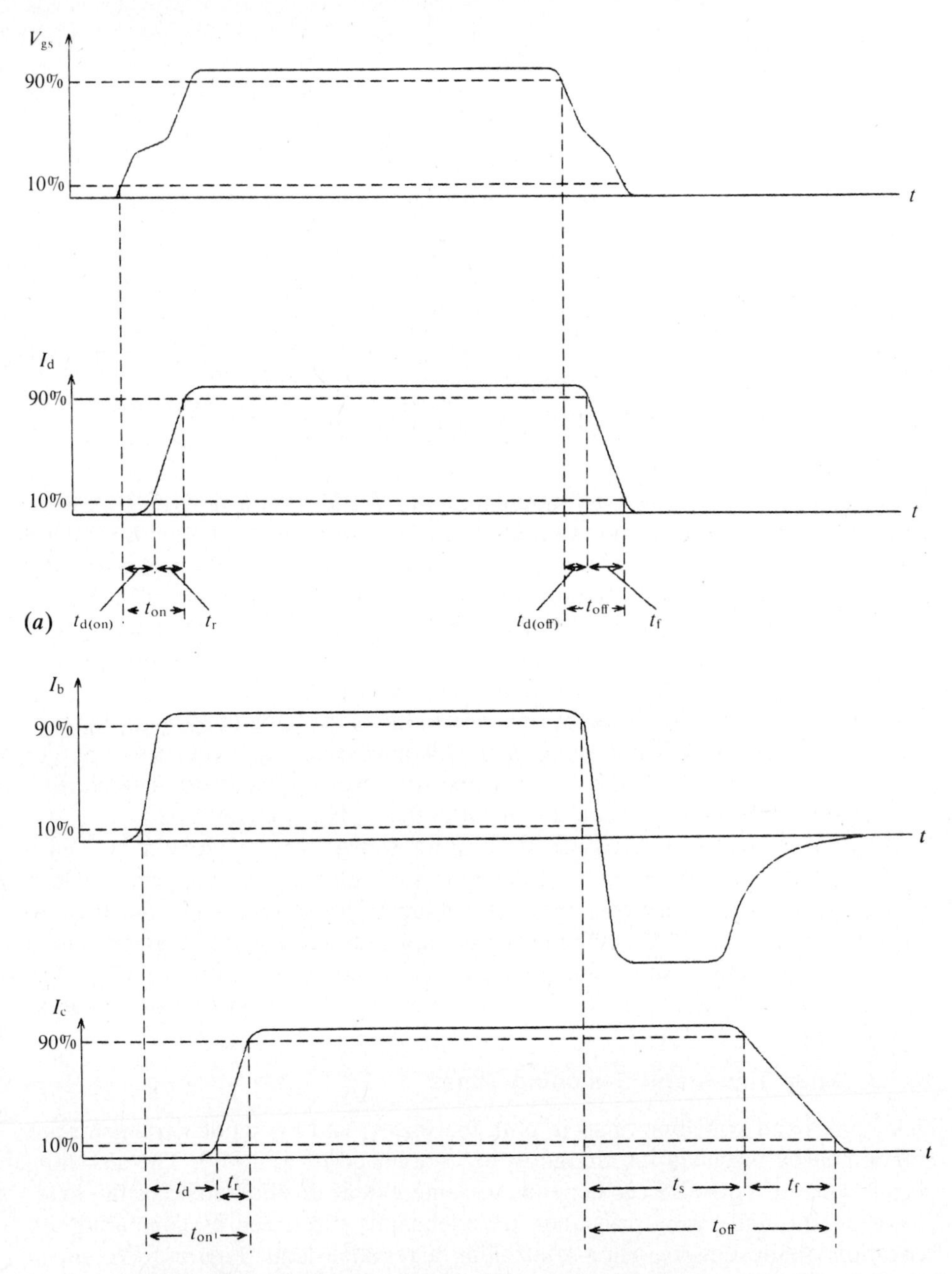

Figure 16.7. Switching waveforms. (*a*) MOSFET; (*b*) bipolar transistor.

is illustrated by these waveforms. Bipolar transistors can have relatively long turnoff delay times due to minority-carrier storage, whereas the MOSFET delay times can be reduced to very low values by hard driving of the gate. The power MOSFET is turned off as soon as sufficient charge has been removed from the gate to take the gate voltage below the threshold. The delay time associated with this transition can be of the order of nanoseconds if the gate-driver impedance is low.

The drain voltage then rises as the load current charges the output capacitance of the MOSFET. In a bipolar transistor charge has to be removed from the base region before the device can block voltage. If the transistor has been operated in saturation and the effective base region has extended beyond the metallurgical base into the collector region, the charge in the collector region will prevent turnoff until a sufficient number of minority carriers have decayed to allow a depletion region to be established. Even when operated in quasisaturation, the rise, fall, and delay times of bipolar transistors tend to be longer than the equivalent parameters for power MOSFETS. Hence the power MOSFET is capable of higher operating frequencies than the bipolar transistor. Since the advent of the power MOSFET, manufacturers of power bipolar transistors have developed techniques to extend their frequency range as well as their SOA characteristics [1]—in particular, by adopting a cellular structure which gives better access to stored charge and increases the resistance of the device to second breakdown [2, 3].

16.1.4. Temperature Dependence

A comparison of the temperature dependence of some important MOSFET and bipolar parameters is shown in Table 16.1.

TABLE 16.1. Effect Produced by an Increase in Temperature

Characteristics	Bipolar	MOSFET
Forward voltage drop	Reduces	Increases
Rise time	Increases	No change
Storage time, turnoff delay time	Increases	No change
Fall time	Increases	No change
V_{BE}, threshold voltage	Decreases	Decreases
Gain, transconductance	Increases	Decreases

An important advantage which the MOSFET has over the bipolar transistor is that the switching times of the MOSFET are essentially independent of temperature. This means that drive strategies for the MOSFET do not have to take account of variations in switching, storage or delay times, as they do with the bipolar transistor. The increase in the forward voltage drop across a MOSFET with increasing temperature is very pronounced, due to the strong relationship between temperature and $R_{DS(on)}$ (see Figure 9 of Appendix 7). While this results in increased losses at elevated temperatures, it does permit power MOSFETS to be connected in parallel without the risk of thermal runaway. In a switching circuit the variation with temperature of the threshold voltage and transconductance of the MOSFET are of little consequence, since the gate voltage alternates between two values representing the fully ON and fully OFF states.

16.1.5. Safe Operating Area

The SOA of the bipolar transistor and that of the power MOSFET differ in a number of important ways.

Firstly, they differ in the definition of SOA. Strictly, the bipolar transistor has two distinct SOAS—the reverse-bias SOA and the forward-bias SOA. Reverse bias and forward bias refer to the bias condition of the base-emitter junction. In a switching cycle during which negative bias is applied to the base to effect turnoff, both SOAS come into play. However, as yet the manufacturers of power MOSFETS have provided only a single SOA curve for all bias conditions, and there does not appear to be any aspect of the MOSFET behavior which would require a different approach. Typical MOSFET and bipolar-transistor SOAS are shown in Figure 16.8.

Secondly, the pulse current rating of the power MOSFET is usually about four times its dc rating, whereas the pulse current rating of a bipolar transistor is often not much greater than its dc rating.

The third important difference between the SOAS of the power MOSFET and the bipolar transistor is that the bipolar SOA has a second breakdown limit, the MOSFET does not (see Figure 16.8). Since the MOSFET is a majority-carrier device, it is free of the thermal runaway and current-focusing phenomena responsible for second breakdown in bipolar transistors. Thus the MOSFET does not require a snubber for load-line shaping, although a snubber may be used to limit dv/dt, to attenuate voltage spikes produced by unclamped inductance, or to reduce switching losses.

16.1.6. Drive Requirements

Generally the drive requirements of a power MOSFET are considerably simpler than those of a bipolar transistor. This is illustrated by Figure 16.9, which shows how the two transistors might be used in a typical power switching application. The bipolar transistor requires substantial drive circuits to provide forward and reverse base current, while the power MOSFET requires only the application or removal of a relatively small amount of charge at turn-on and turnoff. While

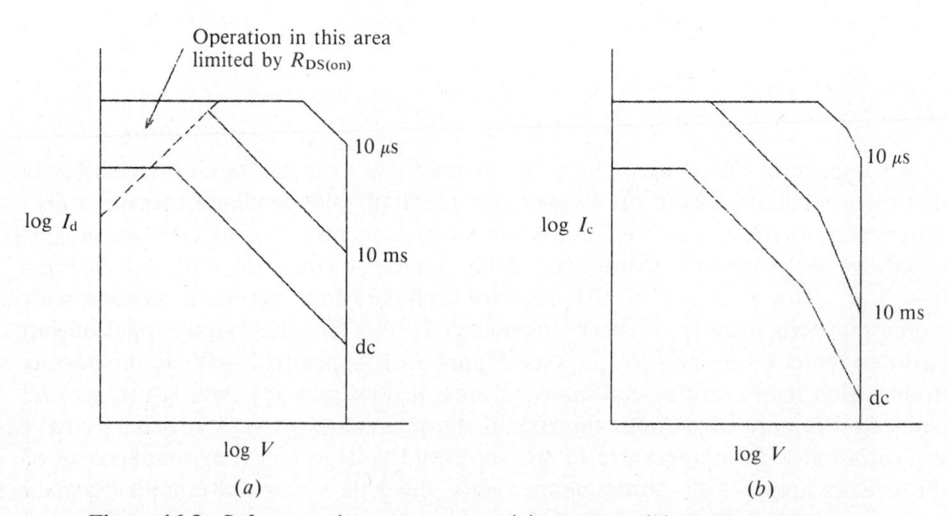

Figure 16.8. Safe-operating-area curves. (*a*) MOSFET; (*b*) bipolar transistor.

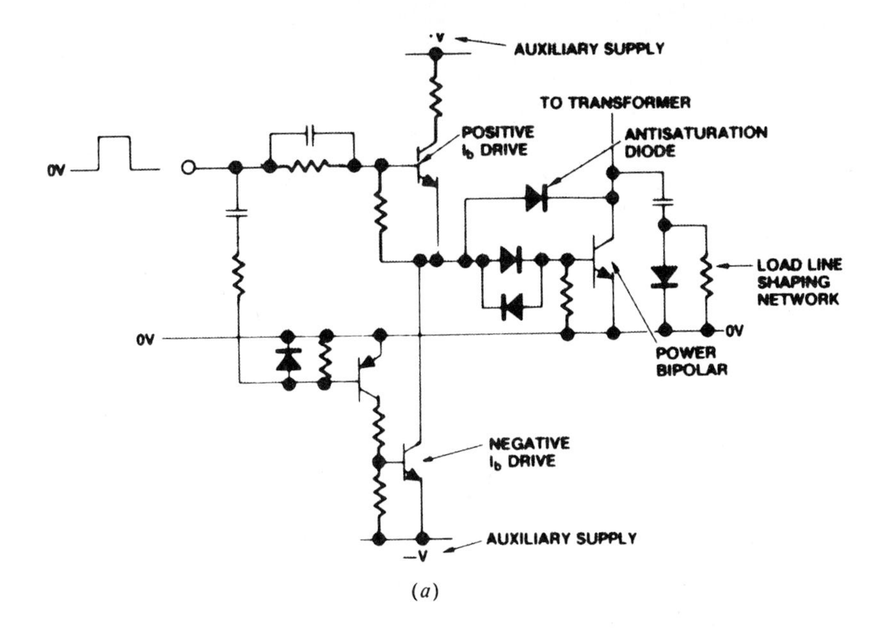

(*a*)

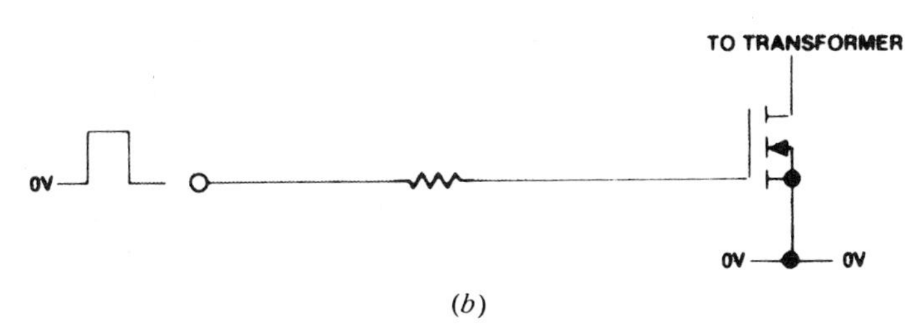

(*b*)

Figure 16.9. Driver circuits. (*a*) For bipolar transistor; (*b*) for MOSFET.

there are more elegant bipolar drive circuits which employ fewer components, there is no doubt that one of the strengths of the power MOSFET is that the gate-drive requirements are generally simple and inexpensive. The advantage is less obvious when the MOSFET is compared with a Darlington bipolar transistor. With a gain of at least 100, the Darlington requires very little base current to maintain it in conduction, although to obtain fast turnoff it may be necessary to provide a negative base drive capable of extracting charge rapidly from the main transistor.

16.1.7. Snubbers, Load-Line Shaping, Antisaturation Diodes

To ensure that the bipolar transistor remains within its safe operating area during switching, it is frequently necessary to employ a snubber or load-line shaping network as shown in Figure 16.10. To ensure fast turnoff of the bipolar transistor, it is necessary to prevent saturation by means of an antisaturation diode arrangement as shown in Figure 16.6.

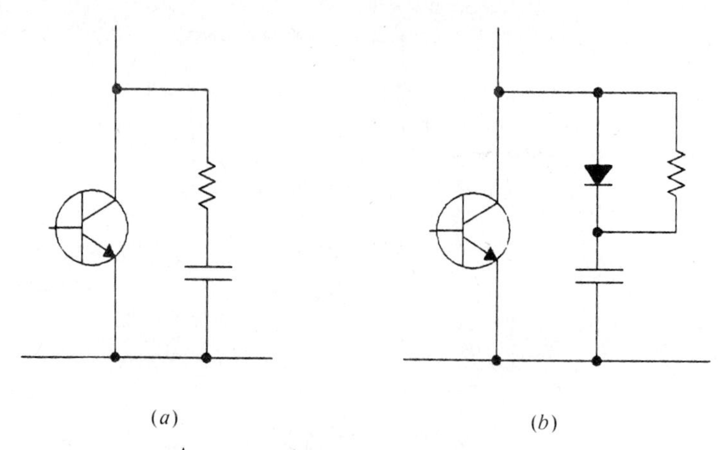

Figure 16.10. Snubber circuits. (*a*) Simple *RC*; (*b*) polarized.

The power MOSFET requires no antisaturation measures, since it is a majority-carrier device. Since the power MOSFET does not experience second breakdown in normal operation, a snubber is not used, as it often is with bipolar transistors, for the purpose of preventing second breakdown. A snubber may however be fitted for a number of other reasons. A MOSFET usually has a safe operating area permitting switching cycles which apply rated voltage to the MOSFET at the same time that it is carrying rated current. However, in high-pulse-current applications it may be necessary to use a snubber in order to keep the instantaneous junction temperature within acceptable limits. A snubber may be used to reduce switching losses in the MOSFET in order to limit the average junction temperature in high frequency applications. A snubber may be required to limit the rate of rise of drain voltage when forward voltage is reapplied immediately after the body–drain diode has been conducting and there is a danger that the dv/dt capability of the device will be exceeded. A snubber may also be required when the load contains some unclamped inductance. Power MOSFETS are capable of extremely rapid switching, and even a small amount of unclamped inductance can cause excessive drain voltage to be developed at turnoff. An alternative to a snubber in this situation is a zener-diode clamp or an active clamp which turns the MOSFET on if a certain level of drain voltage is exceeded (see Figure 9.3). The snubber may also be required to damp unwanted oscillations during switching.

16.1.8. Appropriate Applications

Performance and cost are the two main issues on which the choice between a MOSFET and a bipolar transistor is usually made. In terms of performance, the MOSFET has an advantage when high switching frequencies are required, due to its fast switching times and low switching losses. The ease of drive and freedom from second breakdown are further factors in the MOSFET's favor. The forward voltage drop in a MOSFET can be greater or smaller than that in a bipolar transistor of the same die size, depending on the voltage rating required. The $R_{DS(on)}$ of a MOSFET is approximately proportional to $V^{2.6}$, so that for a given die size, the voltage drop across a high-voltage MOSFET is usually greater than that across a bipolar transistor of the same size, since the voltage drop across a bipolar transistor depends less on the voltage rating. At the other end of the scale, the voltage drop across a low-voltage power MOSFET, say 100 V or below, is less significant, due to the low $R_{DS(on)}$ of the MOSFET at this voltage rating.

If switching losses are taken into consideration, the advantage moves further in the direction of the MOSFET, since switching losses in a MOSFET are generally much lower than those in a bipolar transistor of similar rating. Figure 16.11 shows a comparison of the total losses in a 450-V MOSFET and a 450-V bipolar transistor of comparable current ratings, with switching frequency as a variable [4]. At this voltage rating the MOSFET losses are greater than those of the bipolar transistor up to about 10 kHz. The crossover point of the two curves will be at a lower frequency for lower voltage ratings.

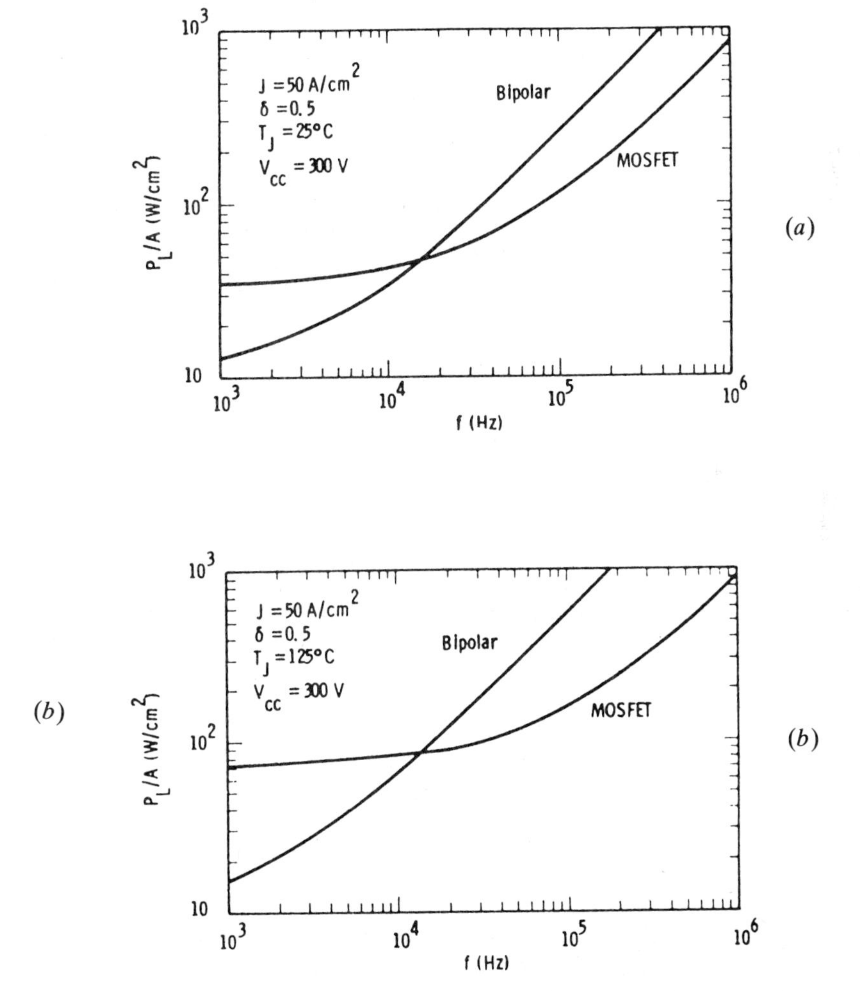

Figure 16.11. Comparison of MOSFET and bipolar losses as a function of frequency at junction temperatures of (a) 25°C (b) 125°C. (From [4]. © IEEE 1980.)

16.1.9. Avalanche Capability

Bipolar transistors are rarely capable of withstanding a significant avalanche current. First breakdown is rapidly followed by second breakdown and the destruction of the transistor. Some devices are available with an avalanche rating (E_{Sb}), but power bipolar transistors are not generally rated for operation in the avalanching regime. The power MOSFET, on the other hand, has a structure which should be capable of supporting a significant amount of current in avalanche provided the parasitic bipolar transistor is not activated. The power MOSFET has a blocking voltage equal to the V_{CES} rating of the parasitic bipolar transistor, since the base of the bipolar transistor is connected to its emitter. If the bipolar transistor in one cell is activated by avalanche current, then the blocking voltage of this cell will drop to $V_{CEO(sus)}$ (see Figure 3.23). This will result in a high local concentration of current and eventual second breakdown of the parasitic bipolar transistor.

16.2. MOSFET-AND-BIPOLAR COMBINATIONS

16.2.1. Introduction

Neither the MOSFET nor the bipolar transistor is an ideal switch—both have their advantages and their limitations. The MOSFET can switch quickly, with no storage time and no second breakdown. Being voltage-controlled, it is easy to drive. However, at high voltage ratings it has a high $R_{DS(on)}$ and consequently a high forward voltage drop. Furthermore the cost of the MOSFET rises steeply as the voltage rating increases. The bipolar transistor, on the other hand, can be made relatively cheaply at high voltages, with a relatively low forward voltage drop. Its disadvantages are that it is prone to second breakdown, has a long storage time when operated in saturation, and requires a continuous base current to maintain it in conduction.

There are several ways in which a MOSFET and a bipolar transistor may be combined, either as discrete components or monolithically, so that the resulting switch has some of the advantages of both the MOSFET and the bipolar transistor. The three most common configurations are the cascade (Darlington), the cascode (series), and the parallel combination.

Another bipolar device, the gate turn-off (GTO) thyristor has also proved suitable for use in combination with the power MOSFET, producing a switch which has some of the useful characteristics of both devices.

16.2.2. Cascade (Darlington) Combination of MOSFET and Bipolar Transistor

The cascade arrangement, when implemented with two bipolar transistors, is more commonly known as the Darlington configuration (see Figure 16.12*a*). The principal advantage of this arrangement is its high beta gain, which is roughly equal to the product of the gains of the two individual transistors. The input

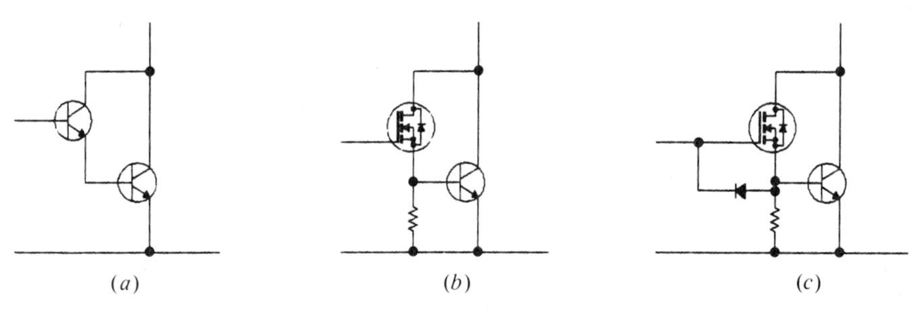

Figure 16.12. Cascade MOSFET-and-bipolar combinations. (*a*) Basic bipolar Darlington configuration; (*b*) cascade MOSFET and bipolar combination with base pull-down resistor; (*c*) cascade MOSFET and bipolar combination with base pulldown resistor and base charge extraction diode.

transistor may be replaced by a MOSFET, as shown in Figure 16.12*b*, to form a MOSFET–bipolar Darlington. This arrangement has the virtue that it is voltage-controlled rather than current-controlled and, particularly for high-voltage devices, is likely to be cheaper than a MOSFET of the same current rating. The MOSFET used in the combination has to have a voltage rating equal to that of the bipolar transistor. A resistor between the base and emitter of the bipolar transistor provides a discharge path for charge stored in the base and for collector leakage current. The presence of this resistor increases the current that must be carried by the MOSFET.

Figure 16.12*c* shows the modification necessary in order to ensure fast turnoff of the bipolar transistor. Active pulldown for the base of the bipolar transistor is achieved by adding a diode between the gate of the MOSFET and the base of the transistor. This permits charge to be extracted from the base of the bipolar transistor when a negative voltage is applied to the gate terminal [5]. However, the switching speed and storage time of the cascade arrangement remain essentially those of the bipolar transistor.

While the MOSFET carries only a relatively small current, it may be necessary to choose a device with a much lower $R_{DS(on)}$ than is dictated by dissipation considerations. From Figure 16.13 it can be seen that the collector–emitter voltage of the bipolar transistor is given by

$$V_{CE} = V_{BE} + I_B R_{DS(on)} \tag{16.1}$$

Therefore, to achieve a low value of V_{CE}, the value of $R_{DS(on)}$ must be correspondingly low. Using a MOSFET with the required value of $R_{DS(on)}$ may not be economic. For example, suppose the bipolar transistor carries a collector current of 10 A with a gain of 10. The base current is therefore 1 A. If V_{CE} is to be limited to 1.7 V, and V_{BE} is 0.7 V, this allows a maximum voltage drop across the MOSFET of 1 V. Since I_B is 1 A, the maximum allowable $R_{DS(on)}$ of the MOSFET is 1 Ω. A power MOSFET with an $R_{DS(on)}$ of 1 Ω is capable of handling between 5 and 8 A, depending on the cooling arrangements. The MOSFET current rating is nearly equal to that of the bipolar transistor, and there is little economic justification for the arrangement. However, it is possible to envisage situations in which this

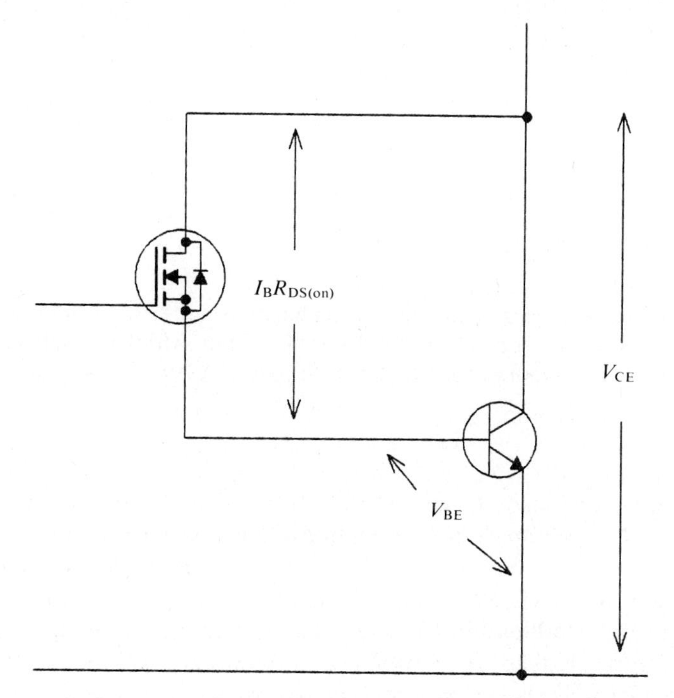

Figure 16.13. MOSFET–bipolar Darlington connection.

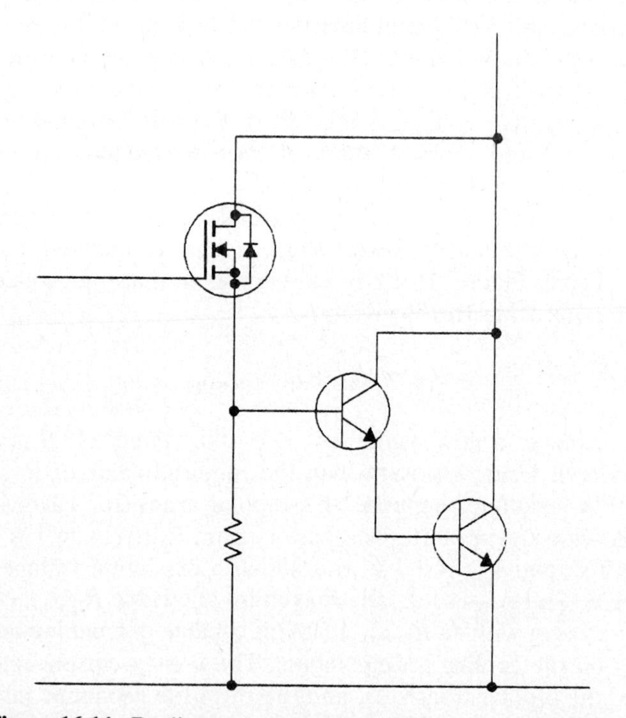

Figure 16.14. Darlington arrangement with MOSFET base drive.

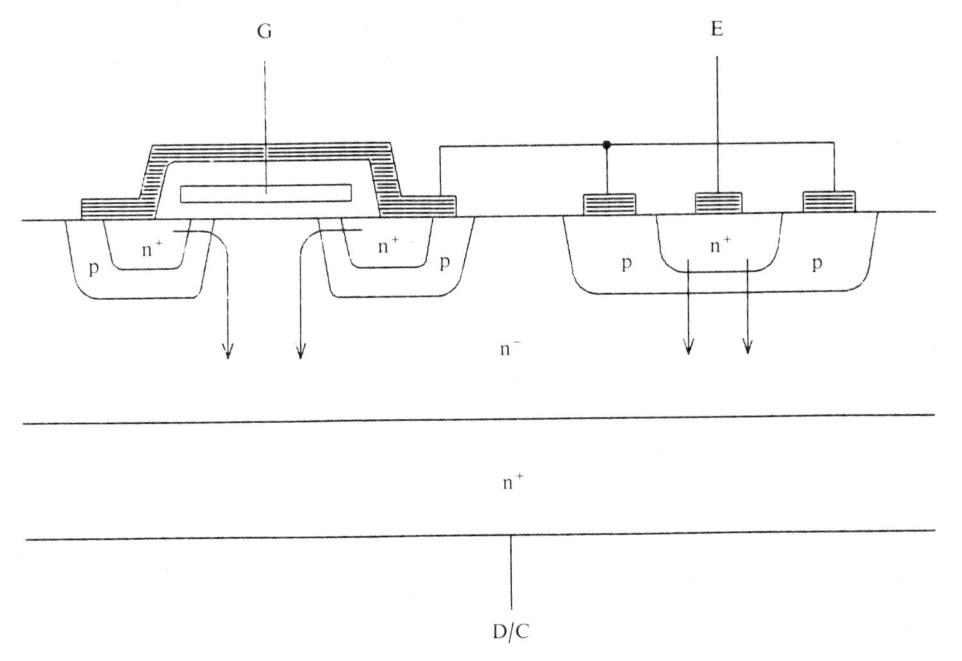

Figure 16.15. Monolithic cascade (Darlington) combination of MOSFET and bipolar transistor.

arrangement can be used to greater benefit, for example in the triple Darlington arrangement shown in Figure 16.14. The base current provided by the MOSFET is reduced by an order of magnitude, allowing the use of a device with correspondingly higher value of $R_{DS(on)}$.

The cascade arrangement may be made as a monolithic device in the manner shown in Figure 16.15. However, as much as 40% of the silicon area may be required for the MOSFET subsection [6].

16.2.3. Cascode (Series) Combination of MOSFET and Bipolar Transistor

The cascode method of connection is illustrated in Figure 16.16. When the MOSFET is off, the emitter of the transistor is isolated. This means that the bipolar transistor has no gain with respect to leakage current, and its blocking capability is equal to its V_{CBO} rating. This is a major advantage of the cascode arrangement, since V_{CBO} is generally considerably greater than V_{CEO}. The MOSFET is required to block only the base driver voltage V_B. The advantages of the cascode combination have been recognized by device manufacturers, who are beginning to make the arrangement available in module form [7].

When the MOSFET turns on, current can flow from the base driver source through the base–emitter junction, therby turning on the bipolar transistor. At turnoff, due to stored charge in its base region, the bipolar transistor does not turn off instantaneously, even though its emitter may be isolated. The zener diode Z1 clamps the base voltage, thereby limiting the voltage applied to the MOSFET. The MOSFET can therefore be a low-voltage, low-cost device.

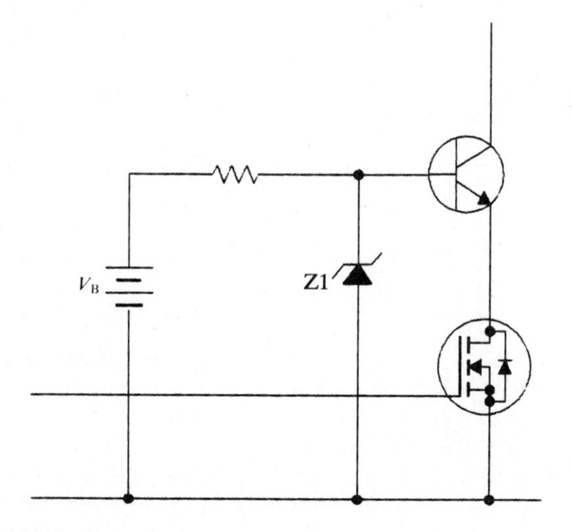

Figure 16.16. Cascode connection of MOSFET and bipolar transistor.

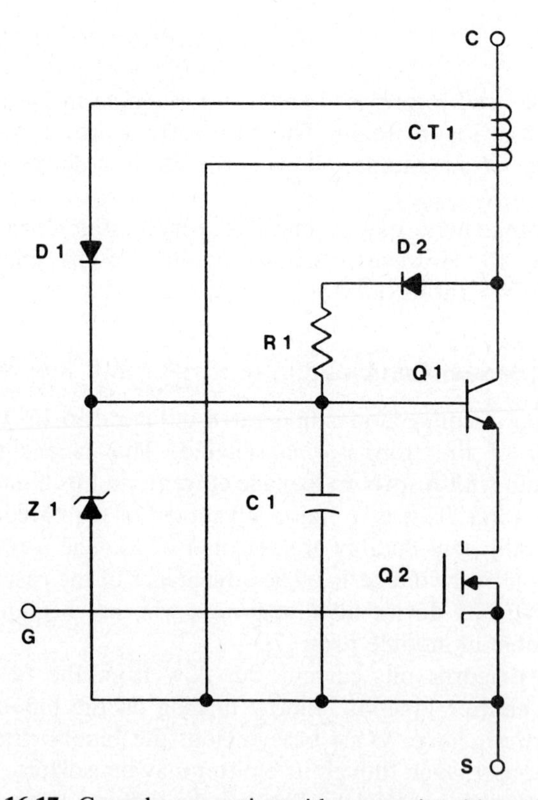

Figure 16.17. Cascode connection with proportional base drive.

The drawback with this circuit is that it is a four-terminal device requiring two inputs. The circuit shown in Figure 16.17 overcomes this difficulty by supplying the bipolar transistor base current from a current transformer, CT1, located in the path of the collector current [8, 9]. This has the additional advantage of providing a proportional drive to the base, thereby enabling the transistor to operate at approximately the same degree of saturation for all values of collector current. D2 and R1 provide the base current to initiate the conduction process. C1 provides a low-impedance source of base current to aid fast turn-on.

Another means by which base drive current can be provided for the bipolar transistor is by the use of a second MOSFET between the collector and the base as shown in Figure 16.18 [10, 11]. An asymmetric drive to the series MOSFET is required to ensure that both MOSFETS do not conduct simultaneously. The drawback of this arrangement is that the base drive MOSFET must be of the same voltage rating as the bipolar transistor.

As well as exploiting the V_{CBO} blocking capability of the bipolar transistor, cascode operation permits rapid switching, since the negative base current at

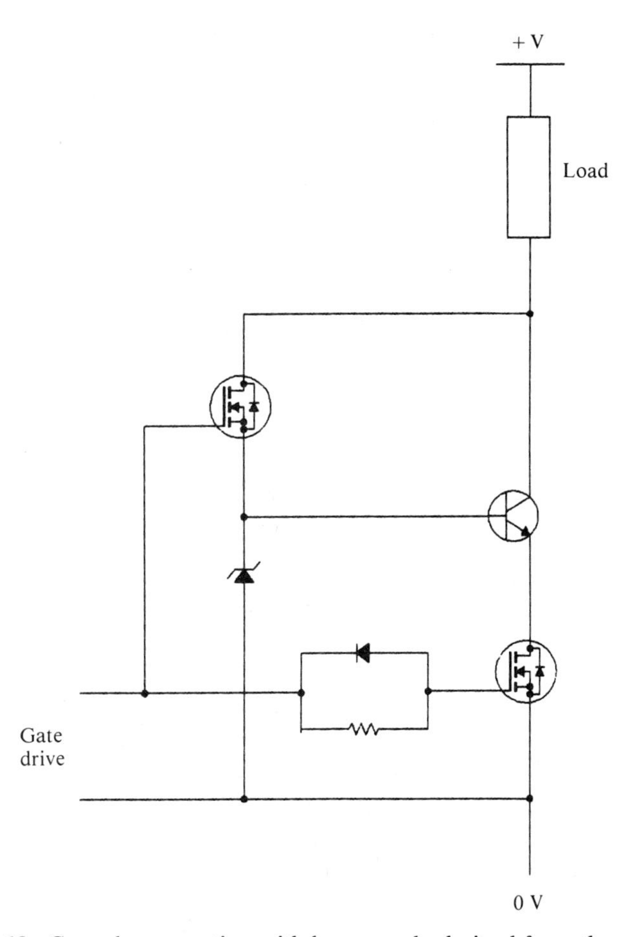

Figure 16.18. Cascode connection with base supply derived from the collector.

turnoff can be equal to the collector current. Switching times are reduced by a factor of 10 to 15. This reduces the storage time and fall time of the bipolar transistor and permits the cascode switch to operate at high switching frequencies [12, 13]. In power-supply applications a switching frequency of up to 500 kHz would appear to be practical [14]. A further advantage of the arrangement is that in the cascode mode of operation the fall time does not vary significantly with temperature [15]. The switching performance of the cascode arrangement has been analyzed in detail by Lorenz and Amann [16].

A further advantage of the cascode circuit is that the possibility of second breakdown in the bipolar transistor can be reduced or eliminated [17]. As Figure

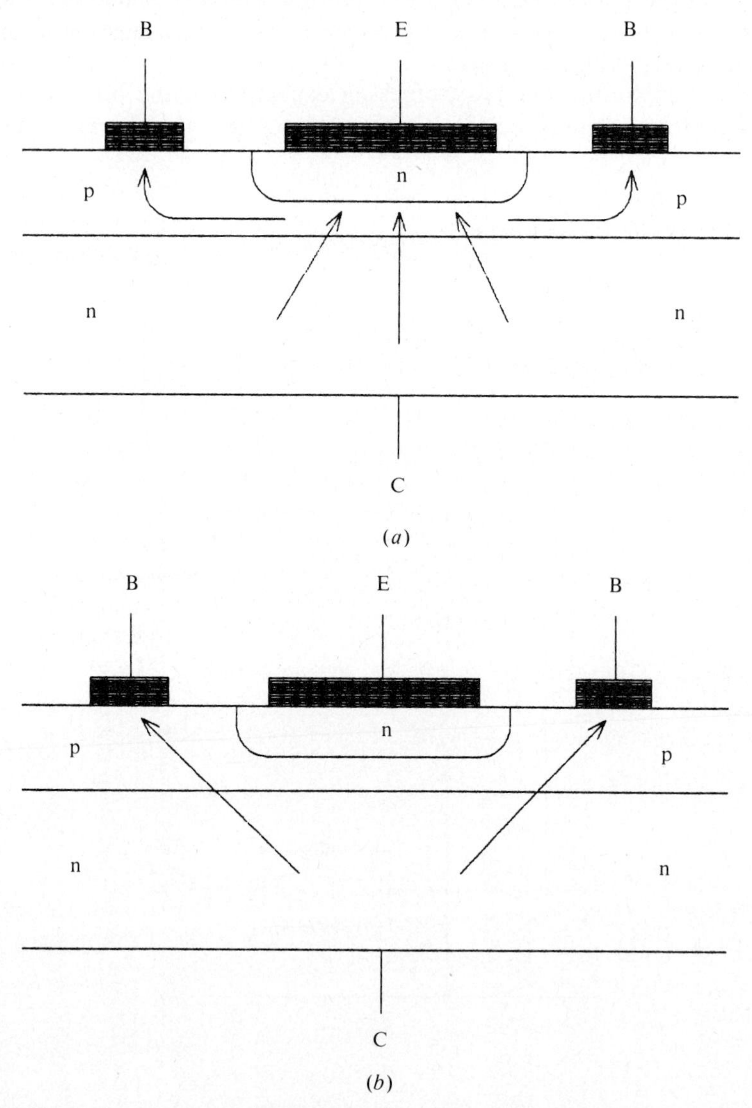

Figure 16.19. Emitter-current defocusing in cascode operation. (*a*) Normal common-emitter operation; (*b*) cascode operation.

16.19 shows, the current-crowding process which occurs at the center of the emitter region in common-emitter operation is reduced, since as soon as the emitter current has been reduced to zero, the collector current flows away from the center of the emitter regions and towards the base connections.

Cascode operation exposes the bipolar transistor to stresses not normally encountered in common-emitter operation. At turnoff, the base connection and base region metallizations are called upon to carry a pulse of current equal to the value of the collector current. Some transistors may not be designed to withstand such levels of base current. Also, energy will be stored in the parasitic inductance in the loop formed by the emitter lead, the zener diode, and the base lead. During very rapid switching, this energy will be dissipated in avalanche breakdown of the base–emitter junction. While this is apparently not deleterious to the life of the transistor [15], it does represent an unusual mode of operation and must be viewed with caution.

The cascode arrangement can incorporate inherent current limiting if the base voltage of the bipolar transistor is limited by appropriate selection of the zener which clamps the base of the bipolar transistor [18, 19]. When the current limit is reached, the voltage drop across the MOSFET approaches the zener clamp voltage, and the voltage available to drive base current in the bipolar is reduced. The current limiting cannot be very accurate, however, because of the wide temperature variation of the $R_{DS(on)}$ of the MOSFET.

16.2.4. MOSFET and Bipolar Transistor in Parallel

This circuit arrangement is illustrated in Figure 16.20. The MOSFET and the bipolar transistor both conduct the load current, but at different times. The MOSFET carries the load current during turn-on and turnoff, and the bipolar transistor carries the load current during steady-state operation. The MOSFET has to have the same voltage rating as the bipolar transistor, but as it conducts for only short periods, its pulse current rating can be exploited. Thus a device with an $R_{DS(on)}$ which would be too high to carry load current continuously can be used to carry the load current at the beginning of a conduction period while charge builds up in the base region of the bipolar transistor, and again during turnoff while charge is being removed from the base region. During the rest of the conduction period the bipolar transistor carries the current with a forward voltage drop that is much

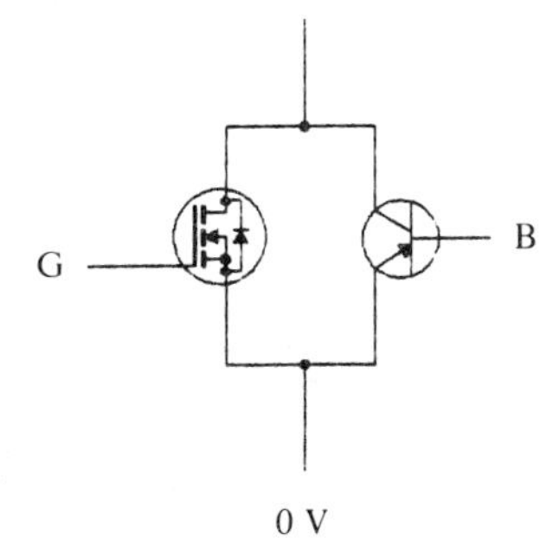

Figure 16.20. Parallel combination of MOSFET and bipolar transistor.

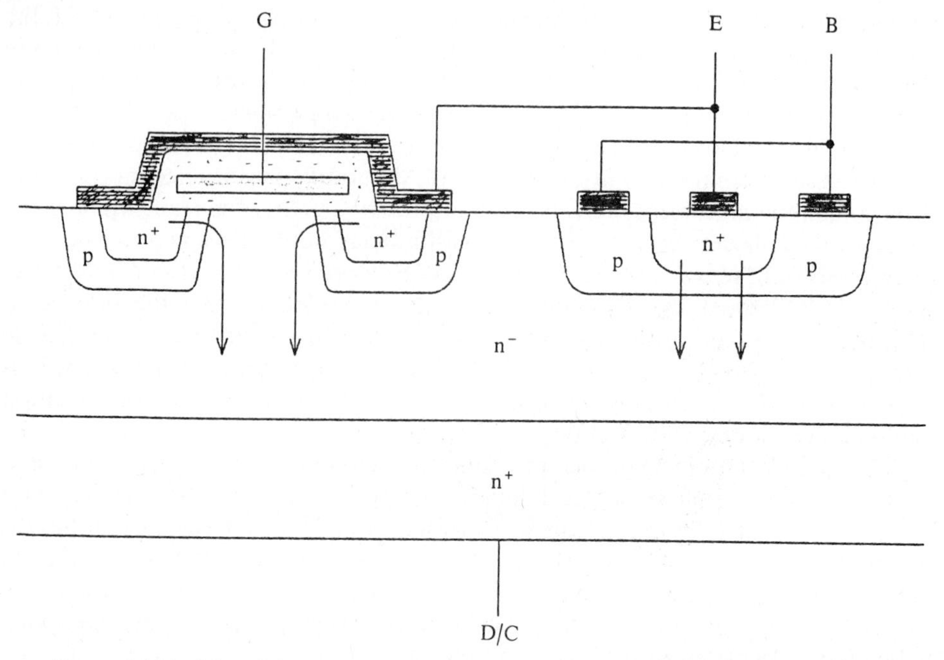

Figure 16.21. Monolithic parallel combination of MOSFET and bipolar transistor.

lower than would be economically feasible with a MOSFET. The MOSFET does the switching quickly with low losses, and the bipolar transistor carries the load current for most of the time with low conduction losses. By allowing the MOSFET to do the switching, the possibility of second breakdown of the bipolar transistor is avoided.

The major drawback of this type of switch is the need for two separate drive signals appropriately spaced in time. However, this is not as much of a disadvantage as it may seem: the MOSFET is inherently easy to drive, and the bipolar transistor, since it is not required to carry the load current during the switching interval, can be driven by a very basic driver circuit without the facilities normally necessary to ensure fast switching.

Monolithic versions of this device have been proposed along the lines shown in Figure 16.21 [20]. Lateral versions have also been proposed [21].

16.2.5. GTO and MOSFET in Cascode

The GTO can also be used in combination with a MOSFET in the cascode configuration as shown in Figure 16.22. This operates in a similar manner to the bipolar-transistor cascode arrangement. The GTO is turned on by the turn-on of the MOSFET. As in the case of the bipolar transistor, a continuous current must be supplied to the base, or gate in this case, to ensure that the GTO remains fully turned on under all conditions. The GTO cascode has an advantage in that this current is likely to be an order of magnitude less than that required in the case of the bipolar transistor and therefore more easily and economically supplied. At

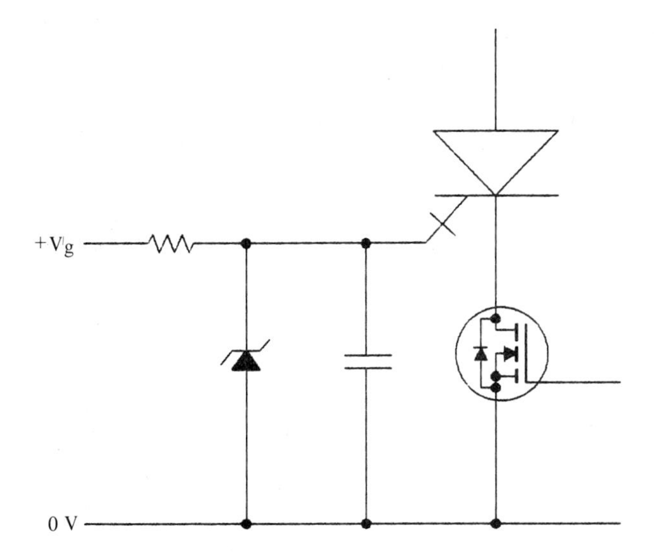

Figure 16.22. Cascode connection of a GTO and a power MOSFET.

turnoff the GTO is at a disadvantage in that even after sufficient charge has been removed from the gate to cause turnoff of the GTO, there will still be charge trapped in the remoter parts of the silicon, which will cause the anode base–emitter junction to inject for some time after turnoff. This will give rise to a tail of anode current which creates losses in the GTO and the gate zener. It is these losses which limit the switching frequency of this arrangement and therefore its usefulness.

A high rate of rise of negative gate current is required for fast turnoff of the GTO. The rate of rise of gate current is essentially determined by the voltage rating of the gate zener clamp and the inductance of the gate–cathode loop. Losses in the zener and the MOSFET voltage rating limit the zener voltage. The loop inductance must be kept very small. This becomes difficult as the current rating and therefore the physical size of the GTO increase. At the same time higher current rating GTOS require even higher values of gate-circuit di/dt. Therefore cascode turnoff tends to become ineffective for GTOS with a turnoff capability greater than 100 A. Furthermore, Wirth [22] has shown that, at least for some GTOS, similar performance to that obtained with cascode turnoff can be obtained with conventional gate turnoff.

The gate drive current can be derived from the anode in a similar manner to that used to derive base current for the bipolar transistor when operated in the cascode mode (see Figure 16.18). As with the bipolar-transistor cascode switch, the factor limiting the blocking voltage in this arrangement is likely to be the voltage rating of the MOSFET.

REFERENCES

1. P. Aloisi, "Bipolar or power MOS." *Proc. Power Convers. Int. Conf.* pp. 328–342 (1981).

2. M. Ichijo, H. Shigekane, and S. Kobayashi, "New high speed, 1200 V, bipolar transistor modules with superior, short-circuit withstand capability." *Proc. Power Convers. Int. Conf.* May, pp. 119–133 (1987).
3. L. Marechal and J. Wojslawowicz, "Bipolar transistor unit cell concept." *Proc. Power Convers. Int. Conf.* May, pp. 134–147 (1987).
4. P. L. Hower, "A comparison of bipolar and field-effect transistors as power switches." *IEEE, Ind. Appl. Soc. Annu. Meet.* pp. 682–688 (1980).
5. H. V. Manjunath, V. T. Ranganathan, and B. S. Ramakrishna Iyengar, "MOSFET–bipolar hybrid Darlington power switches-design. Gate and protection circuits." *IEEE, Ind. Appl. Soc. Anu. Meet.* pp. 733–737 (1986).
6. R. Blanchard, "A new high-power MOS transistor for very high-current, high-voltage switching applications." *Proc. Powercon* **8,** N.1.1–N.1.7 (1981).
7. Y. Kamitani, G. Majumdar, H. Yamaguchi, and S. Mori, "A new 100 A, 1000 V dual cascode BIMOS power module for high power and high frequency inverter applications." *Proc. Power Convers. Int. Conf.* pp. 143–153 (1986).
8. S. Clemente, B. Pelly, R. Ruttonsha, and B. Taylor, "High voltage, high frequency using a cascode connection of bipolar transistor and a power MOSFET." *IEEE, Ind. Appl. Soc. Annu. Meet.* pp. 1395–1405 (1982).
9. N. J. Barabas, "Cascode switching element with proportional drive assures low losses and enhances feasibility." *IEEE, Ind. Annu. Appl. Power Electron. Conf.,* San Diego pp. 37–43 (1987).
10. D. Y. Chen and S. A. Chin, "Design considerations for FET-gated power transistors." *IEEE, Power Electron. Spec. Conf.* pp. 144–149 (1983).
11. E. Hebenstreit, "A new bimos switching stage for 10 kW range." *Proc. Power Convers. Int. Conf.* April, pp. 140–145 (1983).
12. M. S. Adler, "A comparison between bimos device types." *IEEE, Power Electron. Spec. Conf.* pp. 371–377 (1982).
13. V. Farrow and B. Taylor, "A 300 kHz off-line switching supply using a unique BI-MOS switch combination." *Proc. Power Convers. Int. Conf.* pp. 3A.6.1–3A.6.10 (1980).
14. C. M. Penalver and J. Farina, "An improved high-frequency BI-MOS switch application to 300 kHz switch mode power supply." *Eur. Power Electron. Conf.* pp. 553–556 (1987).
15. P. Bardos, "A new switching configuration improves performance of off-line switching converters." *Proc. Powercon* **8,** G.2.1–G.2.7 (1981).
16. L. Lorenz and H. Amann, "Switching behaviour of a BIMOS switching stage." *IEEE, Power Electron. Spec. Conf.* pp. 112–119 (1986).
17. D. Y. Chen and J. P. Walden, "Application of transistor emitter-open turn-off scheme to high voltage power inverters." *IEEE Trans. Ind. Appl. Soc.* **IAS-18**(4), July/August, pp. 411–415 (1982).
18. P. Despagne and P. Davies, "MOS BIP hybrid cascode switch." *Proc. Power Convers. Int. Conf.* June, pp. 303–315 (1986).
19. D. Lafore and D. Delage, "The cascode switch in series-resonant inverters." *High Freq. Power Convers. Conf.* April, pp. 71–81 (1987).
20. J. Meador and N. Zomer, "Using bipolar–MOSFET combinations to optimize the switching transistor function." *Proc. Powercon* **8,** F.4.1–F.4.6 (1981).
21. J. D. Morse and D. H. Navon, "Optimized design of a merged bipolar MOSFET device." *IEEE Trans. Electron Devices* **ED-32**(11), 2277–2281 (1985).
22. W. F. Wirth, "High speed, snubberless operation of GTOS using a new gate drive technique." *IEEE Ind. Appl. Soc. Ann. Meet.* pp. 453–457 (1986).

CHAPTER 17

Packaging, Testing, Reliability, and Handling

17.1. PACKAGING

17.1.1. Discrete Devices

Power MOSFETS are readily available in all the conventional power transistor packages, such as those shown in Figure 17.1. As with all power transistors, the merit of a package is judged by its ability to conduct heat away from the die and by the long-term reliability of the device. In addition, because of the switching speed of which power MOSFETS are capable, the parasitic inductance introduced by the package, in particular source inductance, can be an important consideration. The package dimensions are obviously important in determining the package inductance, but so are other less obvious factors such as whether nickel plating is used on the package [1].

The following table compares some typical characteristics for popular packages:

Type	Also Called	Hermetic	R_{thJC} [a]	R_{thCS} [b]	L_S [c]
TO-220		No	1.0	0.5	7.5
TO-204	TO-3	Yes	0.8	0.1	12.5
TO-247	TO-3P	No	0.7	0.1	13.0

Notes:
[a] Typical minimum value; in K/W or °C/W.
[b] Mounting surface flat, smooth, and greased; in K/W or °C/W.
[c] Measured from the source lead, 6 mm from the package, to the center of the die; in nH.

The largest die that will fit in a TO-220 package is of the order of 25 mm^2. Dice up to 60 mm^2 are regularly used with the TO-3 package. The $R_{DS(on)}$ of low voltage dice of these dimensions can be of the same order of magnitude as the resistance introduced by the package. The package resistance can be reduced by the use of thicker leadthrough pins and then by the use of packages with multiple leads.

The TO-3 is a far from ideal package because it has the electrical connections protruding through the surface that is used for heat extraction. In situations where a hermetic package is not required, the package of choice is often the

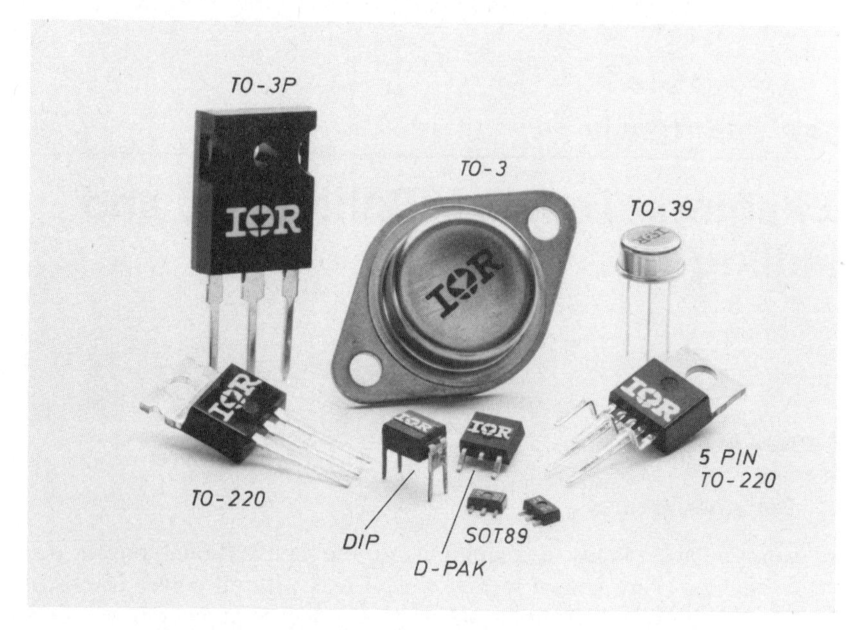

Figure 17.1. Common power-MOSFET package types.

TO-220 because of its low cost and because mounting is easier than for the TO-3. However, the power-handling capability of the TO-220 is considerably less than that of the TO-3. The TO-3P plastic package combines some of the best features of the plastic TO-220 and the metal can TO-3. The TO-3P has the economic construction found in the TO-220 yet has the power handling capability of a TO-3. In some applications, such as battery-powered equipment, forward voltage drop and energy losses must be kept low. Where device dissipation is low, a smaller, cheaper package than the TO-220 can then be used, such as the TO-237 [2].

The advantages of the plastic TO-3P package cannot be enjoyed by military equipment designers requiring hermetic packaging. To meet the military and high-reliability requirements for a package that is hermetic but with the configuration of the TO-220 and the TO-3P, metal can versions of these devices have appeared (Figure 17.2). These packages are seen as resolving the long-standing problem of the difficulty of mounting the TO-3 and the lack of hermetic sealing of the TO-220, so that it is to be expected that commercial-grade MOSFETS will become available in the new packages. The die may be insulated from the metal can by an internal layer of beryllia or alumina [3].

A step forward in convenience is the TO-3P package with the tab encapsulated in plastic on the upper side (TO-247). The insulated screw hole simplifies mounting, since the collets traditionally used to insulate the screw from the metal tab can be omitted. An even greater reduction in mounting hardware can be achieved when the metal tab is totally enclosed (isolated TO-218), so that the only metal parts protruding from the package are the three leads. The traditional mica insulator between the heatsink and the metal face of the package can be dispensed with [4, 5], although the thermal resistance of the insulated package may be greater than that of the noninsulated package with a mica washer.

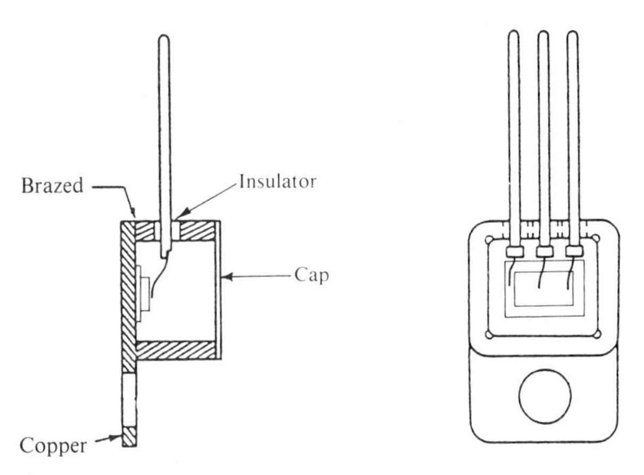

Figure 17.2. Hermetic TO-220 package with heat extraction and electrical connection on different planes. (Courtesy of International Rectifier Corp.)

In many applications in which power MOSFETS are used, sensing of the MOSFET drain current is either already a necessity or a desirable feature. Current-sensing MOSFETS are therefore likely to become widely used, and it is to be expected that the more popular packages will be produced in five-lead form for them. Some of those already developed are shown in Figure 17.3. Staggering of the leads may be necessary to give adequate clearance between PCB tracks.

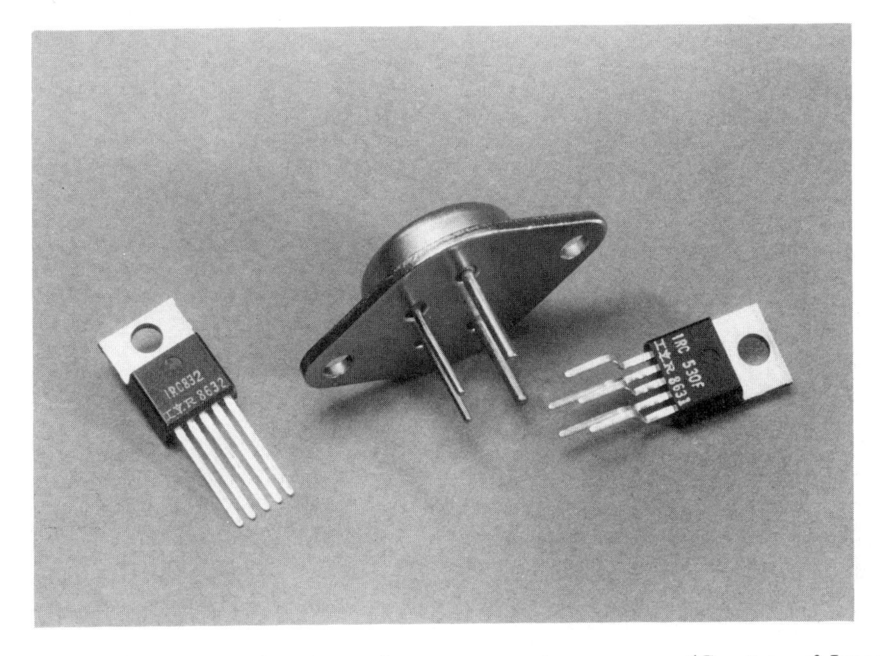

Figure 17.3. Five-terminal packages for current-sensing MOSFETS. (Courtesy of International Rectifier Corp.)

Multilead packages may also be required where the MOSFET die size is large and the current rating high, requiring a number of parallel leads to carry the drain and source current. Reference 6 describes a power MOSFET employing a single square die with a side length of 0.5 in. As well as multiple source and drain leads, a separate source lead for the gate drive is provided to reduce the effect of source inductance on switching time.

17.1.2. Multiple-Die Packages

The power MOSFET is renowned for the ease with which devices may be paralleled. This is a fortunate attribute, since the maximum practical size of a power-MOSFET die is limited by the VLSI manufacturing methods employed and the consequent rapid reduction in yield as the die area increases. High current capability may be obtained by paralleling a number of MOSFET dice in a single package. Figure 17.4 shows an example of such a package. Four MOSFET dice are connected together in a TO-240 package. A thick copper base is required to protect the dice from strain produced by mounting the package on an uneven surface. Current sharing between dice is encouraged by close thermal coupling. Current sharing is also aided if the dice are taken from adjacent sites on the same wafer [7].

As well as connecting all the dice in parallel, packages such as the one shown in Figure 17.4 can be connected in other common circuit configurations such as the half-bridge circuit (Figure 17.5*b*). Two- and three-phase bridge circuits can be fabricated by mounting packages side by side, with bus bars on the top side for

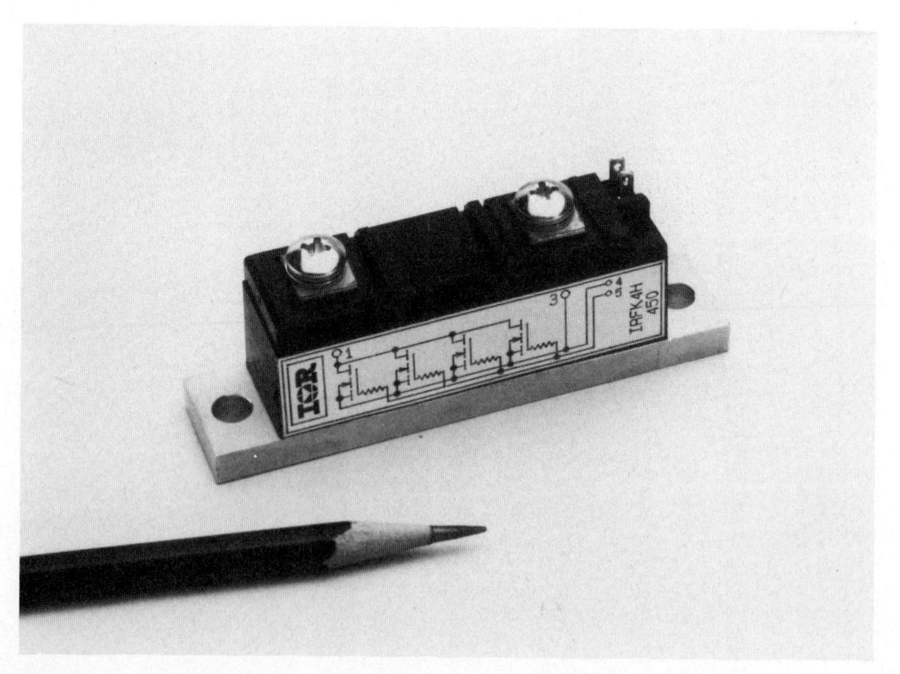

Figure 17.4. A multiple-die power-MOSFET package. (Courtesy of International Rectifier Corp.)

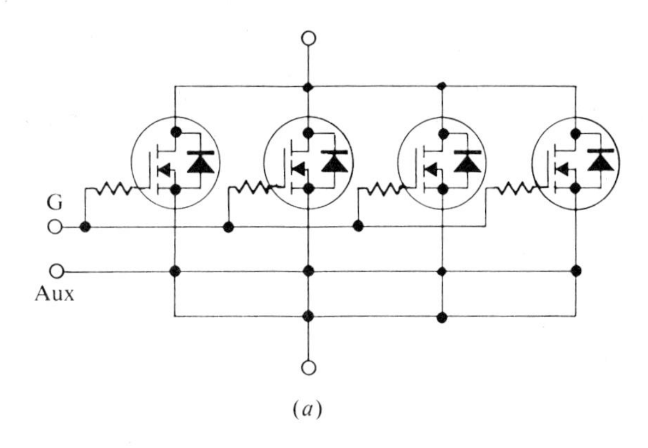

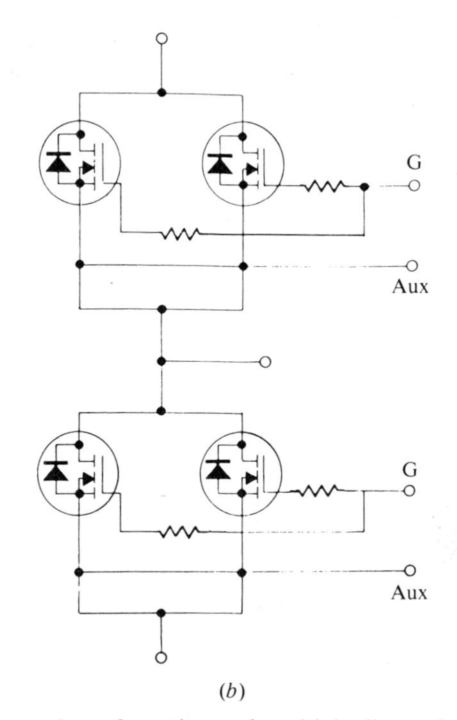

Figure 17.5. Typical internal configurations of multiple-die packages. (*a*) All dice in parallel; (*b*) bridge connection.

interconnection. Dice may be electrically isolated from the base by a layer of alumina or beryllia, providing several kilovolts of isolation. Applications for this kind of device include dc motor control, ac motor drives, uninterruptible power supplies, and high-frequency welders.

Common-source inductance in any power MOSFET limits the switching speed of the device and is therefore undesirable. The amount of common source inductance in multiple die packages tends to be high because of the necessarily large dimensions of the package. However, the common-source inductance can be minimized by the provision of an auxiliary source terminal which is connected to an appropriate point of the internal source connection bus. Some imbalance in the common-source inductance associated with each die is inevitable, so that there may be uneven current sharing during turn-on, although this can be minimized by a judicious choice of the point at which the auxiliary source is connected. Another possible cause of uneven heating is the heat generated in the interconnectors due to the skin effect when switching large currents at a high frequency. These effects can be minimized by good package design and attention to internal layout.

In parallel-die packages it is usual to locate a resistor of about 10 Ω in series with the gate of each die in order to suppress parasitic oscilllation (see Figure 17.5). These resistors are located as close as possible to the gate pad of each die. Alternatively the resistivity of the gate polysilicon may be made high enough so that additional resistance is not required. Because of these series gate resistors, further external gate resistance may not be required to ensure circuit stability.

The value of the drain–gate capacitance tends to be high in packages in which dice are paralleled. Therefore, when high rates of rise of drain voltage are expected, it is advisable to use a low-impedance gate driver, or even to apply a negative gate bias in the OFF condition, in order to avoid spurious turn-on due to feedback between the drain and the gate [8]. Negative gate bias in the OFF state can also shorten crossover times, which tend to be lengthened by the high values of di/dt in conjunction with the parasitic source inductance [9].

The high values of current which can be handled by multiple-die packages require a derating of the device in a manner not normally considered necessary with single-die devices. Firstly, due to the internal voltage drop produced by the

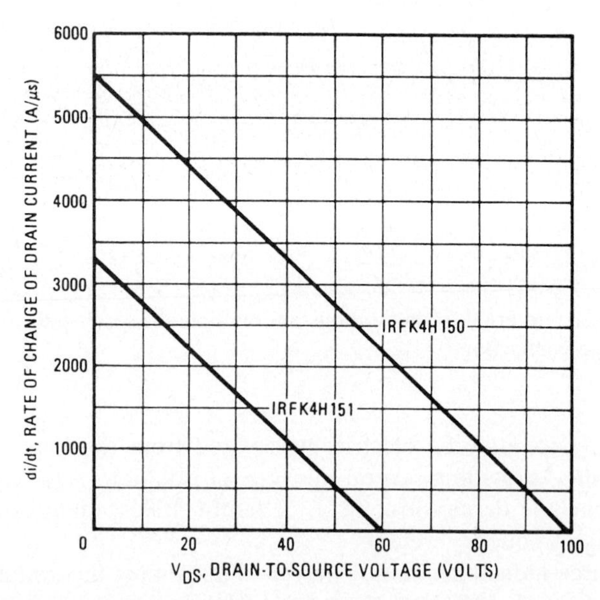

Figure 17.6. Derating curve for multiple-die package, showing reduction in voltage rating with increasing di/dt.

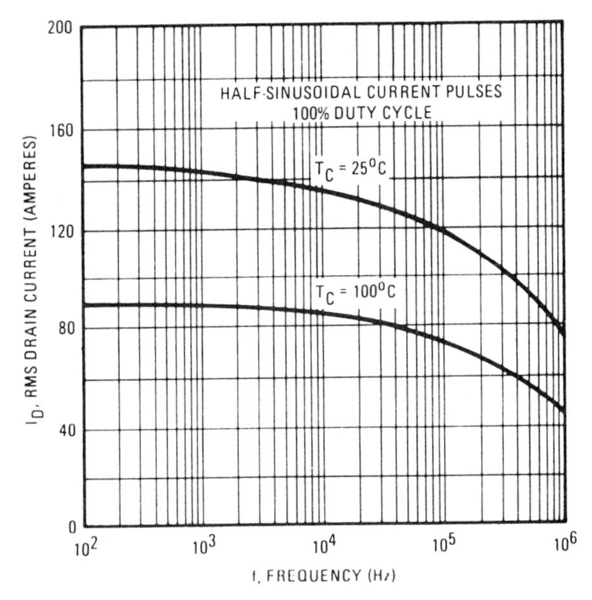

Figure 17.7. Derating curve for multiple-die package, showing reduction in current rating with increasing frequency.

parasitic inductance of the package in combination with high values of d*i*/d*t*, the voltage-blocking capability of the device is reduced as d*i*/d*t* increases (Figure 17.6). Secondly, due to the skin effect and consequent internal conductor heating, the current rating of the device must be reduced as frequency increases. Figure 17.7 shows typical derating curves.

While the economic advantages to the designer of a multiple-die, isolated package can be considerable, the current level at which these devices operate and the rate at which they are able to switch current require that careful attention be paid to circuit layout in order to minimize the generation of EMI and RFI. The potential of a circuit for generating interference may be kept to a minimum by adopting good circuit layout techniques as illustrated in Figure 17.8.

For military and other high-reliability applications hermetically sealed packages are required. These are typically metal can packages with glass-insulated leadthroughs. When the dice are to be electrically isolated from the package, the more expensive beryllia isolation is used in preference to alumina, since beryllia has a much higher thermal conductivity (0.45 Watt-in/in^2 °C for Alumina and 5.2 Watt-in/in^2 °C for Beryllia). Beryllia is also used in some commercial packages, although concerns about the toxicity of beryllia dust, as well as high cost, limit its use.

17.1.3. Hybrid Packages

Equipment manufacturers are constantly seeking to reduce component counts in order to cut assembly and stocking costs and to increase equipment reliability. One method of achieving this is to encapsulate a complete power circuit in a

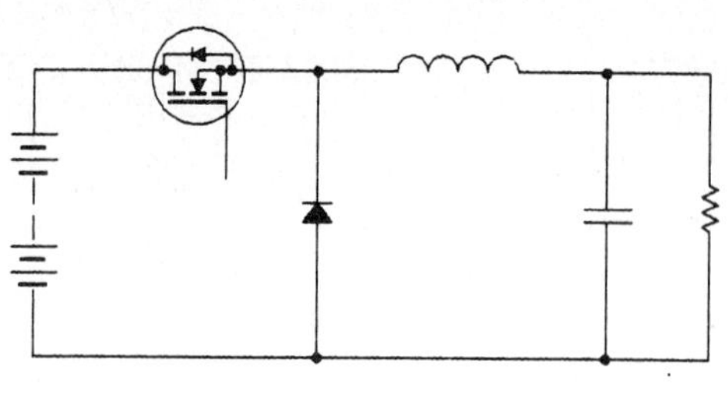

Large loop carrying high di/dt
High level of EMI

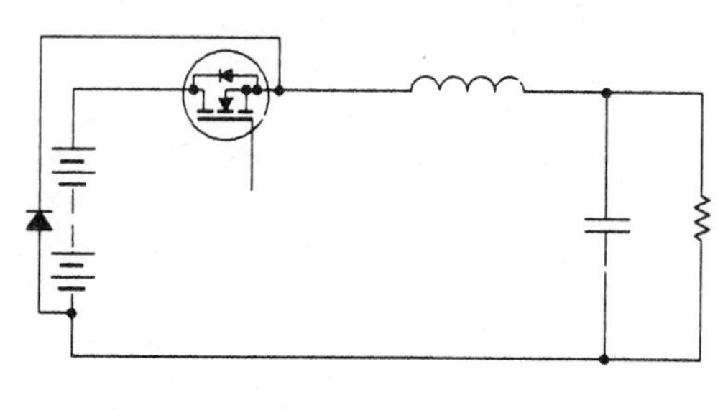

Small loop carrying high di/dt
Low level of EMI

Figure 17.8. Layout for minimum interference when using high-current, multiple-die power MOSFETS.

single package. There are a number of applications with sufficient commonality to merit the provision of the module as a standard item.

A common application for multiple power MOSFETS is the brushless-dc-motor drive. Brushless dc motors are commonly used in fractional- and subfractional-horsepower drives for such applications as disc memory drives, magnetic tape drives, printers, VCRs, and electronic typewriters. In such applications power MOSFETS have become widely used because their SOA characteristics are superior to those of bipolar transistors, they have intrinsic diode structures to handle the stator inductive energy, and they can carry very high peak currents, such as are encountered during starting conditions with high inertia loads. Furthermore, MOSFETS may be driven directly from motor-control logic or a motor-control integrated circuit.

Figure 17.9 shows a typical hybrid motor-drive package. The MOSFETS are mounted on a metallized alumina substrate which can typically provide up to 1.5 kV of isolation between the power devices and the heatsink. The power rating of a package of this type when heat-sunk is approximately 50 W. With 500 V MOSFETS, this package is capable of controlling motors of up to half a horsepower. For low-power drives free air cooling is often adequate, with the package mounted either vertically or horizontally to the printed circuit board. Amongst the considerations that prompt the designer to choose hydrid modules rather than

Figure 17.9. Hybrid MOSFET package for motor control. (Courtesy of International Rectifier Corp.)

discrete power MOSFETS are the space reduction that is possible with modules; the reduction in pin connections, which increases reliability; the simplicity of obtaining device isolation should the use of a heat sink be necessary; low device-to-heatsink capacitance; and a reduced part count.

Better thermal management is often possible while providing the required degree of isolation. In extreme ambient conditions, such as those found under the hood of an automobile, it may be necessary to operate at elevated die temperatures (for example 175°C). While the die may be reliable at these temperatures, this is approaching the temperature limit of some plastic packages. Hybrid packaging, on the other hand, can permit reliable operation at higher junction temperatures.

17.1.4. Low-Inductance Packages for Fast Switching

The inherent switching speed of a power MOSFET is dictated by the carrier transit time, which will be in the sub-nanosecond range. The switching times measured under normal conditions are a result of the time required to charge the parasitic capacitances of the die, and of parasitic inductances in the packaging, which limit the rate of rise of current in the drain circuit and the gate circuit. To achieve very fast switching requires a low-inductance package as well as a low-impedance gate drive to ensure rapid charging and discharging of the input capacitance. Packages of the type shown in Figure 17.10 incorporating parallel devices offer high speed, high current switching capability. For fast switching it is important that the package should incorporate a separate source terminal for the gate return connection to reduce inductance common to the drain–source and the gate–source circuits. By the use of this kind of package, currents of over a hundred amperes have been turned on in a few nanoseconds. Specialist device manufacturers are predicting single-device peak powers of 20 kW, average powers of 5 kW, switching times approaching 1 ns, and switching frequencies in excess of 10 MHz [10]. The use of stripline packaging of power MOSFETS has also proved beneficial in obtaining a good bandwidth in linear applications [11].

17.1.5. Surface Mounting

The electronics industry is continually seeking ways to reduce the size and cost of its products. Surface-mounted components represent a major step in this

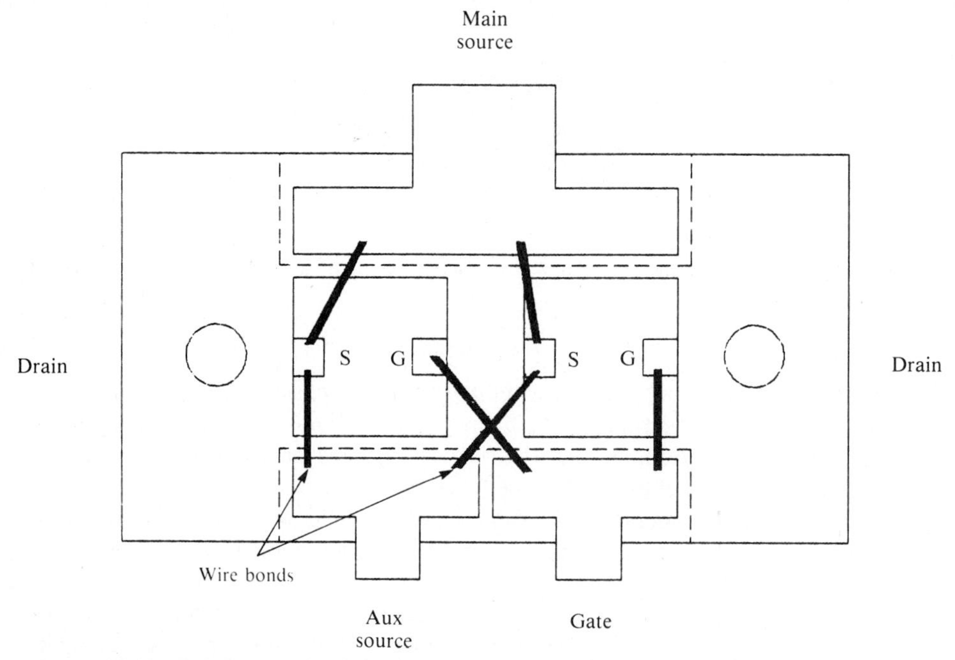

Figure 17.10. Stripline packaging of power MOSFETS and the use of an auxiliary source terminal permit fast switching.

direction. To date, most progress in surface-mount components has been made in integrated circuits. However, with surface-mount technology gaining rapid acceptance in all sectors of the semiconductor industry, there is a growing demand for discrete semiconductor devices in surface-mount packages. To meet this need MOSFETS are becoming available in such packages as the SOT-89 (TO-243AA) and D-Pak (TO-252AA) (see Figure 17.1). Surface-mount devices are held in place with glue or solder paste prior to reflowing the solder in a belt furnace, an infrared oven, or a vapor-phase oven.

17.1.6. Die Attachment

One of the most important aspects of packaging is the method of die attachment, since this has a major influence on the thermal resistance and on the long-term reliability of the device.

The usual method of attaching the silicon die to the package tab or base plate is by soldering. The back side (drain) of the MOSFET die is metallized with a solderable metal, for instance a nickel–silver combination. The die is usually mounted directly onto the copper header or tab. However, the coefficient of expansion of copper differs significantly from that of silicon. To relieve the stress created by differential expansion when the die heats up, a molybdenum wafer may be inserted between the die and the copper base. The coefficient of expansion of molybdenum is closer to that of silicon, so that less strain is created as a result of a rise in temperature. Unfortunately, molybdenum is expensive and is therefore not often used.

In hermetically sealed packages soft solder is used for die mounting. Soft solder (typically 93% lead, 5% tin, 2% silver) is malleable enough to absorb the differential expansion between the die and the copper base plate or header. In nonhermetic packages where oxygen and moisture may be present, soft solder is not generally used because of the possibility of oxidation and corrosion. A hard solder (typically 65% tin, 25% silver, and 10% antimony) is then used. While hard solder is not subject to the corrision problems of soft solder, it is not as good as soft solder at buffering the strain between the die and the base.

An alternative to solder for die attachment is the use of silver-loaded epoxy. The epoxy has adequate thermal and electrical conductivity, although its thermal resistance is somewhat higher than that of solder. The epoxy has the advantage that at high temperatures it softens slightly, thereby relieving the strain caused by differential expansion. Enhanced reliability under thermal cycling is claimed for this method, particularly when used in conjunction with a molybdenum buffer [12].

17.1.7. Active-Area Bonding

The source pad of a power MOSFET can occupy a significant area of the die. This is particularly true of low-voltage devices, which because of their low $R_{DS(on)}$ for a given area, have high current ratings. Either a single thick bonding wire or a number of thinner bonding wires in parallel are required to carry the source current. In either case the bonding pad is large and expensive in terms of silicon real estate. A way of avoiding this loss of active silicon area is to bond the source wires directly onto the active area of the die. No source pad is provided, and the bond is made, usually ultrasonicallly, onto the source metallization covering active cells. The ultrasonic bond applies thermal and mechanical stress to the die, and therefore due attention must be paid to the possibility of damage to the gate oxide. Increasing use of active-area bonding indicates that manufacturers of power MOSFETS have become confident in this technique.

17.2. TESTING

17.2.1. Measuring MOSFET Characteristics

Methods commonly employed for verifying the principal electrical parameters of power MOSFETS are described in Appendix 9. These circuits can be used as the basis for high-volume test equipment, for example at incoming inspection at a production facility. Test procedures and a definition of MOSFET ratings are given in JEDEC Standard 24 [13].

17.2.2. Measuring MOSFET Characteristics with a Curve Tracer

Curve tracers have generally been designed for making measurements on bipolar transistors. While power MOSFETS can be tested satisfactorily on most curve tracers, the controls of these instruments have until recently been labeled with

TABLE 17.1. MOSFET and Bipolar Equivalent Parameters (Approximate)

MOSFET	Bipolar
Drain	Collector
Gate	Base
Source	Emitter
g_{fs}	h_{FE}
BV_{DSS}	BV_{CES}
$V_{GS(th)}$	$V_{BE(on)}$
$V_{DS(on)}$	$V_{CE(sat)}$
I_{DSS}	I_{CES}
I_{GSS}	I_{EBO}

reference to bipolar transistors, and the procedure to follow in the case of MOSFETS is not immediately obvious. References 14 and 15 detail methods for measuring power-MOSFET characteristics on commonly used types of curve tracers.

Testing MOSFETS on a curve tracer is a straightforward procedure, provided the broad correspondence between bipolar-transistor and MOSFET features is borne in mind. Table 17.1 matches some features of MOSFETS with their bipolar counterparts.

The polarity settings for an n-channel MOSFET are the same as those for an npn bipolar transistor, and for a p-channel MOSFET the pnp settings are used. For measurements with currents above about 20 A, or for pulsed tests not controlled by the gate, the use of a pulsed high-current fixture, available for some makes of curve tracer, may be necessary. For tests in which there is significant heating of the MOSFET, a low repetition rate should be used. For tests involving a slow transition through the linear region, a damping resistor of at least 10 Ω should be connected in series with the gate, close to the gate lead, to prevent oscillation.

17.2.3. Using oscilloscopes with Power MOSFETS

When switching power MOSFETS, rise and fall times of the order of ten nanoseconds are not uncommon. To accurately measure and detect all the features in power-MOSFET waveforms it is essential to use an oscilloscope of adequate bandwidth. In fast-switching applications an oscilloscope bandwidth of 100 MHz is usually required, although 60 MHz may be just adequate. With wide-bandwidth oscilloscopes, however, ringing is often visible. This may be a normal feature of the circuit operation, or it may be generated within the parasitic inductances and capacitances of the probe. If the probe is responsible for the ringing, a clean trace can often be obtained, at the expense of using two channels per waveform, by making a differential measurement [16].

17.2.4. Measuring Power-MOSFET Temperatures

The surface temperature of power MOSFETS in a metal package, such as a TO-3, can be measured by removing the top of the can and examining the die with an infrared camera. Where an infrared camera is not available, or if the device is plastic-encapsulated and the package is not to be damaged, the temperature of the die must be determined by electrical means.

Three temperature-sensitive parameters are commonly used. These are the ON-state drain–source resistance, the forward drop of the body–drain diode, and the threshold voltage. The method chosen will depend to some extent on how the device is being used or tested. The nature of the circuit which incorporates the MOSFET may allow one particular parameter to be used more conveniently than the others, or one parameter may provide a more accurate measurement than another. For instance, the threshold voltage appears to give a temperature closest to the peak device temperature, while the body–drain voltage shows least variation in calibration from device to device [17].

For a constant forward current, the sensitivity to temperature change of the voltage drop across the body-drain diode (V_{SD}) is given by:

$$\frac{\mathrm{d}V_{SD}}{\mathrm{d}T} = -\frac{E_{g0}/e - V_{SD}}{T} \tag{17.1}$$

where E_{g0} is the 0-K band-gap energy and e is the electron charge. Typically the sensitivity of V_{SD} is $-2\,\mathrm{mV/°C}$. A negative gate voltage during the measurement would appear to be helpful in minimizing variations in the sensitivity of V_{SD} between devices [18].

The increase of $R_{DS(on)}$ with increasing temperature depends on the device, ranging from an increase of 0.6%/°C for low-voltage devices to 0.9%/°C for high-voltage devices. It must be remembered that $R_{DS(on)}$ is dependent on V_{GS} and to a small extent on I_D.

The relationship between gate voltage (V_{GS}) and temperature for a low value of drain current in the constant current region is given by

$$\frac{\mathrm{d}V_{GS}}{\mathrm{d}T} = \left(\frac{2I_D}{k^3}\right)^{1/2} \frac{\mathrm{d}k}{\mathrm{d}T} + \frac{\mathrm{d}V_T}{\mathrm{d}T} \tag{17.2}$$

and k is given in Chapter 3 as

$$k = \frac{w}{l} C_{ox}\mu_e \tag{17.3}$$

where w is the channel width, l is the channel length, C_{ox} is the gate oxide capacitance per unit area, and μ_e is the electron mobility in the channel inversion layer. This gives a threshold voltage sensitivity of approximately $-6\,\mathrm{mV/°C}$.

For an approximate temperature measurement, data-sheet curves of parameter variation with temperature can be used. For more accurate measurement the device must be calibrated, for instance by immersion in a temperature-controlled bath.

17.3. RELIABILITY

17.3.1. Reliability Testing

Reliability theory in relation to power MOSFETS is discussed in Chapter 5. This present chapter seeks to show some of the ways that the results derived from reliability testing may be applied in circuit design.

A number of tests are used to establish the reliability of power MOSFETS. Some are among those commonly applied to power semiconductor devices, while others are specific to power MOSFETS. Obtaining the necessary number of failures under normal operating conditions for statistically valid results is not usually possible within a reasonable time. Therefore the test is accelerated by increasing the stress conditions above their normal level [19]. For example, the temperature may be raised above the normal maximum allowable value. The failure rate at lower temperatures can be predicted from the Arrhenius model [20]. The model is described by Equation (5.10). It is common practice in the semiconductor industry to scale failure rates using the Arrhenius acceleration factor. The applicability of this model is discussed in Section 5.2.2.

An important aspect of MOSFET reliability is the lifetime of the gate oxide. Like all insulators, the gate oxide degrades at a rate determined by the temperature and the applied voltage. At normal values of gate voltage and device temperature, the life of a power MOSFET gate is likely to be far greater than the lifetime of any equipment in which it is used. Figure 17.11 shows typical gate-oxide life data.

Commonly applied to all power semiconductors, the high-temperature reverse-bias (HTRB) test assesses the long-term reliability of the voltage blocking capability of the device. Forward voltage, typically 80% of the voltage rating, is applied to

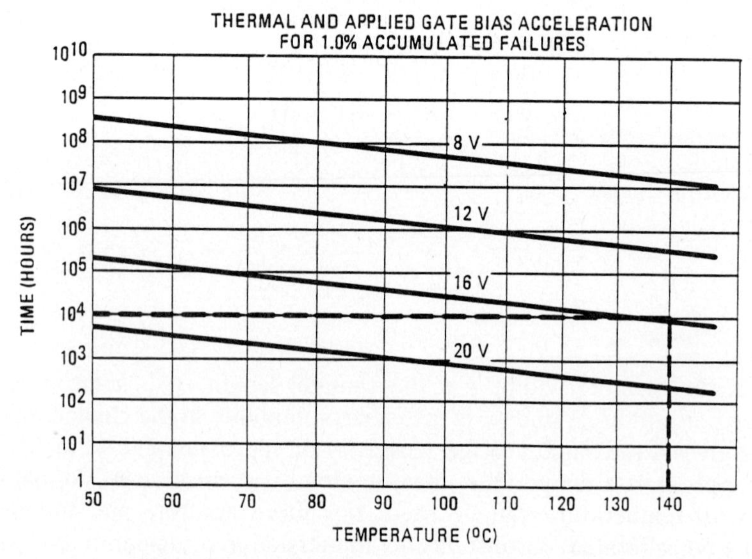

Figure 17.11. Power-MOSFET gate-oxide life data.

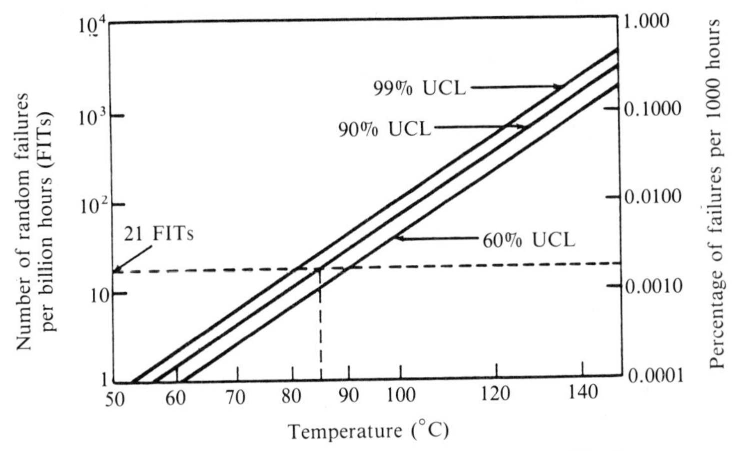

Figure 17.12. Power-MOSFET HTRB life data. The three solid lines represent upper confidence limits based on the number of samples subjected to the lifetest.

the device at an elevated junction temperature. Figure 17.12 shows an example of power-MOSFET HTRB lifetime data.

Other commonly applied tests of reliability include power cycling (thermal cycling with heating by device dissipation), temperature cycling (external heating) and a combined temperature and humidity test [21].

17.3.2. Lifetime Data

Traditionally, reliability results have been presented in terms of *mean time between failures* (MTBF) or *median time to failure.* While these results have their value, they do not necessarily tell the designer what he most needs to know. For example, the MTBF, while providing the equipment designer with a measure of long-term reliability [22], does not give specific information about the all-important first few percent (or fraction of a percent) of device failures, since it is based on the constant-failure-rate section of the classic bathtub curve (see Figure 5.5). Of greatest interest to the designer is the time to failure of a small percentage of devices—say 1% or 0.1%. For example, if it has been decided that one failure per hundred units over five years is an acceptable failure rate for the equipment, and each unit contains one critical component, the designer knows that the time to accumulate 1% failure of that component must be at least five years. If there are ten such components per unit, then no more than 0.1% of the components may fail in five years. Therefore a useful way of presenting MOSFET reliability data is a graph of number of failures expected versus elapsed time.

17.3.3. Gate-Oxide Lifetime Calculation

A gate voltage of 10 or 12 V is adequate to ensure that a MOSFET stays fully turned on in most applications. However, in applications where high peak currents are encountered, a greater gate voltage may be required to ensure that the device does not go into the pinch-off region with a consequent increase in drain-to-source

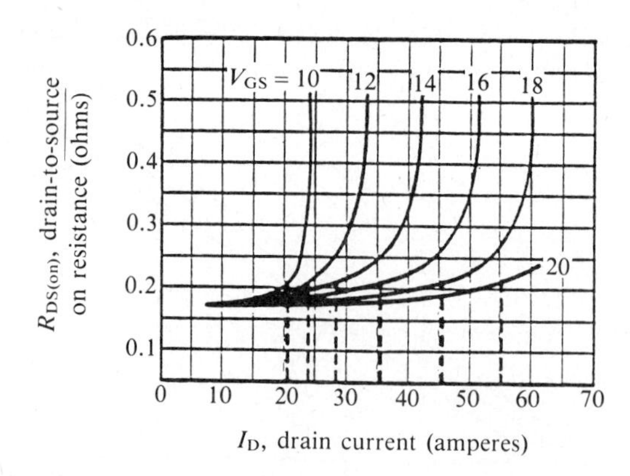

Figure 17.13. Relationship between $R_{DS(on)}$, I_D, and V_{GS} for an IRF 130.

voltage. This is particularly true of low-voltage devices, which have high current ratings compared with their high-voltage counterparts. The channel region is called upon to carry higher peak currents, and a greater gate voltage is required. Under such circumstances, it can be the gate-oxide lifetime which limits the allowable peak drain current. The following example illustrates one way in which gate-oxide reliability information may be used in the selection of a device.

A circuit uses a IRF 130 to switch a current of 35 A with a duty cycle of 10%. The designer wants to know how long it will take to accumulate 1% failures under these conditions. The supply voltage is low with respect to the voltage rating of the device, so that the HTRB failure rate is not significant. The device is to be operated at a maximum junction temperature of 140°C.

Figure 17.13 shows the relationship between $R_{DS(on)}$, the drain current, and the gate voltage for an IRF 130 at $T_J = 25°C$. From this it can be seen that a practical minimum for the gate voltage, taking into consideration device dissipation and the possible spread in threshold and transconductance characteristics, is 16 V. From the gate lifetime curves in Figure 17.11, the time to accumulate 1% failures at $T_J = 140°C$ and $V_{GS} = 16$ V is 10^4 hours. The MOSFET duty cycle is 0.1. Therefore the time to accrue 10^4 hours of exposure to a gate voltage of 16 V is given by

$$t = 10^4/0.1 = 100{,}000 \text{ hours} = 11.4 \text{ years} \tag{17.4}$$

If this time to 1% failure is unacceptable, then a device with a lower $R_{DS(on)}$ must be used so that the required drain current can be obtained with a lower gate–source voltage. A lower $R_{DS(on)}$ will also result in less dissipation and a lower junction temperature with the same heatsink, thereby extending the life of the device. However, as will be seen from Figure 17.11, gate voltage is the most significant factor in determining gate-oxide lifetime.

17.3.4. HTRB Lifetime Calculation

The blocking capability of a semiconductor device gradually deteriorates over its life, due to the migration of ionic contaminants under the influence of the electric field. Particularly when a power semiconductor is operating at a high junction temperature and blocking a high voltage, it is necessary to consult the HTRB reliability data to ensure that the failure rate resulting from HTRB wearout is acceptable. The following example illustrates how HTRB lifetime data may be used to determine if the desired level of equipment reliability can be achieved.

A power supply is to be designed that will provide a continuous output of 250 W to a dedicated load, twenty-four hours per day. The circuit employs two MOSFETS as the power-switching elements. The MOSFETS and their heatsinks are to be chosen so that there are no more than 0.1% accumulated failures over 5 years. This corresponds to one unit per thousand failing in 5 years.

First, the gate-oxide lifetime is checked. The maximum applied gate voltage has been set at 10 V. This is quite adequate to ensure full enhancement of the channel of the MOSFET when the drain current is at its maximum value. The junction temperature is as yet unknown but is estimated as no greater than 90°C. From Figure 17.11 the time to 1% accumulated failures under these conditions is found to be 1141 years. This suggests that gate failure rates are low enough to be ignored. More detailed reliability data [21] confirms this.

Next, the maximum allowable junction temperature is obtained from the HTRB failure graph. Assuming a conduction duty cycle of 45% for each device, the number of device hours (blocking voltage) per 1000 units in 5 years is given by

$$\begin{aligned} t &= 5 \times 365.25 \times 24 \times 1000 \times 2 \times 0.55 \\ &= 4.82 \times 10^7 \text{ device-hours} \end{aligned} \tag{17.5}$$

The failure rate in FITS (failures in 10^9 hours) is given by

$$F = \frac{10^9}{4.82 \times 10^7} = 21 \text{ FIT} \tag{17.6}$$

From the HTRB failure-rate graph (Figure 17.12) the maximum allowable junction temperature is 84°C. Assuming a half-bridge circuit operating from 220 V minus 15% (low line condition) and an efficiency of 80%, the peak current in each device will be 2.7 A at a duty cycle of 45% (assuming a rectangular waveform).

Comparing two different devices which might be used in this application illustrates how the HTRB reliability criterion can be met by a range of MOSFETS. The two devices are:

1. The IRF 430 ($R_{DS(on)max} = 2.4\ \Omega$ at 84°C; junction-to-case thermal resistance = 1.8°C/W).
2. The IRF 440 ($R_{DS(on)max} = 1.36\ \Omega$ at 84°C; junction-to-case thermal resistance = 1.1°C/W).

The next stage of the design is to choose a heatsink for each device that will ensure that the peak junction temperature does not exceed 84°C. For the IRF 430 the conduction losses are given by

$$\begin{aligned} P_C &= I^2 R \times \delta \\ &= (2.7)^2 \times 2.4 \times 0.45 = 7.9 \text{ W} \end{aligned} \tag{17.7}$$

where δ is the conduction duty cycle. Assuming an ambient temperature of 45°C, the temperature rise between the junction and ambient is given by

$$T_{JA} = 84 - 45 = 39°C \tag{17.8}$$

The junction-to-ambient thermal resistance required is given by

$$\begin{aligned} R_{thJA} &= T_{JA}/P_C \\ &= 39/7.9 = 4.9°C/W \end{aligned} \tag{17.9}$$

If the case-to-sink thermal resistance is discounted, the sink-to-ambient thermal resistance is therefore

$$\begin{aligned} R_{thSA} &= R_{thJA} - R_{thJC} \\ &= 4.9 - 1.8 = 3.1°C/W \end{aligned} \tag{17.10}$$

For the IRF 440 the conduction losses are given by

$$\begin{aligned} P_C &= I^2R \times \delta \\ &= (2.7)^2 \times 1.36 \times 0.45 = 4.46 \text{ W} \end{aligned} \tag{17.11}$$

Therefore, from Equation (17.9), the junction-to-ambient thermal resistance required is given by

$$\begin{aligned} R_{thJA} &= T_{JA}/P_C \\ &= 39/4.46 = 8.7°C/W \end{aligned} \tag{17.12}$$

The sink-to-ambient thermal resistance required is therefore

$$R_{thSA} = 8.7 - 1.1 = 7.6°C/W \tag{17.13}$$

Comparing the heatsink thermal resistances required for each device (3.1°C/W for the IRF 430 and 7.6°C/W for the IRF 440) shows that the designer's reliability objectives can be achieved by employing either a low resistance device on a small heatsink or a higher resistance device on a large heatsink. Considerations such as size, cost, and efficiency will determine which is more suitable.

17.3.5. Power Cycling

The die bond in any power semiconductor eventually fatigues if submitted to a sufficient number of thermal cycles. This is due mainly to differential expansion between the silicon and the solder or metal surface to which it is attached. The failure rate is dependent on the temperature excursion in each power cycle. Figure 17.14 shows the relationships between accumulated failures, the number of power cycles, and the temperature excursion for a MOSFET with a die size of 6.5 mm × 6.5 mm. Failure rates are generally greatest for the largest die, so that

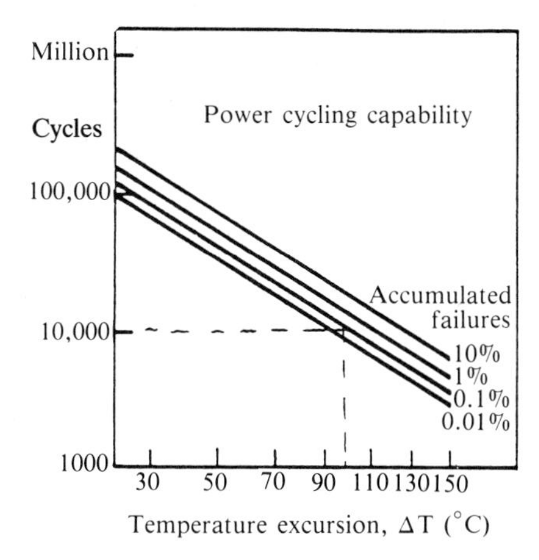

Figure 17.14. Typical power-cycling reliability data.

this information may be used for smaller dice mounted in the same manner to obtain a conservative estimate of their reliability.

As an example of the use of these curves, consider the reliability of a piece of equipment which is turned on and off once an hour. The equipment contains one MOSFET operating at a temperature of 100°C above ambient. The designer wishes to know the time to accumulate 0.1% failures. From Figure 17.14 it can be seen that, for a ΔT of 100°C, the number of cycles to accumulate 0.1% failures is 10,000. Therefore the time to accumulate 0.1% failures is given by

$$t = \frac{10{,}000}{24 \times 365} = 27 \text{ years} \tag{17.14}$$

17.3.6. Induced Failures

Components which are found to be free of infant-mortality failures when tested by the manufacturer can exhibit early failures after mounting in equipment. These failures are usually the result of poor handling or poor mounting.

Electrostatic discharge (ESD) can partially puncture the gate oxide, allowing the device to work sufficiently well for the equipment to pass inspection, but finally causing device failure after a few hours, days, or weeks of operation. The creation of these "walking-wounded" devices can be avoided by the application of ESD protection measures during equipment assembly (see Appendix 8). Partial gate failures can be detected by measuring the time constant of the decay of gate charge [23].

Failures may be induced during mounting by the application of excessive torque to mounting screws or by attempting to mount the device on an uneven surface. This will cause deformation of the metal base on which the die is mounted. The strain will be transmitted to the die so that either the die or the die bond suffers mechanical failure during thermal cycling. The mechanical shock produced by the cropping of unsoldered leads or the bending of leads too close to the package can also damage the die. During soldering the lead temperature and soldering time specified in the data sheet (see Appendix 7) must not be exceeded.

17.4. HANDLING POWER MOSFETs

Handling techniques for power MOSFETS are governed by the necessity to prevent damage to the gate oxide by ESD. Compared with MOS logic circuits, power MOSFETS are more robust because of their larger gate capacitance. Nevertheless, when large numbers of devices are being handled, some failures will occur if proper ESD procedures are not followed, with unfortunate consequences for the purchaser's opinion of the quality of the device [24]. ESD protection measures are described in detail in Appendix 8.

REFERENCES

1. R. Severns, R. Blanchard, A. Cogan, and T. Fortier, "Special features of power MOSFETS in high-frequency switching circuits." *High-Freq. Power Conver. Conf.* May, pp. 133–148 (1986).
2. M. Sciammas and D. Kapur, "A smaller packaged high performance power MOSFET." *Power Electron. Conf.*, San Jose pp. 384–389 (1986).
3. R. Frank, "Isolated, hermetic packages for power MOSFETS." *Powertechnics* April, pp. 26–30 (1988).
4. R. Ruttonsha, "Insulated TO-220 eliminates need for mica insulator without degrading thermal performance." *Powertechnics* March, p. 38 (1986).
5. E. Jansen, 'MOSPOWER transistors with 4000 V electrical isolation." *Siliconix Interface* Winter, pp. 7–9 (1987).
6. A. Cogan, J. Chen, and N. Maluf, "Very high-power, high-voltage DMOS transistors." *Power Electron. Conf.* pp. 142–146 (1987).
7. G. Majumdar, Y. Kamitani, Y. Yu, and H. Iwamoto, "Characteristics of a new 50 A, 500 V power MOSFET module manifested for high frequency inverter applications." *Proc. Power Convers. Int. Conf.* June, pp. 260–274 (1986).
8. B. E. Taylor, "Optimise high current designs with International Rectifiers HEXPAK." *Int. Rectifier Appl. Note* (1985).
9. W. Schultz, "Multichip power MOSFETS beat bipolars at high-current switching." *Electron. Des.* June 14, pp. 223–232 (1984).
10. G. J. Krausse, "Microwave-packaged power MOSFETS allow 10 MHz switching." *Power Convers. Intell. Motion* (*PCIM*) May, pp. 40–45 (1987).
11. H. Granberg, "Get 600 watts from four power FETS." *Motorola Eng. Bull.* **EB-104** (1983).
12. P. Freundel, "A high-current power MOSFET for 45 A and 50 V." *Proc. Power Convers. Int. Conf.* March, pp. 354–364 (1982).
13. JEDEC, "Power MOSFETS—JEDEC Standard No. 24." *Electron. Ind. Assoc. Pub.* (1985).
14. R. Pearce, S. Brown, and D. Grant, "Measuring HEXFET characteristics." *Int. Rectifier Appl. Note* **AN-957** (1986).
15. Motorola Inc., "Chapter 11. Characterization and measurement." *Power MOSFET Transistor Data, Motorola Inc.* pp. A142–A167 (1984).
16. Motorola Inc., "High Frequency measurements of power MOSFETS using oscilloscopes." *Hewlett Packard Appl. Bull.* **35** (1980).
17. D. L. Blackburn and D. W. Berning, "Power MOSFET measurement." *IEEE, Power Electron. Spec. Conf.* pp. 400–407 (1982).

18. D. L. Blackburn and D. W. Berning, "Power MOSFETS thermal resistance measurement." *Proc. Power Convers. Int. Conf.* March/April, pp. 10–14 (1980).
19. J. Eachus, "Failure analysis in brief." *Semicond. Int.* January, pp. 103–110 (1982).
20. J. Aguilar and K. Teasdale, "Power MOSFET reliability testing." *Proc. Power Conver. Int. Conf.* November/December, pp. 17–22 (1983).
21. International Rectifier, "Quarterly HEXFET reliability report no. 19." *Int. Rectifier* (1987).
22. P. Koetsch, "Establishing switcher reliability." *Powertechnics* March, pp. 24–28 (1986).
23. E. Hebenstreit, "Operational disturbances due to power FETS with damaged gate insulation." *Proc. Power Convers. Int. Conf.* June, pp. 294–302 (1986).
24. R. Ghent, "Relationship of device sensitivity to perceived quality." *Prof. Program Sess. Rec. Southcon 85,* Atlanta March, pp. 1–6 (1985).

CHAPTER 18

Modeling Power MOSFETS

18.1. INTRODUCTION

Power-MOSFET modeling has a variety of objectives:

1. Modeling of carrier behavior within the silicon die.
2. Simulation of switching using piecewise analysis.
3. Device modeling for high-frequency linear applications.
4. Circuit simulation using SPICE.

Models of type 1 seek to explain the behavior of power MOSFETS in terms of carrier movement and interaction of carriers with the crystal lattice. Central to these models is a resolution of the equations describing charge distribution and electric field within the device. Such models have been produced which predict power MOSFET behavior at this level with varying degrees of sophistication [1–5].

Models of type 2 seek to predict the behavior of MOSFETS in switching circuits using a set of equations to describe various aspects of the power MOSFETS' behavior. Analysis is usually on a piecewise basis, with separate equations or sets of equations being invoked for different phases of the switching process [6–9]. In some cases it may be convenient to separate the device into separate cells to model device behavior, for instance when determining the effect of gate resistance and capacitance on the spread of conduction across the die [10].

Type-3 models are used for small- or large-signal analysis of power MOSFETS operating in the linear regime, for instance in an audio or rf amplifier [11]. Circuit elements originating in the package as well as in the silicon die are incorporated in the model. In such models the components of the equivalent circuit are assumed linear, and the frequency response, stability, etc. are calculated in the usual manner.

Type-4 models are based on the ubiquitous SPICE computer program for circuit modeling. While SPICE has a built-in MOSFET model, it requires modification to achieve good accuracy when applied to power MOSFETS.

18.2. SPICE MODELING OF POWER MOSFETS

18.2.1. Limitations of Built-In MOSFET Models

The built-in MOSFET models in SPICE II are more attuned to the modeling of low-voltage lateral MOSFETS such as might be encountered in integrated circuits than to vertical DMOS power MOSFETS. The built-in models are unable to simulate some of the features of a power MOSFET, such as the variable drain–gate capacitance, the pinch effect between cells, and the body–drain diode, which are important in determining the response of the device.

In the absence of a suitable built-in model, the power MOSFET is usually simulated by combining further elements with the built-in MOSFET model to enable it to give a more faithful representation of the power MOSFET. An example of such a subcircuit, along with a description of methods for extracting the subcircuit values for data-sheet parameters, is given in Reference 12.

The model described in Reference 12 successfully models the basic switching and dc behavior of the power MOSFET. However, there have been a number of attempts to increase the sophistication of this model [13–17]. Other proposed power-MOSFET models have adopted a significantly different approach [18, 19]. The distinguishing aspects of the model are the ways in which they seek to implement subcircuit representation of various features.

18.2.2. Drain–Gate Capacitance

While the major test of a MOSFET model for small-signal applications is its behavior in the linear regime, assessment of a power MOSFET model is usually based on its ability to accurately reproduce the switching of the device. The drain–gate (Miller) capacitance is a critical factor in determining the switching behavior of a power MOSFET. This capacitance varies with drain–gate voltage, and therefore an accurate power-MOSFET model requires an accurate method of representing the variation in C_{DG} (see Section 3.6.3).

The model described in Reference 12 uses a fixed value for C_{DG}, with the possibility of replacing this with the junction capacitance of a pn junction for increased accuracy. However, the pn-junction method cannot generally cope with the wide variation of C_{DG} found in most power MOSFETS.

An alternative arrangement is shown in Figure 18.1. The method used here is to represent C_{DG} by a polynomial based on the drain–gate voltage [17, 20]. The voltage-dependent capacitor is connected in series with an offset voltage, required to make the capacitance decrease with increasing drain–gate voltage.

Another approach is to use a fixed value of drain–gate capacitance but to modulate the conductivity of the path between the capacitor and the drain by use of a JFET [19]. The JFET serves the double purpose of representing the pinch effect between cells and of controlling the influence of the fixed drain–gate capacitance in order to mimic the action of a variable capacitance. When a large value C_{DG} is effectively isolated by the JFET, only a smaller value C_{DG} is left to represent the drain–gate capacitance. The abrupt change in C_{DG} that occurs as the accumulation layer forms in the drain region under the gate oxide cannot be modeled in a completely satisfactory fashion by this method. Therefore a current-controlled current source is added in parallel with the drain–gate capacitance across the built-in MOSFET model.

Another method of simulating the change in C_{DG} with drain–gate voltage is by the use of the derivative function, where this is available [21]. The drain–gate capacitor is modeled as a voltage dependent current source. This source invokes the DERIV derivative function to implement the relationship between current,

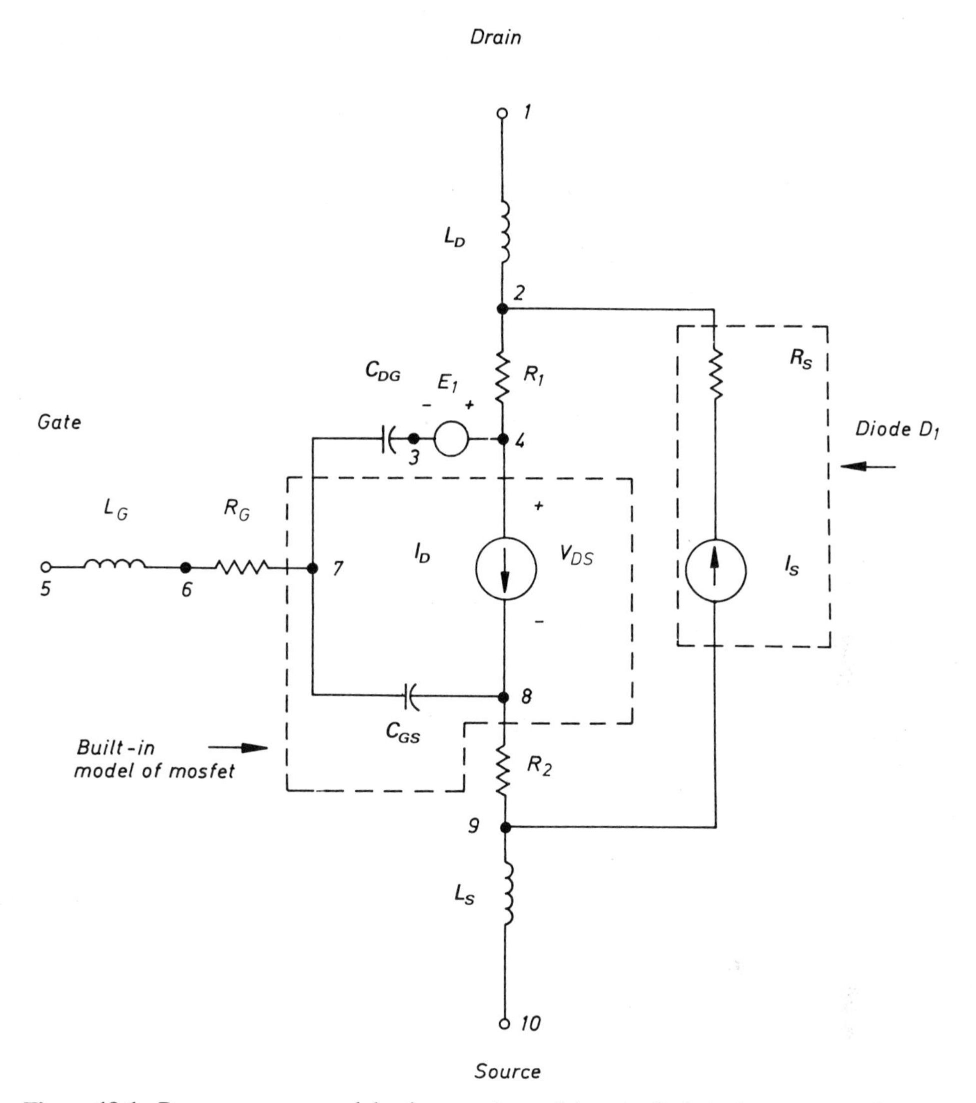

Figure 18.1. Power-MOSFET model using a polynomial-controlled drain–gate capacitance.

capacitance and $\mathrm{d}v/\mathrm{d}t$:

$$i = C\frac{\mathrm{d}v}{\mathrm{d}t} \tag{18.1}$$

The capacitance variation is therefore represented as a ramp between two fixed values (see Figure 18.2). This models the abrupt change in capacitance which occurs as the accumulation layer forms. In Figure 18.2 this corresponds to the point $V_{GD} = -1$ V. This method models the capacitance variation less accurately at low values of capacitance, but for switching considerations accuracy at high capacitance values is more important.

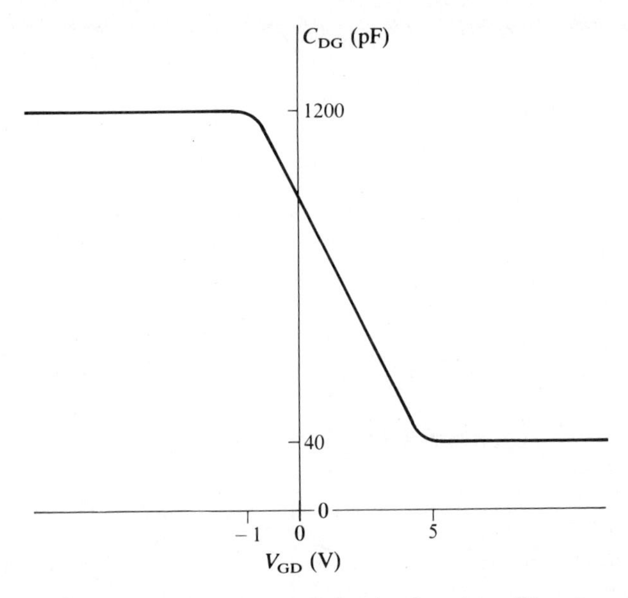

Figure 18.2. Capacitance modeling using a derivative function. (Courtesy AB Associates, Tampa, Florida.)

18.2.3. Computer-Time Requirements

While a MOSFET model must be judged by its accuracy, another important consideration is the computer time required to analyze a circuit containing the model. Simpler models tend to use less computer time. With the more sophisticated models a circuit analysis can use several hours of computer time.

Where circuit components other than the MOSFET dominate the behavior of the circuit, the MOSFET can be replaced, at least for a first try, by the simple model shown in Figure 18.3 in order to reduce computer time. The voltage source is given an arbitrarily high value during the period when the MOSFET is off. This ensures that the diode is reversed biased and no current flows during this time. When the MOSFET is on, the blocking voltage is reduced to zero.

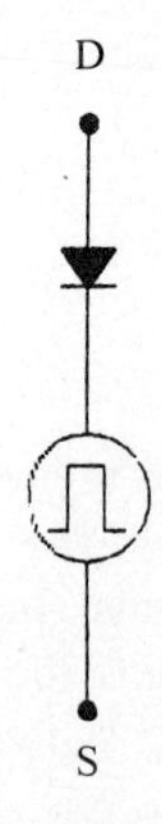

Figure 18.3. Simplified MOSFET model

REFERENCES

1. D. H. Navon and C. T. Wang, "Numerical modeling of power MOSFETS." *Solid-State Electron.* **26**(4) pp. 287–290 (1983).
2. A. W. Wieder, C. Werner, and J. Tihanyi, "2-D analysis of the negative resistance region of vertical power MOS-transistors." *IEEE, Int. Electron Devices Meet.* pp. 95–99 (1980).
3. S. Selberherr, "Modeling static and dynamic behavior of power devices." *IEEE, Int. Electron Devices Meet.* pp. 71–74 (1983).
4. K. Yamaguchi, "A time dependent and two-dimensional numerical model for MOSFET device operation." *Solid-State Electron.* **26,** (9) pp. 907–916 (1983).
5. M. Le Helley and J. P. Chante, "Optimization of BV-R_{ON}-C_{GD} of power MOS transistors." *Eur. Power Electron. Conf.* pp. 397–401 (1987).
6. H. R. Ronan, Jr. and C. F. Wheatley, Jr., "Power MOSFET switching waveforms: A new insight." *Powercon* **11,** C.3.1–C.3.10 (1984).
7. C. L. Rym, P. F. Cools, J. Y. Barral, and P. V. Bury, "Switching power FET structure in hybrid technology: A new approach to performant spaceborne converters." *IEEE, Power Electron. Spec. Conf.* pp. 225–237 (1984).
8. R. Erickson, B. Behen, R. D. Middlebrook, and S. Cuk, "Characterization and implementation of power MOSFETS in switching converters." *Powercon* **7,** D4.1–D4.17 (1980).
9. H. A. Owen, T. H. Sloane, B. H. Rimer, and T. G. Wilson, "Switching interval modeling in very high frequency high power MOSFET converters." *Powercon* **7,** G1.1–G1.13 (1980).
10. J. Thoma and M. Beer, "Switching time limits of paralleled MOSFET devices obtained by modeling the dynamic input behavior." *Proc. Power Convers. Int. Conf.,* Munich pp. 166–181 (1987).
11. M. Alexander, "Frequency response analysis of the MOSFET source-follower." *Siliconix Appl. Note* **AN-83-8** (1983).
12. "SPICE II Computer Models for HEXFETs." International Rectifier HEXFET Data Book 3. (1983).
13. H. A. Nienhaus and J. C. Bowers, "A low power MOSFET computer model." *IEEE, Southeastcon* **80,** pp. 18–21 (1980).
14. H. A. Nienhaus and J. C. Bowers, "A high-power MOSFET computer model." *IEEE, Power Electron. Spec. Conf.* pp. 97–103 (1980).
15. H. Cheng and A. G. Milnes, "Power MOSFET characteristics with modified SPICE modeling." *Solid-State Electron.* **25**(12), 1209–1212 (1982).
16. R. A. Minasian, "Computer-aided simulation of power MOSFET switch-mode convertors." *Electron. Lett.* **20**(11), 454–456 (1984).
17. P. O. Lauritzen and F. Shi, "Computer simulation of power MOSFETS at high switching frequencies." *Proc. Power Convers. Int. Conf.* October, pp. 372–383 (1985).
18. F. X. Timmes, D. T. Dodt, and N. Maluf, "Circuit simulation of power MOSFETS." *IEEE Ind. App. Soc. Ann. Meet.,* pp. 393–395 (1986).
19. G. M. Dolny, H. R. Ronan, and C. F. Wheatley, Jr., "A SPICE II subcircuit representation for power MOSFETS using empirical methods." *Power Electron. Des. Conf.* Anaheim CA. October, pp. 147–153 (1985).
20. International Rectifier, "Spice computer models for HEXFET power MOSFETS." *Int. Rectifier. HEXFET Data Book 4, Appl. Note* **AN-975** (1987).
21. J. C. Bowers, "Extended HEXFET model for I-G Spice." *Appendix, Int. Rectifier Appl. Note* **AN-954A** (1984).

CHAPTER 19

Special-Purpose MOSFETs

19.1. CURRENT-SENSING MOSFETs

19.1.1. Current Sensing

Many power-MOSFET applications incorporate device current sensing to implement control or protection features. Traditional current-sensing techniques suffer from a number of problems. Current-sensing resistors are inefficient, increasing heat removal problems. Current-sense transformers have bandwidth and saturation limitations, and Hall-effect devices need to be sophisticated to overcome accuracy and temperature-dependence problems. Wire-wound resistors and transformers introduce inductance into the circuit, which limits high-frequency operation. Furthermore, all these methods involve extra components that must be sourced and stocked, put some obstacle in the way of automated assembly, use relatively bulky components, and increase equipment cost. Current-sensing MOSFETs have the benefits that special current-sensing components are not essential, automated assembly capability is not affected, and switching performance is unimpaired.

Current-sensing MOSFETs have been developed to meet the need of current monitoring in the power device. In current-sensing MOSFETs a known fraction of the source current is diverted into a second, independent source terminal. This had led to the device being sometimes called a current mirror [1]. The sense current is typically of the order of $\frac{1}{1500}$ of the drain current, and can be converted into a scaled-voltage indication of the drain current in a way that is efficient in terms of power, cost, and board space.

19.1.2. Principle of Operation

A vertical diffused power MOSFET consists of many parallel cells formed on the upper surface of the silicon die. An enhanced channel is formed around the edge of each cell to allow vertical drain–source conduction. When fully inverted, the channel regions act as many similar resistors in parallel, all with equal, positive temperature coefficients. This situation results in current being evenly distributed between cells. It is this even distribution of current that allows current sensing by the detection of the current flowing through a few cells, thereby giving a scaled indication of the total drain–source current.

19.1.3. Structure

The source metallization covering a small number of ordinary cells is isolated from the main source metallization. A separate bonding pad is provided to allow

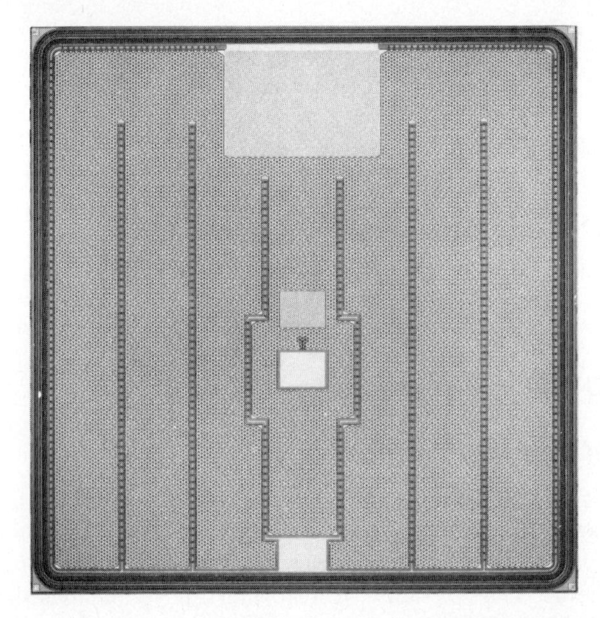

Figure 19.1. Current-sensing MOSFET die. The sense cells are visible just above the central current-sense bonding pad. Just above that is the Kelvin source bonding pad. The main source pad is at the top of the picture; the gate pad is at the bottom. The dots on the source pads are shorts to the underlying p^+ region, used to give the device improved avalanche and dv/dt characteristics. (Courtesy of International Rectifier Corp.)

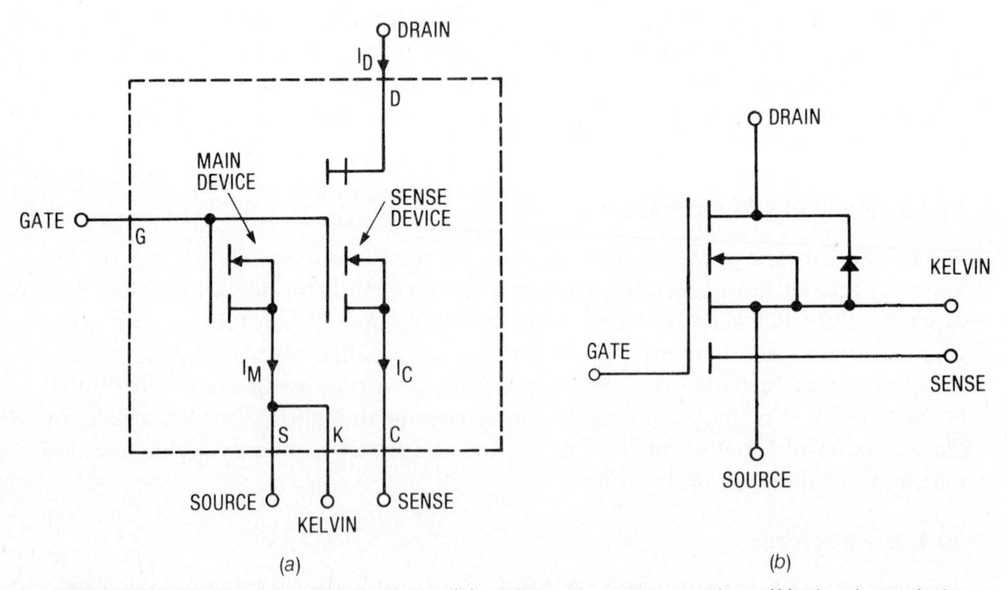

Figure 19.2. Current-sensing MOSFET. (a) Schematic representation; (b) circuit symbol.

connection of the sources of the isolated cells to the current-sense terminal. Two connections are made to the main source metallization via separate bonding pads. One connection is used as the main source terminal and carries the main source current. The other, known as the Kelvin contact, is connected to the main source metallization close to the sense cells. The Kelvin terminal is used to provide a return path for the diverted sense current. The Kelvin contact is located close to the source cells in order to minimize the voltage drop in the main source metallization. Figure 19.1 shows a photograph of a current-sensing MOSFET die. The equivalent circuit of the device is shown in Figure 19.2, along with the circuit symbol for the device. As Figure 19.2 shows, the current-sense MOSFET can be represented by two parallel devices—a main device and a sense device. The ratio of main-device current to sense-device current is termed the sense ratio, and is denoted by r. The intrinsic sense ratio r_0 is that ratio obtained when the effect on current distribution of the measuring system and any parasitic elements is zero.

19.1.4. Static Characteristics

An equivalent circuit for a fully enhanced device is shown in Figure 19.3, with typical element values for a 100-V device and a 500-V device.

With the source and the sense terminals connected together, all cells have essentially equal gate–source bias. The gate voltage must exceed the minimum necessary gate voltage by two or three volts so that the effects of parasitic voltage drops within the device and its package are minimized. Under these conditions

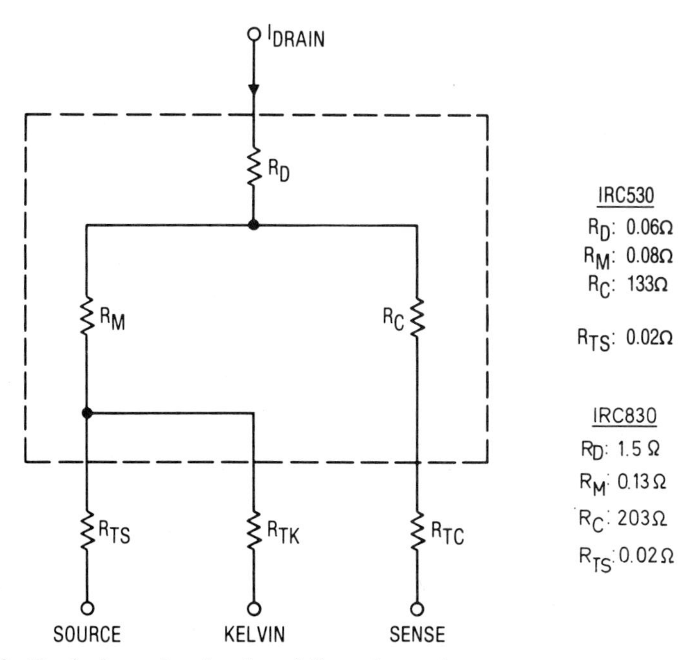

Figure 19.3. Equivalent circuit of a fully enhanced current-sensing MOSFET for static conditions for a 100 V device (IRC 530) and a 500 V device (IRC 830).

the channel resistance can be assumed to be the same for all cells. In Figure 19.3 the equivalent resistance of the parallel cells of the sense device is shown as R_C, and that of the main device cells is shown as R_M. Thus

$$R_C = nR_M \tag{19.1}$$

where n is the cell ratio. R_D is the bulk drain resistance of the device, and R_T represents terminal resistances, including metallization, bonding-wire, and post resistances.

With the sense and source terminals connected, and neglecting parasitic resistances, the ratio of main-cell to sense-cell current is given by

$$\frac{I_M}{I_C} = \frac{R_C}{R_M} = n \tag{19.2}$$

Thus, in the absence of any external resistance, the sense ratio is theoretically equal to the cell ratio. Since R_C and R_M are altered in the same proportion by changes in temperature, drain-current magnitude, and manufacturing spread in $R_{DS(on)}$, the sense ratio remains constant with variation of these conditions. If, however, parasitic resistances external to the silicon die are introduced into the two current paths, the sense ratio will be disturbed from its intrinsic value, as Figure 19.4*a* shows. Furthermore, the external resistances will not behave in the same way as R_C and R_M, and therefore the sense ratio becomes dependent on temperature, current magnitude, and spread in $R_{DS(on)}$. Parasitic impedances can also affect the current ratio. The Kelvin connection is designed to reduce the amount of parasitic resistance and inductance in series with the main cells. The Kelvin connection provides a path for returning the sensing current to the source

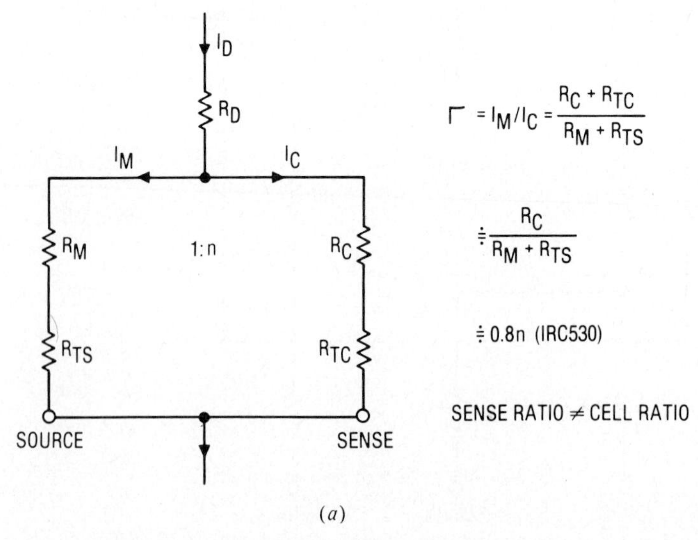

(*a*)

Figure 19.4. Preserving the sense ratio by using the Kelvin source. (*a*) Sense current returned to the source terminal; (*b*) sense current returned to the Kelvin source.

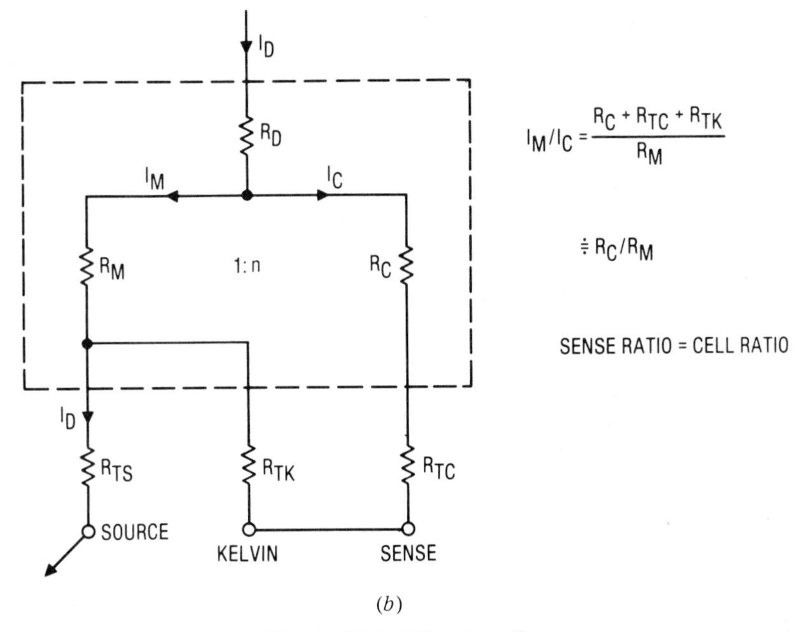

(*b*)

Figure 19.4. (*Continued*)

without including the source wire bond, the source pin, and much of the source metallization. As Figure 19.4*b* illustrates, the sense ratio is then less dependent on parasitic resistances.

The typical data-sheet characteristics shown in Figure 19.5 are based on the virtual-earth current-sensing method (see Section 19.1.7) and therefore reflect variations of the sensing ratio of the packaged device with no external resistance introduced into the sensing circuit. The nominal intrinsic sense ratio of this device is specified as 1665, with $I_D = 14$ A, $V_{GS} = 10$ V, and $T_J = 25$°C. The graphs show a possible manufacturing spread of 2.5% in the sense ratio and a shift in ratio of approximately 0.3% for a change in temperature of 100°C.

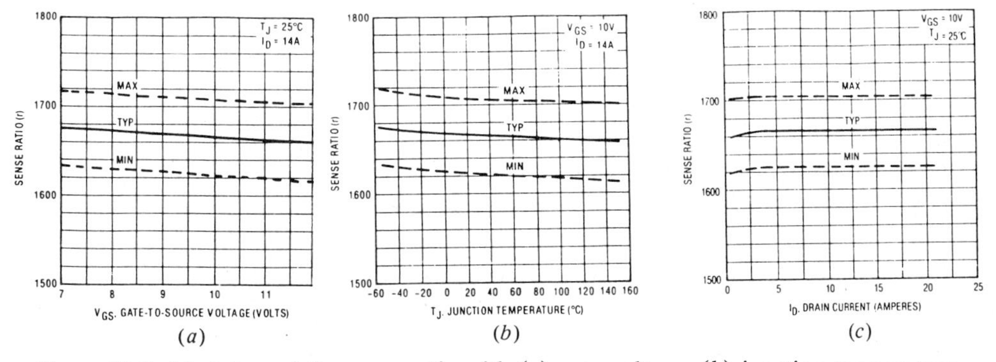

(*a*) (*b*) (*c*)

Figure 19.5. Variation of the sense ratio with (*a*) gate voltage, (*b*) junction temperature, and (*c*) drain current.

19.1.5. Reverse Conduction

With the MOSFET off and current flowing through the integral diode, the current again divides between the main cells and the sense cells. However, this division of the current is not necessarily in the same ratio as for forward conduction, since although the ratio of the channel resistances may be the same, there may be differences in the voltage drops of the diodes. The sense current may therefore only be used to give an approximate indication of the device current. A more accurate current measurement may be obtained if the channel is turned on during diode conduction [2]. Provided the forward drop is below approximately 0.6 V, the current will tend to divide in the ratio of the resistances of the main and sense cells.

During reverse recovery of the MOSFET intrinsic diode, large sense-current spikes will be produced when the drain–source voltage is rising due to the recovery current of the sense-cell diodes.

19.1.6. Dynamic Characteristics

Parasitic capacitances associated with the sense and main devices are not in the same ratio as the intrinsic current sense ratio due to the capacitance of the current sense bondpad. This suggests that at turnoff there will be spurious spikes in the sense signal due to the charging of the output capacitance of the MOSFET as the drain voltage rises. However, in practice, other factors dominate [3].

Current spikes observed at turn-on and turn-off often originate in the measuring equipment. The current sense ratio of a current-sensing MOSFET is typically in the region of 1500 : 1. Signals of the order of a few millamps or a few tens of millivolts are therefore being measured in intimate proximity to circuits carrying tens of amps, with rates of change of current of tens or hundreds of amps per microsecond. Voltages that can swamp the signal are easily generated in the sensing circuit and the ground lead of an oscilloscope probe. Induced errors usually take the form of spikes of opposite polarity at turn-on and turn-off. Even wrapping the ground lead around the probe tip can be insufficient protection. A clean signal may be obtained by removing the probe tip and soldering the sense resistor across the inner tip and the adjacent earth sheath. The resistor is then located directly across the sense and Kelvin leads of the MOSFET package.

Once these precautions have been taken, the sense signal will still contain false peaks at turn-on and turn-off. These peaks are due to a difference in current ratio between the linear operating region and the fully enhanced operating region. In the linear, or constant-current, region the current divides in the ratio of the total channel widths of the main and sense cells. In the fully enhanced region the current division between the main and sense cells is determined mainly by the ratio of the path resistances. Generally, the ratio of the resistance of the drift region of the sense cells to the resistance of the drift region of the main cells is close but not identical to the sense ratio in the linear region. When an external sense resistor is included in the sense path, the ratio of path resistances becomes significantly different.

There is another effect at work during the switching interval that tends to produce spurious signals—transformer action. Even if all magnetic coupling between the main and sense circuits external to the device package is eliminated, there remains the possibility of coupling within the package itself. The drain and source leads form the primary of a transformer, while the sense and Kelvin leads form a secondary. In the usual lead configuration the two loops overlap. Therefore the package constitutes a poorly coupled 1:1 transformer. Since the current ratio in the two circuits is of the order 1500:1, even a low degree of coupling can induce a significant error, and large spikes should be expected in the sense signal during periods of high di/dt. These spikes do not in fact appear during switching since for most of the switching interval the MOSFET is in the linear region and the current in the sense circuit is a function of the transconductance of the sense cells and the gate voltage. However, transformer action will induce spikes in the current sense signal if the current sense MOSFET is maintained on, and its drain current is switched by some external means. The transformer effect clearly has implications for the bandwidth of the current-sensing MOSFET [3].

19.1.7. Virtual-Earth Sensing Method

Two methods are commonly employed for obtaining a scaled voltage indication of drain current. These are the virtual-earth method and the series-resistor method. The virtual-earth method is less sensitive to temperature variations, while the series resistor method is simpler and cheaper.

The virtual-earth method is illustrated in Figure 19.6. The virtual-earth action of the operational amplifier ensures that the sense and Kelvin pins are effectively at the same voltage, so that no voltage which might upset the sense ratio is introduced into the sensing circuit. Operational-amplifier input bias currents do not significantly affect the current ratio. For example, using the device whose characteristics are given in Figure 19.5, an input bias current of 500 nA would result in an error of 0.1% for a drain current of 0.8 A. Amplifiers with bias currents less than 25 nA are readily available. If necessary, bias-current offset

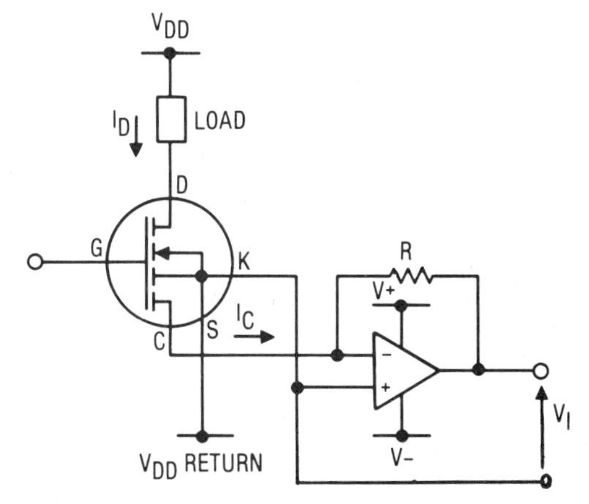

Figure 19.6. Virtual-earth sensing circuit.

may be reduced by locating a resistor between the noninverting terminal and ground. Input offset voltage adds directly to the output voltage. The error it introduces may be significant if a low gain has been chosen. For example, peak-detection circuitry may involve comparing the output with a 200-mV reference. A maximum offset of 2 mV or so would then be required from the operational amplifier.

The operational amplifier must also have a slew rate adequate to allow it to track rapid changes in current. For example, with 0.5-V/A output, and the MOSFET switching 10 A in 100 ns, an amplifier slew rate of 50 V/μs would be required. Many power-supply and motor-control integrated circuits use a current limit threshold voltage of the order of 200 to 300 mV, in which case a slew rate of 2 to 3 V/μs would be adequate.

Capacitance charging currents give rise to sense-current spikes. Current spikes may also be present as part of normal circuit operation, for example diode reverse-recovery spikes. If it is required that these should not trigger overcurrent detection circuitry, filtering or blanking of the spikes may be necessary [4]. An *RC* filter may be incorporated in the operational-amplifier circuit. The operational-amplifier output should not be allowed to be driven into saturation, as the overload recovery time is likely to be long. A zener diode, located across the feedback resistor, may be used to limit the output-voltage swing. Also, the operational amplifier will leave the linear mode if the output becomes slew-rate-limited. Delays in changing slewing directions may cause very short sub-microsecond spikes to appear as pulses 1 to 2 μs wide. These may be difficult to filter and may trigger overcurrent detection circuitry. Therefore slew-rate-limited operation should be avoided.

The disadvantages associated with the virtual-earth circuit are that (a) dual power supplies are required, (b) the current-sense voltage indication is negative, (c) an active device is required to accomplish the current to voltage transformation. These disadvantages are overcome, at the expense of reduced accuracy, with the series-resistor method.

19.1.8. Series-Resistor Sensing Method

The voltage output is obtained by placing a resistor in the path between sense and Kelvin, as shown in Figure 19.7*a*. The voltage across the resistor is a scaled voltage indication of the device current. In some applications the control circuits may of necessity be referenced to the source connection. In this case the source and Kelvin pins have to be connected as shown in Figure 19.7*b*, and the consequent loss in sensing accuracy due to the inclusion of the parasitic source resistance has to be accepted.

The effect of including the series sensing resistor in the sense circuit is to increase the sense ratio. Figure 19.8 shows the equivalent circuit with the sensing resistor included. The sense ratio is then given by

$$r = \frac{R_S}{R_M} + r_0 \tag{19.3}$$

Therefore it is possible to calculate the sense ratio for a given value of R_S, once

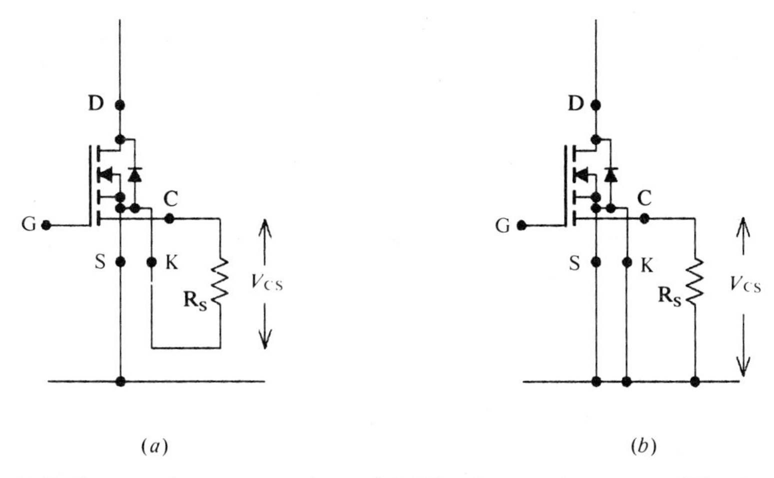

Figure 19.7. Sense-resistor connections. (*a*) Via the Kelvin source; (*b*) via the main source.

R_M and r_0 are known. However, R_M is the main device cell resistance; it is subject to manufacturing spread and is strongly dependent on temperature. Therefore r will vary with these parameters. The variation increases as R_S increases, until the limiting case of $R_S \gg R_C$ is reached when the voltage across R_C is essentially equal to the voltage across the main cell resistance. Then the main device cell resistance is being used to monitor device current, and the current sense output is equal to the drain voltage.

A second disadvantage associated with the resistor sensing method is that the

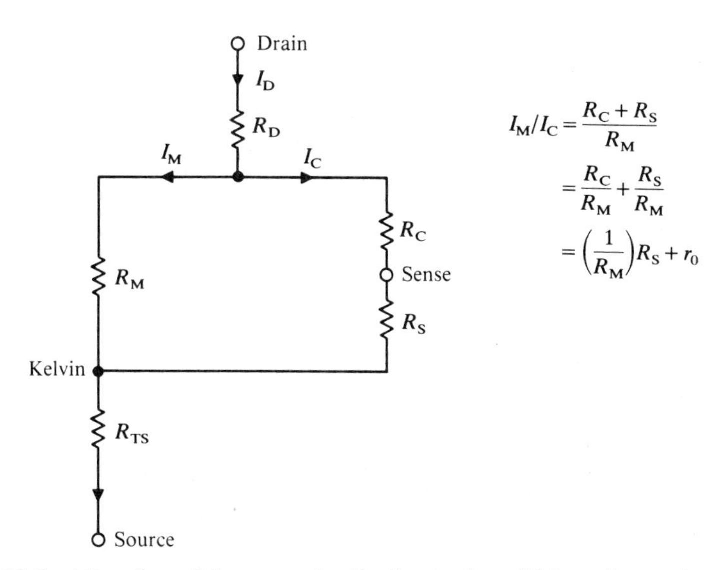

Figure 19.8. Alteration of the current ratio due to the addition of a sensing resistor.

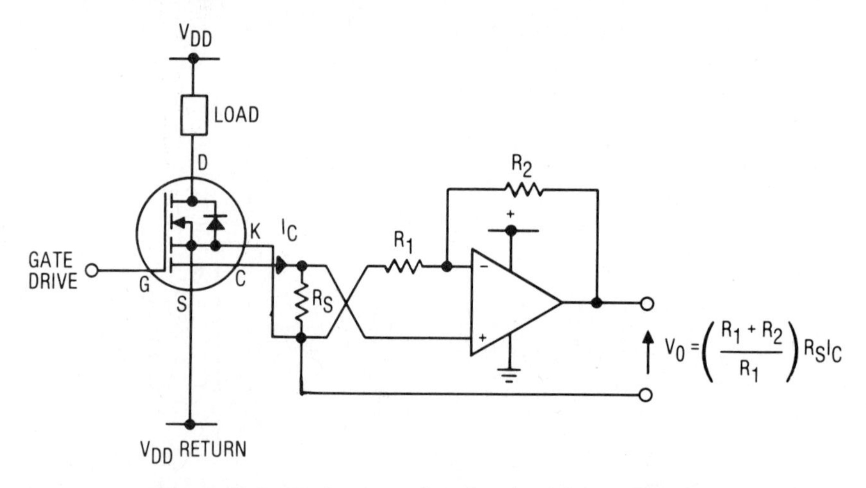

Figure 19.9. Resistor sensing circuit with amplification.

sensitivity range is limited. The circuit sensitivity is given by:

$$\text{sensitivity} = \frac{V_{\text{sense}}}{I_D} = \frac{R_S}{(r+1)} \tag{19.4}$$

As R_S increases, r increases and the maximum available sensitivity approaches a limit. To obtain a higher output voltage, or to allow the use of a low-value sensing resistor in order to reduce the effect of temperature on the sense ratio, an operational amplifier may be used to amplify and buffer the sense voltage as shown in Figure 19.9. The operational amplifier is connected so that a positive output is obtained and only a single positive supply is required. However, the input common-mode voltage range of the amplifier must include the lower power rail.

The static accuracy is dependent on operational-amplifier offsets. The input offset voltage of the amplifier adds directly to the sense voltage and therefore must be kept small relative to V_{sense} at the lowest current value that is of interest. Input bias currents may cause significant error if the noninverting effective source resistance is large.

19.1.9. Applications

Current-sensing MOSFETS find application in all circuits where current is a controlled parameter, for example in motor drives, where current and torque are directly related. Current sensing allows fault protection in power supplies, motor controllers, and other equipment in which overload protection is required. Start-up current limiting in lamp and motor loads can extend system life and reduce component cost. Current-waveform shaping is also possible for such applications as print-hammer drivers, solenoids, and stepper-motor windings.

The Kelvin contact of the current-sensing MOSFET may act as a gate return path in circuits in which very fast switching is required. By referencing the gate drive circuit to the Kelvin pin, the common-mode source inductance associated with the main source lead is eliminated and faster turn-on is possible.

Power supplies often incorporate a current-sensing function which can be implemented with the current-sense MOSFET. In resonant power converters the current-sense signal may be used to implement zero-crossing turn-off. Also, current-mode control of power supplies, which requires sensing of the current in the switching device, is becoming widely used. Examples of this application are given in Chapter 11.

19.2. RADIATION-HARDENED POWER MOSFETs

19.2.1. Introduction

This section deals with the effects of certain kinds of radiation, both naturally occurring and man-made, on the operation of power electronic circuits containing power MOSFETS. Both ionizing and non-ionizing radiation are considered, along with the origins and damage effects of primary radiation interactions and secondary induced interactions. The effects of radiation can be limited by circuit design strategies and the use of radiation-resistant (rad-hard) MOSFETS [5–8].

19.2.2. Types of Radiation

The principal kinds of radiation in the operating environment of an electronic system are gamma rays, X-rays, neutrons, trapped protons and electrons, and cosmic rays. The rate of incidence of gamma rays is of considerable interest when one considers the effects of nuclear weapon bursts on electronics. This rate quantity is usually known as "gamma dot", or the time derivative of the gamma flux, expressed in rad (Si)/s. Another type of radiation, known as electromagnetic pulse (EMP), is generated by nuclear-weapon bursts. While of great interest to circuit and system designers, this does not affect devices and will not be discussed here.

Types of radiation may be broadly classified into two categories: ionizing and non-ionizing. Photon radiation in the gamma and X-ray part of the spectrum is a primary constituent of ionizing radiation. Gamma rays and X-rays are both photons, differing only in their wavelengths. Gamma rays have wavelengths of a fraction of an angstrom, while X-rays have wavelengths of several angstroms. X-rays interact with atoms primarily through the Compton effect, in which a loosely bound electron is scattered along with the incident photon, resulting in ionization of the atom. In a higher-energy interaction, a high-energy photon (gamma ray), upon approaching a nucleus, may be annihilated, resulting in the production of an electron–positron pair. At low photon energies, the matter interaction is via the photoelectric effect, in which an electron is scattered and the incident photon is absorbed. In all cases electrons are scattered. These electrons possess both charge and mass. They therefore readily interact with matter, causing electrical (and chemical) changes to occur. Secondary ionizing radiation is generated when these free electrons collide with the atoms, resulting in bremsstrahlung X-rays.

Trapped protons and electrons are contained by the earth's magnetic field in regions known as the Van Allen belts. These belts form a torus about the earth with the fields converging in the regions of the north and south poles. Additionally, a magnetic anomaly known as the South Atlantic Anomaly (SAA) causes a convergence of the field lines in that area. Protons, being both massive and charged, interact readily with matter, causing ionization damage. Due to their mass, they also cause displacement damage similar to that produced by neutrons. Electrons cause direct ionization damage and produce further secondary interactions via bremsstrahlung radiation.

In addition to protons and electrons, the near-earth radiation environment contains many heavy ions which stream in from the sun and from deep space. These ions (cosmic rays) mainly range from mass numbers around that of iron to simple protons. These particles are responsible for a form of radiation damage known as single-event upset (SEU), in which a single ion can cause a logic circuit to change state or fail. They also have the potential to cause failure in a power MOSFET. Various shielding strategies have been developed to deal with SEU, so that the particles expend their energy in some intervening medium.

The final constituent of the radiation environment is the neutron. The neutron, having no charge, does not cause direct ionization damage. However, it is relatively massive and causes displacement damage within crystal structures. It can also induce reactions which result in the emission of ionizing electrons and photons. Secondary gamma radiation can cause ionization damage even though the incident neutron does not. A similar situation exists when an excited nucleus returns to its ground state by emitting an electron or a photon.

19.2.3. Radiation Effects in MOSFETS

In crystal structures of the sort found in semiconductors, ionizing radiation forms electron–hole charge separations, or "pairs". These electron–hole pairs have two primary effects in a power MOSFET structure, depending upon where they are formed. In the relatively high mobility material which constitutes the MOSFET body and epitaxial diffusions, electron–hole pairs have a very brief existence. They either recombine or are swept out very rapidly. If the circuit configuration permits, a large pulse of drain–source current can result (Figure 19.10). In a

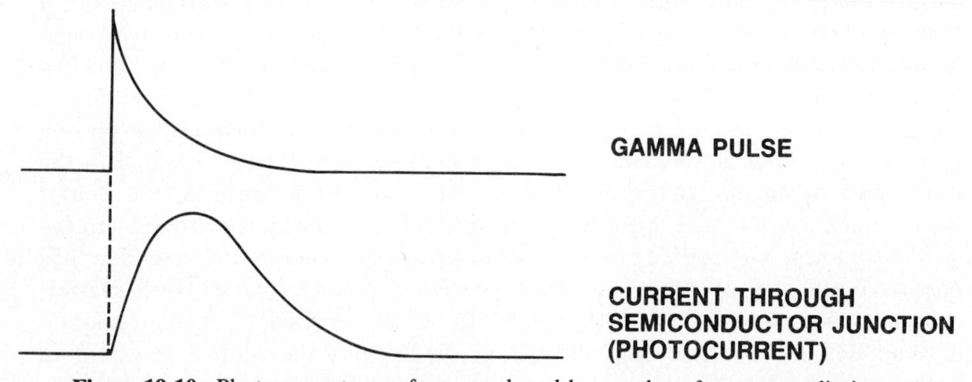

Figure 19.10. Photocurrent waveform produced by a pulse of gamma radiation.

bipolar transistor this may well result in second-breakdown failure. Power MOSFETS in which the parasitic bipolar transistor is effectively muted should be better able to resist damage due to radiation-induced turn-on, provided SOA failure of the parasitic bipolar transistor can be avoided [9–11].

If, however, the electron–hole pairs are generated in the very low-mobility areas of the gate oxide or interface states, recombination is very slow, resulting in persistent charge separations. The negative components (electrons) of these pairs are removed upon reaching the gate structure by virtue of the gate bias voltage applied to the MOSFET. This same gate bias voltage causes the positive components (holes) to collect at the oxide–channel interface, where they act as a permanent bias source. The proximity of these charges to the channel interface enhances this effect. The presence of this charge results in the equivalent V_{GS} of the MOSFET moving in the negative direction and eventually, after sufficient exposure to ionizing radiation, becoming negative. At this point, the MOSFET, which began as an enhancement-mode device, becomes a depletion-mode device (n-channel MOSFET).

The above discussion assumes low to moderate dose rates. At high dose rate [10^8 rad(Si)/s], recombination rates are insufficient to rid the structure of charge separations, resulting in reduced blocking voltage capability and very high drain currents. These photon induced currents (photocurrents) may be many times the steady current rating of the device. The only terrestrial source of such high gamma rates is a nuclear-weapon burst. Various circuit strategies have been developed to deal with this event and are described in Section 19.2.8.

Neutrons are primarily non-ionizing due to their lack of electric charge, and interact with semiconductor crystal structures by causing displacements of the constituent atoms. These defects (called Frankel defects) act to deplete the majority carriers (electrons) in the channel structure, resulting in increased resistivity. This effect is most pronounced in material where dopant concentrations are initially small, such as high-voltage MOSFETS. One may view this process as the canceling of donor atoms in the doped material, resulting in fewer free electrons available as carriers.

The secondary ionizing effects of neutron interactions should not be overlooked, as these effects became pronounced at high neutron fluence levels (10^{10} n/cm^2 s). Gamma rays, bremsstrahlung X-rays, and Compton electrons are generated when neutrons interact with both the semiconductor material and the material used to package the device. The ionizing effects due to the secondary radiation emissions following neutron bombardment are identical to those of incident gamma radiation.

19.2.4. Effects of Ionizing Radiation on the Characteristics of Standard Power MOSFETS

The main effect of ionizing radiation is to introduce charges into the gate oxide which produce a shift in the threshold voltage [12]. The threshold voltage of n-channel MOSFETS decreases with increasing total dose (Figure 19.11) while for p-channel MOSFETS the threshold voltage increases (Figure 19.12). This change of threshold is essentially independent of dose rate and depends only on the total

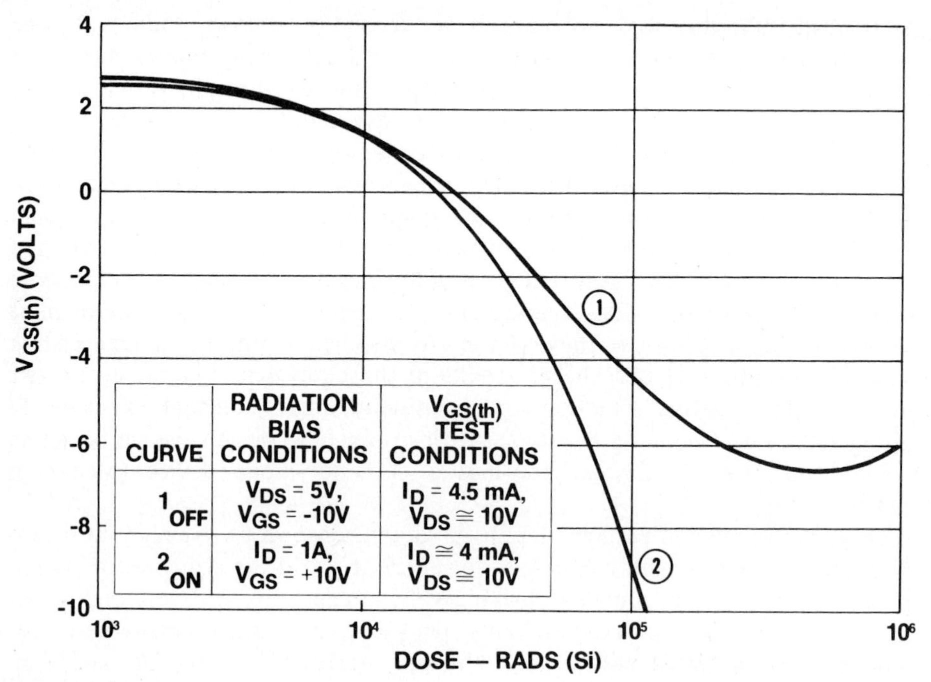

Figure 19.11. Gate threshold voltage versus total dose for a conventional n-channel power MOSFET (IRF 130).

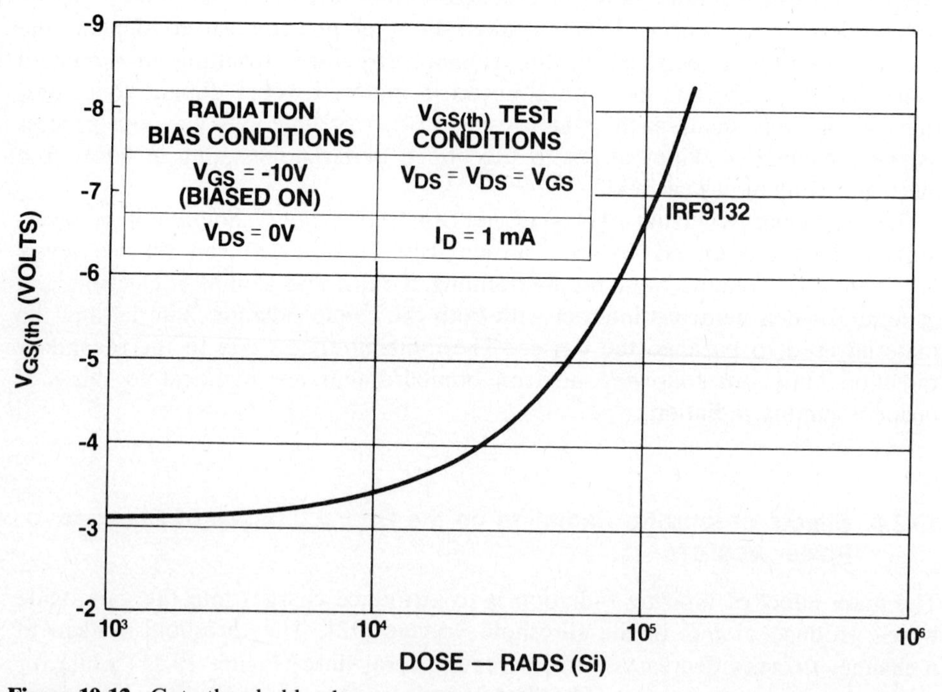

Figure 19.12. Gate threshold voltage versus total dose for a conventional p-channel power MOSFET (IRF 9132).

dose, except when the temperature excursions of the die are sufficient to accelerate recombination of this charge to a point where significant decay occurs.

As can be seen from Figure 19.11, the variation in threshold voltage is roughly independent of bias conditions for doses of up to about 30 krad. Thereafter the threshold voltage decreases much more rapidly for the biased-on condition. The threshold shift appears to be essentially independent of frequency and drain–source voltage [13]. Gate drive circuitry can be designed to nullify the threshold shift by overriding it with an appropriate level of biasing. For very heavy dose levels this may not be possible if the bias voltage required exceeds the maximum allowable gate–source voltage. Application of power to the main circuit may have to be delayed at switch-on in order to allow the gate bias-generating circuitry to become operative.

19.2.5. Effects of Ionizing Radiation on Leakage Current

Figure 19.13 shows the effect on drain–source leakage current of total radiation dose for a p-channel power MOSFET. Since for a p-channel MOSFET the threshold voltage becomes more negative with increasing total dose, the increased leakage current cannot be due to the threshold voltage approaching the gate voltage. Leakage current increases with radiation dose due to the trapping of charge in the oxide covering the field termination structure. This charge causes a local increase in electric field and leads to increased leakage current or, conversely, breakdown at a lower voltage. However, experiments on p-channel MOSFETS suggest that this explanation is simplistic and in fact the mechanisms at work are more complex [14]. Generally, the decrease in breakdown voltage increases with the applied drain–source voltage during irradiation [15]. The fall in blocking voltage will

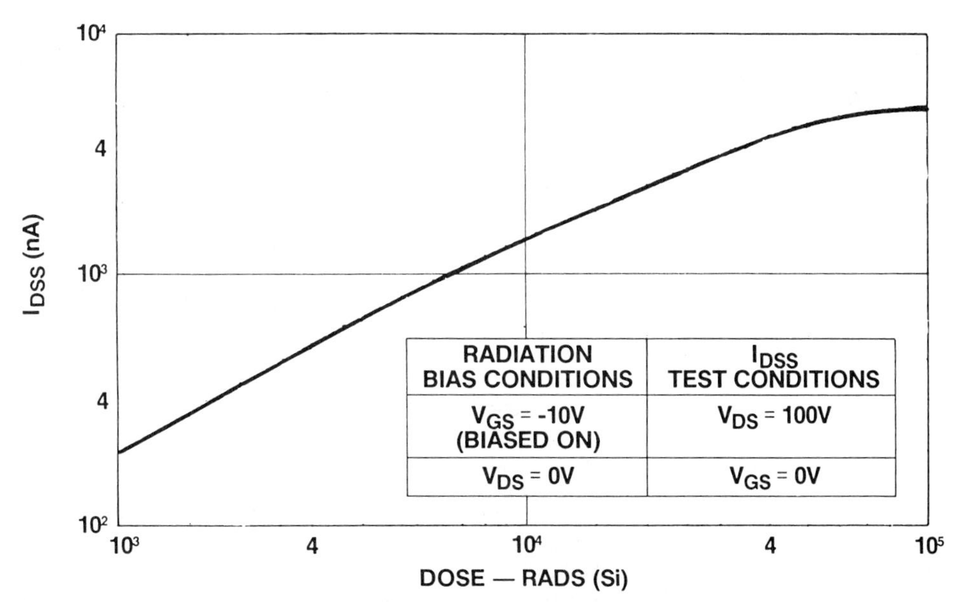

Figure 19.13. Drain–source leakage current versus total dose for a conventional p-channel power MOSFET (IRF 9132).

require the selection of a device with a voltage rating high enough to accommodate the fall in blocking voltage produced by the radiation. This results in a higher $R_{DS(on)}$ and a device that is more susceptible to a shift in $R_{DS(on)}$ due to neutron radiation.

19.2.6. Effects of Non-ionizing Radiation on the Characteristics of Standard Power MOSFETS

The effect of neutron radiation is to increase the $R_{DS(on)}$ of the power MOSFET. An empirically determined expression [16] gives the relationship between resistivity and neutron fluence:

$$\rho_n = \rho_0 \exp\left(\frac{\Phi}{444 n_0^{0.77}}\right) \tag{19.5}$$

where ρ_n is the resistivity after exposure, ρ_0 is the resistivity before radiation, Φ is the neutron fluence, and n_0 is the starting dopant density. The increase of $R_{DS(on)}$ depends on the resistivity of the silicon, and therefore on the MOSFET voltage rating [17]. For 100-V MOSFETS the increase is small, but at 400-V it is significant (Figure 19.14). The increase of ON resistance will result in increased conduction losses in switching applications but will have little effect in linear applications. In switching applications allowance may be made for the increased $R_{DS(on)}$ by selecting a device with an appropriately low initial value of $R_{DS(on)}$. Alternatively, the increase in $R_{DS(on)}$ can be accepted and allowance made for the increased power dissipation by use of an appropriate-sized heatsink. Neutron

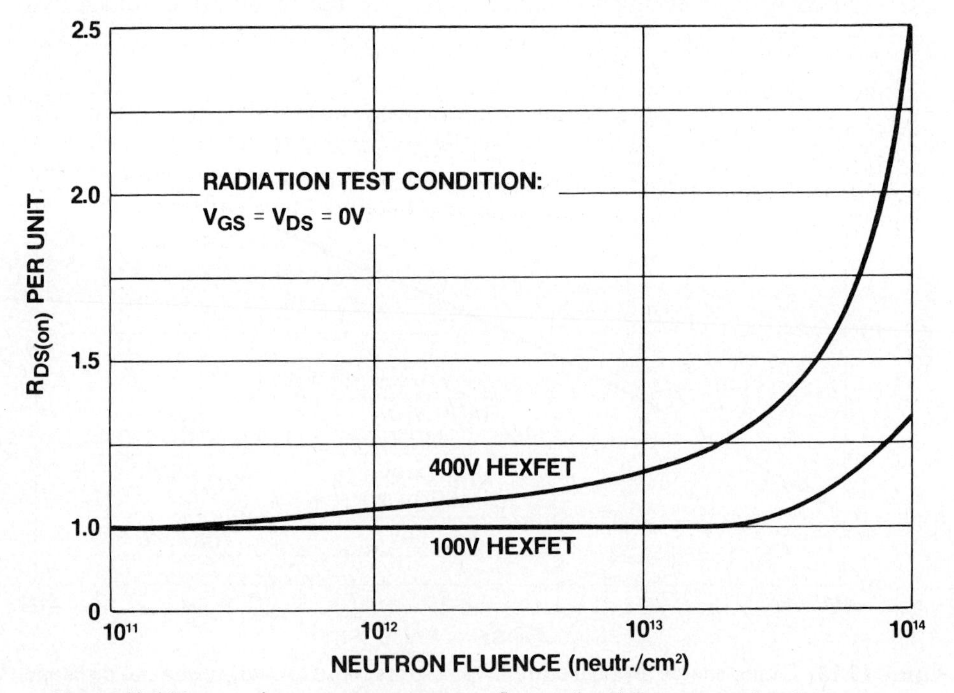

Figure 19.14. ON resistance versus neutron fluence level for conventional MOSFETS.

radiation also has a major effect on the forward voltage drop of the IGBT (see Section 20.2). This is due more to the effect of the radiation on minority-carrier lifetime than to an increase in the resistivity of the silicon [18].

19.2.7. Radiation Countermeasures

In order to construct electronic circuits capable of operating reliably in hostile radiation environments, strategies must be adopted to protect the semiconductor devices. These strategies fall into three broad categories:

1. System shielding.
2. Radiation-tolerant circuit topologies.
3. Radiation-hardened semiconductors.

In system shielding a suitable material is interposed between the radiation source and the electronics. This material is chosen to deplete the energy of the incident radiation by collisions with the material atoms or absorption in nuclear reactions [19]. The degree to which this is used depends on the application. In stationary land-based applications, such as in nuclear power plants, lead shielding is practical but in space applications weight restrictions severely limit the amount of shielding that is possible. Similarly, more shielding is possible in land vehicles and ships than in aircraft.

19.2.8. Circuit Designs to Protect MOSFETs against the Effects of Radiation

Radiation-resistant circuit topologies are ones in which there is reduced sensitivity of the circuit to abnormal behavior of the semiconductor components, the objective being to prevent permanent damage to any circuit component. For example, circuits may be designed so that passive components limit the current which flows between the power rails if all semiconductor components lose their blocking capability [20]. Conventional buck converter and boost converter circuits are unsuitable in this respect, as Figure 19.15 shows. Push–pull and resonant

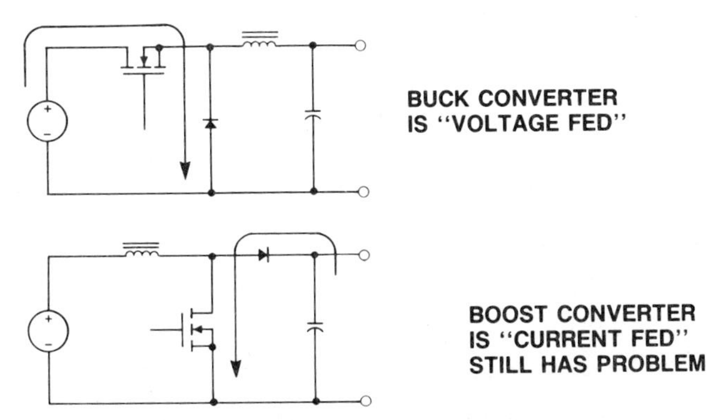

Figure 19.15. Conventional converter circuits experience short circuits between the power rails when semiconductors lose their blocking capability due to gamma radiation.

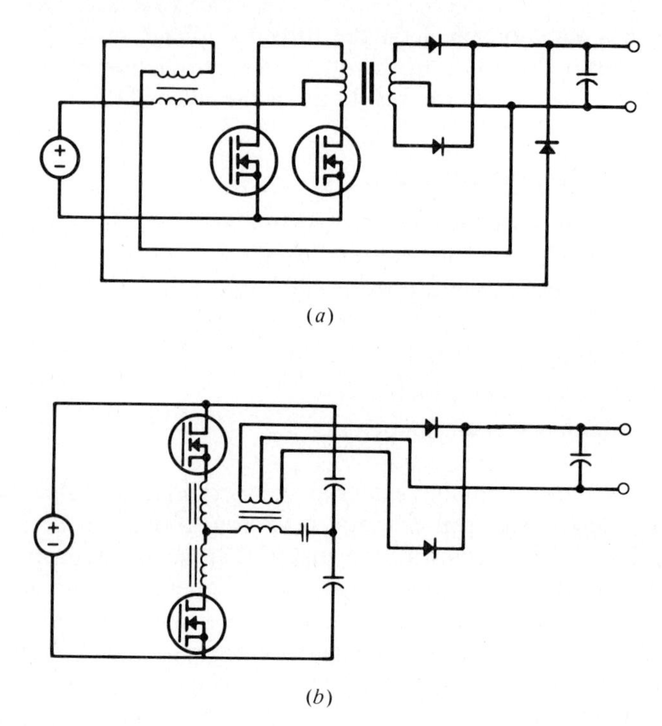

Figure 19.16. Power converter circuits capable of surviving prompt gamma pulses. (*a*) Weinberg converter; (*b*) resonant converter.

circuits, such as those shown in Figure 19.16, rely on inductance to absorb power-rail voltage for the brief period during a gamma-ray pulse when all semiconductor devices are likely to be conducting. Alternatively a radiation detector, such as a p–i–n diode, can initiate shutdown of the equipment with crowbarring of supply rails, as illustrated in Figure 19.17. As well as crowbarring, all FET channels can be turned on to divert current away from the parasitic bipolar transistor, thereby reducing the chance of second breakdown.

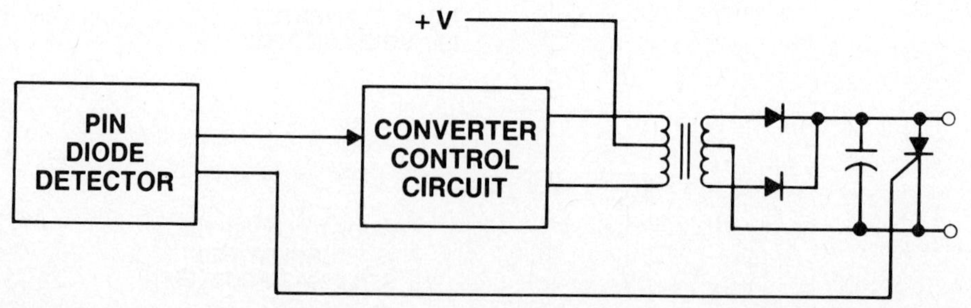

Figure 19.17. Circuit protection by radiation detection and shutdown. Circuits can be protected against damage if the gamma radiation is detected by a p-i-n diode which shuts down the circuit and triggers the crowbar.

19.2.9. Radiation-Resistant Power MOSFETS

Radiation hardened ("rad-hard") power MOSFETS have been successfully produced and continue to be a focus of intense effort. Manufacturers are as yet reluctant to discuss the methods used to obtain resistance to radiation, in part because of the obvious military applications for these devices, although some work has been reported [21, 22]. It is clear that the gate-oxide structure traditionally employed in power MOSFETS requires modification to reduce the threshold shift resulting from ionizing radiation [23]. There can be no doubt that presently available rad-hard MOSFETS are capable of withstanding large doses of radiation without suffering disabling loss of tranconductance or a major shift in threshold voltage.

Assessing the quality of radiation-resistant devices poses some difficulties, since the device is altered by radiation and radiation sources are not always readily available. Nondestructive testing of some characteristics is possible [24]. Where verification of product quality involves irreversible changes in the device tested, the product has to be sampled. The number of samples required to obtain statistically valid results may be reduced by testing the samples beyond normal limits [25]. Bias conditions applied to the device in the minutes after exposure to radiation influence the degree of permanent threshold shift and therefore may affect the accuracy of the measurements [26].

19.3. LOW-THRESHOLD-VOLTAGE MOSFETs

In an attempt to make power MOSFETS more compatible with 5-V logic, some manufacturers have introduced MOSFETS with a threshold voltage that is typically half that of standard ones. They can therefore be driven directly from TTL or other 5-V logic families without the need for level shifting. The reduction in threshold voltage is usually achieved by a reduction in the thickness of the gate oxide, typically from 1000 to 500 Å [27]. Since the gate oxide is thinner, less voltage is required to produce the same electric field in the silicon under the gate and the same inversion depth. Halving the thickness of the gate oxide halves the threshold voltage and doubles the transconductance of the MOSFET. The threshold voltage may also be reduced by a reduction of the doping density of the body region [28], but considerations such as the need to avoid punchthrough restrict the adjustment to the threshold voltage that can be made by this means.

Since low threshold voltage and high transconductance are generally desirable features, it is worth considering why the oxide thickness in ordinary power MOSFETS is 1000 Å and not 500 Å. The primary reason is that it becomes more difficult to grow an oxide layer of the required integrity as the oxide thickness decreases. Defects and variation in thickness become a more serious problem as the oxide thickness approaches the dimension of the defect. Furthermore, hot-electron effects, ion migration, and a lower activation energy for gate-failure phenomena all make it more difficult to maintain an acceptable lifetime for the gate. However, despite these difficulties, power MOSFETS with a threshold voltage in the range of 1 to 2 V can be made and are widely available.

Halving the gate oxide thickness increases gate-source and drain–source capacitances. An increase in these capacitances will tend to increase switching times, while the lowering of the threshold voltage has the opposite effect. For most of the turn-on period when the drain voltage is falling, the drain region is separated from the gate by the parasitic JFET between cells (see Section 3.6.2). However, the transconductance of the device is higher, giving an enhanced Miller effect. There are thus a number of opposing influences on the switching time. A direct comparison of the switching speed of low-threshold MOSFETS with that of ordinary MOSFETS is difficult to make, since the gate-drive conditions are necessarily different for the two types of device. If the gate drive is assumed to be a voltage generator in series with a resistor, the value of the voltage generator should clearly be halved for the low-threshold device (for a device with half the threshold voltage), while the series resistor may either be kept the same, be halved to give the same initial value of gate current, or be reduced by a factor of four to give the same initial gate power. If one maintains the same initial gate power, the switching time of the low-threshold MOSFET is almost half that of an equivalent conventional device [29].

Applications for low-threshold power MOSFETS include interfacing with 5-V logic circuits and use in battery-operated equipment where insufficient voltage may be available to fully turn on normal-threshold voltage devices. However, in many interface applications, where the load is an LED, a lamp, a relay, or a solenoid, the fast switching capability of the power MOSFET is not of particular benefit, and the choice between the power MOSFET and the bipolar Darlington transistor in this application tends to hinge on cost. It should not be assumed that full enhancement of the logic-level MOSFET is achievable when driven from any 5-V logic. For example, the guaranteed minimum high-level output voltage of low-power Schottky logic (2.4 V) is inadequate for full enhancement of most logic-level power MOSFETS [30].

REFERENCES

1. N. Zommer and J. Biran, "Power current mirror devices and their applications." *Proc. Power Convers. Int. Conf.* June., pp. 275–283 (1986).
2. C. E. Cordonnier and D. L. Zaremba, "Application considerations for SenseFET power MOSFET devices." *Proc. Power Convers. Int. Conf.* Munich, May, pp. 47–65 (1987).
3. D. A. Grant and R. Pearce, "The dynamic performance of current sensing power MOSFETS." *Electron. Lett.* **24**(18), 1129–1131 (1988).
4. D. L. Zaremba, Jr., "Current sense power MOSFETS." *Power Electron. Conf.* pp. 1–7 (1988).
5. G. C. Messenger and M. S. Ash, *The Effects of Radiation on Electronic Systems.* Van Nostrand-Reinhold, New York, 1986.
6. J. F. Kircher and R. E. Brown, *Effects of Radiation on Materials and Components.* Reinhold, New York, 1964.
7. U. S. Defense Dept., *The Effects of Nuclear Weapons.* Dept. of Energy, U.S. Gov. Printing Office, Washington, DC.

8. N. Rudie, *Principles and Techniques of Radiation Hardening*. Western Periodical Co., Hollywood, CA, 1976.
9. H. Volmerange and A. Witteles, "Radiation effects on MOSPOWER transistors." *IEEE Trans. Nucl. Sci.* **NS-29**(6), 1565–1568 (1982).
10. W. R. Dawes, Jr., T. A. Fischer, C. C.-C. Huang, W. J. Meyer, C. S. Smith, R. A. Blanchard, and T. J. Fortier, "Transient Hardened Power MOSFETS." *IEEE Trans. Nucl. Sci.* **NS-33**(6), 1425–1427 (1986).
11. A. E. Waskiewicz, J. W. Groninger, V. H. Straham, and D. M. Long, "Burnout of power MOS transistors with heavy ions of californium-252." *IEEE Trans. Nucl. Sci.* **NS-33**(6), 1710–1713 (1986).
12. D. L. Blackburn, D. W. Berning, J. M. Benedetto, and K. F. Galloway, "Ionizing radiation effects on power MOSFETS during high-speed switching." *IEEE Trans. Nucl. Sci.* **NS-29**(6), 1555–1558 (1982).
13. S. S. Seehra and W. J. Slusark, "The effect of operating conditions on the radiation resistance of VMOS power FETS." *IEEE Trans. Nucl. Sci.* **NS-29**(6), 1559–1564 (1982).
14. R. D. Pugh, A. H. Johnston, and K. F. Galloway, "Characteristics of the breakdown voltage of power MOSFETS after total dose irradiation." *IEEE Trans. Nucl. Sci.* **NS-33**(6), 1460–1464 (1986).
15. D. L. Blackburn, J. M. Benedetto, and K. F. Galloway, "The effect of ionizing radiation on the breakdown voltage of power MOSFETS." *IEEE Trans. Nucl. Sci.* **NS-30**(6), 4116–4121 (1983).
16. M. G. Buehler, "Design curves for predicting fast-neutron-induced resistivity changes in silicon." *Proc. IEEE* **56**(10), 1741–1743 (1968).
17. D. Blackburn, T. C. Robbins, and K. F. Galloway, "VDMOS power transistor drain-source radiation dependence." *IEEE Trans. Nucl. Sci.* **28**(6), 4354 (1981).
18. A. R. Hefner, D. L. Blackburn, and K. F. Galloway, "The effect of neutrons on the characteristics of the insulated gate bipolar transistor (IGBT)." *IEEE Trans. Nucl. Sci.* **NS-33**(6), 1428–1434 (1986).
19. A. Holmes-Siedle and R. F. A. Freeman, "Improving radiation tolerance in space-borne electronics." *IEEE Trans. Nucl. Sci.* **NS-24**(6), 2259–2265 (1977).
20. W. E. Abare and W. K. Martindale, "Dose rate tolerant HEXFET power supply." *IEEE Trans. Nucl. Sci.* **NS-28**(6), 4380–4383 (1981).
21. G. B. Roper and R. Lowis, "Development of a radiation hard N-channel power MOSFET." *IEEE Trans. Nucl. Sci.* **NS-30**(6), 4110–4115 (1983).
22. J. M. McGarrity, "Considerations for hardening MOS devices and circuits for low radiation doses." *IEEE Trans. Nucl. Sci.* **NS-27**(6), 1739–1744 (1980).
23. G. Singh, K. F. Galloway, and T. J. Russell, "Radiation-induced interface traps in power MOSFETS." *IEEE Trans. Nucl. Sci.* **NS-33**(6), 1454–1459 (1986).
24. H. E. Boesch, Jr. and J. M. McGarrity, "An electrical technique to measure the radiation susceptibility of MOS gate insulators." *IEEE Trans. Nucl. Sci.* **NS-26**(6), 4814–4818 (1979).
25. A. I. Namenson, "Hardness assurance and overtesting." *IEEE Trans. Nucl. Sci.* **NS-29**(6), 1821–1826 (1982).
26. P. V. Dressendorfer, J. M. Soden, J. J. Harrington, and T. V. Nordstrom, "The effects of test conditions on MOS radiation hardness results." *IEEE Trans. Nucl. Sci.* **NS-28**(6), 4281–4287 (1981).
27. C. F. Wheatley and H. R. Ronan, "Switching waveforms of the L^2FET: A 5-volt gate-drive power MOSFET." *IEEE Power Electronics Spec. Conf.* pp. 238–246 (1984).

28. M. Sciammas and D. Kapur, "Low threshold power FETS: A sensible solution." *Powertechnics* June, pp. 28–31 (1986).

29. L. Marechal, "The logic level FET at switching." *Proc. Power Convers. Inter. Conf.* pp. 197–208 (1984).

30. J. E. Wojslawowicz., "L^2FETS challenge bipolar power devices." *Power Convers. Intell. Motion* (*PCIM*) January, pp. 27–30 (1988).

CHAPTER 20

Other MOS Power Devices

20.1. INTRODUCTION

The power MOSFET has spawned a variety of power semiconductor devices in which the current is carried by many cells in parallel. The techniques proven to be effective in the power MOSFET have been applied in other power devices, even where a fine cellular structure is not essential. For example, designers of bipolar transistors have adopted the cellular structure as a means of obtaining better SOA characteristics [1]. However, the bipolar transistor is not a field-effect device and therefore is not dealt with here.

The essential principles employed in the power MOSFET are:

- Current divided between many cells
- Current flow controlled by a low-voltage field-effect switch
- High voltage held off by pinch effect between cells
- Vertical current flow through the die (even if the cell structure is lateral)

Devices based on these principles include the insulated-gate bipolar transistor (IGBT), the MOS-controlled thyristor (MCT), the static induction transistor (SIT), and the static induction thyristor (SITH). The SIT (also called field-controlled diode or gridistor) is a junction field-effect transistor and therefore does not come within the scope of a book on MOS devices. Similarly, the SITH (also called field-controlled thyristor), a four-layer latching version of the SIT, is not discussed. However, the IGBT and the MCT are MOS cellular devices and therefore have a place in a book on power MOS devices. A comparison of their characteristics shows that there are some areas of application where these devices may be more appropriate than a power MOSFET [2, 3].

20.2. THE IGBT

20.2.1. Principle of Operation

The insulated-gate bipolar transistor developed from the power MOSFET. Devices of the IGBT type have been given various trade names such as IGT, COMFET (conductivity-modulated FET), and GEMFET (gain-enhanced MOSFET), but we refer to it here by the JEDEC preferred term IGBT. The basic structure of the n-channel IGBT is shown in Figure 20.1. A power MOSFET is formed in an n-type

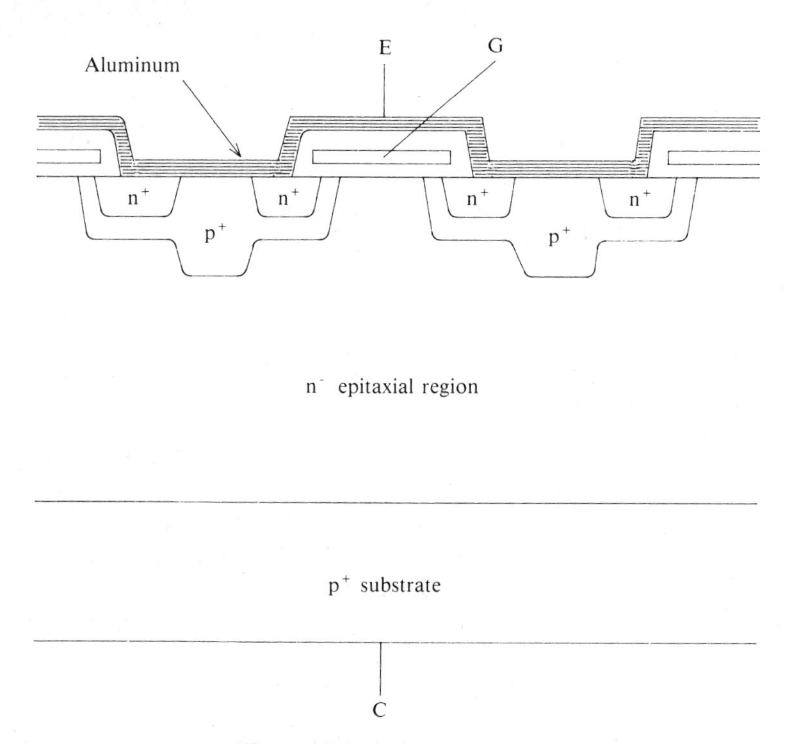

Figure 20.1. IGBT construction.

epitaxial layer in the conventional manner. However, the substrate on which the epitaxial layer has been grown is p-type rather than n-type [4, 5].

Current flow across the junction between the n-type epitaxial layer and the p-type substrate results in minority-carrier injection into the epitaxial layer. This results in conductivity modulation of the epitaxial layer, reducing its resistance. This permits operating current densities up to five times that of equivalent bipolar transistors and twenty times that of power MOSFETs to be achieved [6]. The IGBT symbol (see Figure 20.5 below) reflects both the MOS gate input and the bipolar transistor action resulting from minority-carrier injection. Like the MOSFET and the bipolar transistor, the IGBT may be viewed as a charge-control device [7] (see Section 16.1.2).

A variant of the IGBT structure has been proposed in which the p-type substrate is partly replaced by a Schottky rectifier junction. The potential benefit of this arrangement is an improved tradeoff between switching speed and forward voltage drop [8]. A considerable reduction in IGBT forward voltage drop has also been demonstrated when the traditional surface DMOS-type gate structure is replaced by vertical trenches [9]. Silicon-wafer direct bonding is another technology that appears to be appropriate to IGBT manufacture in that it facilitates the fabrication of buffer layers for optimizing switching performance [10, 11].

20.2.2. Characteristics

Conductivity modulation of the epitaxial region is particularly useful in high-voltage devices in which the broad epitaxial n-base region would otherwise

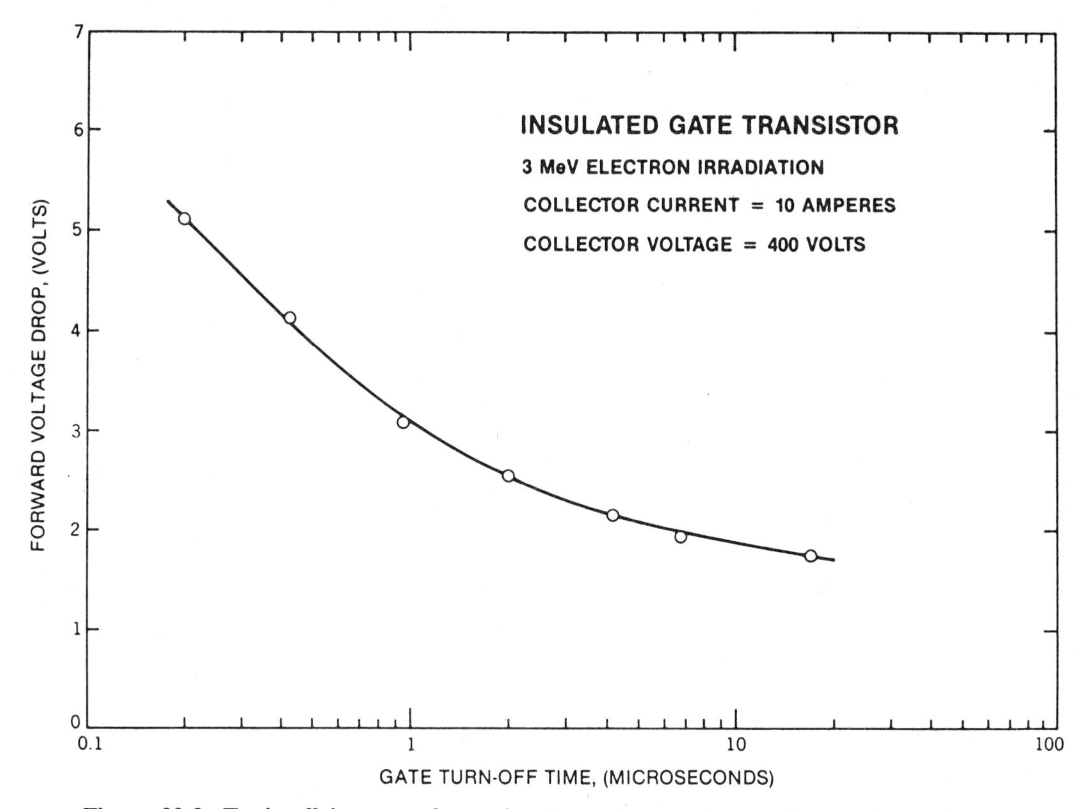

Figure 20.2. Trade-off between forward voltage drop and turn-off time for n-channel IGBTs. From Baliga [12]. © IEEE 1984.

present a high resistance to current flow. However, there is a penalty in the form of stored minority-carrier charge, the effect of which is to introduce delay in the turnoff process. The operating frequency of the IGBT is thereby restricted. If the minority-carrier lifetime is reduced, for example by electron irradiation, the turnoff time can be reduced at the expense of increased forward voltage drop, as Figure 20.2 shows [12]. With little lifetime killing, IGBTs are limited to low-frequency applications such as line-supplied phase-control circuits. With heavier radiation doses, operation up to 10 kHz has proved feasible, and it is expected that the IGBT will become an important device at 20 kHz and above.

Parallel operation of IGBTs is possible, although paralleling requires the precautions that are normally taken both with MOSFETs and with bipolar transistors, such as series gate resistors and emitter resistor ballasting [13].

Except in the case of asymmetric devices (see Section 20.2.5), the presence of a diode junction between the substrate and the epitaxial region allows the device to block reverse voltage, a useful feature in some applications. However, the field termination at the edge of the die also needs to be designed to block reverse voltage. IGBTs are usually made by the planar process on square dies which are scribed from the wafer. If high reverse blocking capability is required, the junction between the p^+ substrate and the n^- epitaxial region requires passivation.

A useful feature of IGBTS is that n- and p-channel devices of a given die area and a given voltage rating have similar forward voltage drops and therefore a similar current-carrying capability [14, 15]. This is in contrast to power MOSFETS, in which the p-channel device has a considerably greater $R_{DS(on)}$ than an n-channel device of the same size and voltage rating. The conductivity modulation present in IGBTS masks the difference in resistivity of the epitaxial regions of the p and n devices. The availability of complementary devices facilitates the design of high-voltage bridge circuits, since both devices can operate with the emitter connected to a supply rail, while the gate drive voltage is derived from the other supply rail.

The transconductance of the IGBT is typically twice that of the equivalent power MOSFET. This is due to the gain of the pnp transistor composed of the p-type substrate, the n-type epitaxial region, and the p-type MOSFET body region. A convenient model for understanding IGBT behavior is shown in Figure 20.3. In this model the MOSFET current constitutes the base current of the pnp transistor. It has become common practice to name the terminals of the IGBT by analogy with the bipolar transistor, so that the substrate terminal is called the collector and the source region of the MOSFET is called the emitter. Clearly this conflicts with the model in which only the MOSFET and the pnp transistor are shown. Another school of thought calls the substrate the anode and the source the cathode, by analogy with the thyristor, since the IGBT is effectively a four-layer device.

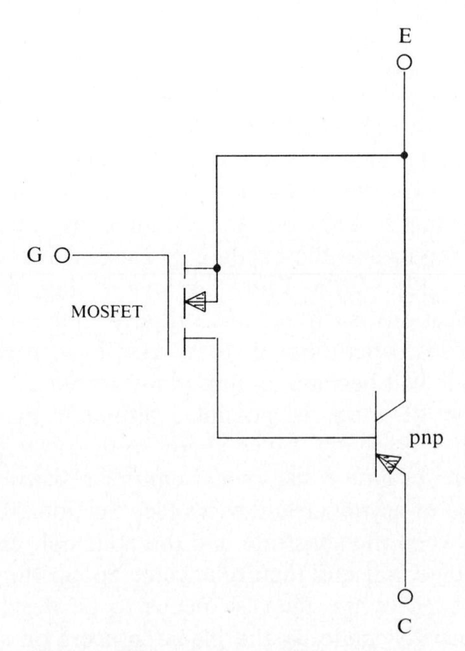

Figure 20.3. IGBT equivalent circuit.

20.2.3. Latchup

The IGBT may be viewed as a four-layer device and therefore, as the equivalent circuit (Figure 20.4) shows, embodies a pair of regeneratively connected npn and pnp transistors. The device is therefore capable of thyristor action, and one of the major objectives of IGBT design is to prevent latchup and consequent loss of gate control. Latchup results when the base–emitter junction of the npn transistor of the equivalent circuit (the junction between body and source of the MOSFET structure) becomes forward-biased. Forward bias of this junction will occur when the voltage drop produced by lateral current flow in the p-type body-channel region exceeds approximately 0.7 V. In practice latching is produced when a certain value of load current is exceeded, since some of the injected hole current will flow laterally in this region. Latchup is detectable by a fall in the ON-state forward voltage drop to below the normal level.

The value of load current at which latchup occurs depends on the junction temperature. At higher temperatures the sheet resistance of the body region is elevated while the V_{BE} of the npn transistor section is reduced. Both these effects make it easier for the base–emitter junction to become forward-biased, so that latchup occurs at a lower value of load current. Latchup in IGBTS also occurs more readily during turnoff, so that the dynamic latchup current is less than the static latchup current. The faster the device turns off, the lower the value of the dynamic latchup current. Therefore asymmetric gate driver circuits like that shown in Figure 20.5 are required to slow turnoff without compromising turn-on speed [16, 17].

Much effort has been concentrated on improving the characteristics of the IGBT

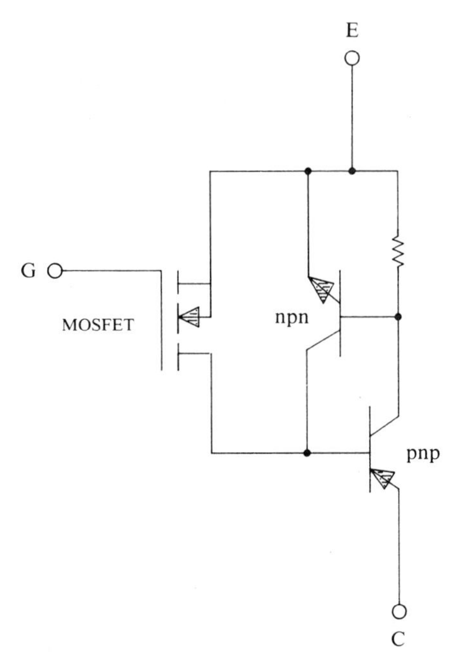

Figure 20.4. IGBT equivalent circuit including parasitic thyristor structure.

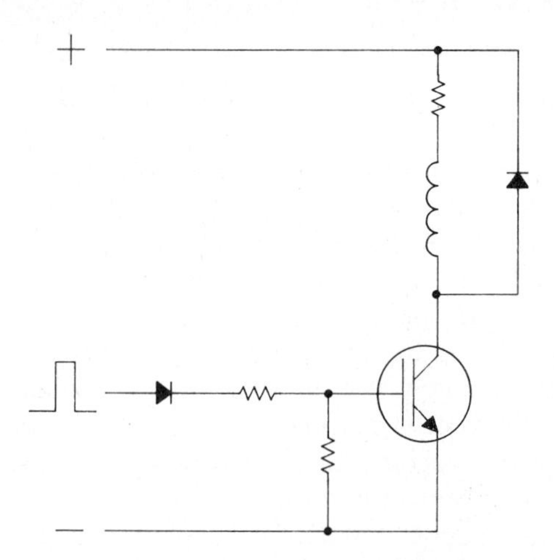

Figure 20.5. Polarized IGBT gate drive to prevent latchup on turnoff.

to the extent that the latching current may lie outside the SOA of the device and therefore in theory latching will never be encountered [18, 19]. One design incorporates sufficient source resistance to have a ballasting effect, so that under high-current conditions the gain of the npn transistor section is reduced sufficiently to prevent latching [20]. (In the MOS SCR, which has an almost identical structure to the IGBT, latching is allowed to occur at a low value of the anode current, and turnoff can only be achieved by reversing the load current [21]. A triac with MOS gate turn-on has also been produced [22].)

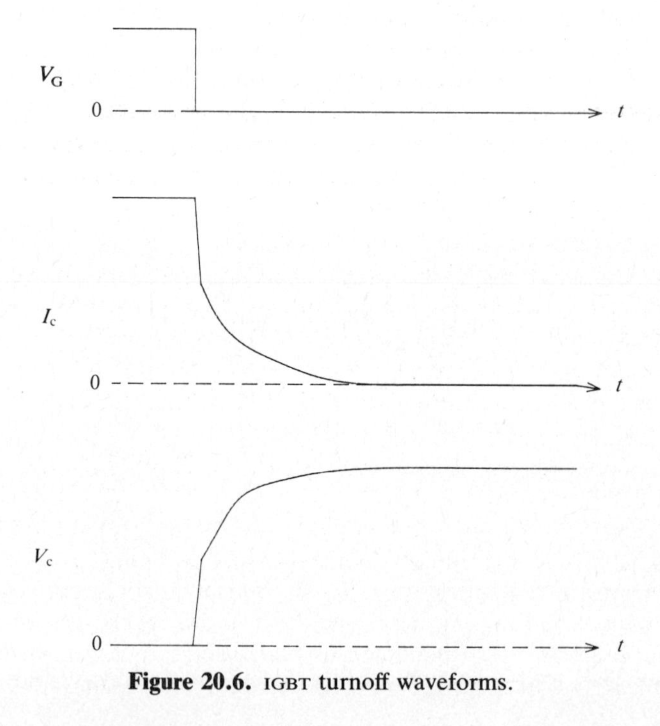

Figure 20.6. IGBT turnoff waveforms.

20.2.4. Switching Waveforms

The waveforms associated with turnoff of the IGBT are shown in Figure 20.6. The most salient feature is the rapid fall in load current to an intermediate value, followed by the gradual decay of the remaining current. The initial rapid fall is caused by the turnoff of the MOSFET current. As the gate voltage drops below the threshold voltage, electron flow through the channel ceases. The pnp base current in the model (Figure 20.4) therefore ceases. However, the high concentration of minority carriers present in the n^- region continues to support hole current. The pnp transistor continues to conduct until the charge in its base region recombines or is swept out.

20.2.5. Reducing the Turnoff Time

A power device becomes more useful when it can operate efficiently at a frequency above the range of human hearing—typically taken as 20 kHz. To achieve this with an IGBT the storage and turnoff times must be reduced below the values which would result from the normal MOSFET process. This is commonly achieved in two ways. The minority-carrier lifetime can be reduced by increasing the concentration of recombination centers in the epitaxial silicon layer. Heavy-metal doping is not favored in this type of device, because of the risk of gate-oxide contamination. Electron, proton, or neutron irradiation [23] is the preferred method, since damage to the oxide which the radiation produces can largely be annealed out by heating the die. Lifetime killing results in a faster device, but also increases the forward voltage drop. The static latching current is increased, since the irradiation reduces the gain of the npn and pnp sections, but their dynamic latching current may be reduced [6].

A second method of reducing the storage time involves the introduction of an n^+ epitaxial layer beneath the n^- epitaxial layer, as shown in Figure 20.7. By adjusting the dopant concentration of this layer and its thickness the turnoff time can be reduced [24, 25]. The presence of the n^+ layer also reduces the emitter injection efficiency of the p^+ substrate. This reduces the loop gain of the npn-and-pnp transistor pair, thereby raising the value of the latching current [26–28]. At the same time the reverse blocking capability of this junction is reduced, but since the field termination is not usually able to support a high reverse voltage, this is not a significant loss. The presence of the n^+ buffer region allows the doping of the n^- region to be reduced, since the heavily doped n^+ region will act as a stopper to the depletion region, which would otherwise reach through to the substrate. The more uniform electric field which this produces in the n^- layer allows the width of this region to be reduced. However, this does not necessarily mean a reduction in forward voltage drop, in view of the reduced injection efficiency of the substrate–epilayer junction, although an improvement in switching speed can be obtained [29]. Much research interest has been focused on the reduction of the turnoff time by the use of multiple epitaxial layers with differing lifetimes and doping densities [30–33] and by the use of collector and emitter shorts [34–36]. It is therefore to be expected that IGBT performance will continue to improve as these technologies make their way into the market.

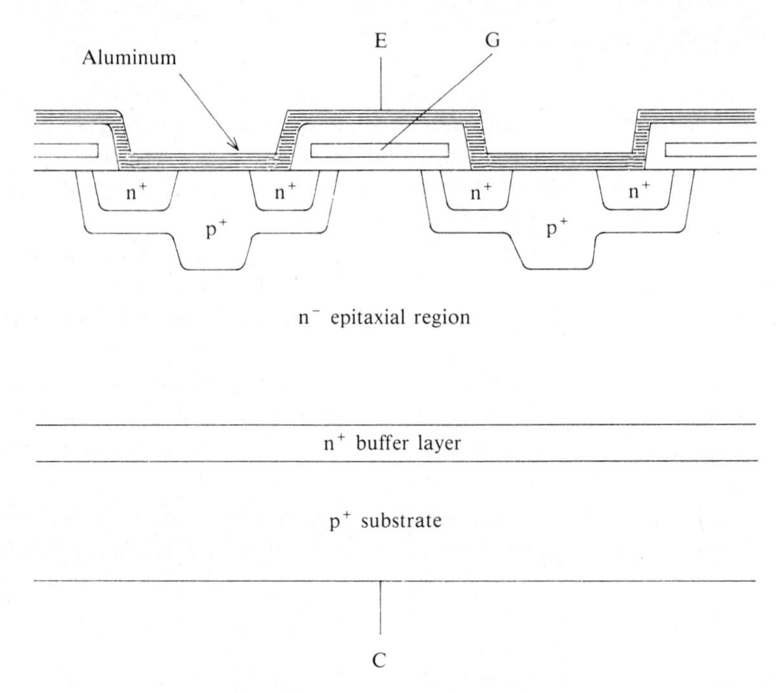

Figure 20.7. IGBT with buffer layer.

20.2.6. Applications

In high-voltage applications the low forward drop of the IGBT compared with that of the MOSFET makes the IGBT a candidate for off-line applications where the switching frequency is below 10 kHz. The principal use for the IGBT is therefore likely to be in motor drives operating from 240 V ac (or 120 V ac with a voltage-doubler input stage) where 1-kHz switching is required for six-step operation or where the PWM frequency is of the order of a few kilohertz [37]. The faster IGBTS may also be used in 20-kHz "silent" PWM inverters [38, 39] and switched-mode power supplies [40, 41].

The IGBT has also been suggested as a suitable device for use in inductive-discharge automobile electronic ignition systems [42]. In this application the switching device is required to operate at a relatively low frequency, to block high voltage, and to have a low forward voltage drop. While the forward voltage drop may not be as low as that obtainable with a bipolar transistor, the SOA may allow active clamping (see Section 9.3) to be used to absorb stored energy in the coil when the secondary is open circuit.

20.3. MOS-CONTROLLED THYRISTOR (MCT)

20.3.1. Principle of Operation

The MCT [43, 44], also known as the MOS-GTO [45], is a four-layer latching device, the turnoff and turn-on of which are under the control of MOS gates. The basic

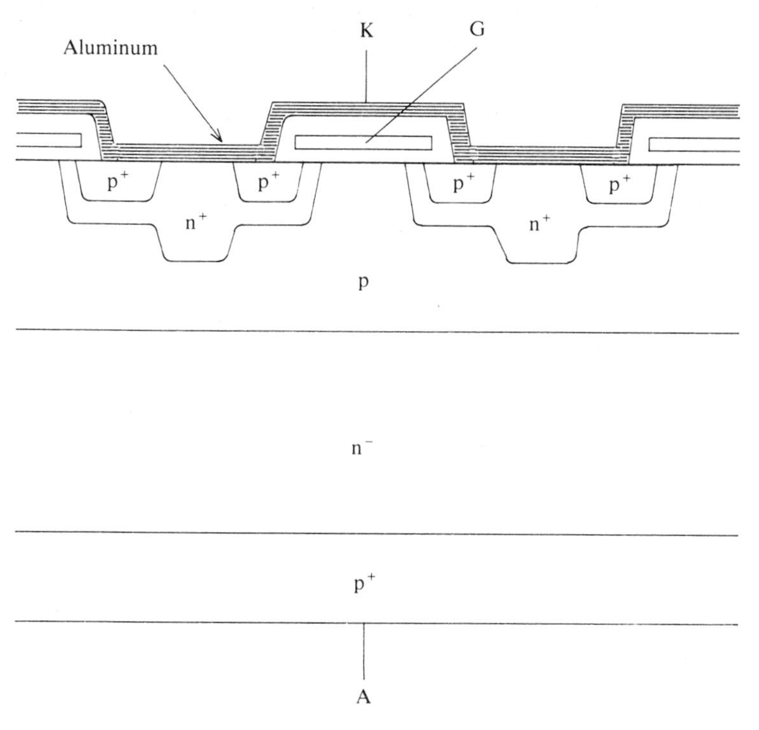

Figure 20.8. MCT structure.

MCT structure is shown in Figure 20.8. The structure resembles that of the IGBT except that a further layer has been added, so that it is inherently a four-layer device and thus merits the name thyristor rather than transistor. An equivalent-circuit model is given in Figure 20.9. The p-base is connected to the n-cathode via a MOSFET channel and the surface metallization. When this channel is inverted, the base–emitter junction of the npn section of the thyristor structure is shorted, thereby suppressing the regenerative switching action of the four layers and turning the device off. When the gate voltage is at the same potential as the cathode, the channel does not conduct and there is no cathode shorting of the thyristor structure. The loop gain of the thyristor npn-and-pnp equivalent transistor pair is such that the device easily latches when the shorting is removed.

20.3.2. Triggering

The stimulus for latching may be provided in several ways. One proposal is for a separate MOS structure at selected sites on the die [45]. In this case conduction has to propagate from these sites across the die in the way normally associated with SCR triggering. Another proposal is to trigger the device with capacitive currents generated by the step in gate voltage at turn-on [46]. To turn on the p-channel MCT shown in Figure 20.8 the gate voltage goes from negative to positive, since the p-channel MOSFET must be turned off. The positive-going leading edge can be used to induce a capacitive current in the lateral p region of sufficient magnitude

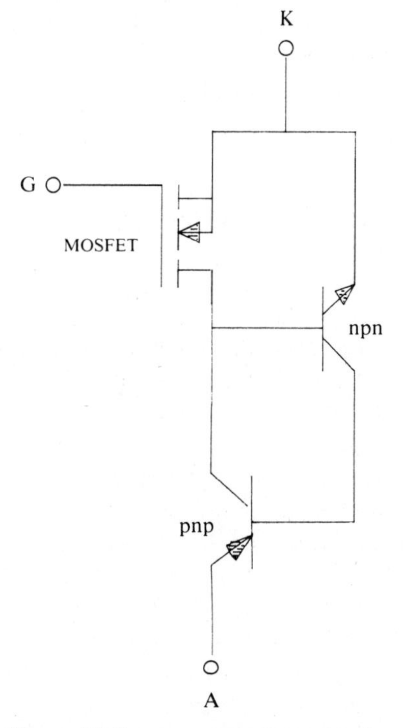

Figure 20.9. MCT equivalent circuit.

to trigger the thyristor. An alternative structure in which the same gate controls turn-on as well as turnoff is shown in Figure 20.10.

Due to regenerative thyristor action, the MCT can be operated at a current density several times higher than is possible with the IGBT. However, it may not be possible to turn the MCT off satisfactorily at such high current densities. Turn-on of the device will be fast due to the regenerative turn-on action, but turn

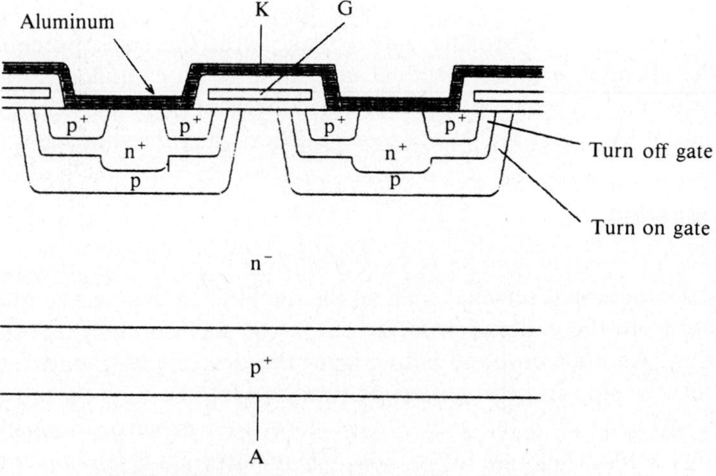

Figure 20.10. MCT structure with gate turn-on.

off is slowed by minority-carrier storage, and lifetime killing may be required to obtain a short storage time and a small current tail. Applications for this device are similar to those for the IGBT, in particular, motor drives [47].

20.4. POWER INTEGRATED CIRCUITS (PICs)

20.4.1. Introduction

The power MOSFET is manufactured using techniques similar to those used to fabricate logic circuits. Optimum MOSFET cell densities appear to be in the range of 1 to 2 million cells per square inch, with feature sizes of a few micrometers. This is similar to the dimensions of the features of early CMOS integrated circuits. It is therefore a logical step to seek to incorporate control functions on the power-MOSFET die, so that the power MOSFET can become more versatile or the cost of expensive external control circuitry can be eliminated and equipment reliability improved. However, this requires low-voltage logic circuits to operate on the same die as a relatively high-voltage power transistor, so that a means of isolating the different structures is central to the construction of a PIC. In addition, to obtain efficient use of the silicon area, the power transistor must operate at a temperature of the order of 150°C, subjecting the logic elements to temperatures and temperature cycling beyond the range normally experienced by integrated circuits.

Power integrated circuits vary in the complexity of the on-board intelligence, and various categories of device may be identified. Simplest of all is the device that is essentially a power transitor with some on-board control functions such as thermal shutdown, overcurrent limiting, shaping of the control signal, and device-status feedback [48–51]. A function commonly required in PICS is a high-side drive circuit which permits an n-channel MOSFET to operate with its drain tied to the positive supply rail [52]. The complexity of the PIC increases when timing and signal-processing functions are incorporated—as, for example, in bridge-type power-supply circuits, which require level shifting for the upper device, timing logic to avoid overlap, PWM waveform generation, and control circuits [53]. At the other extreme, the PIC may be principally a logic circuit with one or more power devices to give the outputs the capability of driving external loads. There is obviously a whole spectrum of complexity between the extremes, and the manufacturing technology employed depends to some extent on the category of device [54].

A market sector which is attractive to the manufacturers of power integrated circuits is the automotive market. As the number of electrically operated accessories on cars has increased, so has the weight, cost, and complexity of the wiring. When a large number of these accessories are operated from door-mounted switches, the problem of accommodating a bundle of wires within the door and at the door hinge may be insurmountable. An attractive solution to this problem is the use of a single-wire distribution system with multiplexed signaling to power integrated circuits which control the electrical loads. A number of designs of such switches have been developed [55–57].

20.4.2. PIC Technology

Some PIC technologies are illustrated in Figure 20.11.

In the self-isolated process shown in Figure 20.11*a* isolation is achieved by surrounding the high-voltage drain region with the source region, which is at low voltage. Because the lateral structure occupies a large silicon area, this type of construction is used for low-power devices [58].

The process shown in Figure 20.11*b* employs dielectric isolation of sections of the circuit. Grooves are etched in the silicon surface, oxidized, and filled with polysilicon. The apexes of the grooves are lapped away to leave isolated islands of epitaxial material in which the power device can be fabricated [59]. Vertical power MOSFETS can be created if the wells are first lined with a heavy n^+-diffusion as shown in Figure 20.11*b*.

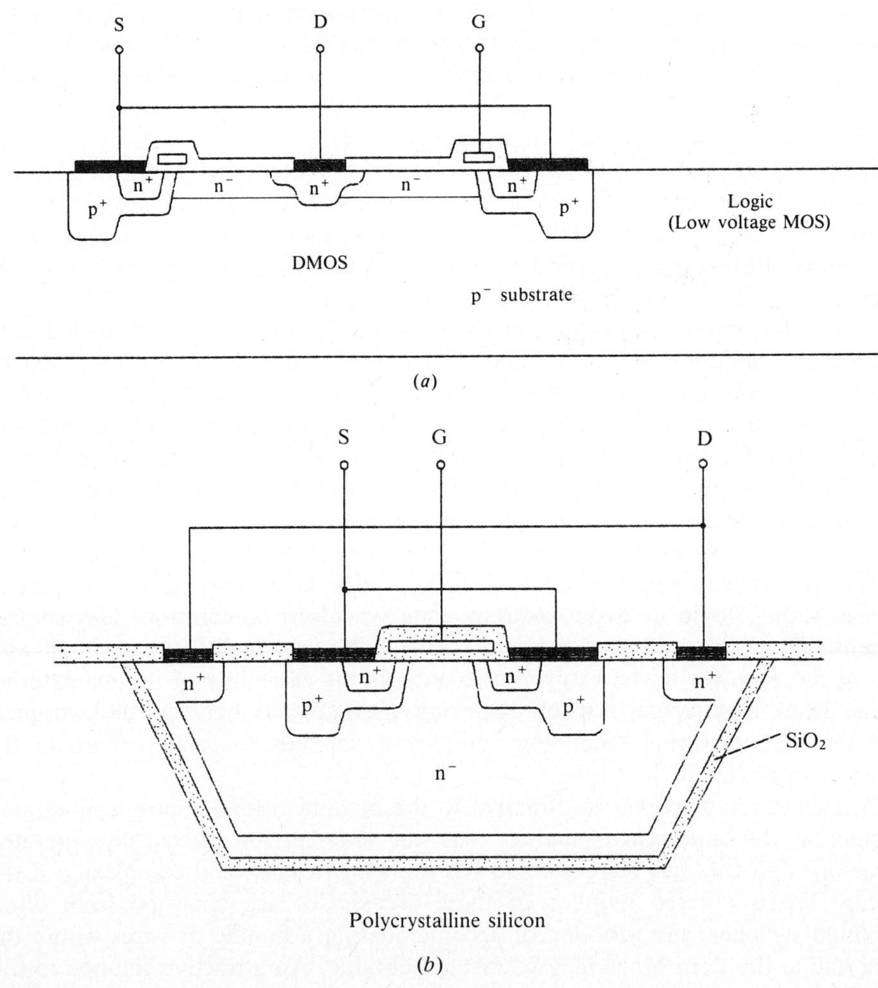

Figure 20.11. Power-integrated-circuit technologies. (*a*) Self-isolated; (*b*) dielectrically isolated; (*c*) junction isolation with lateral power MOSFET; (*d*) junction isolation with vertical power MOSFET.

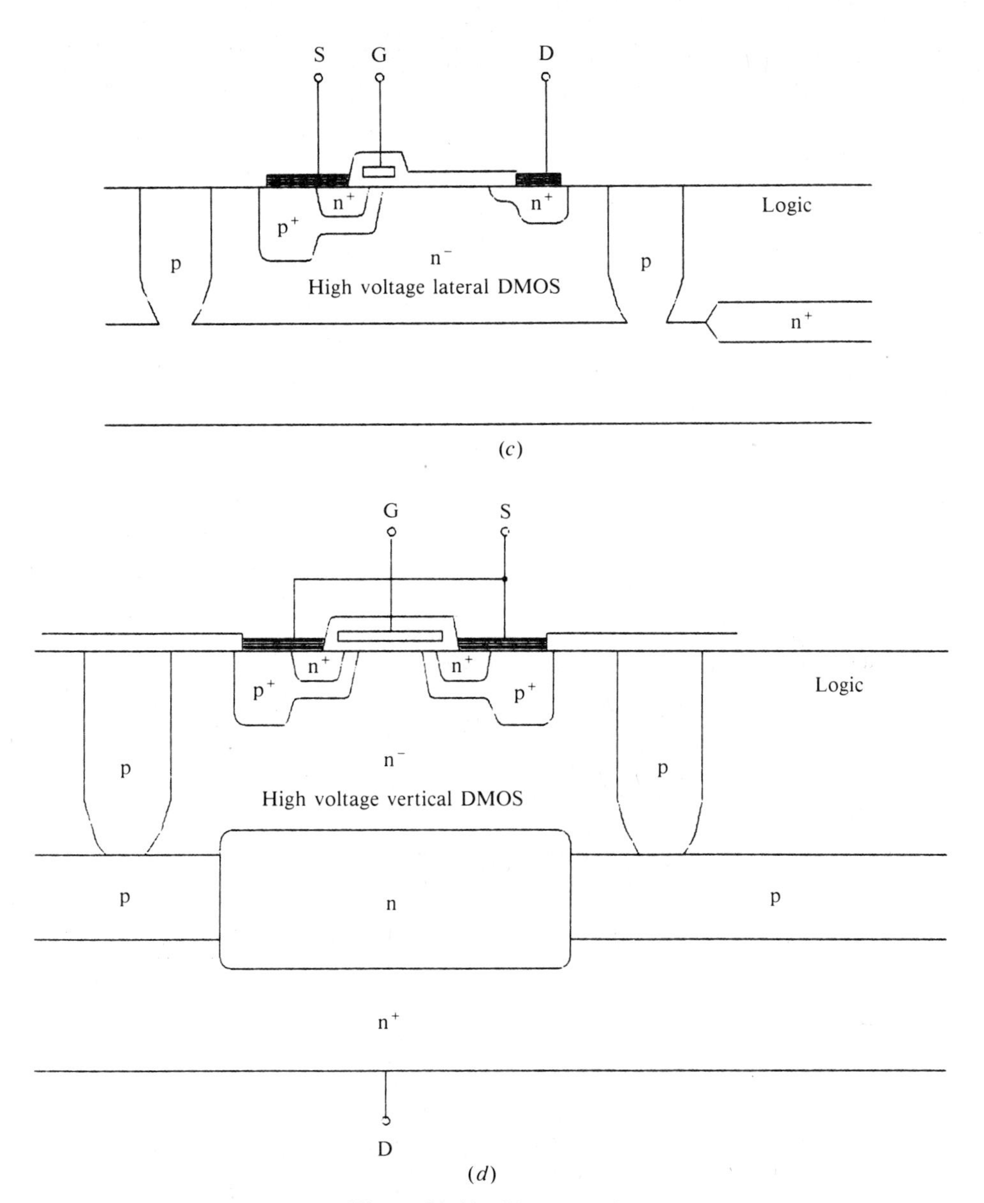

Figure 20.11. (*Continued*)

Both lateral and vertical power MOSFETS are possible with junction isolation, as shown in Figure 20.11*c* and *d*. In this process the high-voltage devices are built in pockets of n-type material isolated from low-voltage regions by wells of p^+ diffusions. Figure 20.11*c* shows a lateral version that allows the construction of more than one power device on a die. Figure 20.11*d* shows a version in which the power MOSFET is a vertical device, allowing higher current densities. The drain current is allowed to flow vertically to the back side of the die through a low-resistivity n^+ region that penetrates the p-type horizontal isolating layer. This type of construction only allows one power device per die.

20.4.3. PIC Example

An example of a power integrated circuit is shown in Figure 20.12. This is a low-power (5-W), 15-V-output switching regulator capable of operating from a wide range of input voltages (100 to 450 V) [60]. The five-pin, junction-isolated device uses a lateral power MOSFET for power control and a mixture of bipolar and MOS logic and interface circuits. The structure includes double polysilicon depositions, high-voltage crossovers, pnp and npn bipolar devices, and lateral DMOS power devices. High-value thin-film resistors are used for level shifting and for processing high-voltage signals [61]. High blocking capability is achieved by

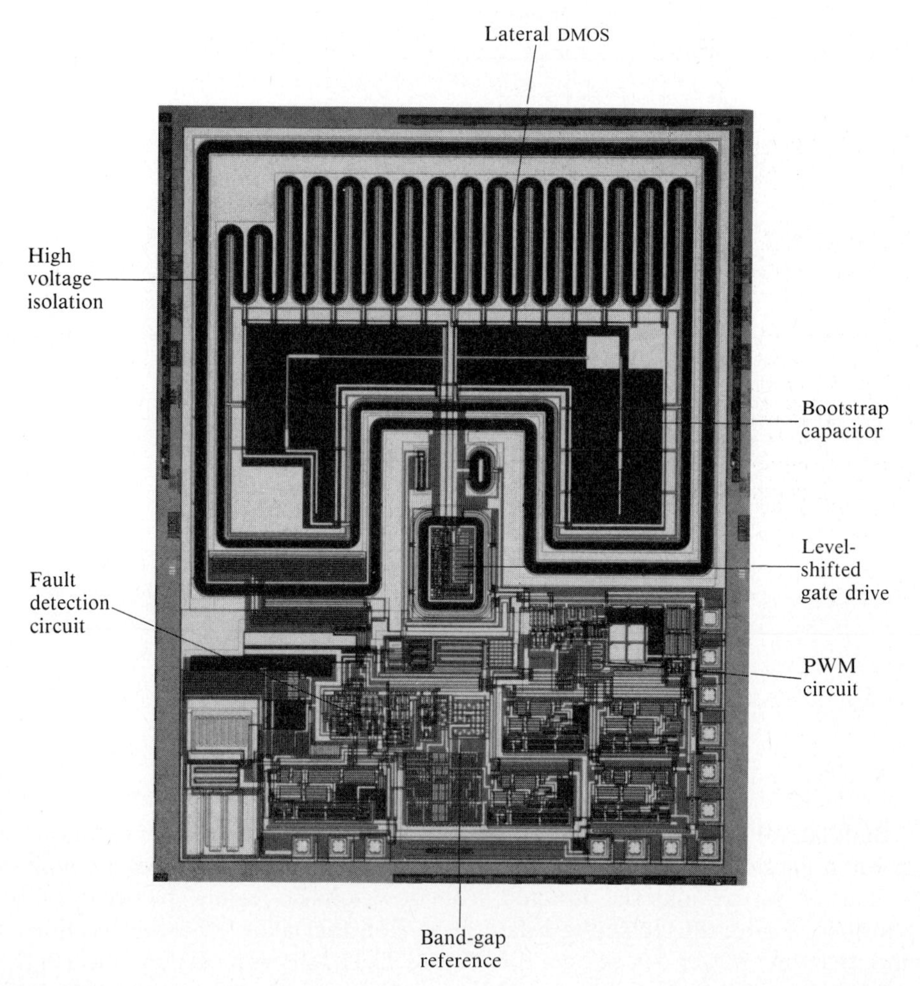

Figure 20.12. Photograph of buck converter PIC die. (Courtesy of International Rectifier Corp.)

two-dimensional charge control (RESURF). Only a diode, an inductor, and a capacitor are required with the PIC to form a complete buck regulator.

Internal power for the chip is supplied from the output, with an auxiliary supply from the input supplying essential functions during startup and under fault conditions. The PIC includes undervoltage lockout and overtemperature shutdown. Full enhancement of the n-channel power MOSFET is obtained with a bootstrap circuit fed from the outpt voltage in order to obtain high efficiency. Switching times are of the order of 30 ns. The device can be used in a wide variety of switched-mode power-supply circuits, both isolated and non-isolated [61], demonstrating that flexibility is not necessarily sacrificed when a power device and control circuit are monolithically integrated.

20.4.4. Basic PICs

The level of complexity required for a device to qualify as a power integrated circuit is open to debate. In a sense every power MOSFET is an integrated circuit in that it contains thousands of integrated transistors, all connected in parallel, although this would clearly be a trivial use of the term. Less trivial examples include devices in which transistors, separate from the main power transistor, perform some function distinct from power switching.

One such example is the power MOSFET with integral gate drive [62] (see Figure 20.13). To achieve maximum efficiency, resonant converters operating at switching frequencies of the order of 10 MHz or greater require fast turnoff of the power switch. An integral gate drive greatly reduces the inductance of the gate capacitance discharge path, thereby making it possible to turn the MOSFET off in a few nanoseconds.

The self-thermal-protecting power MOSFET [63] is another example of a device which could be called a PIC by virtue of incorporating a gate control function. The device utilizes pn junctions built into the polysilicon layer as thermal sensors. At

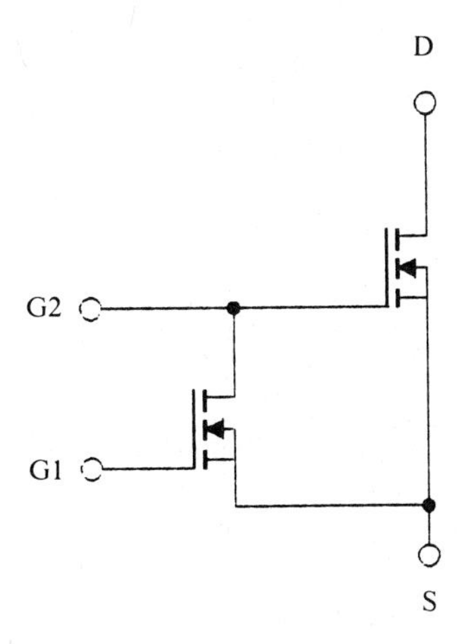

Figure 20.13. Power MOSFET with integrated gate driver.

a critical temperature the forward voltage drop of these diodes falls, activating a bistable latch which turns off the main transistor.

REFERENCES

1. L. Marechal and J. Wojslawowicz, "Bipolar transistor unit cell concept." *Proc. Power Convers. Int. Conf.* May, pp. 134–147 (1987).
2. B. J. Baliga, "Revolutionary innovations in power discrete devices." *IEEE Int. Electron Devices Meet.* pp. 102–105 (1986).
3. E. K. Behr and R. Hanitsch, "Contribution to the comparison of power MOSFET and conductivity enhanced MOSFET." *Eur. Power Electron. Conf.* pp. 99–104 (1987).
4. J. P. Russell, A. M. Goodman, L. A. Goodman, and J. M. Neilson, "The COMFET—a new high conductance MOS-gated device." *IEEE Electron Devices Lett.* **EDL-4**(3), 63–65 (1983).
5. B. J. Baliga, M. S. Adler, R. P. Love, P. V. Gray, and N. D. Zommer, "The insulated gate transistor: A new three-terminal MOS-controlled bipolar power device." *IEEE Trans. Electron Devices* **ED-31**(6), 821–828 (1984).
6. B. J. Baliga, *Modern Power Devices.* Wiley (Interscience), New York 1987.
7. J. G. Fossum and R. J. McDonald, "Charge-control analysis of the COMFET turn-off transient." *IEEE Trans. Electron Devices* **ED-33**(9), 1377–1382 (1986).
8. J. K. O. Sin and C. A. T. Salama, "Hybrid Schottky injection MOS-gated power transistor." *Electron. Lett.* **22**(19), 1003–1005 (1986).
9. H. R. Chang, B. J. Baliga, J. W. Kretchmer, and P. A. Piacente, "Insulated gate bipolar transistor (IGBT) with a trench gate structure," *IEEE, Int. Electron Devices Meet.* pp. 674–677 (1987).
10. H. Ohashi, K. Furukawa, M. Atsuta, A. Nakagawa, and K. Imamura, "Study of Si-wafer directly bonded interface effect on power device characteristics." *IEEE, Int. Electron Devices Meet.* pp. 678–681 (1987).
11. A. Nakagawa, K. Watanabe, Y. Yamaguchi, H. Ohashi, and K. Furukawa, "1800 V bipolar-mode MOSFETS: A first application of Silicon Wafer Direct Bonding (SDB) technique to a power device." *IEEE, Int. Electron Devices Meet.* pp. 122–125 (1987).
12. B. J. Baliga, "Switching speed enhancement in insulated gate transistors by electron irradiation." *IEEE Trans. Electron Devices* **ED-31,** 1790–1795 (1984).
13. S. R. Korn, "Parallel operation of the insulated gate transistor in switching operations." *Proc. Power Convers. Int. Conf.* June, pp. 218–234 (1986).
14. T. Schulz, "P-channel COMFET—the first true complement." *Proc. Power Convers. Int. Conf.* June, pp. 235–246 (1986).
15. M. F. Chang, G. C. Pifer, H. Yilmaz, R. F. Dyer, B. J. Baliga, T. P. Chow, and M. S. Adler, "Comparison of N and P channel IGTS." *IEEE Int. Electron Devices Meet.* pp. 278–281 (1984).
16. H. R. Ronan and C. F. Wheatley, "Circuit influences on COMFET (IGT) dynamic latching current." *IEEE Power Electronic Spec. Conf.* pp. 73–79 (1986).
17. M. W. Smith, "Applications of insulated gate transistors." *Proc. Power Convers. Int. Conf.* April, pp. 121–131 (1984).
18. D. J. MacIntyre, "Motor control applications of second generation IGT power transistors." *Proc. Power Convers. Int. Conf.* October, pp. 206–215 (1985).

19. A. Nakagawa, H. Ohashi, M. Kurata, H. Yamaguchi, and K. Watanabe, "Non-latch-up 1200 V 75 A bipolar-mode MOSFET with large ASO." *IEEE, Int. Electron Devices Meet.* pp. 860–861 (1984).
20. N. Zommer, R. Yu, and G. Chen, "Characteristic and applications of an improved MOS insulated gate transistor that handles 800 V, 50 A." *Proc. Power Convers. Int. Conf.* October, pp. 225–231 (1986).
21. A. Pshaenich, "The MOS SCR, a new thyristor technology." *Motorola Eng. Bull.* **EB-103** (1982).
22. C. J. Heron, "A new triac device utilizing integral FET drive." *Proc. Powercon* **8,** N2.3.1–N2.3.4 (1981).
23. A. M. Goodman, J. P. Russell, L. A. Goodman, C. J. Neuse, and J. M. Neilson, "Improved COMFETS with fast switching speed and high-current capability." *IEEE Int. Electron Devices Meet.* pp. 4.3, 79–82 (1983).
24. A. R. Hefner and D. L. Blackburn, "Performance trade-off for the insulated gate bipolar transistor buffer layer versus base lifetime reduction." *IEEE, Power Electron Spec. Conf.* pp. 27–38 (1986).
25. A. Nakagawa and H. Ohashi, "600- and 1200-V bipolar-mode MOSFETS with high current capability." *IEEE Electron Devices Lett.* **EDL-6**(7), 378–380 (1985).
26. H. Yilmaz, J. L. Benjamin, R. F. Dyer, L. S. Chen, W. R. Van Dell, and G. F. Pifer, "Comparison of the punch-through and non-punch-through IGT structures." *IEEE, Ind. Appl. Soc. Annu. Meet.* pp. 905–908 (1985).
27. M. F. Chang, G. C. Pifer, B. J. Baliga, M. S. Adler, and P. V. Gray, "25 Amp, 500 Volt insulated gate transistors." *IEEE, Int. Electron Devices Meet.* Technical Digest 4.4, pp. 83–86 (1983).
28. L.-S. Chen, H. Yilmaz, and M. F. Chang, "Process optimization of 1200 V N-channel IGT." *IEEE, Ind. Appl. Soc. Annu. Meet.* pp. 909–911 (1985).
29. D. S. Kuo and C. Hu, "Speed-conductance optimization for power bipolar-MOS transistor." *IEEE Ind. Appl. Soc. Annu. Meet.* pp. 396–403 (1986).
30. A. Caruso, P. Spirito, G. Vitale, G. Busatto, G. Cocorullo, and G. Ferla, "Design criteria for a bipolar mode JFET (BMFET) with high blocking voltages." *IEEE, Ind. Appl. Soc. Annu. Meet.* pp. 350–355 (1986).
31. G. Ferla, S. Musumechi, G. Busatto, P. Spirito, and G. Vitale, "Switching characteristics of a high voltage BMFET." *Proc. Inter. Conf. Solid-State Devices Mater., 18th*, Tokyo pp. 93–96 (1986).
32. D.-S. Kuo and C. Hu, "An analytical model for the power bipolar-MOS transistor." *Solid-State Electron.* **29**(12), 1229–1237 (1986).
33. D.-S. Kuo and C. Hu, "Optimization of epitaxial layers for power bipolar–MOS transistor." *IEEE Electron Devices Lett.* **EDL-7**(9), 510–512 (1986).
34. D. Ueda, K. Kitamura, H. Takagi, and G. Kano, "A new injection suppression structure for conductivity modulated power MOSFETS." *Ext. Abst. Conf. Solid State Devices Mater., 18th,* Tokyo pp. 97–100 (1986).
35. A. Nakagawa, K. Imamura, and K. Furukawa, "1800 V bipolar-mode MOSFETS". *Toshiba Rev.* **161,** 34–37 (1987).
36. T. P. Chow, B. J. Baliga, H. R. Chang, P. V. Gray, W. Hennessy, and C. E. Logan, "P-channel, vertical insulated gate bipolar transistors with collector short." *IEEE, Int. Electron Devices Meet.* pp. 670–673 (1987).
37. V. Sukumar and D. Y. Chen, "IGT/COMFET latching characteristics and application to brushless DC motor drive." *IEEE Trans. Aerosp. Electron. Syst.* **AES-22**(5), 540–544 (1986).

38. R. Bayerer, "New IGBT-modules (half-bridge) with ultra-fast free-wheeling diodes in isolated package." *Proc. Power Convers. Int. Conf.* pp. 254–265 (1987).
39. R. Bayerer and J. Teigelkotter, "IGBT-diode half-bridge modules operate at 20 kHz." *Power Convers. Intell. Motion* (*PCIM*) December, pp. 21–35 (1987).
40. H. W. Becke, C. E. Harm, R. T. Lee, H. R. Ronan, Jr., and C. F. Wheatley, Jr., "Applications of COMFETS (IGT) to 40 kHz off-line switcher." *IEEE, 1st Annu. Appl. Power Electron. Conf.* pp. 1–7 (1986).
41. R. Rangan, D. Y. Chen, J. Yang, and J. Lee, "Application of IGT/COMFET to zero-current switching resonant converters." *IEEE, Power Electron. Spec. Conf.* pp. 55–60 (1987).
42. K. Gauen, "The GEMFET—a new option for power control." *Motorola Appl. Note* **AN-934** (1985).
43. B. J. Baliga and J. P. Walden, "Enhancement and depletion mode vertical MOS gated thyristors." *Electron. Lett.* **15**(20), 645–647 (1979).
44. V. A. K. Temple, "MOS Controlled Thyristors (MCT's)". *IEEE, Int. Electron Devices Meet.* pp. 282–285 (1984).
45. M. Stoisiek and H. Strack, "MOS GTO—a turn-off thyristor with MOS-controlled emitter shorts." *IEEE, Int. Electron Devices Meet.* pp. 158–161 (1985).
46. M. Stoisiek and D. Theis, "Turn-on principles of MOS-GTO." *IEEE, Power Electron. Spec. Conf.* pp. 87–93 (1986).
47. J. G. Mansmann and D. L. Zaremba, "An H-bridge configuration using MOS-SCRS and power MOSFETS." *Proc. Motor-Con.* October pp. 181–188 (1986).
48. M. Glogolja and J. Tihanyi, "Smart-SIPMOS, an intelligent power switch." *IEEE, Ind. Appl. Soc. Annu. Meet.* pp. 429–433 (1986).
49. P. Antognetti, *Power Integrated Circuits.* McGraw-Hill, New York, 1986.
50. P. Brauschke and P. Sommer, "Smart SIPMOS: Intelligent power semiconductors." *Siemens Components* **22**(5) 182–185 (1987).
51. J. Wojslawowicz, "Monolithic current-limiting MOSFETS." *Power Electron. Conf.* pp. 297–300 (1988).
52. C. Contiero, A. Andreini, P. Galbiati, and S. Storti, "Design of a high side driver in multipower-BCD and VIPOWER technologies." *IEEE, Int. Electron Devices Meet.* pp. 766–769 (1987).
53. R. L. Steigerwald, M. H. Kuo, G. S. Claydon, and K. C. Routh, "A high-voltage integrated circuit for power supply applications." *IEEE, 2nd Annu. Appl. Power Electron. Conf.*, San Diego pp. 221–229 (1987).
54. A. Foster, "Overview of smart power." *IEE Colloq. Integr. Intell. Devices* March, pp. 1.1–1.5 (1987).
55. S. Mizutani and K. Tamaki, "Power MOS device and its future." *IEE, Int. Conf. Automot. Electron., 6th* pp. 185–189 (1987).
56. T. E. Record, "Smart power (the million $ automotive question)." *IEE, Int. Conf. Automot. Electron. 6th* pp. 180–184 (1987).
57. B. Saby and B. C. Nadd, "Smart power devices in automotive multiplex wiring systems." *IEE, Int. Conf. Automot. Electron. 6th* pp. 195–199 (1987).
58. R. Frank and R. Janikowski, "Trends in power IC development." *Power Convers. Intell. Motion* (*PCIM*) April, pp. 26–28 (1986).
59. H. W. Becke, "Approaches to isolation in high voltage integrated circuits." *IEEE, Int. Electron Devices Meet.* Technical Digest 30.1, pp. 724–727 (1985).

60. S. Clemente, D. Kinzer, and D. Tam, "An integrated dc–dc converter for off-line operation." *Proc. Power Convers. Int. Conf.* May, pp. 290–302 (1987).
61. C. E. Harm, K. J. Timm, D. Kinzer, and D. Tam, "A universal input, fixed output, solid state, dc-to-dc converter." *IEEE, Power Electron. Spec. Conf.* pp. 76–84 (1987).
62. J. B. Bernstein, S. Bahl, and M. F. Schlect, "A low capacitance power MOSFET with an integral gate drive." *IEEE, Power Electron. Spec. Conf.* pp. 61–68 (1987).
63. Y. Tsuzuki, M. Yamaoka, and K. Kawamoto, "Self-thermal protecting power MOSFETS." *IEEE, Power Electron. Spec. Conf.* pp. 31–36 (1987).

APPENDIX 1

Basic Electrostatic Theory of the Depletion Layer

At its simplest, the electrostatic theory of the pn junction assumes an abrupt interface between uniformly doped n-type and p-type regions. The contact potential V_{CP} between the p and n regions (see Section 2.2), together with any applied bias voltage V_0, causes the regions on either side of the junction to be depleted of their majority carriers. This gives rise to a local space-charge density proportional to the local doping concentration. This is positive in the n-type region and negative in the p-type region, as shown in Figure A1.1. Let the depletion layer extend a distance l_p, into the p-type region, and a distance l_n into the n-type region. Then, the charge per unit area, Q_D, stored in the depletion layer on either side of the junction is given by

$$|Q_D| = en_A l_p = en_D l_n \tag{A1.1}$$

where n_A is the acceptor concentration in the p region and n_D is the donor concentration in the n region.

Set the x-coordinate perpendicular to the junction, with the origin at the junction and the p region on the positive side, and apply Poisson's law:

$$\nabla^2 V = -\rho/\epsilon_0 \epsilon_{Si} \tag{A1.2}$$

where V is the local potential, ρ the local charge density, ϵ_{Si} the relative permittivity of silicon (=11.9), and ϵ_0 the permittivity of free space. In our one-dimensional system, this reduces to

$$\frac{d^2V}{dx^2} = \frac{en_A}{\epsilon_0 \epsilon_{Si}} \tag{A1.3}$$

for $x > 0$, while for $x < 0$ we have

$$\frac{d^2V}{dx^2} = \frac{-en_D}{\epsilon_0 \epsilon_{Si}} \tag{A1.4}$$

The electric field strength, $E = -dV/dx$, is assumed to be zero for all $x < -l_n$ and for all $x > l_p$. Its maximum value, E_0, occurs at the junction, $x = 0$. The local potential V is defined to be zero at $x = 0$, for convenience.

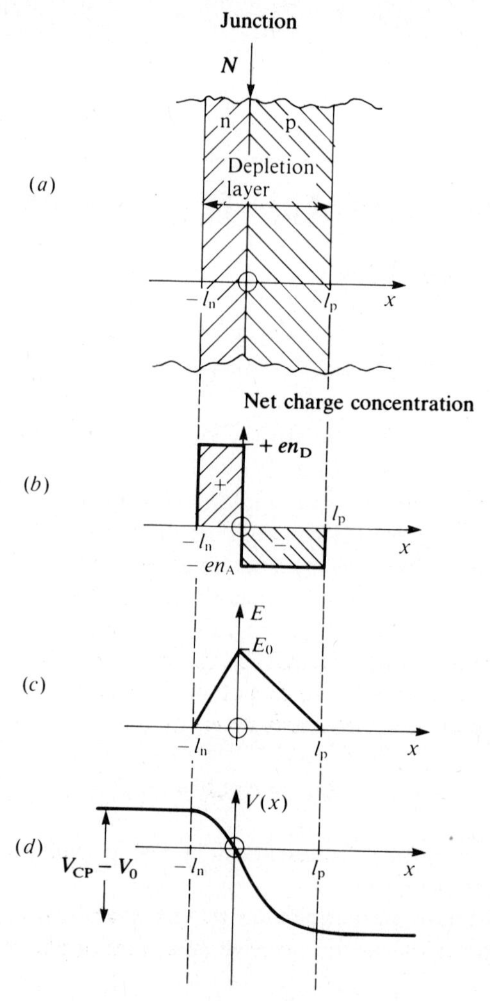

Figure A1.1. Depletion layer at pn junction. (*a*) Showing greater extent into the more lightly doped region, in this case the p-type region; (*b*) net charge concentration assuming uniform doping and negligible free carrier concentration in the depletion layer; (*c*) electric field strength; (*d*) potential variation.

Applying these boundary conditions, Equation (A1.3) integrates to give

$$\frac{\mathrm{d}V}{\mathrm{d}x} = \frac{en_A(x - l_p)}{\epsilon_0 \epsilon_{Si}} \tag{A1.5}$$

for $l_p > x > 0$, and Equation (A1.4) integrates to give

$$\frac{\mathrm{d}V}{\mathrm{d}x} = -\frac{en_D(x + l_n)}{\epsilon_0 \epsilon_{Si}} \tag{A1.6}$$

for $-l_n < x < 0$. Integrating each expression again, we obtain

$$V(x) = \frac{en_A/\epsilon_0\epsilon_{Si}}{\frac{1}{2}x^2 - l_p x} \tag{A1.7}$$

for $l_p > x > 0$, and

$$V(x) = -\frac{en_D/\epsilon_0\epsilon_{Si}}{\frac{1}{2}x^2 + l_n x} \tag{A1.8}$$

for $-l_n < x < 0$. These relationships are plotted in Figure A1.1.

The total change of potential across the depletion layer is

$$V_{CP} - V_0 = V(-l_n) - V(l_p)$$

$$= \frac{en_D l_n^2}{2\epsilon_0\epsilon_{Si}} + \frac{en_A l_p^2}{2\epsilon_0\epsilon_{Si}} \tag{A1.9}$$

$$= \frac{Q_D^2/2e\epsilon_0\epsilon_{Si}}{1/n_D + 1/n_A} \tag{A1.10}$$

where we have substituted Equation (A1.1).

Thus,

$$|Q_D| = \left(\frac{2e\epsilon_0\epsilon_{Si}(V_{CP} - V_0)}{1/n_D + 1/n_A}\right)^{1/2} \tag{A1.11}$$

$$l_p = \frac{|Q_D|}{en_A}$$

$$= \frac{1}{n_A}\left(\frac{2\epsilon_0\epsilon_{Si}(V_{CP} - V_0)}{e(1/n_D + 1/n_A)}\right)^{1/2} \tag{A1.12}$$

$$l_n = \frac{|Q_D|}{en_D}$$

$$= \frac{1}{n_D}\left(\frac{2\epsilon_0\epsilon_{Si}(V_{CP} - V_0)}{e(1/n_D + 1/n_A)}\right)^{1/2} \tag{A1.13}$$

Note that the total depletion layer width l_D is given by

$$l_D = l_p + l_n$$

$$= \left(\frac{2\epsilon_0\epsilon_{Si}(V_{CP} - V_0)(1/n_D + 1/n_A)}{e}\right)^{1/2} \tag{A1.14}$$

and that

$$E_D = \frac{en_A l_p}{\epsilon_0\epsilon_{Si}} = \frac{en_D l_n}{\epsilon_0\epsilon_{Si}} = \frac{|Q_D|}{\epsilon_0\epsilon_{Si}} \tag{A1.15}$$

$$= \left(\frac{2e(V_{CP} - V_0)}{\epsilon_0\epsilon_{Si}(1/n_D + 1/n_A)}\right)^{1/2} \tag{A1.16}$$

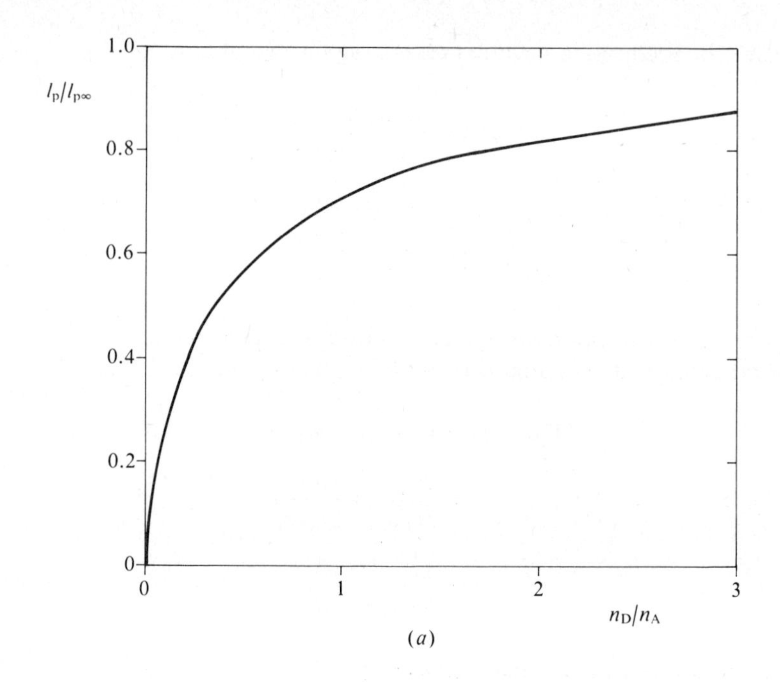

(*a*)

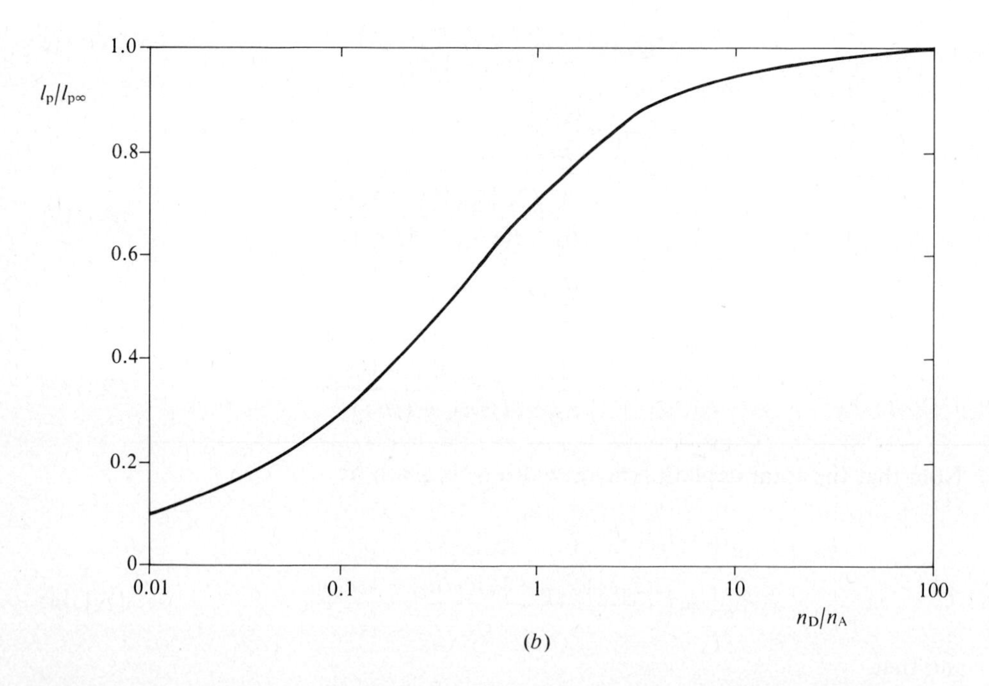

(*b*)

Figure A1.2. Reduction of depletion layer in p-type region as doping concentration in n-type region is reduced. (*a*) Linear scales; (*b*) log scale for (n_D/n_A). As $(n_D/n_A) \rightarrow \infty$, $l_p \rightarrow l_{p\infty}$.

The variation of l_p with n_D, assuming that V_0 and n_A remain constant, is shown in Figure A1.2. The very small variation of V_{CP} with n_D is neglected on the assumption that V_0 is large and negative. It can be seen that for a given applied voltage V_0, reducing n_D reduces the penetration of the depletion layer into the p region. Of course, the converse is also true.

The importance of this theory in the design of high-voltage MOSFETS is explained in Chapters 1 and 2. In order to minimize channel-length modulation caused by the drain–source voltage, the channel region should be more heavily doped than the drain drift region. For exactly similar reasons the base region of a bipolar junction transistor should be more heavily doped than the collector region. This minimizes base-width modulation by the collector voltage, the *Early effect.*

Another important quantity that affects several aspects of device performance is the total static self-capacitance per unit area of the pn junctions. We shall write this as C_T, and it is given by $C_T = |Q_D|/V_0$. Often of greater significance is the incremental self-capacitance, $C_J = d(|Q_D|)/dV_0$. When $(-V_0) \gg V_{CP}$, we can neglect the contact potential in evaluating C_T and C_J. Then

$$C_T = \left(\frac{2e\epsilon_0\epsilon_{Si}}{(V_{CP} - V_0)(1/n_D + 1/n_A)}\right)^{1/2}$$

$$\approx \left(\frac{2e\epsilon_0\epsilon_{Si}}{-V_0(1/n_D + 1/n_A)}\right)^{1/2} \tag{A1.17}$$

and

$$C_J = \left(\frac{e\epsilon_0\epsilon_{Si}}{2(V_{CP} - V_0)(1/n_D + 1/n_A)}\right)^{1/2}$$

$$\approx \left(\frac{e\epsilon_0\epsilon_{Si}}{2(-V_0)(1/n_D + 1/n_A)}\right)^{1/2} \tag{A1.18}$$

These expressions are obtained directly from Equation (A1.11).

Often in modeling the behavior of MOSFETS, situations arise where V_{CP} is not negligible in comparison with $-V_0$ and the full expressions have to be used. It is then convenient to express the effective junction capacitance in terms of the incremental capacitance per unit area at zero applied bias C_{J0}. This is obtained by setting $V_0 = 0$ in the full expression for C_J. Thus

$$C_{J0} = \left(\frac{e\epsilon_0\epsilon_{Si}}{2V_{CP}(1/n_D + 1/n_A)}\right)^{1/2} \tag{A1.19}$$

Then

$$C_J = C_{J0}\left(\frac{V_{CP}}{V_{CP} - V_0}\right)^{1/2} \tag{A1.20}$$

The incremental capacitance C_J is a function of the change in Q_D and the change in l_D brought about by a change in V_0. It is therefore a function of the properties of the semiconductor at the depletion-layer edges. In the case of an abrupt, one-sided junction, the variation of C_J with V_0 can be used to obtain the doping-level profile. For a p^+n junction, it can easily be shown that the donor doping level at a distance x perpendicular to the junction, $n_D(x)$, is related to C_J and V_0 as

$$\begin{aligned} n_D(x) &= \frac{C_J^3}{e\epsilon_0\epsilon_{Si}}\left(\frac{dC_J}{dV_0}\right)^{-1} \\ &= -\frac{2}{e\epsilon_0\epsilon_{Si}}\left[\frac{d}{dV_0}\left(\frac{1}{C_J^2}\right)\right]^{-1} \end{aligned} \tag{A1.21}$$

Of course, a similar expression can be obtained for an n^+p junction. This theory forms the basis of the C–V profiling method.

A transition between two bias voltages, V_1 and V_2, involves a change in the charge stored at the junction given by

$$\begin{aligned} \Delta Q_D &= \left(\frac{2e\epsilon_0\epsilon_{Si}}{1/n_D + 1/n_A}\right)^{1/2}[(V_{CP} - V_2)^{1/2} - (V_{CP} - V_1)^{1/2}] \\ &= 2C_{J0}(V_{CP})^{1/2}[(V_{CP} - V_2)^{1/2} - (V_{CP} - V_1)^{1/2}] \end{aligned} \tag{A1.22}$$

For that particular voltage change, the junction then appears to present an effective capacitance

$$\begin{aligned} (C_J)_{eff} &= \frac{\Delta Q_D}{V_2 - V_1} \\ &= 2C_{J0}(V_{CP})^{1/2}\frac{(V_{CP} - V_2)^{1/2} - (V_{CP} - V_1)^{1/2}}{V_2 - V_1} \end{aligned} \tag{A1.23}$$

Although they are algebraically very cumbersome, these expressions do make arithmetic calculations more straightforward, especially in digital applications where the transition potentials, V_1 and V_2, are usually well defined. They are often 0 and 5 V, respectively, or perhaps 0 and 10 V in the case of power devices. The parameters V_{CP} and C_{J0} are known from the manufacturing process and the device design.

APPENDIX 2

The Formation of Depletion, Inversion, and Accumulation Layers at the Silicon Surface

A depletion layer can be formed in the silicon surface region as well as at a pn junction, if the majority carriers are expelled from the surface layer by externally applied electric fields. As in the case of the pn junction, the remaining ionized dopant atoms set up a local space charge, and Poisson's law has to be satisfied. In a planar geometry this again reduces to a one-dimensional problem. Consider the case of p-type material, and let the x-coordinate be directed into the semiconductor, perpendicular to the surface. The coordinate origin is taken to be at the surface, as shown in Figure A2.1. The depletion layer extends a distance l_p into the semiconductor, at which point $dV/dx = 0$. We set $V = 0$ for $x \geq l_p$. Thus, for $0 < x \leq l_p$,

$$\frac{d^2V}{dx^2} = \frac{en_A}{\epsilon_0 \epsilon_{Si}} \qquad \text{(A1.3)}$$

$$\frac{dV}{dx} = -\frac{en_A}{\epsilon_0 \epsilon_{Si}}(l_p - x) = -E \qquad \text{(A2.1)}$$

$$V = \frac{en_A}{2\epsilon_0 \epsilon_{Si}}(l_p - x)^2 \qquad \text{(A2.2)}$$

Thus, the voltage dropped across the depletion layer is

$$\Delta V = V(0) - V(l_p) = V(0) = en_A l_p^2 / 2\epsilon_0 \epsilon_{Si} \qquad \text{(A2.3)}$$

and the electric field strength at the surface is

$$E(0) = en_A l_p / \epsilon_0 \epsilon_{Si} = E_{Si} \qquad \text{(A2.4)}$$

Assume that the semiconductor surface is covered with a coating of oxide. In the semiconductor the relative permittivity is $\epsilon_{Si} = 11.9$, whereas in the oxide it is $\epsilon_{ox} = 3.9$. Flux continuity across the oxide–semiconductor interface requires that

$$\epsilon_0 \epsilon_{Si} E_{Si} = \epsilon_0 \epsilon_{ox} E_{ox} + Q_{SS} \qquad \text{(A2.5)}$$

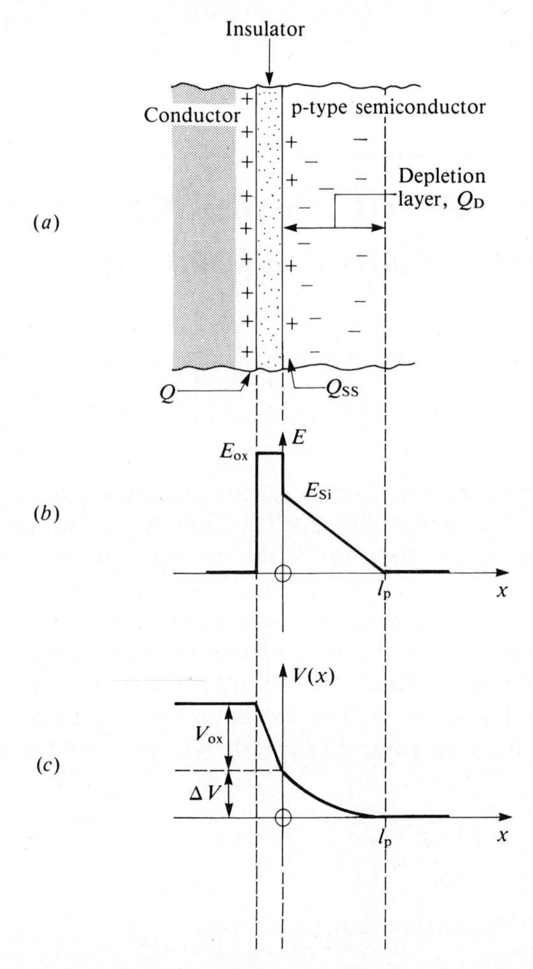

Figure A2.1. Formation of depletion layer at semiconductor surface. (*a*) Surface region; (*b*) electric field strength; (*c*) variation of internal potential. Note in (*b*) that the discontinuity in E at the oxide/semiconductor interface is the net effect of the change in ϵ_r and the presence of the interface charge. The total change in internal potential shown in (*c*) comprises the contact potential, V_{CP}, plus any applied voltage between gate and bulk semiconductor, V_{GB}. Thus, $V_{GB} + V_{CP} = V_{ox} + \Delta V$, in agreement with Equation (2.5).

where E_{ox} is the electric field strength in the oxide and Q_{SS} is the positive surface charge density present at the interface, as discussed in Section 2.2. Two results follow from this. The first is that the static charge per unit area stored in the depletion layer is

$$\begin{aligned} Q_D &= en_A l_p = \epsilon_0 \epsilon_{Si} E_{Si} \\ &= (2e\epsilon_0\epsilon_{Si} n_A \,\Delta V)^{1/2} \qquad \text{(A2.6)} \end{aligned}$$

using Equations (A2.3) and (A2.4). The second is that the voltage dropped across the gate oxide is

$$V_{ox} = E_{ox}t_{ox} = \frac{(\epsilon_0\epsilon_{Si}E_{Si} - Q_{SS})t_{ox}}{\epsilon_0\epsilon_{ox}}$$

$$= \frac{Q_D - Q_{SS}}{C_{ox}} \tag{A2.7}$$

where $C_{ox} = \epsilon_0\epsilon_{ox}/t_{ox}$ is the gate-oxide capacitance per unit area. Here we have used Equation (A2.5).

In Figure 2.6 it can be seen that the voltage applied between the gate and the bulk semiconductor, which is represented on the diagram by the difference in the Fermi levels in the two regions, is given by

$$V_{GB} = V_{ox} + \Delta V - V_{CP} \tag{A2.8}$$

where V_{CP} is the contact potential between the gate and the bulk semiconductor,

$$V_{CP} = \frac{kT}{e}\ln\frac{n_A n_D}{n_i^2} \tag{2.6}$$

In Section 2.2 it is explained that an inversion layer is formed at the silicon surface, in addition to the depletion layer, when

$$\Delta V = \Delta V_{Th} = 2(-V_i)_{body}$$

$$= \frac{2kT}{e}\ln\frac{n_A}{n_i} \tag{2.9}$$

Thus at threshold, $V_{GB} = V_T$, and $\Delta V = \Delta V_{Th}$, so that putting these values into equation (A2.8) we obtain

$$V_T = V_{ox} + \Delta V_{Th} - V_{CP}$$

$$= \frac{Q_D - Q_{SS}}{C_{ox}} + \Delta V_{Th} - V_{CP} \tag{A2.9}$$

where Q_D is given by Equation (A2.6) with $\Delta V = \Delta V_{Th}$. In Equation (A2.9) we have neglected as small the voltages dropped across the inversion layer and across the depleted surface of the poly-Si gate.

In p-type material the inversion layer is characterized as a very thin surface layer of additional negative charge comprising free electrons. This is illustrated in Figure A2.2. It can most easily be treated as having a negative surface charge density, $-Q_{IL}$, which subtracts from Q_{SS}, the fixed positive interface charge. Because the depletion layer is sensibly unchanged above threshold, Q_{IL} increases the electric field in the gate oxide and hence increases V_{ox}. Equations (A2.5) and (A2.7) then become

$$\epsilon_0\epsilon_{Si}E_{Si} + Q_{IL} = \epsilon_0\epsilon_{ox}E_{ox} + Q_{SS} \tag{A2.10}$$

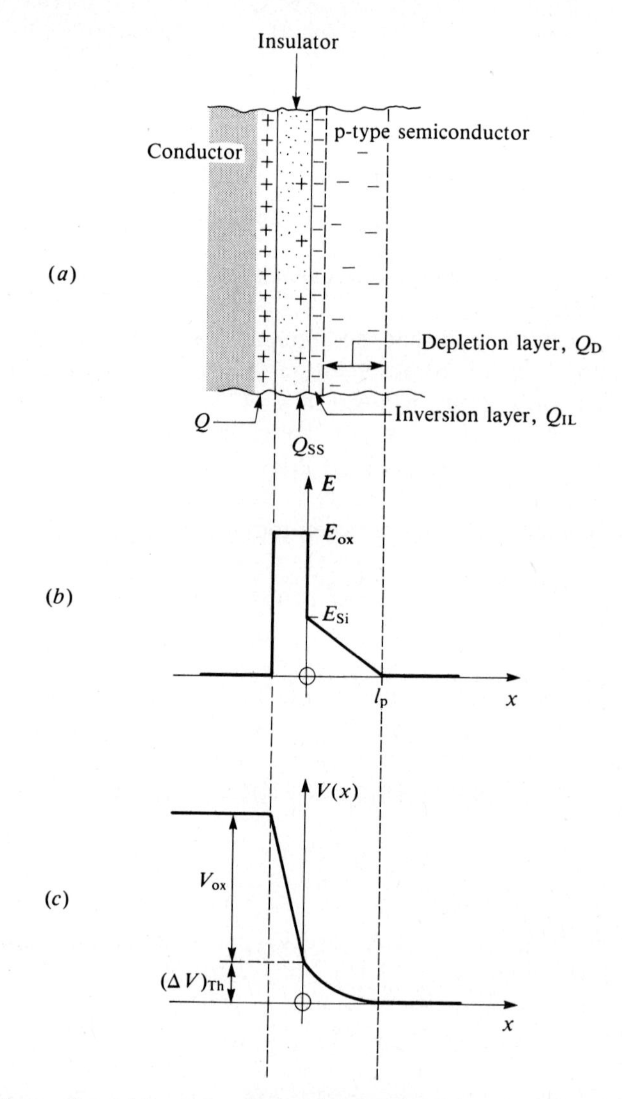

Figure A2.2. Formation of inversion layer. (*a*) Charge layers at semiconductor surface; (*b*) electric field strength; (*c*) internal potential.

and

$$V_{ox} = \frac{Q_D + Q_{IL} - Q_{SS}}{C_{ox}} \tag{A2.11}$$

Thus,

$$\begin{aligned} V_{GB} &= V_{ox} + \Delta V_{Th} - V_{CP} \\ &= \frac{Q_D + Q_{IL} - Q_{SS}}{C_{ox}} + \Delta V_{Th} - V_{CP} \\ &= V_T + \frac{Q_{IL}}{C_{ox}} \end{aligned} \tag{A2.12}$$

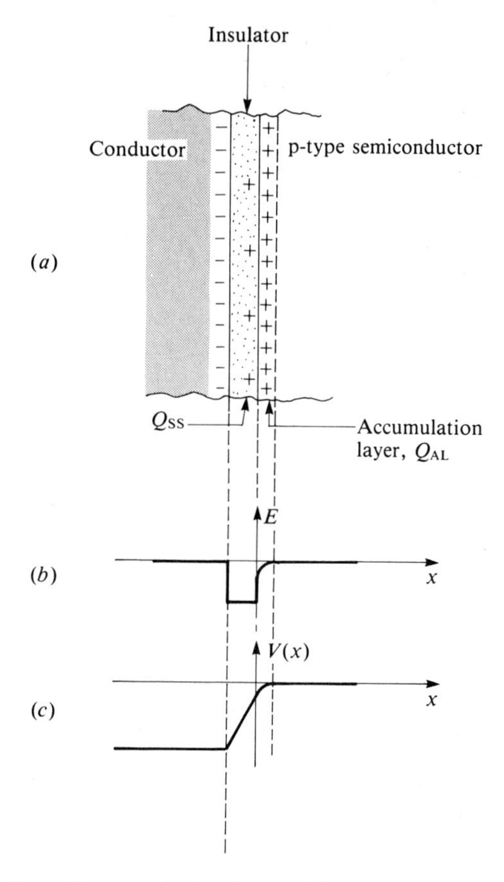

Figure A2.3. Formation of accumulation layer. (*a*) Charge distribution; (*b*) electric field strength; (*c*) internal potential.

Rearranging, we obtain

$$Q_{IL} = C_{ox}(V_{GB} - V_T) \quad \text{(A2.13)}$$

In other circumstances the surface of the p-type semiconductor may be at a more negative potential than the bulk material. Then an accumulation layer forms, as shown in Figure A2.3. Because the hole concentration increases exponentially with the change in potential, the accumulation layer is very thin and can be treated as having a positive surface charge density, Q_{AL}, which adds to Q_{SS}. Thus, in this case, Equation (A2.5) should be modified to

$$\epsilon_0\epsilon_{Si}E_{Si} = \epsilon_0\epsilon_{ox}E_{ox} + Q_{SS} + Q_{AL} = 0 \quad \text{(A2.14)}$$

on the assumption that beyond the accumulation layer the field in the bulk semiconductor is zero.

With an n-type bulk semiconductor region the polarities required to produce depletion, inversion, and accumulation layers are reversed, except that Q_{SS} remains positive, and V_{CP} becomes small if the gate is also doped n-type. Thus, in region Ⓑ of the VDMOS FET shown in Figure 2.1, a positive gate potential raises the potential at the surface of the n-type bulk silicon and so tends to form an accumulation layer under the gate oxide.

APPENDIX 3

More Rigorous Theory of the Formation of Inversion and Accumulation Layers

The theory presented in Appendices 1 and 2 assumed that at all places the bulk semiconductor was in one of two possible states: either neutral or fully depleted. When accumulation and inversion layers formed, they were represented as thin layers having a certain surface charge density. The validity of these approximations is confirmed by the more rigorous solution of Poisson's equation in semiconductor material given here. The analysis is again one-dimensional and assumes uniform doping and complete lateral uniformity. The x-coordinate is taken to be perpendicular to the silicon surface, and Poisson's equation becomes

$$\frac{d^2V_i}{dx^2} = \frac{e}{\epsilon_{Si}\epsilon_0}(n - p + n_A - n_D) \tag{A3.1}$$

In Equation (A3.1), as in the main text, V_i is the midband potential (that is, the potential corresponding to the position of the Fermi level in intrinsic material) measured with respect to the actual potential of the Fermi level. The concentrations of free electrons, n, and holes, p, at any point are then given by

$$n = n_i \exp(eV_i/kT) \tag{A3.2}$$

$$p = n_i \exp(-eV_i/kT) \tag{A3.3}$$

where n_i is the intrinsic carrier concentration, which in silicon at room temperature (298 K) has a value of about $1.5 \times 10^{16}\,\mathrm{m}^{-3}$. At this temperature $kT/e = 0.0257$ V. The other variables in Equation (A3.1) have their usual meanings: $\epsilon_0 = 8.854 \times 10^{-12}$ F/m is the permittivity of free space; $\epsilon_{Si} = 11.9$ is the relative permittivity of silicon; n_A is the local concentration of acceptor impurities; n_D is the local concentration of donors. In p-type material we may put $n_D = 0$.

It is convenient to normalize V_i against kT/e by putting $\psi = eV_i/kT$; then Equation (A3.1) becomes

$$\frac{kT}{e}\frac{d^2\psi}{dx^2} = \frac{e(n - p + n_A)}{\epsilon_{Si}\epsilon_0} \tag{A3.4}$$

Substituting for n and p from Equations (A3.2) and (A3.3) gives

$$\frac{d^2\psi}{dx^2} = \frac{e^2}{kT\epsilon_{Si}\epsilon_0}(n_i e^{\psi} - n_i e^{-\psi} + n_A)$$

$$= \frac{n_A e^2}{kT\epsilon_{Si}\epsilon_0}\left(2\frac{n_i}{n_A}\sinh\psi + 1\right) \quad \text{(A3.5)}$$

It is also convenient to normalize distance to the Debye length $L_D = (kT\epsilon_0\epsilon_{Si}/n_A e^2)^{1/2}$ by putting $X = x/L_D$. If in addition we put $K = 2n_i/n_A$, Equation (A3.5) reduces to

$$\frac{d^2\psi}{dX^2} = K\sinh\psi + 1 \quad \text{(A3.6)}$$

We expect the potential to change smoothly from its value at the surface, ψ_S, to its value deep in the bulk semiconductor, ψ_∞. Equation (A3.6) may be solved numerically, subject to these boundary conditions. Some solutions are shown in Figure A3.1.* They are compared with the solutions from the usual, simple theory which omits the sinh ψ term in the equation. For the simple theory the origin of the X-coordinate is positioned to coincide with the boundary of the depletion layer and the neutral bulk material. Then, the equations reduce to

$$\frac{d^2\psi}{dX^2} = 1$$

$$\frac{d\psi}{dX} = X, \qquad \text{so that } \frac{d\psi}{dX} = 0 \quad \text{at } X = 0 \text{ and for all } X > 0$$

$$\psi - \psi_\infty = \tfrac{1}{2}X^2, \qquad \text{giving } \psi = \psi_\infty \text{ at } X = 0 \text{ and for all } X > 0$$

This solution is represented by the dotted curves in Figure A3.1. The semiconductor surface on this simple theory is positioned where $\psi = \psi_S$, that is, at $-[2(\psi_S - \psi_\infty)]^{1/2}$. Furthermore, $\psi = 0$ at

$$X = X_0 = -(-2\psi_\infty)^{1/2} \quad \text{(A3.7)}$$

In order to facilitate comparison between the full solution and the simple theory, the point $\psi = 0$, $X = X_0$ is taken to be a fixed point where both solutions are coincident. For the full solution we know that at $X = \infty$, $\psi = \psi_\infty$ and $d\psi/dX = d^2\psi/dX^2 = 0$. Putting these values into Equation (A3.6) gives

$$\psi_\infty = \sinh^{-1}(-1/K) \approx -\ln(-2/K) \quad \text{(A3.8)}$$

* The authors are grateful to Dr. I. B. Stewart of Bristol University Computer Centre for carrying out these calculations, which are similar to those first published by R. H. Kingston and S. F. Neustadter, *J. Appl. Phys.*, **26**, pp. 718–720 (1955).

Equation (A3.6) may be integrated directly to give

$$\frac{d\psi}{dX} = -[2K(\cosh \psi - \cosh \psi_\infty) + 2(\psi - \psi_\infty)]^{1/2} \tag{A3.9}$$

With the given fixed point of $\psi = 0$ at $X = X_0$, Equation (A3.9) may be integrated numerically to find the position of the semiconductor surface, X_S, given that the surface potential is ψ_S:

$$X_S = X_0 + \int F(\psi)\, d\psi \tag{A3.10}$$

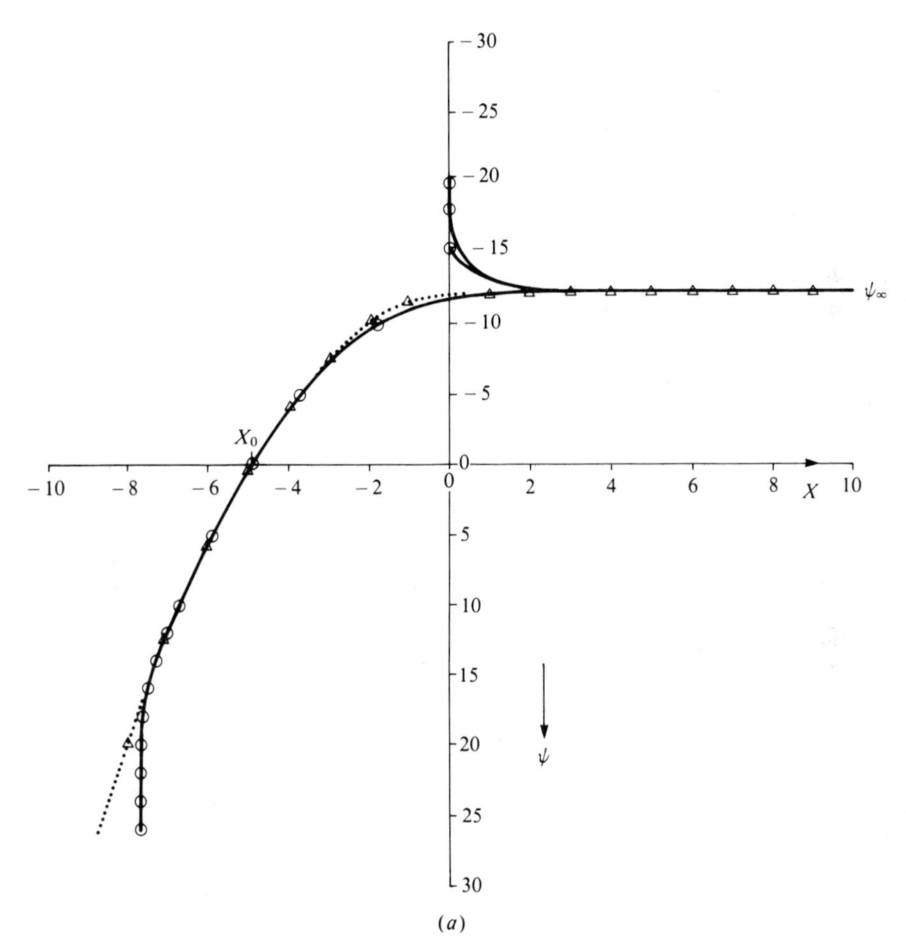

Figure A3.1. Potential variation under the semiconductor surface for different values of surface potential. (*a*) $K = 10^{-5}$; (*b*) $K = 10^{-6}$; (*c*) $K = 10^{-7}$. Normalized units (ψ, X) are used for potential and distance. The curves are centered on $\psi = 0$, $X = X_0$. The position $X = 0$ corresponds to the boundary between the depletion layer and the neutral, bulk semiconductor under the conditions of the simple theory. The circles represent the locus of the relative surface position, X_S, for different values of the surface potential, ψ_S. For $\psi_S < \psi_\infty$, an accumulation layer forms, and the surface is assumed to be positioned at $X = 0$.

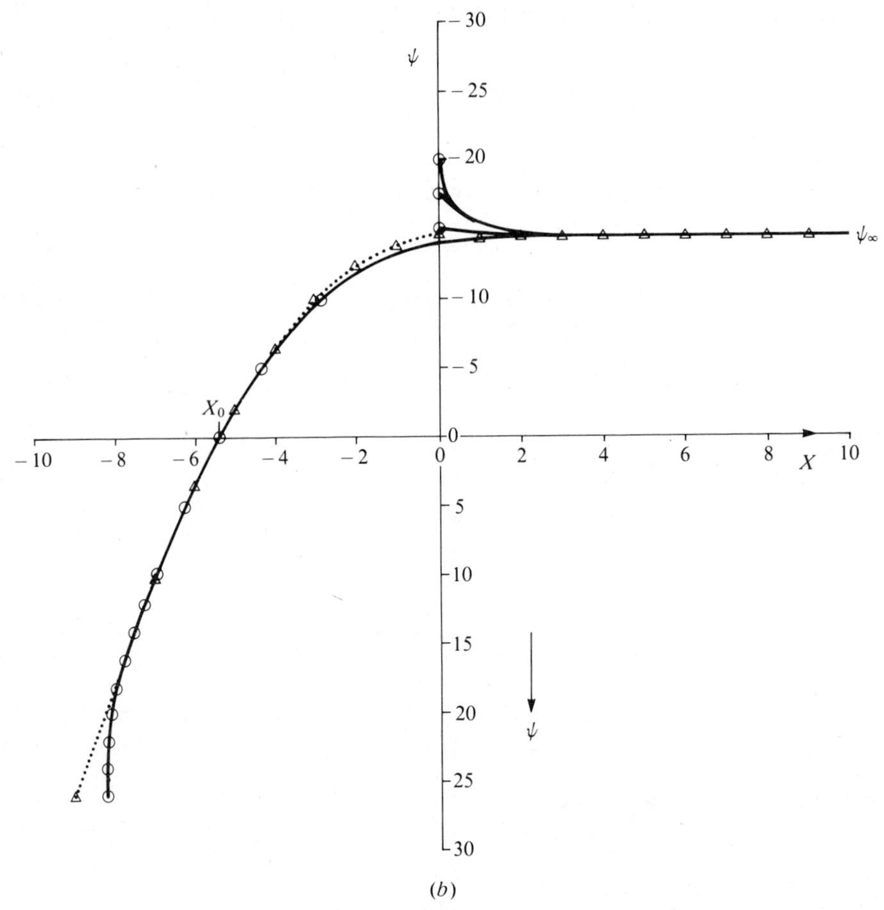

(*b*)

Figure A3.1. (*Continued*)

where

$$F(\psi) = \sqrt{\tfrac{1}{2}}\,[K(\cosh\psi - \cosh\psi_\infty) + (\psi - \psi_\infty)]^{-1/2} \tag{A3.11}$$

The full solution from X_S to $+\infty$ (taken as $X = +10$) may then be obtained using a finite-difference routine. The locus of the points (ψ_S, X_S) lies close to the set of curves given by the full solutions, which are shown as the full curves in Figure A3.1. The pinning of the depletion layer with the formation of the inversion layer at large positive values of ψ_S can be seen, as can the slight divergence of the two solutions around $X = 0$.

The total negative surface charge density in the depletion–inversion layer, $Q_D + Q_{IL}$, is given directly by $(d\psi/dX)_{X_S}$ in units normalized against $Q_0 = (n_A kT\epsilon_0\epsilon_{Si})^{1/2}$. The fixed negative space charge of the depleted region, Q_D, may

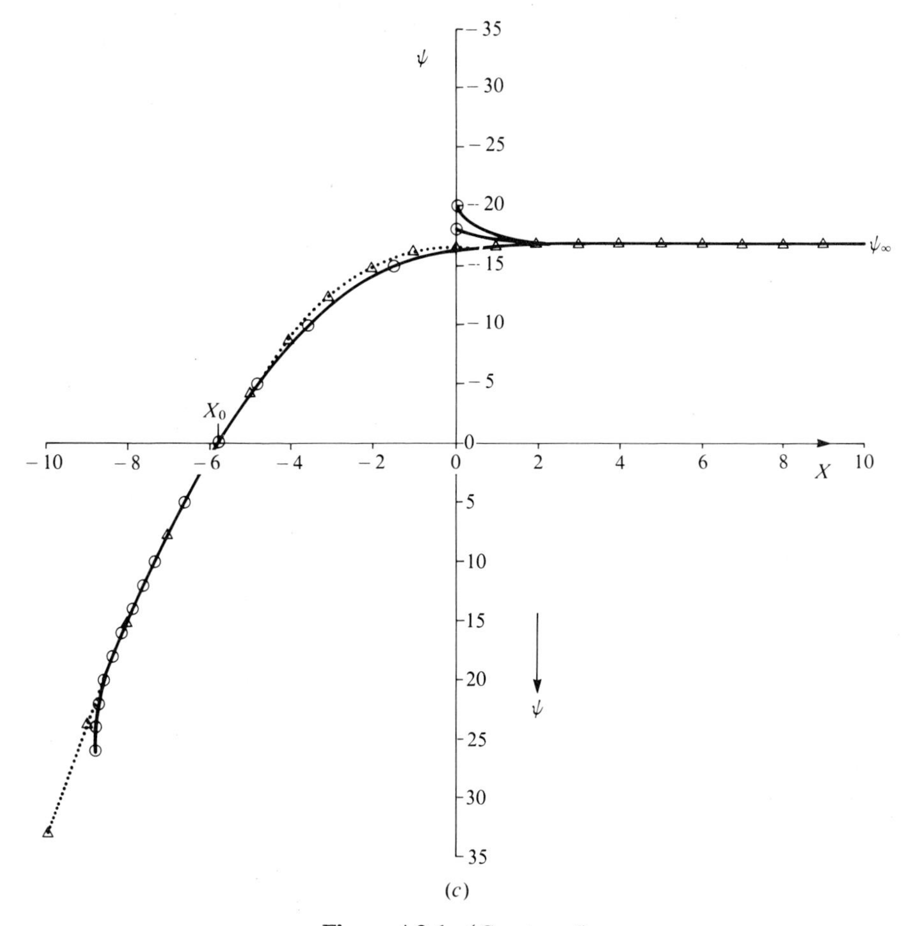

Figure A3.1. (*Continued*)

be calculated from

$$\frac{Q_D}{Q_0} = \left(\frac{d\psi}{dX}\right)_{X_0} + (X - X_0) \qquad \text{(A3.12)}$$

Any free-electron charge in a surface inversion layer is then given by

$$\frac{Q_{IL}}{Q_0} = \left(\frac{d\psi}{dX}\right)_{X_S} - \frac{Q_D}{Q_0} \qquad \text{(A3.13)}$$

In Figure A3.2 we have plotted X_{IL}, the depth of the inversion layer as normally defined, and the potential difference across it, $(\Delta\psi)_{IL}$, against the inversion-layer charge Q_{IL}. Thus, $X_{IL} = X_S - X(-\psi_\infty)$ and $(\Delta\psi)_{IL} = \psi_S - (-\psi_\infty)$, on the usual assumption that the inversion layer forms where $\psi = -\psi_\infty$. There is, of course, some free-electron charge at the surface before this point is reached, as

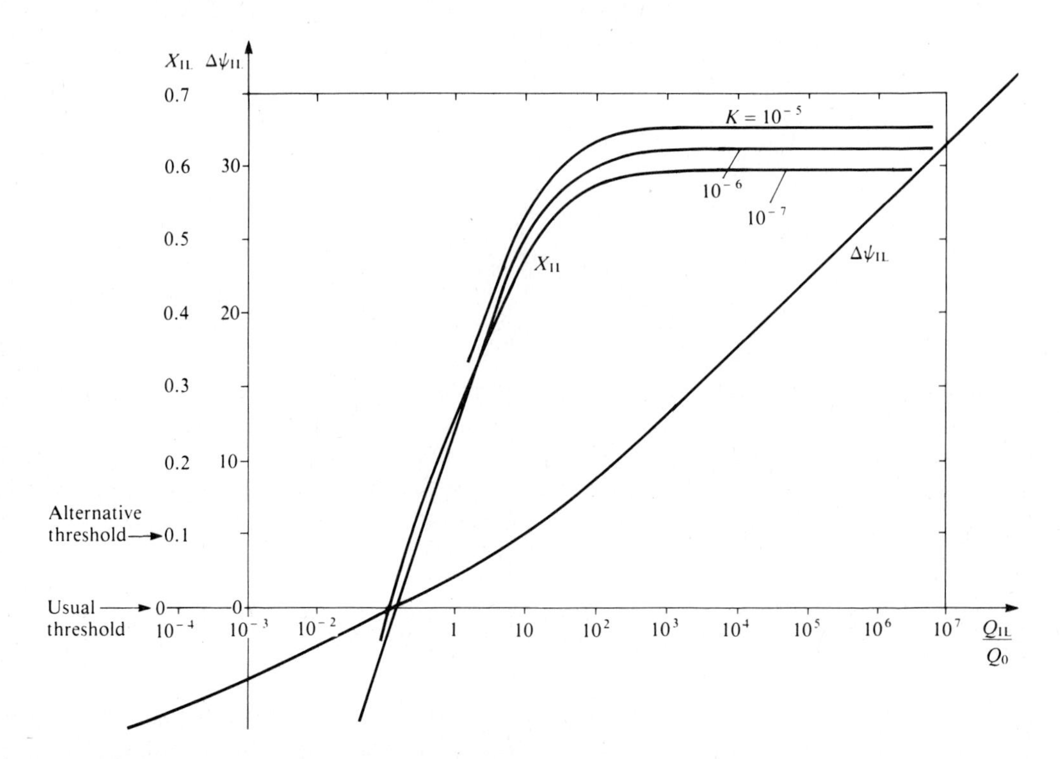

Figure A3.2. Potential dropped across the inversion layer, $\Delta\psi_{IL}$, and its width, X_{IL} as a function of the inversion layer charge, Q_{IL}/Q_0. Normalized units.

can be seen in the figure. This accounts for the so-called subthreshold current normally observed in MOSFET devices and discussed in Section 3.3. The theoretical definition of "threshold" is thus, to a degree, arbitrary and is chosen to give a simple physical definition.

The rapid increase in Q_{IL} for a small increase in ψ_{IL} that was discussed in Section 2.2 is made clear in Figure A3.2.

At room temperature and under normal operating conditions the mobile inversion layer charge is limited to $Q_{IL}/Q_0 = 10$ to 100. Thus, $X_{IL} \approx 0.6$ and $(\Delta\psi)_{IL} \approx 5$ to 10. In comparison, the thickness of the depletion region, as normally defined, and the fixed charge contained in it are both given in normalized units by $2[\ln(n_A/n_i)]^{1/2}$, which is typically in the region of 7 to 8. This means that the thickness of the inverted region is some 8% of that of the depletion layer that separates it from the bulk semiconductor, and some 0.1 to 0.2 V is dropped across it.

We have extended the graph shown in Figure A3.2 to illustrate what would happen at larger values of ψ_S and $(\Delta\psi)_{IL}$. Although Q_{IL} grows exponentially as the potential increases, the inversion-layer thickness remains almost constant. The increased free charge occupies a very thin surface layer. In fact, the lower limit to the thickness is set by quantum theory. This essentially two-dimensional structure can be produced at low temperatures in MOS devices and leads to a number of interesting and unusual physical properties.

Figures A3.1 and A3.2 are plotted for values of the doping parameter K of 10^{-5}, 10^{-6}, and 10^{-7}. At room temperature the corresponding values of the boundary and normalizing parameters are

K	n_A (m^{-3})	ψ_∞	X_0	L_D (nm)	Q_0 ($C\,m^{-2}$)
10^{-5}	3×10^{21}	-12.2	-4.94	74	3.56×10^{-5}
10^{-6}	3×10^{22}	-14.5	-5.39	23.4	1.125×10^{-4}
3×10^{-7}	1×10^{23}	-15.7	-5.61	13.0	2.08×10^{-4}
10^{-7}	3×10^{23}	-16.8	-5.80	7.4	3.56×10^{-4}

It is instructive to re-evaluate these for a temperature of 125°C (398 K) when $n_i \approx 6 \times 10^{18}\,m^{-3}$, and again at -55°C (218 K) when $n_i \approx 5 \times 10^{12}\,m^{-3}$.

In Figure A3.1 curves have been included for cases where $\psi_S < \psi_\infty$, so that an accumulation layer forms at the surface. In these cases the surface has been taken to be at $X = 0$. It can be seen that the effective normalized thickness of the accumulation layer is in the region of 1 to 2, with the bulk of the excess charge confined to a layer that is thin compared to the thickness of a normal depletion layer (7 to 8). The use of the simple model of Appendix 2 is thus largely vindicated for both accumulation and inversion layers.

APPENDIX 4

Channel Transconductance

A4.1. THE TURNED-ON CONDITION

In Appendix 2 we show that the charge per unit area in the channel inversion layer, Q_{IL}, is

$$Q_{IL} = C_{ox}(V_{GS} - V_T) \tag{A2.13}$$

where $C_{ox} = \epsilon_0\epsilon_{ox}/t_{ox}$ is the capacitance per unit area of the gate oxide, and V_T is a threshold voltage, which depends on the channel doping concentration. This equation assumes that the source and the bulk semiconductor are at the same potential: $V_{GS} = V_{GB}$. When significant drain current flows through the inverted channel, an ohmic voltage, V_{Ch}, develops along the channel. As a result, the depletion-layer charge density varies along the length of the channel, as does Q_{IL}. Equation (A2.13) has to be modified. It is no longer possible for the electrons in the inversion layer to maintain thermodynamic equilibrium with the source electrons. However, charge conservation has to be satisfied, as do the laws of electrostatics.

To deal with this problem, we use the system of coordinates defined in Figure 3.3*a*. First consider an element of the channel, dy, at a distance y from the source. Assume that at this point, the local channel potential has increased to $V(y)$ above the source potential. The local value of the inversion-layer charge density, $Q_{IL}(y)$, is diminished for two reasons: first, because of the increased voltage dropped across the depletion layer, $\Delta V_{Th} + V(y)$, and secondly, because of the extra fixed charge contained in the expanded depletion layer, $Q_D(y)$. This becomes

$$Q_D(y) = \{2e\epsilon_0\epsilon_{Si}n_A[\Delta V_{Th} + V(y)]\}^{1/2} \tag{A4.1}$$

The potential difference between the gate and the bulk semiconductor, which was previously given by Equation (A2.8), now becomes

$$V_{GS} = V_{ox} + \Delta V_{Th} + V(y) - V_{CP} \tag{A4.2}$$

The first term on the right-hand side, which was previously given by Equation (A2.11), is now

$$V_{ox} = \frac{Q_D(y) + Q_{IL}(y) - Q_{SS}}{C_{ox}} \tag{A4.3}$$

with $Q_D(y)$ given by Equation (A4.1). The expression for $Q_{IL}(y)$ thus takes the form

$$Q_{IL}(y) = Q_{SS} + C_{ox}[V_{GS} + V_{CP} - \Delta V_{Th} - V(y)] - \{2e\epsilon_0\epsilon_{Si}n_A[\Delta V_{Th} + V(y)]\}^{1/2} \quad \text{(A4.4)}$$

To obtain Equation (A4.4) we have rearranged Equation (A4.3) and substituted for $Q_D(y)$ using Equation (A4.1).

What we have really done here is to apply Poisson's equation in one dimension in the direction perpendicular to the semiconductor surface. This is on the assumption that the rate of change of electric field strength along the channel (in the y-direction) is much smaller than its rate of change through the channel in the x-direction. This approximation is the equivalent of Shockley's "gradual channel" approach in the theory of the junction FET. It ceases to be valid at the drain end of the channel as pinchoff is aproached.

In order to satisfy charge continuity along the channel, we require that:

$$I_D = Q_{IL}(y)wv_d = -Q_{IL}(y)w\mu_e E_y = \text{constant} \quad \text{(A4.5)}$$

Here,

v_d = the electron drift velocity,
μ_e = the electron mobility,
$E_y = -dV(y)/dy$ is the electric field along the channel.

Note that we have defined Q_{IL} to be >0, even though in an n-channel device it represents free electrons. Likewise, under normal operating conditions $I_D > 0$ and $dV(y)/dy > 0$. Thus,

$$\frac{I_D}{w\mu_e} = Q_{IL}(y)\frac{dV(y)}{dy} = \text{constant} \quad \text{(A4.6)}$$

The total voltage along the channel may be obtained by rearranging Equation (A4.6) and integrating y from 0 to l, and V from 0 to $V(l) = V_{Ch}$:

$$\frac{I_D}{w\mu_e}\int_0^l dy = \int_0^{V_{Ch}} Q_{IL}(y)\, dV \quad \text{(A4.7)}$$

Substituting for $Q_{IL}(y)$ using Equation (A4.4) yields

$$\begin{aligned}\frac{I_D l}{w\mu_e} &= \int_0^{V_{Ch}} \{Q_{SS} + C_{ox}(V_{GS} + V_{CP} - \Delta V_{Th}) - C_{ox}V(y) \\ &\quad - (2e\epsilon_0\epsilon_{Si}n_A)^{1/2}[\Delta V_{Th} + V(y)]^{1/2}\}\, dV \\ &= \{Q_{SS} + C_{ox}(V_{GS} + V_{CP} - \Delta V_{Th})\}V_{Ch} - \tfrac{1}{2}C_{ox}V_{Ch}^2 \\ &\quad - \tfrac{2}{3}(2e\epsilon_0\epsilon_{Si}n_A)^{1/2}\{(\Delta V_{Th} + V_{Ch})^{3/2} - \Delta V_{Th}^{3/2}\}\end{aligned} \quad \text{(A4.8)}$$

This expression may be simplified by substituting C_D as defined in Equation (3.18). This represents the depletion layer capacitance at $y = 0$ at threshold, when $\Delta V = \Delta V_{Th}$:

$$C_D = (e\epsilon_0\epsilon_{Si}n_A/2\,\Delta V_{Th})^{1/2} \tag{A4.9}$$

We may also substitute V_T for the threshold voltage, as defined by Equation (2.10):

$$V_T = \frac{Q_D(0) - Q_{SS}}{C_{ox}} + \Delta V_{Th} - V_{CP} \tag{2.10}$$

$$= \frac{(2e\epsilon_0\epsilon_{Si}n_A\,\Delta V_{Th})^{1/2} - Q_{SS}}{C_{ox}} + \Delta V_{Th} - V_{CP}$$

$$= \Delta V_{Th}\left(1 + 2\frac{C_D}{C_{ox}}\right) - \frac{Q_{SS}}{C_{ox}} - V_{CP} \tag{A4.10}$$

Then,

$$\frac{I_D l}{\mu_e w C_{ox}} = (V_{GS} - V_T)V_{Ch} - \tfrac{1}{2}V_{Ch}^2 + \frac{2C_D}{C_{ox}}\Delta V_{Th}^2\left[\frac{2}{3} + \frac{V_{Ch}}{\Delta V_{Th}} - \frac{2}{3}\left(1 + \frac{V_{Ch}}{\Delta V_{Th}}\right)^{3/2}\right] \tag{A4.11}$$

For very small values of V_{Ch} such that $V_{Ch} \ll \Delta V_{Th}$, we may use a binomial expansion of the final term of Equation (A4.11) to give

$$\frac{I_D l}{\mu_e w C_{ox}} = (V_{GS} - V_T)V_{Ch} - \tfrac{1}{2}V_{Ch}^2 + \frac{2C_D}{C_{ox}}\Delta V_{Th} V_{Ch}$$

$$- \frac{4}{3}\frac{C_D}{C_{ox}}\Delta V_{Th}^2\left(1 + \frac{3}{2}\frac{V_{Ch}}{\Delta V_{Th}} + \frac{3}{8}\frac{V_{Ch}^2}{\Delta V_{Th}^2} \cdots - 1\right)$$

$$\approx (V_{GS} - V_T)V_{Ch} - \tfrac{1}{2}V_{Ch}^2\left(1 + \frac{C_D}{C_{ox}}\right) \tag{A4.12}$$

At low drain current, when $V_{Ch} \ll V_{GS} - V_T$, Equation (A4.12) approximates to:

$$I_D l / w\mu_e C_{ox} = (V_{GS} - V_T)V_{Ch} \tag{A4.13}$$

as does Equation (3.10) of the simple theory. But at higher currents, the V_{Ch}^2 term becomes significant and its coefficient is increased by the factor $1 + C_D/C_{ox}$ in comparison with the simple theory. For VDMOS devices this is very significant because of the high channel doping concentration n_A. Using the values $t_{ox} = 100$ nm and $n_A = 1 \times 10^{23}\ \text{m}^{-3}$, we have $C_D = 3C_{ox}$. Then, Equation (A4.12) becomes

$$I_D = \mu_e \frac{w}{l} C_{ox}[(V_{GS} - V_T)V_{Ch} - 2V_{Ch}^2] \tag{A4.14}$$

Remember, this applies only for $V_{Ch} \ll \Delta V_{Th}$—in this case, for $V_{Ch} \ll 0.81$ V.

An alternative way of expressing Equation (A4.8) is in terms of an effective threshold voltage V_{T1} that varies slowly with V_{Ch}:

$$I_D l/\mu_e w C_{ox} = (V_{GS} - V_{T1})V_{Ch} - \tfrac{1}{2}V_{Ch}^2 \tag{A4.15}$$

This has the same form as Equation (3.9), but now

$$V_{T1} = \frac{\langle Q_D \rangle - Q_{SS}}{C_{ox}} + \Delta V_{Th} - V_{CP} \tag{A4.16}$$

where $\langle Q_D \rangle$ is an average value of the depletion-layer charge density along the channel, taken with respect to the channel voltage:

$$\langle Q_D \rangle = \frac{\int_0^{V_{Ch}} (2e\epsilon_0\epsilon_{Si}n_A)^{1/2}[\Delta V_{Th} + V(y)]^{1/2}\, dV}{\int_0^{V_{Ch}} dV}$$

$$= \tfrac{2}{3}(2e\epsilon_0\epsilon_{Si}n_A)^{1/2}\frac{(\Delta V_{Th} + V_{Ch})^{3/2} - \Delta V_{Ch}^{3/2}}{V_{Ch}} \tag{A4.17}$$

Clearly, $\langle Q_D \rangle$ and V_{T1} are larger than Q_D and V_T, respectively. As a result, I_D and $I_{D(Sat)}$ are smaller than the values predicted by the simple theory given in Section 3.2 of Chapter 3.

Differentiation of Equation (A4.7) shows that

$$\frac{\partial I_D}{\partial V_{Ch}} = \frac{w}{l}\mu_e Q_{IL}(l)$$

so that, at pinchoff, $Q_{IL}(l) = 0$ and $\partial I_D/\partial V_{Ch} = 0$. The drain current thus makes a smooth transition into the constant current regime.

A4.2. AFTER PINCHOFF

In Section 3.2 we suggest that pinchoff occurs and the constant-current region starts when $Q_{IL}(l) = 0$, and this has been corroborated by the theory just presented. However, a closer examination of this theory shows it to be riddled with self-contradictions. For example, with $Q_{IL}(l) = 0$, current continuity requires that $v_d = \infty$ at the drain end of the channel. Even if E_y became infinite at $y = l$, we have seen that that the carrier drift velocity saturates at a value of 9.2×10^4 m/s (v_s) for fields higher than about 5×10^6 V/m (E_s). In Section 3.2 we also discuss what happens when the electric field along the channel becomes so large that the carriers travel at their saturation drift velocity along most of the length of the channel. This leads to Equation (3.14), but it brings further problems. Adapting Equation (A4.5) to this situation would imply

$$I_D = Q_{IL}(y) w v_s = \text{constant} \tag{A4.18}$$

With v_s fixed, Q_{IL} must be independent of y. But that is not possible if the channel voltage, $V(y)$ is to vary, which it must in order to provide the field, E_y, in the first place. So what really happens?

We start to try to unravel this problem by simplifying the assumed variation of drift velocity with electric field. We take the mobility to remain constant for $0 < E < E_s$, and assume that the drift velocity saturates at v_s for $E > E_s$. This is shown in Figure 3.5, where it can be seen that it represents an overestimate of μ_e at high fields.

As long as

$$Q_{IL}(y) > I_D/wv_s \tag{A4.19}$$

everywhere in the channel, the assumptions that led to Equation (A4.8) remain valid. Consider first a situation in which V_{GS} is held constant and V_{DS} is increased, thereby increasing V_{Ch} and I_D, according to Equation (A4.8). Note that $Q_{IL}(y)$ is smallest at the drain end of the channel, and that $Q_{IL}(l)$ decreases as I_D and V_{Ch} increase:

$$\begin{aligned} Q_{IL}(l) = {} & Q_{SS} + C_{ox}(V_{GS} + V_{CP} - \Delta V_{Th} - V_{Ch}) \\ & - [2e\epsilon_0\epsilon_{Si}n_A(\Delta V_{Th} + V_{Ch})]^{1/2} \end{aligned} \tag{A4.20}$$

We have obtained Equation (A4.20) by substituting $y = l$ and $V(l) = V_{Ch}$ into Equation (A4.4).

When the limiting condition of (A4.19) is reached at the drain end of the channel—that is,

$$Q_{IL}(l) = \frac{I_D}{wv_s} = \frac{I_{D(Sat)}}{wv_s} \tag{A4.21}$$

—the drain current I_D ceases to increase with further increase in V_{DS}, remaining constant at $I_D = I_{D(Sat)}$. The potential variation along the channel, $V(y)$, and with it the charge distribution $Q_{IL}(y)$, becomes "pinned". They are not significantly influenced by further increase in V_{DS}. Any such increase simply sets up the familiar space-charge layer at the channel–drain junction. Conditions in the channel are not affected unless the space-charge layer causes a significant reduction in the channel length l. With the high channel doping levels used in VDMOS FETS, this should not occur.

We now have a new definition of pinchoff. The drain current, I_D, is given by Equation (A4.8), the inversion layer charge density at the drain end of the channel. $Q_{IL}(y)$, is given by Equation (A4.20), and at pinchoff they are related by Equation (A4.21). To calculate $I_{D(Sat)}$ we have to solve these equations simultaneously. The effect of the new theory is to cause a further reduction in the predicted value of $I_{D(Sat)}$, compared with the simple theory of Chapter 3 and with Equation (A4.11).

It is interesting to see how conditions in the channel become "pinned" in the way we have described. Once v_s is reached at the drain end, larger values of I_D would require an increase in $Q_{IL}(l)$. But that is not possible without reducing

$V(l) = V_{Ch}$ and hence I_D itself. Thus V_{Ch} becomes fixed, and the drain current saturates.

We have seen that, just above threshold, $I_{D(Sat)}$ and hence $Q_{IL}(l)$ are proportional to $(V_{GS} - V_T)^2$. But $Q_{IL}(0)$ is proportional to $V_{GS} - V_T$. With larger gate voltages applied and hence larger drain currents, $Q_{IL}(l)$ becomes an appreciable fraction of $Q_{IL}(0)$. Then, both are constrained to increase together in direct proportion to $V_{GS} - V_T$, and so $I_{D(Sat)}$ also becomes proportional to $V_{GS} - V_T$.

At the drain end of the channel the gradual channel approximation no longer applies. Indeed, in order to account for the current and voltage distributions here, we may have to invoke two new effects. One is that the carrier concentration gradient may become sufficiently high for diffusion to account for a significant proportion of the total current. The other is that the potential gradients parallel and perpendicular to the channel may become comparable in magnitude, with the result that Poisson's equation has to be satisfied in two dimensions.

The assumed discontinuity in the variation of v_d with E_y gives rise to a discontinuity in each I_D versus V_{DS} curve. It is clear from Figure 3.1 that no such sharp transitions occur in practice. The use of a device model that implies this can lead to difficulties in some circuit modelling and simulation packages.

To model an actual device more closely, two further effects have to be considered. First, the correct relationship of drift velocity versus electric field should be used. This will tend to reduce the value of I_D at higher values of V_{Ch}, and to smooth out the onset of saturation. Higher values of E_y are needed to support a given current than those predicted by the idealized curve. Secondly, due notice should be taken of the dopant concentration gradient along the channel from source to drain. This is a consequence of the double diffusion process used to define the channel. It means that $Q_D(y)$ increases with $V(y)$ at a rate less than that predicted by Equation (A4.1). As a result, $Q_{IL}(y)$ decreases less than the amount predicted by Equation (A4.4), and the reduction in $I_{D(Sat)}$ compared to the simple theory is less than Equation (A4.8) would suggest.

For more precise modeling, both effects may be put into the theory just discussed. We may use the appropriate function for $n_A(y)$, derived from the double diffusion process. This needs to be introduced into Equation (A4.4) to obtain $Q_{IL}(y)$. It is then necessary to use numerical methods to solve Equation (A4.7) for I_D as a function of V_{Ch}. We may also find a more satisfactory way to define the field-dependent mobility, $\mu_e(E_y)$, or the drift velocity, $v_d(E_y)$. The empirical expression proposed in Ref. 2 of Chapter 3 is

$$v_d(E_y) = \frac{\mu_e E_y}{[1 + (\mu_e E_y / v_s)^{\alpha}]^{1/\alpha}} \tag{A4.22}$$

The saturation drift velocity, v_s, and the index, α, were determined to give a best fit. Their values were 92,300 m/s and 1.92, respectively. The low field mobility, μ_e, depends on the normal field, E_x:

$$\mu_e = \frac{\mu_0}{[1 + (E_x/E_0)^c]} \tag{A4.23}$$

where $E_0 = 30.5$ V/μm and $c = 0.657$.

Neither of these expressions is very convenient for analysis, and they were obtained from carefully prepared specimens having a much lower doping level than that of a typical power MOSFET. The effect of using $\alpha = 2$ and $\alpha = 1$ in Equation (A4.22) is shown in Figure 3.5. Putting $\alpha = 1$ does significantly underestimate the drift velocity, just as our earlier approximation overestimated it. But it does permit a straightforward analytical solution of the device equations to be obtained. Equation (A4.5) becomes:

$$I_D = \frac{-Q_{IL} w \mu_e E_y}{(1 + E_y/E_s)} \tag{A4.24}$$

where $E_s = v_s/\mu_e$. Thus,

$$I_D + \frac{I_D E_y}{E_s} = -Q_{IL} w \mu_e E_y$$

$$\therefore \quad I_D = (Q_{IL} w \mu_e - I_D/E_s)\frac{dV}{dy} \tag{A4.25}$$

Integrating the left hand side from 0 to l and the right hand side from 0 to V_{Ch}, as before, we obtain:

$$I_D(l + V_{Ch}/E_s) = \int Q_{IL} w \mu_e \, dV \tag{A4.26}$$

Carrying out the integration as before, yields Equation (A4.8) or (A4.11), with l replaced by $l + V_{Ch}/E_s$. The effect of the reduced high field mobility is simply to cause an apparent increase in the length of the channel of V_{Ch}/E_s. With $E_s = 1.85\ \mathrm{V}/\mu\mathrm{m}$, and $l = 1\ \mu\mathrm{m}$, the effective length of the channel is doubled when $V_{Ch} = 1.85\ \mathrm{V}$.

The electron mobility at low longitudinal fields also varies with the mean normal field in the inversion layer, E_x, which can easily be related to the gate voltage:

$$E_x = \frac{(Q_D - \frac{1}{2}Q_{IL})}{\epsilon_0 \epsilon_{Si}} \tag{A4.27}$$

$$\frac{Q_D}{\epsilon_0 \epsilon_{Si}} = \left(\frac{2en_A \Delta V_{Th}}{\epsilon_0 \epsilon_{Si}}\right)^{1/2} \approx 16\,[\mathrm{V}/\mu\mathrm{m}] \tag{A4.28}$$

$$\frac{Q_{IL}}{2\epsilon_0 \epsilon_{Si}} = \frac{C_{ox}(V_{GS} - V_T)}{2\epsilon_0 \epsilon_{Si}} = \frac{\epsilon_{ox}}{\epsilon_{Si}} \frac{V_{GS} - V_T}{2t_{ox}}$$

$$\approx 1.6(V_{GS} - V_T)[\mathrm{V}/\mu\mathrm{m}] \tag{A4.29}$$

Then,

$$E_x/[V/\mu\mathrm{m}] \approx 16[1 + 0.1(V_{GS} - V_T)] \tag{A4.30}$$

Equation (A4.23) is also not a very convenient form for obtaining analytical solutions, although an even more complex, semi-empirical, expression has been proposed [1]. This fits a wide range of experimental results of different workers, and takes account of several expected carrier scattering mechanisms and their variation with temperature, field, doping concentration, and surface condition. The surface charge density, Q_{SS}, is expected to give rise to significant scattering, even though it is shielded by electrostatic screening at high carrier densities. This has been confirmed by experiment [2].

It is most convenient if the data for the variation of μ_e with E_x can be fitted to Equation (A4.23) with $c = 1$. Then,

$$\mu_e = \frac{\mu_0}{(1 + E_x/E_0)} \tag{A4.31}$$

Approximate values for μ_0 and E_0 under the conditions obtaining in a power MOSFET are $0.06\,\mathrm{m^2/Vs}$ and $30\,\mathrm{V/\mu m}$, respectively. Thus, substituting Equation (A4.30), we obtain:

$$\mu_e = \frac{\mu_{e0}}{[1 + \theta(V_{GS} - V_T)]} \tag{A4.32}$$

where $\mu_{e0} = 0.052\,\mathrm{m^2/Vs}$ and $\theta = 0.014\,\mathrm{V^{-1}}$.

Putting all these effects together, the equation for the linear region of the characteristics becomes:

$$I_D = \frac{\mu_{e0}(w/l)C_{ox}\left\{(V_{GS} - V_T)V_{Ch} - \frac{1}{2}V_{Ch}^2 + \frac{2C_D}{C_{ox}}\Delta V_{Th}^2\left[\frac{2}{3} + \frac{V_{ch}}{\Delta V_{Th}} - \frac{2}{3}\left(1 + \frac{V_{Ch}}{\Delta V_{Th}}\right)^{3/2}\right]\right\}}{[1 + \theta(V_{GS} - V_T) + \mu_{e0}V_{Ch}/lv_s]} \tag{A4.33}$$

The drain current saturates when $\partial I_D/\partial V_{Ch} = 0$, so there is no discontinuity at that point.

To represent our typical IRF 540 device, we have assumed the following parameters, using slightly larger values for μ_{e0} and θ:

$\mu_{e0} = 0.06\,\mathrm{m^2/Vs}$ $\quad w/l = 1.4 \times 10^6$
$C_{ox} = 3.45 \times 10^{-4}\,\mathrm{F/m^2}$ $\quad C_D = 10.2 \times 10^{-4}\,\mathrm{F/m^2}$
$\theta = 0.02\,\mathrm{V^{-1}}$ $\quad lv_s/\mu_{e0} = 1.3\,\mathrm{V}$
$V_T = 3.5\,\mathrm{V}$ $\quad \Delta V_{Th} = 0.81\,\mathrm{V}$

In SI units, Equation (A4.33) then becomes:

$$I_D = \frac{30\{(V_{GS} - 3.5)V_{Ch} - \frac{1}{2}V_{Ch}^2 + 3.9[\frac{2}{3} + 1.24V_{Ch} - \frac{2}{3}(1 + 1.24V_{Ch})^{3/2}]\}}{[1 + 0.02(V_{GS} - 3.5) + 0.93V_{Ch}]}$$

and

$$V_{DS} = V_{Ch} + I_D R_{DS(on)} \tag{A4.34}$$

It is Equation (A4.34) that is plotted in Figure 3.7 and the corresponding values of $I_{D(Sat)}$ are shown in Figures 3.4 and 3.6.

REFERENCES

1. S. A. Schwarz and S. E. Russek, "Semi-empirical equations for electron velocity in silicon: Part II—MOS inversion layer." *IEEE Trans. on Electron Devices,* **ED-30,** pp. 1634–1639 (1983).
2. S. C. Sun and J. D. Plummer, "Electron mobility in inversion and accumulation layers on thermally oxidised silicon surfaces." *IEEE Trans. on Electron Devices,* **ED-27,** pp. 1497–1508 (1980).

APPENDIX 5

Cell Geometry

In Section 3.4 the device transconductance g_m, and hence the current-controlling capability of the MOSFET, are shown to be proportional to the gate width w. As a result, an important design objective must be to obtain as wide a gate as possible on a given area of silicon.

In principle, VDMOS FET gates may be laid out as linear arrays, interdigitated with the source, as shown in Figure 3.9*a*. Then, the gate width per unit area is simply $w/A = 1/a$, where a is the pitch of the array. Photolithographic line definition and mask registration set a lower limit to the pitch size.

Aside from these considerations, it is also important that spacing between the channel diffusions, b, should always exceed a certain minimum value. The reason for this is that built into the VDMOS structure is a parasitic JFET in series with the drain. This is described in Section 3.6. It is most likely to have a serious effect in high-voltage devices, in which the epitaxial layer is more lightly doped. What happens is that the conduction path below the gate is pinched off by the depletion layers that form around the junction with the p-type regions.

The alternative to the linear gate geometry is a cellular array. There are several reasons why, in practice, the latter is normally preferred. By comparison, the linear structures have a lower packing density (w/A), they give rise to a higher device capacitance, and they usually require rather more complicated photolithographic masks. The long gate fingers give rise to significant resistance when a polycrystalline silicon gate is used, and this causes turn-on delays. However, one manufacturer advocates the use of a linear metal gate structure, in conjunction with a simple four-mask fabrication process, for producing inexpensive high-voltage devices. Another, quite sophisticated interdigitated design has been made for high-frequency applications [1].

Several different cellular structures have been proposed, and of these, a few remain in widespread production by different manufacturers. Designs have used cells of hexagonal, square, and even triangular shape, and they have been based on square or hexagonal lattices. Some of the options are illustrated in Figure 3.9. Technological advances have led to a reduction of the cell pitch, and hence to larger values of gate width for a given chip size. More than three million cells per square inch have been obtained, giving gate widths exceeding 1 m on a 10-mm^2 chip (100×150 mil). Reducing cell dimensions also gives rise to a more uniform current distribution in lower-voltage devices, for reasons discussed in Section 3.5, and this brings the further benefit that it lowers $R_{DS(on)}$.

Let us assume that the need to avoid the effects of the parasitic JFET, together with the level of technology used in fabrication, sets a lower limit on the spacing between adjacent channel diffusions under the gate oxide. Note that this distance, b, is about 5 μm less than the breadth of the polysilicon arms forming the gate web. (It must be clearly distinguished from the gate *width*, w, which depends on the total length of these arms.) Then, for each of the possible cell structures there is an optimum cell pitch that maximizes the value of w/A, the gate width per unit area. Consider the case shown in Figure 3.9b, where hexagonal cells of side s lie on a hexagonal lattice of pitch a. Each cell occupies an area $\sqrt{3}\,a^2/2$ and provides a gate width of $6s$. The geometry of the configuration means that $a = b + \sqrt{3}\,s$. Thus,

$$\frac{w}{A} = \frac{12s}{\sqrt{3}\,(b + \sqrt{3}\,s)^2} \tag{A5.1}$$

Differentiating with respect to s and setting the derivative equal to zero shows that w/A is a maximum when $s = b/\sqrt{3}$, and hence $a = 2b$. Thus,

$$\left(\frac{w}{A}\right)_{\max} = \frac{1}{b} = \frac{1}{\sqrt{3}\,s} = \frac{2}{a} \tag{A5.2}$$

Results for other cell patterns may be derived similarly and are set out in Table A5.1. On this argument the hexagon-on-hexagon, square-on-square, and triangle-on-hexagon arrangements are equally good, and have a slight advantage over the others.

Clearly, the smaller b can be made, and with it a, the larger is $(w/A)_{\max}$. However, the limit on b is far from being the only constraint. In particular, the need to make a low-resistance ohmic contact to the source means that the cell size

TABLE A5.1. Optimum Cell Geometries

Figure	Configuration	Constraint	Condition for $(w/A)_{\max}$	$(w/A)_{\max}$
3.9(a)	Linear	$a = b + s$	$s \to 0$	$1/a = 1/b$
3.9(b)	Hexagons on hexagonal lattice	$a = b + \sqrt{3}\,s$	$s = b/\sqrt{3}$	$1/b = 1/\sqrt{3}\,s = 2/a$
3.9(c)	Hexagons on square lattice	$a = b + 2s$	$s = b/2$	$3/4b = 3/8s = 3/2a$
3.9(d)	Squares on square lattice	$a = b + s$	$s = b$	$1/b = 1/s = 2/a$
3.9(e)	Squares on an offset square lattice	$a = b + \sqrt{2}\,s$	$s = b/\sqrt{2}$	$1/\sqrt{2}\,b = 2/s = \sqrt{2}/a$
3.9(f)	Squares on a hexagonal lattice	$a \approx (2/\sqrt{3})(b + s)$	$s \approx b$	$\sqrt{3}/2b \approx \sqrt{3}/2s \approx 2/a$
3.9(g)	Triangles on hexagonal lattice	$a = \sqrt{3}(\sqrt{3}\,b + s)$	$s = \sqrt{3}\,b$	$1/b = \sqrt{3}/s = 6/a$

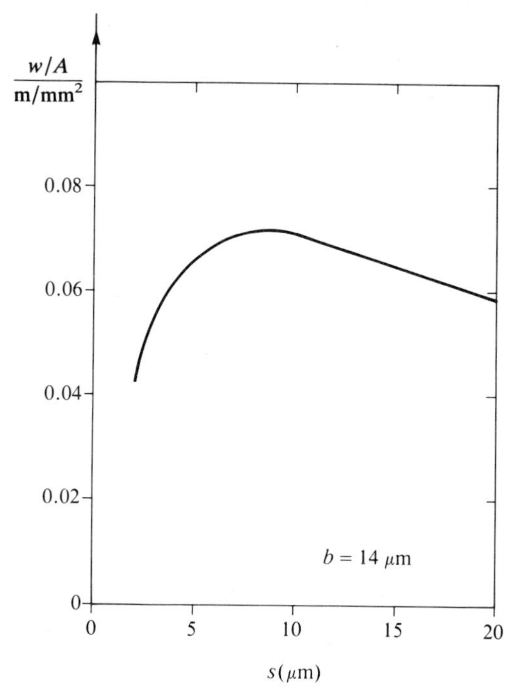

Figure A5.1. Variation of gate width per unit area with cell size. Hexagons on a hexagonal lattice with $b = 14\ \mu$m.

s may not be made too small either. As a result, it is normal practice for the cell pitch a to be made rather larger (for a given value of b) than the optimum cell pitch predicted by the simple theory and set out in Table A5.1. The important parameter $(w/A)_{max}$ is not very sensitive to such variations, as can be seen in Figure A5.1. Thus, the HEXFET shown in Figure 3.10*a* has $b = 14\ \mu$m but $a = 38\ \mu$m ($s = 14\ \mu$m). Even so, the gate width per unit area is 0.067 m/mm^2, compared with the optimum value of 0.071 m/mm^2 when $a = 28\ \mu$m ($s = 8\ \mu$m).

There are two other important considerations. The first is concerned with the added resistance that results from the higher current density under the gate oxide between the p diffusions. This is less when the shortest distance between adjacent cells lies between their vertices rather than their sides. This goes some way to redress the slight disadvantage of the arrangements of Figures 3.9*c*, *e*, and *f*. The second consideration is the need for an even distribution of current around the whole width of the channel. This is obtained more easily with the more obtuse vertices of the hexagonal cell than it is with square or triangular cells.

Overall it can be said that the theoretical advantages of one cell arrangement over another are at most marginal. What matters is the exploitation of their respective merits by the development of sophisticated, specialized manufacturing techniques.

REFERENCE

1. D. Fuoss, "Vertical DMOS power field-effect transistors optimized for high-speed operation." *Proc. IEEE Int. Electron Devices Meet.* **82,** 250–253 (1982).

APPENDIX 6

ON-Resistance and Breakdown Voltage

A6.1. BASIC EQUATIONS

Here we give a simple theoretical treatment of the expected functional relationship between the avalanche breakdown voltage V_B and the basic ohmic resistance R of a planar $p^+n^-n^+$ diode structure. It should be understood that R is the resistance presented by the n^- layer to a uniformly distributed current of majority electrons, when there is no conductivity modulation by minority carriers. We shall show how an optimum doping profile may be determined, in theory. However, of greater significance is the form of the variation of R with V_B.

The resistance of such a diode of area, A, is given by

$$RA = \int_0^d \rho(x)\,\mathrm{d}x = \int_0^d \frac{\mathrm{d}x}{n_0(x)e\mu_e} \tag{A6.1}$$

In Equation (A6.1) the x-coordinate is defined to be perpendicular to the junctions, with its origin at the p^+n^- junction. The total thickness of the n^- region is d. The value of $n_D(x)$, the donor concentration, may vary locally, and with it so does the local value of the resistivity ρ. As usual, e is the electronic charge and μ_e the electron mobility. At 25°C and low doping levels, $\mu_e = 0.15\ \mathrm{m^2/V\,s}$. This value becomes $0.07\ \mathrm{m^2/V\,s}$ at 125°C and $0.32\ \mathrm{m^2/V\,s}$ at −55°C. The electron mobility also varies with n_D, which is a matter we return to.

Under conditions of reverse bias the electric field is determined by Poisson's equation, as shown in Appendix 1. The maximum field occurs at the p^+n^- junction, as shown in Figure A6.1. Two situations are illustrated. Both assume that n_D is uniform. In the first case the depletion layer has not quite *reached through* to the n^-n^+ junction. Raising the applied voltage increases the field by the same amount everywhere until the breakdown field E_B is reached at the p^+n^- junction at the voltage V_B. In the case illustrated this has caused reachthrough. At low values of n_D, $E_B = 2.2 \times 10^7\ \mathrm{V/m}$. This value also varies slowly with n_D, as we discuss in due course. The total voltage dropped across the reverse-biased structure is

$$V_B = E_B d - \int_0^d \left[\int_0^x \frac{e n_D(x')}{\epsilon_0 \epsilon_{Si}}\,\mathrm{d}x' \right] \mathrm{d}x \tag{A6.2}$$

Note that we have neglected as very small the voltages dropped in the p^+ and n^+

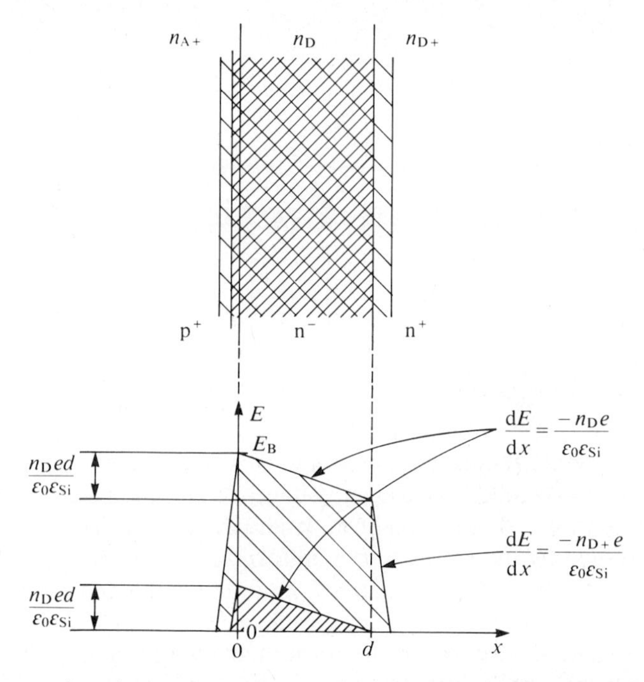

Figure A6.1. The $p^+n^-n^+$ junction under reverse bias. Heavy cross-hatching shows extent of depletion layer at low voltage when it just reaches the n^-n^+ junction. The lighter hatching shows the depletion layer at breakdown when E reaches E_B.

regions, which are represented by the areas shaded in those regions. They are usually quite negligible compared to the voltage across the n^- region.

A6.2. FUNCTIONAL RELATIONSHIP

Ideally we wish to find the doping profile $n_D(x)$ that minimizes RA in Equation (A6.1). However, the functional relationship between RA and V_B can be established without going through the full minimization procedure. We do this first for the simpler case when n_D is uniform. It is important to understand that this does not affect the functional relationship, only the optimum value of RA.

With uniform doping, Equation (A6.1) becomes:

$$RA = d/n_D e\mu_e \tag{A6.3}$$

and Equation (A6.2) becomes

$$V_B = E_B d - \frac{n_D e d^2}{2\epsilon_0 \epsilon_{Si}} \tag{A6.4}$$

When n_D is very small, the field remains constant at E_B and $V_B \approx E_B d$, as shown in Figure A6.2*a*. When n_D is such that the depletion layer just reaches

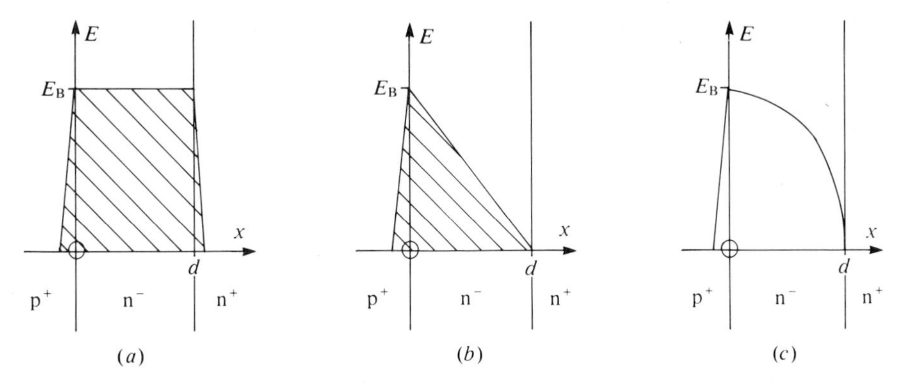

Figure A6.2. Effect of doping profile on electric field distribution. (*a*) With very lightly doped n^- region; (*b*) more heavily doped n^- region, such that breakdown and reachthrough occur at the same voltage; (*c*) field distribution corresponding to the optimum doping profile of Equation (A6.26). Note that V_B is represented by the hatched areas under the curves of E_x, and that $dE/dx = (n_A - n_D)e/(\epsilon_0\epsilon_{Si})$. In (*a*), $E_B \approx$ constant, so that $V_B \approx E_B d$. In (*b*), $V_B \approx \frac{1}{2}E_B d$.

through to d, as shown in Figure A6.2*b*, $V_B \approx \frac{1}{2}E_B d$. In general we may write

$$V_B = E_B d/a \tag{A6.5}$$

where a is a constant between 1 and 2. Then, eliminating d from Equation (A6.4),

$$n_D = 2\left(1 - \frac{1}{a}\right)\frac{\epsilon_0\epsilon_{Si}E_B^2}{eV_B} \tag{A6.6}$$

so that

$$RA = \frac{d}{n_D e\mu_e} = \frac{a^3V_B^2}{2(a-1)\mu_e\epsilon_0\epsilon_{Si}E_B^3} \tag{A6.7}$$

Thus we have the result that $RA \propto V_B^2$.

This relationship needs to be modified to allow for the dependences of E_B and μ_e on the doping concentration n_D. These are shown in Figures A6.3 and A6.4. Clearly, the dependence of E_B has much greater effect because of the third-power relationship in Equation (A6.7).

In order to keep the analysis general, put $\mu_e \propto n_D^{-x}$ and $E_B \propto n_D^y$. Then Equation (A6.3) becomes

$$RA = k_1 d n_D^{-(1-x)} \tag{A6.8}$$

and Equation (A6.6) becomes, after rearrangement,

$$V_B = k_2 n_D^{2y}/n_D = k_2 n_D^{-(1-2y)} \tag{A6.9}$$

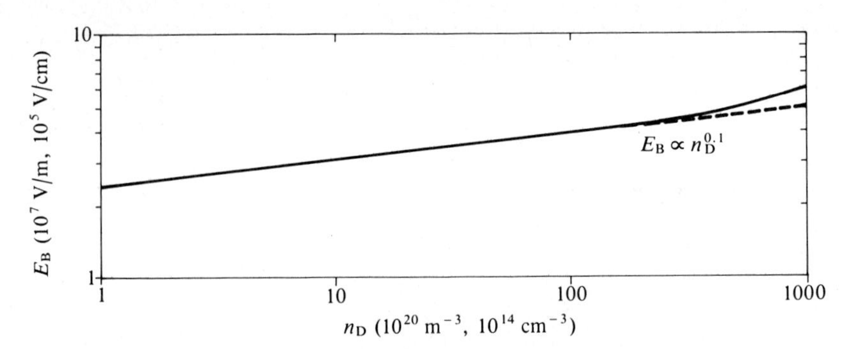

Figure A6.3. Variation of breakdown field with doping concentration. The curve shown is for an abrupt, asymmetric, plane p$^+$n junction.

Substituting Equations (A6.5) and (A6.9) into Equation (A6.8) gives

$$RA = k_1 \frac{aV_B}{E_B} n_D^{-(1-x)} = k_2 V_B n_D^{-(1-x+y)}$$

$$= k_3 V_B^{\alpha} \qquad \text{(A6.10)}$$

where

$$\alpha = \frac{2-x-y}{1-2y} \qquad \text{(A6.11)}$$

We have introduced the coefficients k_1, k_2, and k_3 in order to simplify the constant terms in the expressions.

From Figures A6.3 and A6.4, it can be seen that over a narrow range of values of n_D, in the region of 10^{21} m^{-3} (the region of interest to us), $x \approx 0.06$ and $y \approx 0.1$. These values give $\alpha = 2.3$. This would be different at other values of n_D. Indeed, we should expect α to increase at higher doping levels. This helps to explain why several different values for α are to be found qouted in the literature. For

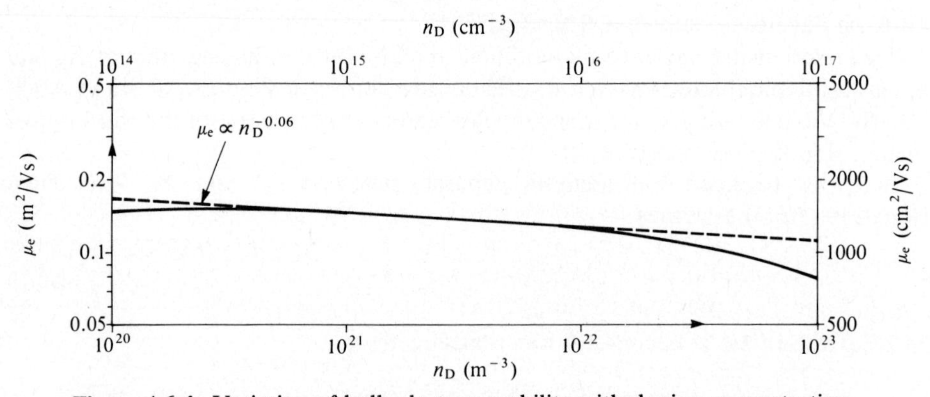

Figure A6.4. Variation of bulk electron mobility with doping concentration.

example, a well-established empirical relationship [1] between V_B and n_D for silicon p^+n diodes that do not "reach through" is (in SI units)

$$V_B = 60(n_D/10^{22})^{-3/4} \tag{A6.12}$$

If we use Equation (A1.15) from Appendix 1 in conjunction with this by putting $E_B = E_0$, $V_B = V_0 - V_{CP}$, and $1/n_A = 0$, we obtain

$$E_B = \left(\frac{2en_D V_B}{\epsilon_0 \epsilon_{Si}}\right)^{1/2} \tag{A6.13}$$

$$= k_4 n_D^{1/8} \tag{A6.14}$$

That is, $y = 0.125$. With $x = 0$, this gives $\alpha = 2.5$. Another quoted approximation [2] is

$$E_B = 78 \times 10^6 V_B^{-0.2} \tag{A6.15}$$

Substituting Equation (A6.12) gives

$$E_B = k_5 n_D^{0.15} \tag{A6.16}$$

Using $x = 0$ and $y = 0.15$ in Equation (A6.11) gives the frequently quoted value $\alpha = 2.6$. With the constants evaluated,

$$(RA)_{\text{optimum}} \approx 4 \times 10^{-13} V_B^{2.6}\ \Omega\ \text{m}^2 \qquad (V_B \text{ in volts}) \tag{A6.17}$$

A6.3. OPTIMIZATION WITH UNIFORM DOPING

We wish to minimise RA (that is, d/n_D), subject to the constraint imposed by Equation (A6.4). This equation may be arranged as

$$\left(\frac{d}{n_D}\right)^2 - \frac{2\epsilon_0\epsilon_{Si}E_B}{en_D}\left(\frac{d}{n_D}\right) + \frac{2\epsilon_0\epsilon_{Si}}{en_D^3}V_B = 0 \tag{A6.18}$$

We are seeking the value of n_D which gives the lowest value of d/n_D consistent with Equation (A6.18). To simplify the algebra, put $y = d/n_D$, $b = 2\epsilon_0\epsilon_{Si}E_B/en_D^2$, and $c = 2\epsilon_0\epsilon_{Si}V_B/en_D^3$. Then Equation (A6.18) may be written as

$$f(y) = y^2 - by + c = 0 \tag{A6.19}$$

One way of solving this type of problem is to introduce a Lagrange undetermined multiplier λ and seek the minimum with respect to n_D of $y + \lambda f(y)$. This requires that

$$\frac{\partial}{\partial n_D}[y + \lambda f(y)]_{y=\text{constant}} = \lambda\left(-y\frac{db}{dn_D} + \frac{dc}{dn_D}\right) = 0$$

That is,

$$
\begin{aligned}
y = y_{\min} &= \frac{dc/dn_{\mathrm{D}}}{db/dn_{\mathrm{D}}} \\
&= \frac{2\epsilon_0\epsilon_{\mathrm{Si}} V_{\mathrm{B}}}{e}\left(\frac{-3}{n_{\mathrm{D}}^4}\right)\frac{e}{2\epsilon_0\epsilon_{\mathrm{Si}} E_{\mathrm{B}}}\left(\frac{n_{\mathrm{D}}^3}{-2}\right) \\
&= 3V_{\mathrm{B}}/2n_{\mathrm{D}}E_{\mathrm{B}} \qquad \text{(A6.20)}
\end{aligned}
$$

Substitution of Equation (A6.20) back into Equation (A6.19) or (A6.18) yields the required optimum value for n_{D}:

$$
\frac{9V_{\mathrm{B}}^2}{4n_{\mathrm{D}}^2E_{\mathrm{B}}^2} - \frac{2\epsilon_0\epsilon_{\mathrm{Si}}E_{\mathrm{B}}}{en_{\mathrm{D}}^2}\frac{3V_{\mathrm{B}}}{2n_{\mathrm{D}}E_{\mathrm{B}}} + \frac{2\epsilon_0\epsilon_{\mathrm{Si}}V_{\mathrm{B}}}{en_{\mathrm{D}}^3} = 0
$$

Therefore,

$$
\frac{9V_{\mathrm{B}}^2}{4n_{\mathrm{D}}^2E_{\mathrm{B}}^2} = \frac{\epsilon_0\epsilon_{\mathrm{Si}}V_{\mathrm{B}}}{en_{\mathrm{D}}^3} \qquad \text{(A6.21)}
$$

so that

$$
n_{\mathrm{D}} = 4\epsilon_0\epsilon_{\mathrm{Si}}E_{\mathrm{B}}^2/9eV_{\mathrm{B}} = (n_{\mathrm{D}})_{\mathrm{opt}} \qquad \text{(A6.21a)}
$$

and

$$
\begin{aligned}
y &= \frac{d}{n_{\mathrm{D}}} = \frac{3V_{\mathrm{B}}}{2E_{\mathrm{B}}}\frac{9eV_{\mathrm{B}}}{4\epsilon_0\epsilon_{\mathrm{Si}}E_{\mathrm{B}}^2} \\
&= \frac{27eV_{\mathrm{B}}^2}{8\epsilon_0\epsilon_{\mathrm{Si}}E_{\mathrm{B}}^3} \qquad \text{(A6.21b)}
\end{aligned}
$$

Hence,

$$
RA = \frac{d}{n_{\mathrm{D}}e\mu_{\mathrm{e}}} = \frac{27V_{\mathrm{B}}^2}{8\epsilon_0\epsilon_{\mathrm{Si}}\mu_{\mathrm{e}}E_{\mathrm{B}}^3} \qquad \text{(A6.22)}
$$

At 25°C, with $\mu_{\mathrm{e}} = 0.15\ \mathrm{m^2/V\,s}$ and $E_{\mathrm{B}} = 2.2 \times 10^7\ \mathrm{V/m}$,

$$
RA = 2 \times 10^{-11}V_{\mathrm{B}}^2\ \Omega\,\mathrm{m}^2 \qquad (V_{\mathrm{B}} \text{ in volts}) \qquad \text{(A6.23)}
$$

Although this may be quite a neat piece of analysis, it clearly suffers from a number of serious shortcomings. In the first place, there is the assumption of a uniform doping profile. Next is the neglect of the variation of E_{B} with n_{D}. Finally there is the neglect of the variation of μ_{e} with n_{D}. However, the analysis may readily be extended to deal with each of these factors, as shown in the next section.

A6.4. OPTIMUM DOPING PROFILE

The optimization procedure using Lagrange multipliers may be extended so as to obtain the optimum doping profile required to minimize RA [2]. When n_D is a function of position,

$$RA = \int \frac{dx}{n_D e \mu_e} = \int_0^d \frac{-dx}{\epsilon_0 \epsilon_{Si} \mu_e (dE/dx)} \tag{A6.24}$$

and

$$V_B = \int -E \, dx \tag{A6.25}$$

a solution of which was given as Equation (A6.2). The optimum profile [2] is found to be

$$n_D(x) = \frac{\epsilon_0 \epsilon_{Si} E_B^2}{3eV_B(1 - 2E_B x/3V_B)^{1/2}} \tag{A6.26}$$

That is, the donor doping density is lowest just before the p^+n^- junction and increases with x, becoming large as $x \to d = 3V_B/2E_B$. The corresponding electric field distribution is shown in Figure A6.2(c) This profile gives

$$(RA)_{min} = 3V_B^2/\epsilon_0 \epsilon_{Si} \mu_e E_B^3 \tag{A6.27}$$

This represents an 11% improvement over the case of uniform doping, as given by Equation (A6.22).

Still further refinement to the theory could be obtained by putting the empirical variations of E_B with μ_e and n_D into the analysis at the beginning of the optimization procedure. To the knowledge of the authors this analysis has not been published, and the problem is left to the interested reader with time to spare. However, this detail is academic in view of the lateral nonuniformities in actual devices caused by the source and channel diffusions. And it does not affect the predicted functional relationship between RA and V_B in optimized devices. This should involve a power law with an index in the region of 2.5, and a coefficient in SI units of about 6×10^{-13}. Thus,

$$(RA)_{opt} \approx 6 \times 10^{-13} V_B^{2.5} \ \Omega \ m^2 \qquad (V_B \text{ in volts}) \tag{A6.28}$$

Equation (A6.28) is plotted as the "theoretical limit" in Figure 3.14. As explained in Section 3.5, more important considerations associated with the lateral nonuniformity of the current distribution in real devices determine the best doping profile to use in practice.

REFERENCES

1. S. M. Sze, *The Physics of Semiconductors,* 2nd ed. Wiley, New York, 1981.
2. C. Hu, "Optimum doping profile for minimum ohmic resistance and high breakdown voltage." *IEEE Trans. Electron Devices,* **ED-26,** 243–244 (1979).

APPENDIX 7

A Power-MOSFET Data Sheet

A7.1. INTRODUCTION

Figure A7.1 shows a typical power-MOSFET data sheet. Data sheets from different manufacturers initially were very similar, especially where they used the same industry-standard part numbers. However, as new ratings have been introduced, manufacturers have adopted different rating methods and styles of presentation. Some of these alternatives are identified in the notes that follow. The notes refer to sections in the specimen data sheet which have the appropriate number alongside in the left-hand margin.

A7.2. NOTES ON DATA-SHEET FEATURES

1. *Power-MOSFET Symbol.* Some manufacturers have chosen to use a zener-diode symbol to represent the integral body–drain diode in order to draw attention to the repetitive avalanche capability. Others continue to use an ordinary diode symbol.

2. *Product-Summary I_D Rating.* The value given for the I_D rating in product summaries and brief device descriptions is usually the I_D rating with the case temperature held at 25°C. In practice it is impossible for the case temperature to be held at 25°C while the device is carrying this current unless it is cooled by exotic means. However, it has become common practice in the power-semiconductor industry to use the $T_J = 25$°C value for the current rating of the MOSFET. A more usable current rating is given by the I_D rating with the case temperature held at 100°C, given later in the data sheet. The use of a plainly optimistic value for the MOSFET current rating has its parallel in bipolar-transistor data sheets, where it is common to assign an I_C rating to the transistor which is in practice not usable because of the low gain of the transistor at that value.

3. *Voltage Downgrades.* Devices are usually selected for voltage breakdown after packaging. Manufacturers commonly sell two voltage grades of the same device—"prime" devices which meet the higher voltage specification, and "downgrade" devices which meet a lower voltage specification. Since the thickness of the epitaxial drain region is the same in both cases, the $R_{DS(on)}$ is the same, so that the lower-voltage device does not have an optimally low value of $R_{DS(on)}$. Devices may also be graded according to $R_{DS(on)}$.

4. *Device Marking.* As well as the part number, manufacturers commonly

INTERNATIONAL RECTIFIER 

REPETITIVE AVALANCHE AND dv/dt RATED*

LOWER ON STATE RESISTANCE, 175°C OPERATING TEMPERATURE

HEXFET® TRANSISTORS

IRF540
IRF541
IRF542
IRF543

N-CHANNEL POWER MOSFETs

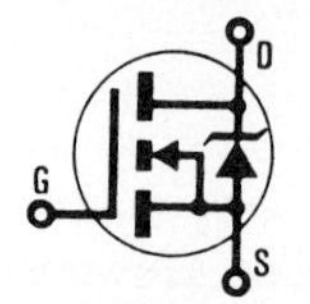

100 Volt, 0.077 Ohm HEXFET
TO-220AB Plastic Package

The HEXFET® technology is the key to International Rectifier's advanced line of power MOSFET transistors. The efficient geometry and unique processing of this latest "State of the Art" design achieves: very low on-state resistance combined with high transconductance; superior reverse energy and diode recovery dv/dt capability.

The HEXFET transistors also feature all of the well established advantages of MOSFETs such as voltage control, very fast switching, ease of paralleling and temperature stability of the electrical parameters.

They are well suited for applications such as switching power supplies, motor controls, inverters, choppers, audio amplifiers and high energy pulse circuits.

Product Summary

Part Number	BV_{DSS}	$R_{DS(on)}$	I_D
IRF540	100	0.077	28
IRF541	80	0.077	28
IRF542	100	0.100	25
IRF543	80	0.100	25

FEATURES:

- Repetitive Avalanche Ratings
- Dynamic dv/dt Rating
- Simple Drive Requirements
- Ease of Paralleling

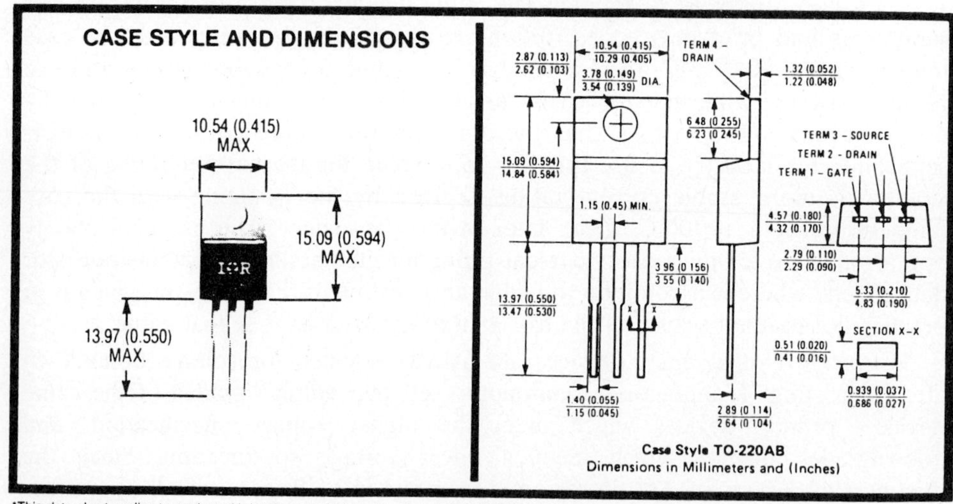

*This data sheet applies to product with batch codes that begin with a digit, ie. 2A3B

Figure A7.1. Power-MOSFET data sheet.

Absolute Maximum Ratings

		Parameter	IRF540, IRF541	IRF542, IRF543	Units
5	I_D @ T_C = 25°C	Continuous Drain Current	28	25	A
6	I_D @ T_C = 100°C	Continuous Drain Current	20	17	A
7	I_{DM}	Pulsed Drain Current ①	110	100	A
8	P_D @ T_C = 25°C	Max. Power Dissipation	150		W
9		Linear Derating Factor	1.0		W/K ⑤
10	V_{GS}	Gate-to-Source Voltage	±20		V
11	E_{AS}	Single Pulse Avalanche Energy ②	230 (See Fig. 14)		mJ
12	I_{AR}	Avalanche Current ① (Repetitive or Non-Repetitive)	28 (See E_{AR})		A
13	E_{AR}	Repetitive Avalanche Energy ①	15 (See I_{AR})		mJ
14	dv/dt	Peak Diode Recovery dv/dt ③	5.5 (See Fig. 17)		V/ns
15	T_J T_{STG}	Operating Junction Storage Temperature Range	−55 to 175		°C
		Lead Temperature	300 (0.063 in. (1.6mm) from case for 10s)		°C

Electrical Characteristics @ T_J = 25°C (Unless Otherwise Specified)

		Parameter	Type	Min.	Typ.	Max.	Units	Test Conditions	
16	BV_{DSS}	Drain-to-Source Breakdown Voltage	IRF540 IRF542	100	—	—	V	V_{GS} = 0V, I_D = 250 μA	
			IRF541 IRF543	80					
17	$R_{DS(on)}$	Static Drain-to-Source On-State Resistance ④	IRF540 IRF541	—	0.060	0.077	Ω	V_{GS} = 10V, I_D = 17A	
			IRF542 IRF543	—	0.080	0.100			
18	$I_{D(on)}$	On-State Drain Current ④	IRF540 IRF541	28	—	—	A	$V_{DS} > I_{D(on)} \times R_{DS(on)}$ Max. V_{GS} = 10V	
			IRF541 IRF543	25					
19	$V_{GS(th)}$	Gate Threshold Voltage	ALL	2.0	—	4.0	V	$V_{DS} = V_{GS}$, I_D = 250μA	
20	g_{fs}	Forward Transconductance ④	ALL	8.7	13	—	S (℧)	$V_{DS} \geq$ 50V, I_{DS} = 17A	
21	I_{DSS}	Zero Gate Voltage Drain Current	ALL	—	—	250	μA	V_{DS} = Max. Rating, V_{GS} = 0V	
				—	—	1000		V_{DS} = 0.8 x Max. Rating V_{GS} = 0V, T_J = 150°C	
22	I_{GSS}	Gate-to-Source Leakage Forward	ALL	—	—	500	nA	V_{GS} = 20V	
23	I_{GSS}	Gate-to-Source Leakage Reverse	ALL	—	—	−500	nA	V_{GS} = −20V	
24	Q_g	Total Gate Charge	ALL	—	40	60	nC	V_{GS} = 10V, I_D = 28A	
25	Q_{gs}	Gate-to-Source Charge	ALL	—	8.8	13	nC	V_{DS} = 0.8 x Max. Rating See Fig. 16	
26	Q_{gd}	Gate-to-Drain ("Miller") Charge		—	16	25	nC	(Independent of operating temperature)	
27	$t_{d(on)}$	Turn-On Delay Time	ALL	—	15	23	nS	V_{DD} = 50V, $I_D \approx$ 28A, R_G = 9.1Ω	
28	t_r	Rise Time	ALL	—	72	110	nS	R_D = 1.7Ω	
29	$t_{d(off)}$	Turn-Off Delay Time	ALL	—	40	60	nS	See Fig. 15	
30	t_f	Fall Time	ALL	—	50	75	nS	(Independent of operating temperature)	
31	L_D	Internal Drain Inductance	ALL	—	4.5	—	nH	Measured from the drain lead, 6mm (0.25 in.) from package to center of die.	Modified MOSFET symbol showing the internal inductances.
32	L_S	Internal Source Inductance	ALL	—	7.5	—	nH	Measured from the source lead, 6mm (0.25 in.) from package to source bonding pad.	
33	C_{iss}	Input Capacitance	ALL	—	1500	—	pF	V_{GS} = 0V, V_{DS} = 25V	
34	C_{oss}	Output Capacitance	ALL	—	500	—	pF	f = 1.0 MHz	
35	C_{rss}	Reverse Transfer Capacitance	ALL	—	90	—	pF	See Fig. 10	

Figure A7.1. (*Continued*)

Source-Drain Diode Ratings and Characteristics

		Parameter	Type	Min.	Typ.	Max.	Units	Test Conditions
35 A	I_S	Continuous Source Current (Body Diode)	ALL	—	—	28	A	Modified MOSFET symbol showing the integral Reverse p-n junction rectifier.
36	I_{SM}	Pulsed Source Current (Body Diode) ①	ALL	—	—	110	A	
37	V_{SD}	Diode Forward Voltage ④	ALL	—	—	2.5	V	T_J = 25°C, I_S = 28A, V_{GS} = 0V
38	t_{rr}	Reverse Recovery Time	ALL	97	210	460	ns	T_J = 25°C, I_F = 28A, di/dt = 100 A/μs
39	Q_{RR}	Reverse Recovery Charge	ALL	0.42	0.95	2.1	μC	
40	t_{on}	Forward Turn-On Time	ALL	Intrinsic turn-on time is negligible. Turn-on speed is substantially controlled by $L_S + L_D$.				

Thermal Resistance

41	R_{thJC}	Junction-to-Case	ALL	—	—	1.0	K/W ⑤	
42	R_{thCS}	Case-to-Sink	ALL	—	0.50	—	K/W ⑤	Mounting surface flat, smooth, and greased
43	R_{thJA}	Junction-to-Ambient	ALL	—	—	80	K/W ⑤	Typical socket mount

① Repetitive Rating; Pulse width limited by maximum junction temperature (see figure 5) Refer to current HEXFET reliability report

② @ V_{DD} = 25V, Starting T_J = 25°C, L = 440 μH, R_G = 25Ω, Peak I_L = 28A.

③ $I_{SD} \leq$ 28A, di/dt ≤ 170A/μs, $V_{DD} \leq BV_{DSS}$, $T_J \leq$ 175°C Suggested R_G = 9.1Ω

④ Pulse width ≤ 300 μs; Duty Cycle ≤ 2%

⑤ K/W = °C/W W/K = W/°C

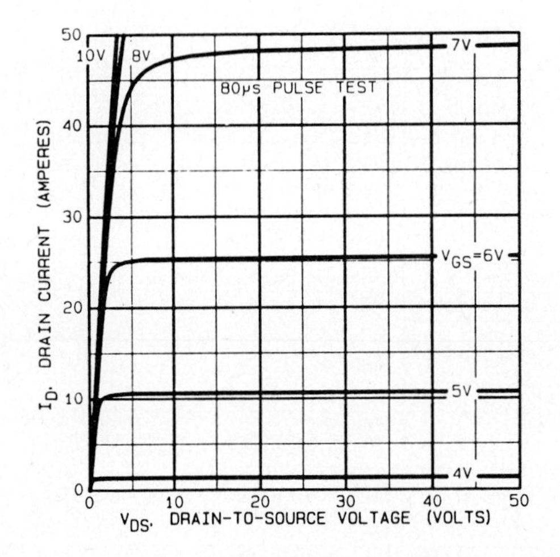

Fig. 1 — Typical Output Characteristics
note (44)

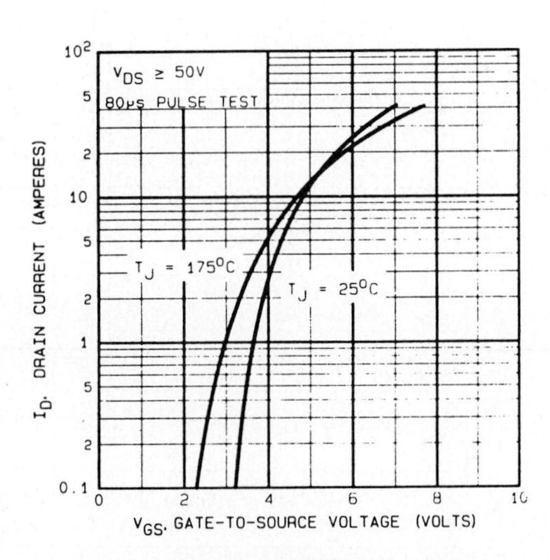

Fig. 2 — Typical Transfer Characteristics
note (45)

Figure A7.1. (*Continued*)

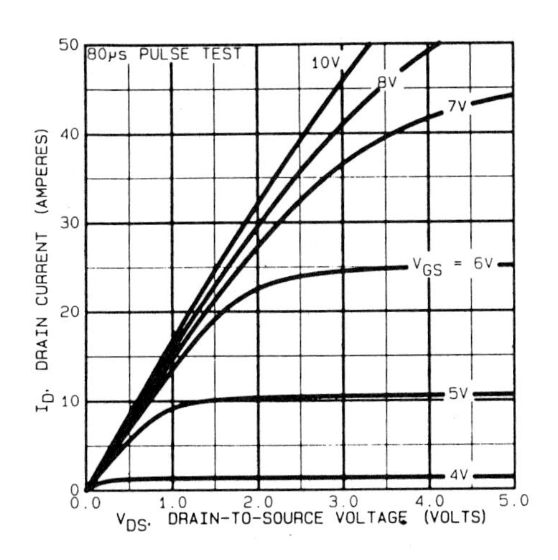

Fig. 3 — Typical Saturation Characteristics

note (46)

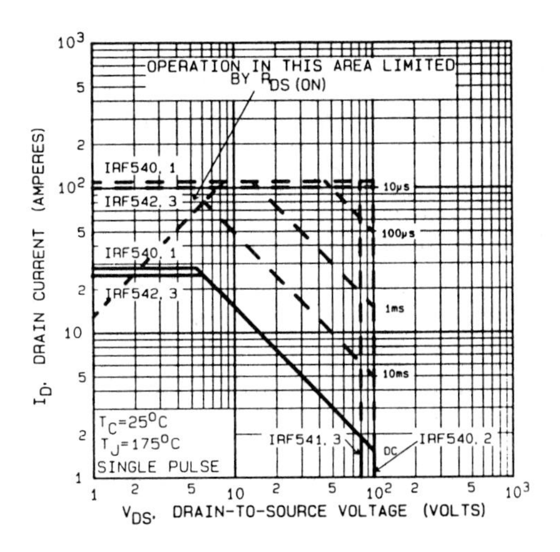

Fig. 4 — Maximum Safe Operating Area

note (47)

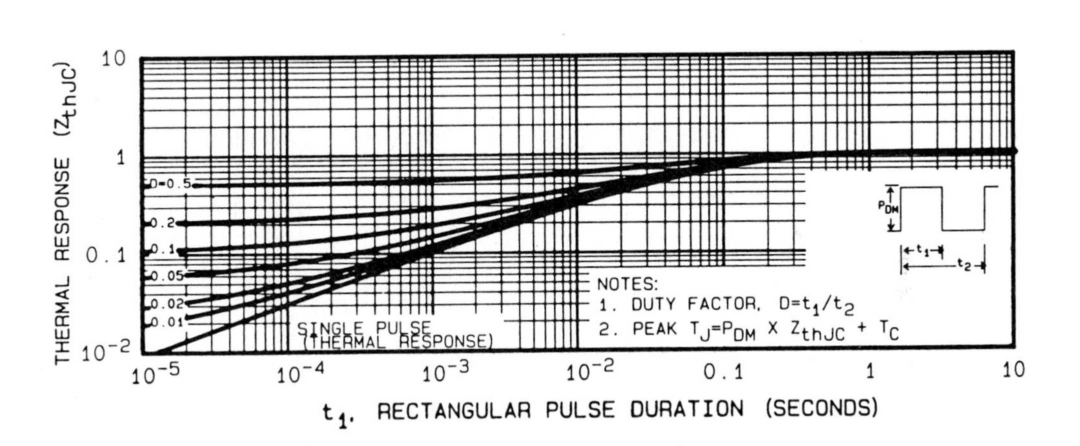

Fig. 5 — Maximum Effective Transient Thermal Impedance, Junction-to-Case Vs. Pulse Duration

note (48)

Figure A7.1. (*Continued*)

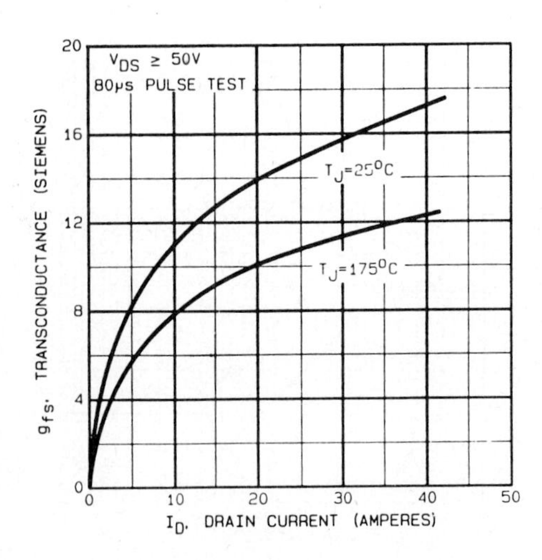

Fig. 6 — Typical Transconductance Vs. Drain Current
note (49)

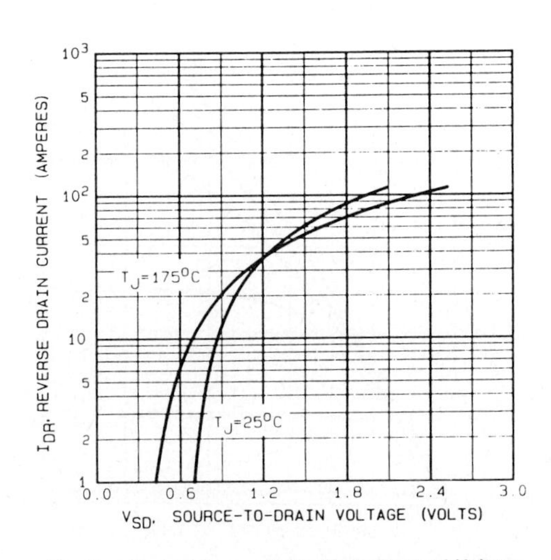

Fig. 7 — Typical Source-Drain Diode Forward Voltage
note (50)

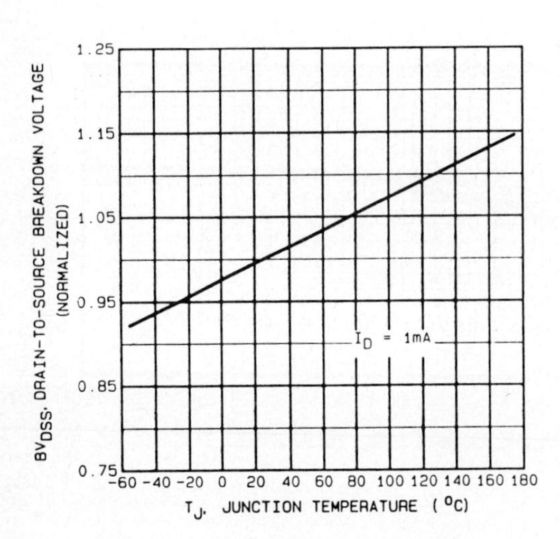

Fig. 8 — Breakdown Voltage Vs. Temperature
note (51)

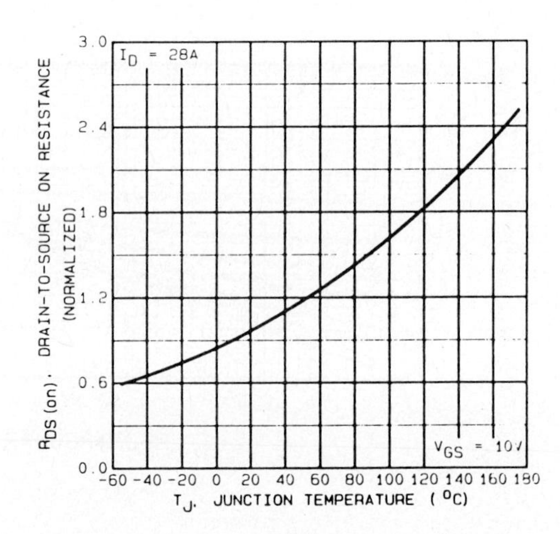

Fig. 9 — Normalized On-Resistance Vs. Temperature
note (52)

Figure A7.1. (*Continued*)

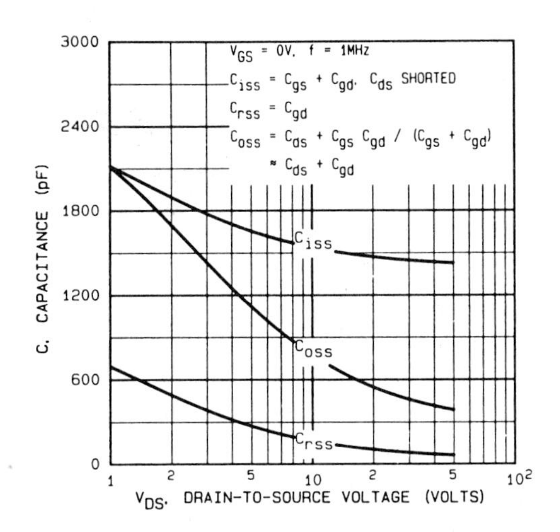

Fig. 10 — Typical Capacitance Vs. Drain-to-Source Voltage
note (53)

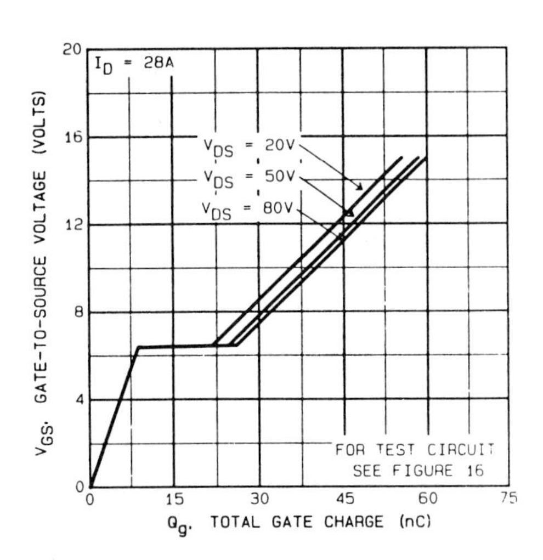

Fig. 11 — Typical Gate Charge Vs. Gate-to-Source Voltage
note (54)

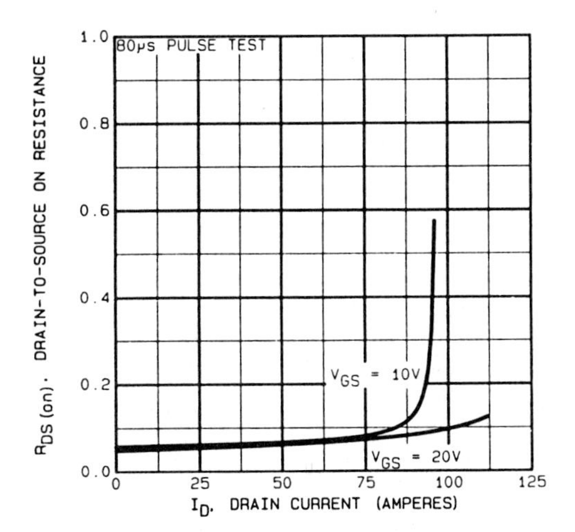

Fig. 12 — Typical On-Resistance Vs. Drain Current
note (55)

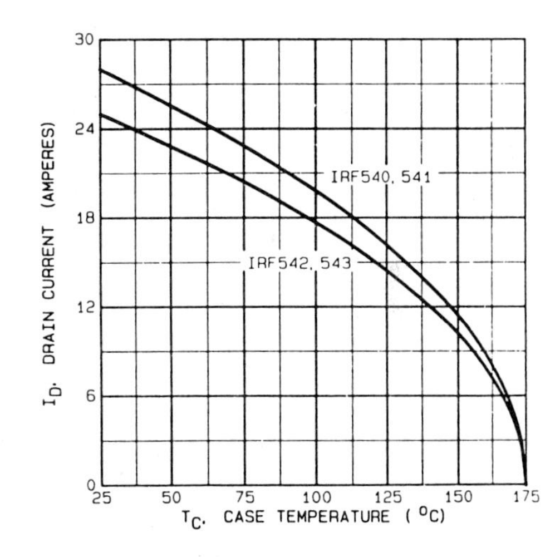

Fig. 13 — Maximum Drain Current Vs. Case Temperature
note (56)

Figure A7.1. (*Continued*)

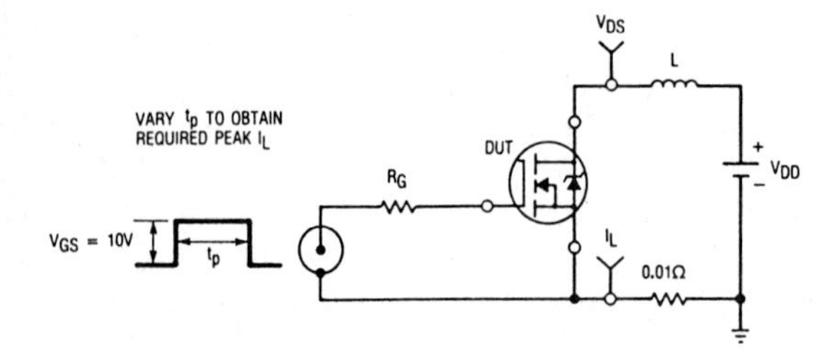

Fig. 14a — Unclamped Inductive Test Circuit

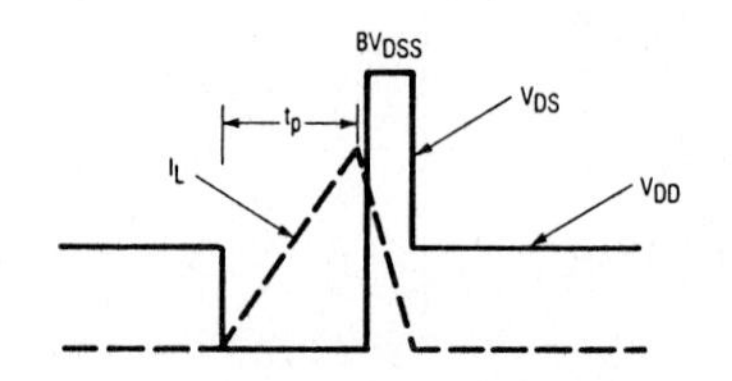

Fig. 14b — Unclamped Inductive Waveforms

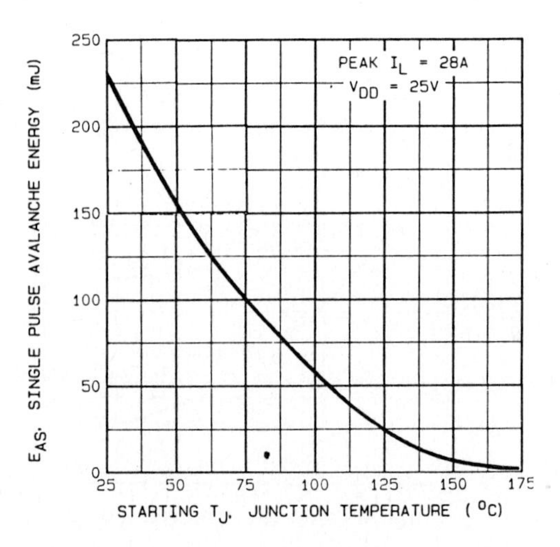

Fig. 14c — Maximum Avalanche Energy Vs. Starting Junction Temperature

note (57)

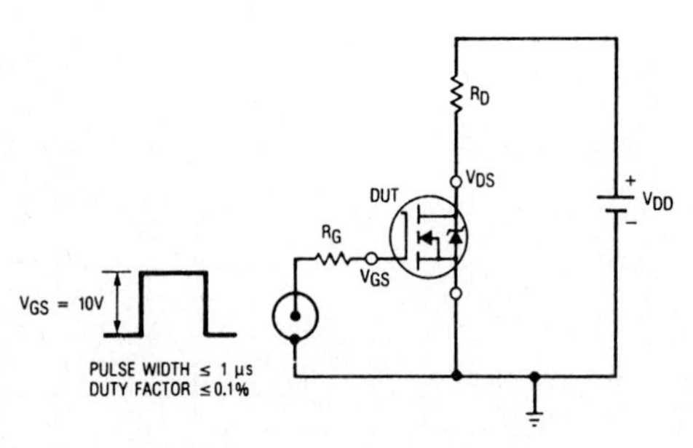

Fig. 15a — Switching Time Test Circuit

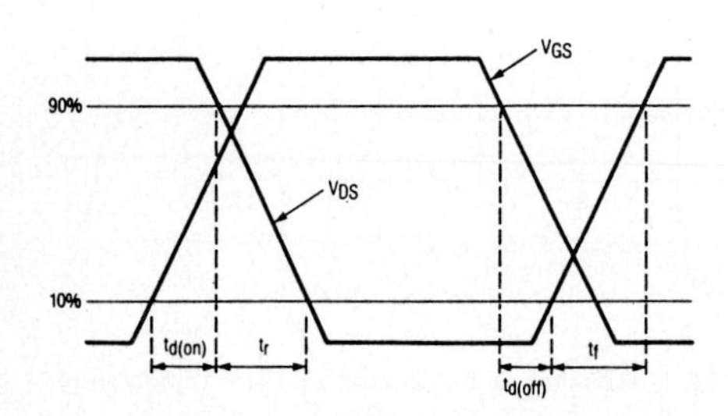

Fig. 15b — Switching Time Waveforms

note (58)

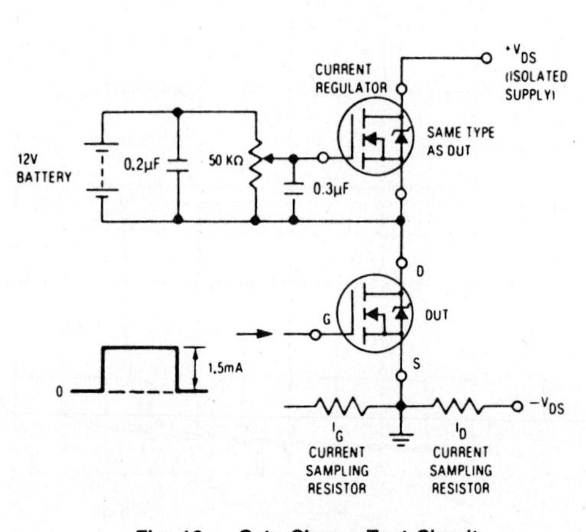

Fig. 16 — Gate Charge Test Circuit

note (59)

Figure A7.1. (*Continued*)

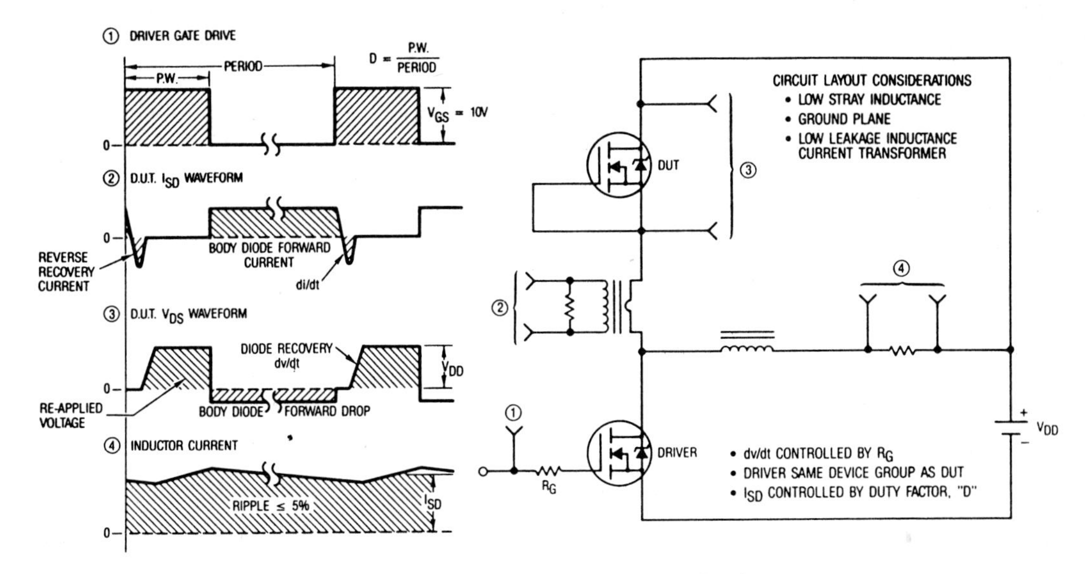

Fig. 17 — Peak Diode Recovery dv/dt Test Circuit

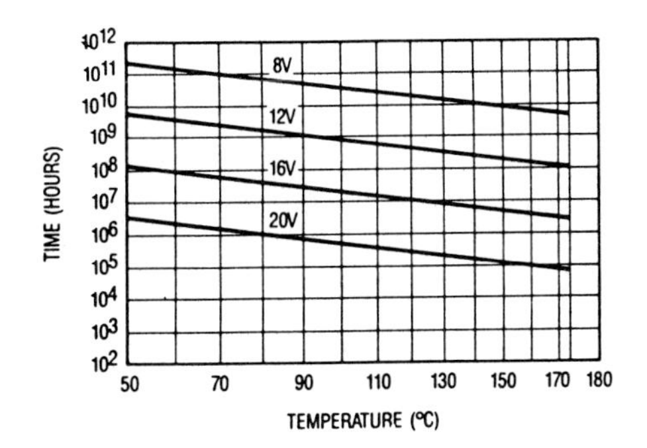

•Fig. 18 — Typical Time to Accumulated 1% Gate Failure

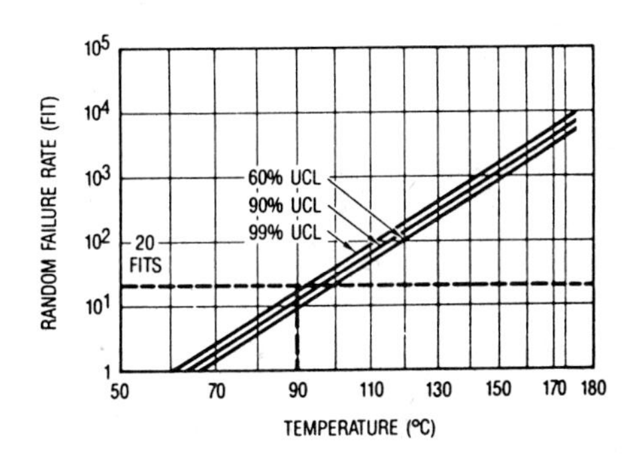

•Fig. 19 — Typical High Temperature Reverse Bias (HTRB) Failure Rate

•The data shown is correct as of Jan 15, 1987. This information is updated on a quarterly basis; for the latest reliability data, please contact your local IR field office.

Figure A7.1. (*Continued*)

include the date of manufacture. Typically this consists of a four-digit number in which the first two digits represent the year and the second two the week. Thus a device marked 8723 would have been made in the 23rd week of 1987. Some manufacturers use their own code which is not readily interpretable by the user.

5. *Current Rating at* $T_C = 25°C$. (See note 2.) This rating is the value of the drain current which will raise the die temperature to its maximum allowable value while the case is held at 25°C. The maximum allowable dissipation is 150 W at a die temperature of 175°C. The $R_{DS(on)}$ of the device is 0.077 Ω at a die temperature of 25°C. From the graph shown in Figure 9 of the data sheet, the resistance of the die at 175°C is 2.5 times its resistance at 25°C. Therefore, since the power dissipation in the die is given by

$$P_d = I_D^2 R_{DS(on)} \tag{A7.1}$$

the current rating at $T_C = 25°C$ is then

$$\begin{aligned} I_D &= \sqrt{(P_d/R_{DS(on)})} \\ &= \sqrt{\frac{150}{0.077 \times 2.5}} \\ &= 27.9 \text{ A} \end{aligned} \tag{A7.2}$$

This agrees with the data-sheet value to within the accuracy to which the graph of $R_{DS(on)}$ versus temperature may be read, and the data sheet is self-consistent.

6. *Current Rating at* $T_C = 100°C$. (See note 2.) I_D may be calculated in the manner described in note 5 except that P_d will be reduced because of the higher case temperature. The allowable power dissipation is given by

$$P_d = \frac{T_{J(max)} - 100}{R_{thJC}} \quad (T \text{ in °C}) \tag{A7.3}$$

7. *Pulsed Drain Current.* This is the maximum allowable value of the drain current and is typically four times the 25°C average-current rating. The dissipation is such that this level of current can only be sustained for a short time without exceeding the maximum allowable die temperature. The SOA graph (Figure 4 of the data sheet) gives an indication of the allowable length of a current pulse for a given current magnitude. The value of I_{DM} is mainly determined by reliability considerations relating to such factors as bonding-wire fatigue and metal migration. Power MOSFETS will generally survive pulses of current in excess of I_{DM} provided the die temperature does not reach a harmful level, although repetitive exposure to overcurrents will compromise the reliability of the device.

8. *Maximum Allowable Power Dissipation.* This is the power dissipation which will raise the die temperature to the maximum allowable value when the

case temperature is held at 25°C. Therefore in this case

$$P_d = \frac{T_{J(max)} - 25°C}{R_{thJC}}$$

$$= \frac{175 - 25}{1} = 150\ W \qquad (A7.4)$$

9. *Linear Derating Factor.* Enables P_d to be adjusted in accordance with the case temperature.

10. *Maximum Allowable Gate–Source Voltage.* This is determined by the thickness of the gate oxide and its dielectric qualities. ±20 V is almost an accepted standard for this rating, except that at least one manufacturer offers ±40 V, while low-threshold devices can be expected to have lower limits. The margin between the maximum allowable value and the voltage at which the oxide breakdown will occur varies from manufacturer to manufacturer. For standard power MOSFETS with a 20-V limit, a well-controlled manufacturing process should be able to deliver gate-oxide breakdown voltages of the order of 50 V with a spread of a few volts. In a last resort this gives the user some security against accidental gate overvoltage, but, as the graph in Figure 18 of the data sheet shows, exposure to voltages above the allowed limit can rapidly cause degradation of the gate oxide. The gate may also carry a rating which guarantees that the oxide will not rupture below a given voltage [1].

11. *Single-Pulse Avalanche Energy.* This is the maximum energy dissipation allowed during avalanche breakdown for a single pulse of avalanche current. The test circuit is described in Figure 14*a* and *b* of the data sheet, while the conditions of the rating are given in note ②. The value of avalanche energy given here is valid only when the starting temperature of the die is 25°C. The graph given in Figure 14 shows how this value must be reduced for other starting temperatures. Other manufacturers use a different method of specifying avalanche energy. The topic is discussed more fully in Section 3.6.4.

12. *Avalanche Current.* The maximum allowable current in avalanche breakdown. Note ① of the data sheet is a reminder that avalanche current produces die heating and maximum junction temperature limitations must be observed however heating occurs. While some manufacturers prefer to specify avalanche capability in terms of an unclamped inductive load test [2], there is a school of thought which prefers to specify avalanche rating as an extension of the conventional SOA graph [3].

13. *Repetitive Avalanche Energy.* The repetitive avalanche energy rating is the highest value of repetitive avalanche current that can be allowed (see data-sheet note ①). The duration and frequency of the pulses are limited by the requirement not to exceed the maximum junction temperature.

14. *Diode-recovery* dv/dt. The maximum rate of rise of drain–source voltage permitted upon recovery of the body–drain diode. The rating conditions are given in data-sheet note ③. Other manufacturers have different names for this rating and different methods of specification. See Section 3.6.4 for a full discussion of the topic.

15. *Maximum Junction Operating Temperature.* This is determined primarily by reliability considerations. The maximum die temperature is that temperature which gives an acceptable device lifetime when blocking rated voltage. With the passage of time, ion migration leads to an increase in leakage current. At low operating temperatures this effect is negligible, but at high operating temperatures the device eventually may no longer meet the specification and may even fail due to a thermal runaway. A typical maximum junction-temperature rating for a power MOSFET is 150°C, although, as in this case, some lower-voltage devices may be rated at a higher temperature where reliability tests show that this is merited.

16. *Drain–Source Breakdown Voltage.* The maximum allowable drain current at the rated breakdown voltage is 250 μA. The temperature relevant to this condition is 25°C. This is stated in the table heading—"Electrical Characteristics @ $T_C = 25$°C . . .".

17. *Static Drain–Source On-State Resistance.* Usually measured using a low duty cycle pulsed drain current to reduce heating of the die.

18. *On-State Drain Current.* This is not a widely used rating. It represents a guarantee of the transconductance of the device at rated current. It ensures that the device will conduct rated current with $V_{GS} = 10$ V.

19. *Gate Threshold Voltage.* The gate voltage required to produce 250 μA of drain current is guaranteed to be not greater than 4.0 V. At the same time it is guaranteed that at a gate voltage of 2.0 V the drain current will be less than 250 μA. For most power MOSFETS a drain current of 250 μA is, in theory, likely to be in the subthreshold region (Section 3.3).

20. *Forward Transconductance.* A benchmark point on the transconductance curve (see Figure 6 of the data sheet). Note the importance of temperature in this rating.

21. *Zero-Gate-Voltage Drain Current.* This is the corollary of the drain–source breakdown voltage rating in that the maximum allowable drain current in the OFF state is the condition used for specifying the breakdown voltage. The maximum current is also given for a higher-temperature, lower-voltage condition.

22. *Forward Gate-Source Leakage Current.* The maximum gate–source current that can be expected if the gate is "healthy". A greater leakage current than this implies that the gate oxide has been damaged. A device with some degree of gate-oxide damage may continue to function but early failure is a high probability. Typical gate leakage currents are much lower than the figure given here.

23. *Reverse Gate-Source Leakage Current.* See note 22.

24. *Total Gate Charge.* This is another benchmark rating. It gives the maximum charge drawn by the gate when its voltage is changed from 0 to +10 V to turn on a current of 28 A from a supply of 0.8 times the voltage rating of the device. The magnitude of the drain current has little influence on the total charge-transfer value, but, as can be seen from Figure 11 of the data sheet, the supply voltage does affect the total charge, due to differences in the charge transferred through the drain–gate capacitance. The maximum value of the total charge for a given gate voltage can be obtained by interpolating from the

maximum value quoted in the numerical data and the shape of the curve given in Figure 11 of the data sheet. The total gate charge is an important quantity, since the charge multiplied by the switching frequency gives the average current passing through the gate driver circuit due to charging and discharging of the gate [4]. The total gate charge can also be used to calculate the turn-on time of the MOSFET (see Section 4.4). Designers are usually interested in the amount of charge that is transferred during switching rather than capacitance values. Therefore manufacturers have moved from guaranteeing maximum values of input, Miller, and output capacitances to guaranteeing a maximum value of total gate charge, gate-to-source charge, and Miller charge.

25. *Gate-to-Source Charge.* This is the charge required to raise the gate voltage to the threshold voltage. It represents a maximum value for the area under the first sloping section of the graph shown in Figure 11 of the data sheet.

26. *Gate-to-Drain Charge.* This is the charge required during the period when the MOSFET is turning on and the drain voltage is falling. Feedback from the drain to the gate through the drain–source capacitance keeps the gate voltage approximately constant during this phase of turn-on. The gate-to-drain charge represents the maximum area under the flat part of the curve shown in Figure 11 of the data sheet.

27. *Turn-on Delay Time, Rise Time, Turnoff Delay Time, Fall Time.* These times are dependent on the parasitic capacitances of the MOSFET, the package inductance, and to a very large extent the test circuit used to make the measurements. These ratings are usually included in imitation of the delay-time, rise-time, storage-time, and fall-time ratings given on bipolar-transistor data sheets. Time constants associated with majority-carrier flow in a power MOSFET are negligibly short compared with the effects associated with the interaction between the gate drive circuit, the load, and the parasitic capacitances and inductances of the MOSFET. However, the switching times of power MOSFETS are often quoted as a yardstick of comparison even though they are quite meaningless without a description of the test circuit. Manufacturers make comparison more difficult by using a wide variety of test circuits.

28. See note 27.

29. See note 27.

30. See note 27.

31. *Internal Drain Inductance.* Of interest to designers requiring a precise knowledge of circuit inductances. Drain inductance is a function of the package style, with TO-220 packaged devices having a low inductance and devices in a TO-3 steel can having a relatively high inductance. Except in very low-inductance circuits, the drain inductance is usually swamped by the circuit wiring inductance.

32. *Internal Source Inductance.* This is a significant parameter, since it places a limitation on the ultimate switching speed that can be attained. The source inductance is common to the drain–source current path as well as the gate circuit. At turn-on a voltage is developed across the source inductance by the source current di/dt. This voltage opposes the voltage applied to the gate, thereby retarding turn-on (see Chapter 4).

33. *Input Capacitance.* This is a typical value, since gate charge values have

become the guaranteed figures of merit with respect to switching speed and gate current. The input capacitance is formed from several elements, and the equation showing how the input capacitance is derived from measured values of these elements is shown in Figure 10 of the data sheet. Figure 10 shows how capacitance varies with drain–source voltage. The input capacitance is specified at a drain–source voltage of 25 V. The fact that the capacitance curve still has a significant slope at this point and that the shape of the capacitance curve varies between manufacturers has led to the replacement of capacitance values by charge values as guarantees of performance and yardsticks of comparison. The curves show capacitance values for the condition $V_{GS} = 0$ with V_{DS} varying. The curves may be continued to the left by fixing V_{DS} at zero and increasing V_{GS}, since it is the gate drain voltage which is the significant factor in the variation of these capacitance values (see Figure 3.18). Capacitance measurements are made using a small-amplitude 1-MHz signal.

34. *Output Capacitance.* See note 33.

35. *Reverse Transfer Capacitance.* Also called the Miller capacitance. The smallest of the capacitances, but the one having the greatest influence at turn-on, due to the Miller effect. See note 33.

35A. *Continuous Source Current.* The forward drop across the body–drain diode when conducting a current equal to the I_D-rating of the MOSFET is usually less than the forward drop when the MOSFET is conducting in the forward direction. This is due to conductivity modulation of the drain drift region by minority-carrier injection during diode conduction. The significance of this is that the diode can safely be given a source-current rating equal to the I_D-rating of the device, since the dissipation will be less than that produced by a drain current of the same magnitude. Therefore it is common practice to give the I_S and the I_D ratings the same value. In many cases I_S could be made greater than I_D if required, but there are few applications in which average diode current is likely to be greater than average forward MOSFET current. Large low-voltage devices with very low values of $R_{DS(on)}$ could have a lower forward drop in the forward direction than when carrying the same current in the reverse direction through the body–drain diode. In that case it might not be possible to give I_S the same value as I_D.

36. *Pulsed Source Current.* The pulsed-source-current rating is determined by the same reliability considerations that fix the pulsed-drain-current rating and is therefore the same.

37. *Diode Forward Voltage.* This is the guaranteed maximum forward drop of the body–drain diode at the specified value of source current. Figure 7 of the data sheet shows the relationship between diode current and diode voltage. p-channel devices may well have a much larger diode forward drop than n-channel devices of the same voltage rating and die area. The cause of this is to be found in the metal–silicon interface between the source metallization and the n-type body region. A poor-quality Schottky barrier diode with a low breakdown voltage is formed, which contributes significantly to the total voltage drop.

38. *Reverse-Recovery Time.* This is the reverse-recovery time of the body–drain diode under the specified conditions of source current, temperature, and di/dt. A wide spread in this parameter is to be expected, given that the

body–drain diode is an incidental product of the DMOS process and manufacturers do not generally seek tight control over its characteristics, preferring instead to optimize the process to yield consistently to the MOSFET specifications.

39. *Reverse-Recovery Charge.* See note 38.

40. *Forward Turn-on Time.* Any voltage spikes appearing across the device as the diode is forward biased are likely to be a result of di/dt and the package inductance rather than the forward-recovery characteristics of the body–drain diode.

41. *Junction-to-Case Thermal Resistance.* The term "kelvins per watt" instead of "degrees C per watt" is widely used, particularly in Europe.

42. *Case-to-Sink Thermal Resistance.* A parameter very dependent on mounting technique.

43. *Junction-to-Ambient Thermal Resistance.* The thermal resistance in free air with no heatsink.

44. *Figure* 1. *Typical Output Characteristics.* All graphs are assumed to be for the prime device at a temperature of 25°C except where temperature is one of the variables. This is the graph from which it can be determined whether a MOSFET will be in the fully ON state for given values of drain current and gate voltage, or whether it will be in the constant-current region. This is a typical curve, so that the difference between typical and guaranteed maximum values as well as the effect of temperature on threshold voltage (approximately a reduction of 6 mV/K) must be taken into account.

45. *Figure* 2. *Typical Transfer Characteristics.* Here shown on a log–linear graph, but often shown on a linear-linear graph.

46. *Figure* 3. *Typical Saturation Characteristics.* This shows expanded detail of the transfer characteristics of Figure 2. These curves give an indication of the forward voltage drop across the MOSFET at various values of gate voltage when the device is in the saturated or fully ON condition. It shows that a significant reduction can be made in forward drop if a marginal gate voltage can be increased in some way.

47. *Figure* 4. *Safe Operating Area.* The top left-hand corner of the area is in fact inaccessible because the current is limited by the $R_{DS(on)}$ of the device and the low drain–source voltage. The 10-μs line encloses a square area delineated by the voltage rating of the device and its I_{DM}-rating. Since switching in most applications typically takes no more than 1 μs, the device is capable of switching its rated peak pulsed drain current at rated voltage. This is true of most power MOSFETS. As would be expected, this graph shows no second-breakdown restriction of the SOA.

48. *Figure* 5. *Transient Thermal-Impedance Curves.* These curves allow calculation of the peak junction temperature for repetitive intermittent conduction using the values of the duty cycle, pulse duration, and peak pulse power (not average power). P_{DM} is calculated from the current magnitude and $R_{DS(on)}$ (or V_{DS}), the duty cycle is calculated from the waveform, and Z_{thJC} is read from the graph. $T_{J(max)}$ is then calculated using the formula given in Figure 5.

49. *Figure* 6. *Transconductance versus Drain Current.* A large proportion of the operating current range of typical power MOSFETS is in the square-law region,

and not until the drain current approaches the rated average current does the device enter the region of velocity saturation and show approximately constant transconductance.

50. *Figure* 7. *Typical Diode Forward Voltage Drop*. See note 37.

51. *Figure* 8. *Breakdown Voltage versus Temperature.* The increase in breakdown voltage with temperature is advantageous, given that power semiconductors heat up when operated. It should however be remembered that the breakdown voltage of a device operating in a cold environment may be less than its rated value if no heating of the die has yet occurred.

52. *Figure* 9. *Normalized* $R_{DS(on)}$ *versus Temperature.* It is important to remember that the vertical axis is normalized with respect to the $R_{DS(on)}$ of the device at 25°C. It is easy to forget this and misread the scale as ohms, especially when $R_{DS(on)}$ is close to 1 Ω.

53. *Figure* 10. *Typical Capacitance versus Drain–Source Voltage.* See note 33.

54. *Figure* 11. *Typical Gate Charge versus Gate–Source Voltage.* See note 24.

55. *Figure* 12. *Typical* $R_{DS(on)}$ *versus Drain Current.* Provided the gate voltage is adequate to maintain the MOSFET in saturation, there is a small increase in $R_{DS(on)}$ with drain current. The sudden increase exhibited by the $V_{GS} = 10$ V line is due to the device entering the constant-current region.

56. *Figure* 13. *Maximum Continuous Drain Current versus Temperature.* This curve assumes a die temperature of 175°C. If average current is substituted for continuous current, cyclic variations in die temperature will reduce the allowable average current value.

57. *Figure* 14. *Unclamped Inductive Load Test.* See note 11.

58. *Figure* 15. *Switching-Time Test Circuit.* See note 27.

59. *Figure* 16. *Gate-Charge Test Circuit.* See note 24.

REFERENCES

1. W. Atkins and R. Frank, "Gate rupture voltage—the new criteria for power MOSFETS." *Powertechnics* December, pp. 29–33 (1987).
2. J. Wojslawowicz and P. J. Carlson, "Ruggedized transistors emerging as power MOSFET standard bearer." *Powertechnics* January, pp. 29–32 (1988).
3. K. Gauen, "Specifying power MOSFET avalanche stress capability." *Powertechnics* January, pp. 34–38 (1987).
4. K. Gauen, "Understanding the power MOSFET's input characteristics." *Power Electron. Conf.* pp. 127–131 (1987).

APPENDIX 8

Electrostatic-Discharge Protection

Reproduced from "Protecting Power MOSFETs from ESD" by kind permission of International Rectifier Corp.

A8.1. INTRODUCTION

Most power-MOSFET users are familiar with the warning that MOSFETs must be protected against electrostatic discharge (ESD). However, familiarity may breed contempt, especially if one has never destroyed a power MOSFET by improper handling. Statistically, it is unlikely that a particular MOSFET will be destroyed by ESD. However, when thousands of MOSFETs are handled, even a statistically small number of failures may be significant [1]. In view of the fact that power-MOSFET manufacturers generally seek an outgoing quality level better than one defective device per 10,000 shipped, it is evident that destroying 1 or 2 parts per 1000 during incoming handling will have a significant impact on the perceived quality of the units.

Users of power MOSFETs should therefore adopt the following measures to avoid ESD damage [2]:

1. Always store and transport MOSFETs in closed conductive containers.
2. Remove MOSFETs from containers only after grounding at a static-control work station.
3. Personnel who handle power MOSFETs should wear a static dissipative outer garment and should be grounded at all times.
4. Floors should have a grounded static-dissipative covering or treatment.
5. Tables should have a grounded static-dissipative covering.
6. Avoid insulating materials of any kind.
7. Use antistatic materials in one-time applications only.
8. Always use a grounded soldering iron to install MOSFETs.
9. Test MOSFETs only at a static-controlled work station.
10. Use all of these protective measures simultaneously and in conjunction with trained personnel.

A8.2. THE NATURE OF ESD

ESD is the discharge of static electricity. Static electricity is an excess or deficiency of electrons on one surface with respect to another surface or to ground. A surface exhibiting an excess of electrons is negatively charged, and an electron-deficient surface is positively charged. Static electricity is measured in terms of voltage (volts) and charge (coulombs).

When a static charge is present on an object, the molecules have an electrical imbalance. ESD takes place when an approach to equilibrium occurs through the transfer of electrons between one object and another that is at a different potential. When an ESD-sensitive device, such as a power MOSFET, becomes part of the discharge path, or is brought within an electrostatic field, it can be permanently damaged.

A8.3. GENERATION OF STATIC ELECTRICITY

The most common way of generating static electricity is by tribo-electrification. Rubbing two materials together will cause tribo-electrification, as will bringing two materials together and then separating them. The magnitude of the charge is highly dependent upon the propensity of the material to giving up or taking on electrons. Dissimilar materials are particularly susceptible, especially if they have high surface resistivity.

Another way of placing a static charge on a body is by induction. Induction can be caused, for example, by placing a body in close proximity to a highly charged object.

A8.4. MOSFET FAILURE MODE

The high input resistance (typically $> 4 \times 10^9\ \Omega$), which is one of the biggest operating advantages of the power MOSFET, can also be a disadvantage with respect to ESD. The gate of the power MOSFET may be considered to be a low-voltage, low-leakage capacitor. The capacitor plates consist primarily of the polysilicon gate and the underlying silicon. The capacitor dielectric is the silicon dioxide gate oxide layer.

ESD destruction of the MOSFET occurs when the gate-to-source voltage is high enough to puncture the gate dielectric. The breakdown voltage of the oxide is typically two to three times the rated maximum gate–source voltage. The arc burns a microscopic hole in the gate oxide, which results either in an immediate

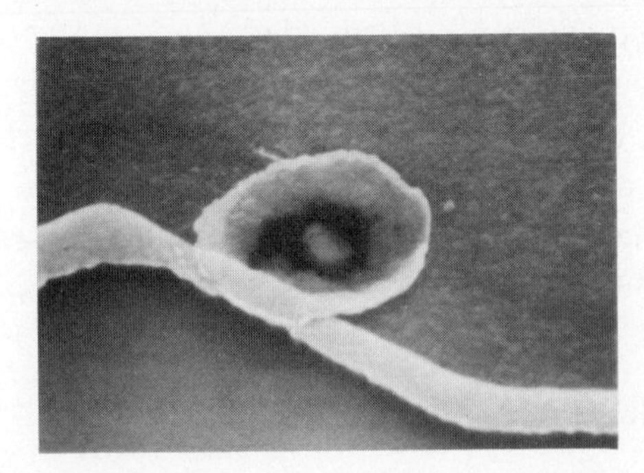

Figure A8.1. Typical ESD failure site.

gate-to-source short circuit or in a site of high leakage current which eventually breaks down into a full short circuit. Like any capacitor, the gate of a power MOSFET must be supplied with a certain amount of charge to reach a particular voltage. Since larger devices have greater capacitance, they require more charge per volt and are therefore less susceptible to ESD than smaller MOSFETS.

A typical ESD destruction site can be seen in Figure A8.1. This was caused by a

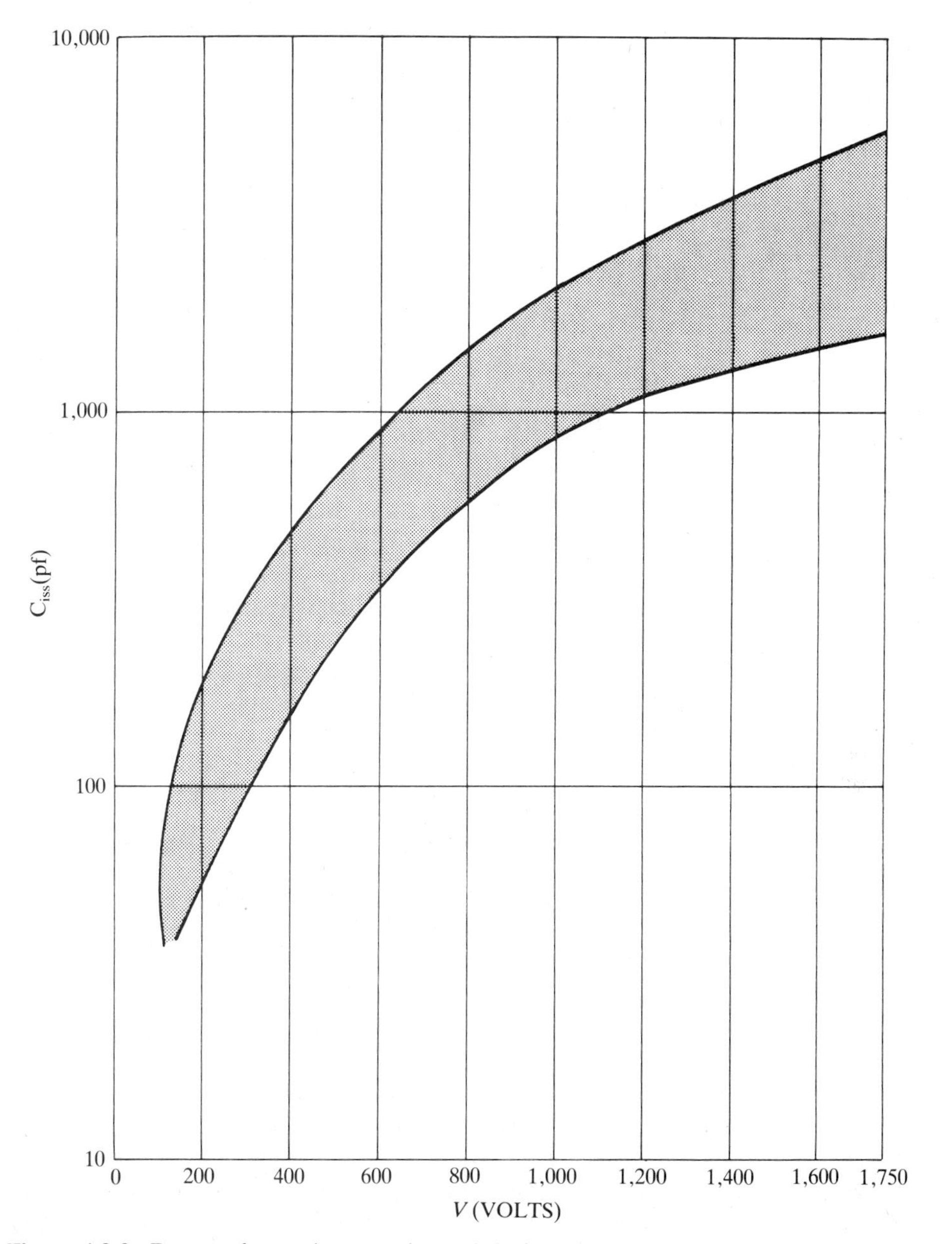

Figure A8.2. Range of capacitance values of devices damaged by contact with a human body charged to the voltage V.

human body model charged to 700 volts being discharged into the gate of the device. This photograph was taken with a scanning electron microscope after stripping the surface of the die down to the polysilicon layer. The failure site shown in Figure A8.1 is about 8 μm in diameter. The electrical symptom of the ESD failure is a low resistance or a zener effect between gate and source with less than 20 V applied.

The voltages required to induce ESD damage can be 1000 V or higher (depending upon the die size). This is due to the fact that the capacitance of the body carrying the charge tends to be much lower than the C_{iss} of the MOSFET, so that when the charge is transferred, the resulting common voltage will be much lower than the original body voltage. The graph of Figure A8.2 shows the relationship between chip size and the body voltage required to induce ESD damage.

Electrostatic fields can also destroy the power MOSFET. The failure mode is ESD, but the effect is caused by placing the unprotected gate of the FET in a corona discharge path. Corona discharge is caused by a positively or negatively charged surface discharging into small ionic molecules in the air (CO_2^+, H^+, O_2^-, OH^-).

A8.5. COMPLETE STATIC PROTECTION

Direct protection of the MOSFET relies on methods which limit the gate–source voltage, such as shorting the gate to the source or applying zener protection from gate to source. While effective for in-circuit or small-quantity situations, the direct method is usually impractical in the manufacturing environment because of the large number of MOSFETS involved. The practical approach in such situations is complete static protection [3]. The basic concept of complete static protection for the power MOSFETS is the prevention of static buildup where possible and the quick, reliable removal of any charge that does build up.

Materials in the environment can either help or hinder static control. These may be placed into four categories with respect to surface resistivity:

Insulating	$>10^{14}$ Ω/square
Antistatic	10^{9}–10^{14} Ω/square
Static-dissipative	10^{5}–10^{9} Ω/square
Conductive	$<10^{5}$ Ω/square

Ideally, to protect the MOSFET, one should have only grounded conductive bodies in the facility. Additionally, all personnel involved in the manufacturing process should, from the point of view of device protection, be connected solidly to ground. However, people so grounded would be vulnerable to electrocution by faulty electrical equipment. Therefore earthing leads should incorporate resistance which limits to a safe value any current flowing through the body of the wearer. When people are required to be mobile and an earthing lead is impractical, protective materials must be relied on to give ESD protection.

A8.6. INSULATING MATERIAL

Because of their propensity for storing static charge and the difficulty of discharging them, it is imperative to keep objects made of insulating materials away from power MOSFETs and, if possible, out of the environment entirely. Since electric current cannot flow through an insulator, electrical connections from an insulator to ground are ineffective in controlling static charges.

Insulating materials include: ceramics, mica, many organic materials and most plastics, in particular, polyethylene (found in ordinary plastic bags), polystyrene (found in plastic foam cups and packing granules), Mylar and vinyl. When plastic products must be used in a power-MOSFET-handling facility, only ones impregnated with a conductive material or treated with antistatic compounds should be used.

A8.7. ANTISTATIC MATERIAL

Antistatic material is resistant to the generation of tribo-electric charges, but does not provide a shield from electric fields. Corona discharge will pass through an antistatic enclosure, possible destroying any MOSFETs inside. Because of its high surface resistivity, grounding antistatic material is not very effective for removing a charge.

Some plastic insulators can be treated with antistatic agents which chemically reduce their susceptibility to tribo-electrification and lower their surface resistivity. Most antistatic agents require high relative humidity to be effective. Therefore the relative humidity of facilities where power MOSFETs are handled should be kept above 40%. Also, antistatic agents tend to wear off or wear out after a period of time, and most of them use reactive ionic chemicals which can be corrosive to metal. Antistatic plastics should be limited to short-term use in one-time-only situations, such as in DIP and TO-3 tubes, and packing materials for shipping components.

A8.8. STATIC-DISSIPATIVE MATERIAL

Static-dissipative materials are effective in application on any surface to facilitate the removal of static charges by conducting them to ground. It is possible to generate tribo-electric charges in static-dissipative materials, but the charges will be dissipated throughout the material and can easily be discharged to ground. Static-dissipative material is suitable for use in or on floors, table tops, and clothing.

A8.9. CONDUCTIVE MATERIAL

Conductive materials are suitable for use in the construction of enclosures for storing or transporting power MOSFETs. Like static-dissipative materials, conductive materials are susceptible to tribo-electrification, but can easily be discharged to ground. Plastic, though normally highly insulating, can be made conductive when manufactured from carbon- or metal-impregnated base material. Conductive tote bins and bags are constructed from these materials. Containers should be constructed so that the conductive elements will not separate, migrate, or otherwise contaminate the environment.

Power MOSFETS contained in closed conductive containers are safe from corona discharge, as electric current is conducted around the container. The contents of the container are shielded.

A8.10. IONIZERS

In addition to passive static control by proper material selection, active controls may sometimes be necessary or advantageous. Ionizers are a form of active static control.

Ionizers are designed to produce large and equal quantities of positive and negative ions. When used to neutralize a specific object, the object tends to attract only those ions necessary for neutralization. Excess positive or negative ions either tend to neutralize each other or flow to ground. There are three basic types of ionizers: ac, dc and nuclear.

Among dc-powered ionizers, the balanced type is preferred. Unbalanced dc ionizers can create imbalances in the amount of ions of different polarities, creating exactly the type of dangerous condition one is trying to eliminate. Solid objects, and even the air, can sometimes acquire a charge from an ionizer. Consequently, ionizers of any type should be avoided for general room area or direct personnel neutralization. Ionizers should never be used where there is possible contact with moisture. Ionizers may be installed where necessary for dedicated applications. For example, it may be advisable to use an ionizer on printed-circuit boards where MOSFETS are to be mounted.

A8.11. FLOORS

The foremost consideration is to prevent the generation of tribo-electrically induced static charges in the first place. A good place to start is with the use of grounded static-dissipative floor coverings or treatments. Conductive floor tiles are the most permanent solution, but their cost tends to be prohibitive unless the installation is on a new floor. Floor mats, conductive or static-dissipative, can be used, but may constitute a safety hazard with curled-up edges or corners.

One suitable and cost-effective method is to use a static-dissipative floor finish. Such a finish can be aesthetically pleasing and have a surface resistivity that remains well within the static-dissipative range after about two months of normal pedestrian traffic. Grounding of a static-dissipative floor finish is usually accomplished without special effort. Since it is applied as a liquid, it tends to make contact with grounding rods or other static control ground points.

A8.12. TABLE TOPS

Grounded, static-dissipative tabletops should be used at every work station where power MOSFETS are handled either in or out of protective containers. As with floors, the most permanent solution is static-dissipative tabletop laminant (see Figure A8.3). New benches can be ordered with a static-dissipative surface, and old benches can be resurfaced. Alternatively, benches can be covered with soft, static-dissipative mats (see Figure A8.4). However, mats should be avoided where they may be exposed to heat or chemicals.

Metal tabletops should not be used in place of static-dissipative ones, as they

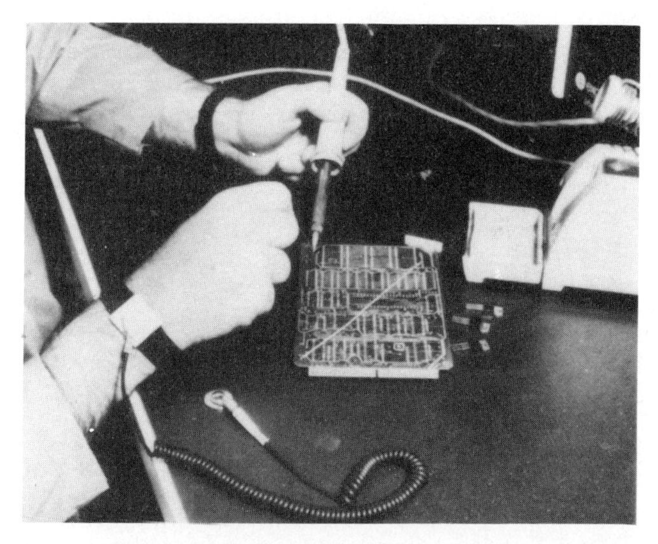

Figure A8.3. Static-dissipative mat with wriststrap grounding.

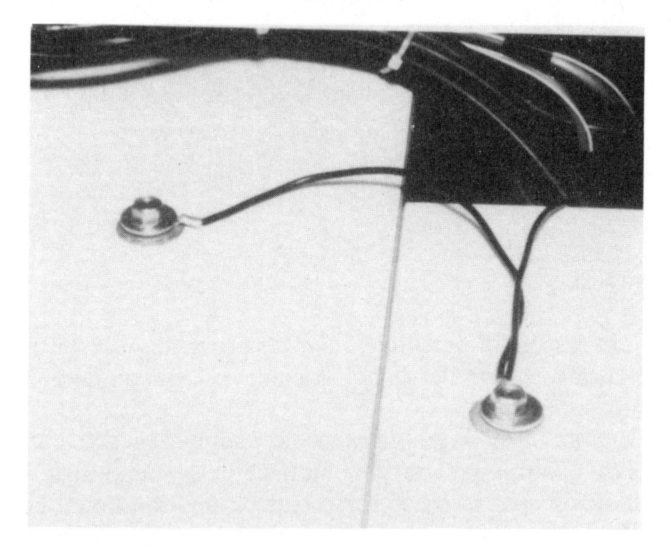

Figure A8.4. Static-dissipative table-top laminate and typical grounding.

are far too conductive and therefore present a shock hazard where electrical equipment is used. The ideal work surface should fall within the static-dissipative range.

A8.13. CONTAINERS

Power MOSFETS, even if they are already contained in antistatic tubes or bags, should always be stored and transported in closed conductive bags or containers.

If power MOSFETS are to be stored in a dry atmosphere, such as nitrogen, the gas should be ionized as it goes into the bag or dessicator to prevent static buildup in the container. Conductive bags and containers should never by opened except at a static-controlled work station and only after the bag or container has been placed on a grounded static-dissipative surface (Figure A8.5).

Figure A8.5. Handling power MOSFETS.

A8.14. PERSONNEL

Static protection in respect of personnel involves prevention of static buildup and dissipation of existing static. Personnel training is also an important part of effective ESD control.

Materials found in most outer garments constitute an ESD hazard. Typical laboratory coats of a cotton–polyester blend have been found to store charges of up to 5000 V. Static-dissipative laboratory coats or smocks should be supplied to employees, as this will shield the environment from personal clothing. Gloves should be worn only when necessary for cleanliness, since the surface resistivity of the human body falls within the static-dissipative or conductive range.

Removal of existing charges can be effected with grounded wriststraps, which should be worn whenever physically possible. Ground straps and grounded table-top surfaces should have at least 1 MΩ of resistance to ground to prevent shock hazard.

A8.15. GROUNDING

Earth ground rods for ESD protection should be solid copper or copper-jacketed steel and should be driven six to eight feet into the earth beyond the building slab, with approximately six inches exposed above the floor for making connections (Figure A8.6). Dry soil conditions may require a copper sulfate drip. Electrical grounds should be isolated from static control grounds. Water pipes should never be used to terminate static control grounds, since they may not be connected to ground.

These grounding methods may seem excessive, since the ground rod may be in series with 1 MΩ or more of resistance. However, these techniques are for minimizing the difference of potential between separate grounds and not for reducing the ohmic resistance to earth.

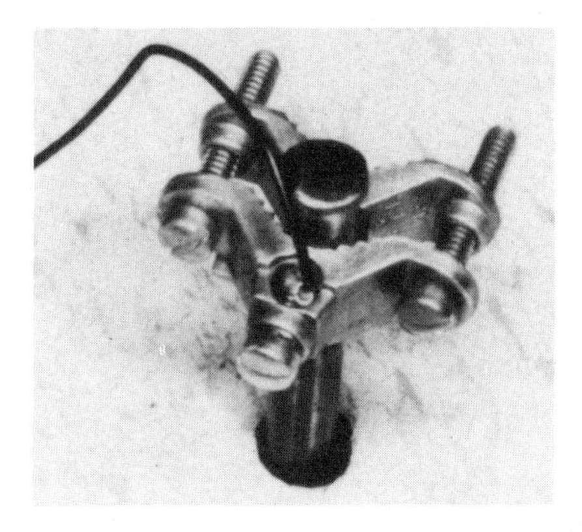

Figure A8.6. Static-control rod.

A8.16. TEST EQUIPMENT

Test equipment should be placed on grounded static-dissipative floors or table tops. Operators should wear a static-dissipative or antistatic garment and a ground strap at all times. Automatic testers and handlers should have antistatic feed paths and receptables for the MOSFETS.

A8.17. COMPLETE ESD PROTECTION

The most effective protection from ESD occurs when the total environment is under control. Changing only the floors or the tabletops is not enough. When all of the appropriate ESD protection devices are used simultaneously and in conjunction with trained personnel, ESD damage can be reduced to negligible levels.

REFERENCES

1. R. M. Ghent, "Relationship of device sensitivity to perceived quality." *Prof. Program Sess. Rec., Southcon 85,* Atlanta, pp. 2/2, 1–6 (1985).
2. S. Brown and B. Ghent, "Protecting power MOSFETS from ESD." *Int. Rectifier Appl. Note* **AN-955** (1984).
3. R. M. Ghent, "A working model of a passive ESD protection shceme." *1st Ann. E.O.E. Tech. Conf.* San Jose, CA April pp. 1–12 (1984).

APPENDIX 9

Power-MOSFET Test Circuits

A9.1. INTRODUCTION

This appendix gives details of ways in which some MOSFET data-sheet parameters may be measured [1]. Where device-dependent component values are given in the circuits, these are the values that would be used for an IRF 630. It may be necessary to use different component values for other devices.

A9.2. BV_{DSS}: DRAIN–SOURCE BREAKDOWN VOLTAGE

The test circuit is shown in Figure A9.1. The current source will typically consist of a power supply with an output voltage capability of about $3BV_{DSS}$ in series with a current-defining resistor of the appropriate value. When testing high-voltage power MOSFETS it may not be practical or safe to use a supply of $3BV_{DSS}$, in which case some other form of constant-current source must be used.

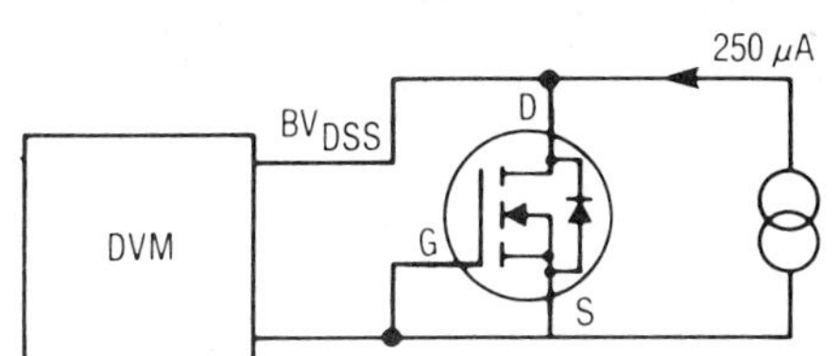

Figure A9.1. Test circuit for BV_{DSS}.

A9.3. $V_{GS(th)}$: THRESHOLD VOLTAGE

The test circuit is shown in Figure A9.2. The 10-kΩ gate resistor is required to suppress potentially destructive oscillations at the gate. The current source may be derived from a voltage source equal to the voltage rating of the MOSFET and a series resistor.

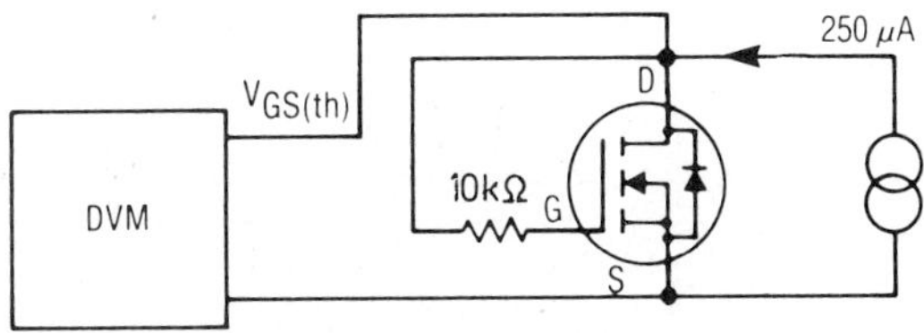

Figure A9.2. Test circuit for $V_{GS(th)}$.

A9.4. $R_{DS(on)}$: SATURATED ON RESISTANCE

The test circuit is shown in Figure A9.3. The pulse width should be 300 μs at a duty cycle of less than 2%. The value quoted is at a junction temperature of 25°C. $R_{DS(on)}$ is calculated by dividing $V_{DS(on)}$ by I_D. The ground of the gate supply should be connected as close to the source lead as possible.

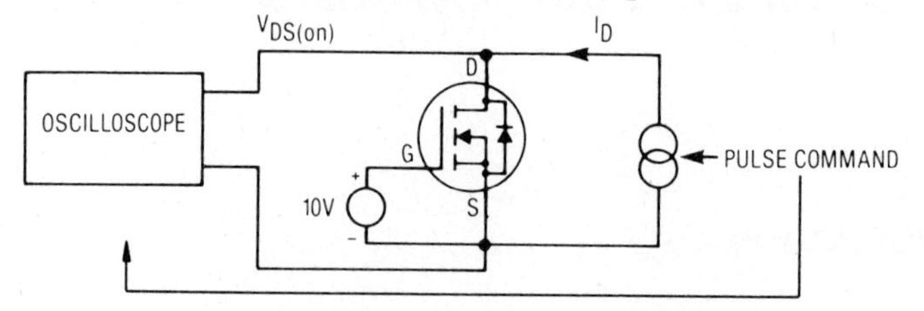

Figure A9.3. Test circuit for $R_{DS(on)}$.

A9.5. C_{iss}, C_{oss}, AND C_{rss}: INPUT, OUTPUT, AND REVERSE TRANSFER CAPACITANCES

The test circuits are shown in Figures A9.4, A9.5, and A9.6. A 1-MHz capacitance bridge is used for all these tests. The capacitance to be measured is connected in series with a 1-μF capacitor to provide dc isolation.

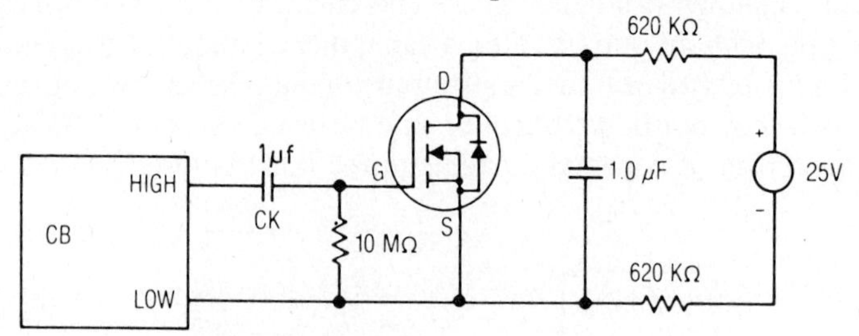

Figure A9.4. Test circuit for C_{iss}.

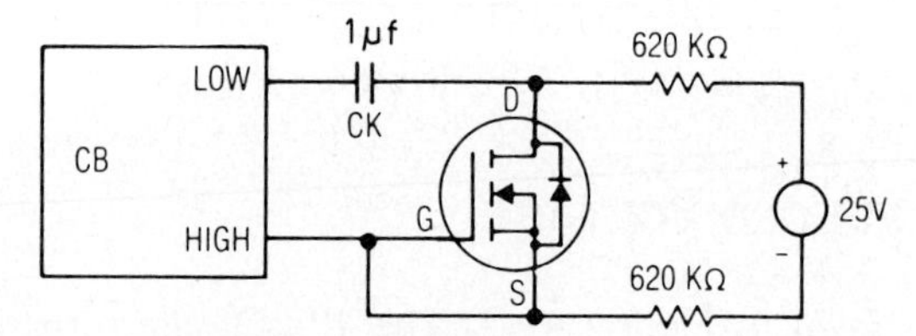

Figure A9.5. Test circuit for C_{oss}.

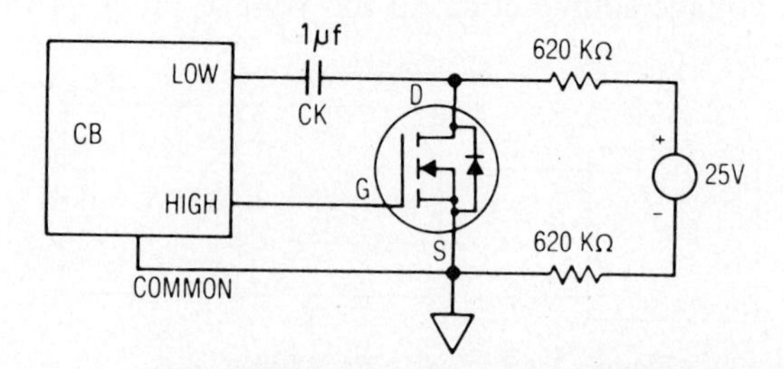

Figure A9.6. Test circuit for C_{rss}.

A9.6. $t_{d(on)}$, t_r, $t_{d(off)}$, AND t_f: TURN-ON DELAY TIME, RISE TIME, TURNOFF DELAY TIME, AND FALL TIME

The test circuit is shown in Figure A9.7. The gate pulses should be just long enough to achieve complete turn-on, with a duty cycle of the order of 0.1%. The series resistor is chosen according to the die size. The definitions of the rise, fall, and delay times are given in Figure A9.8.

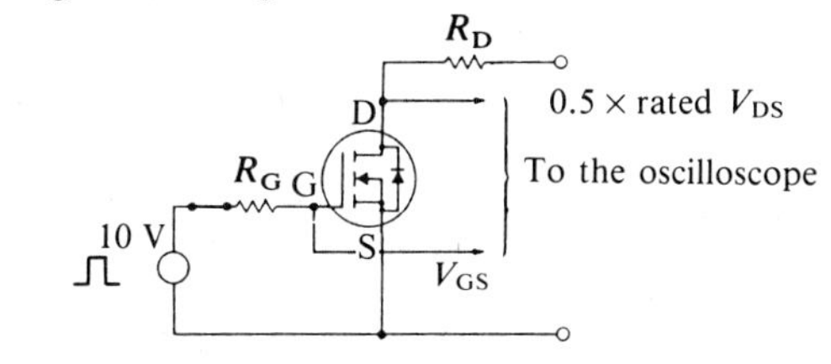

Figure A9.7. Test circuit for $t_{d(on)}$, t_r, $t_{d(off)}$, and t_f.

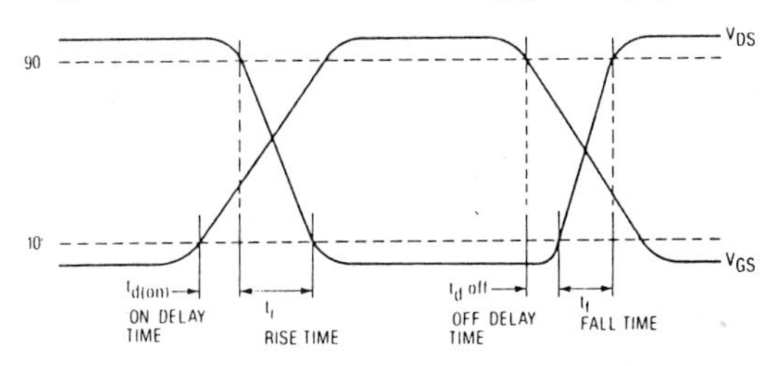

Figure A9.8. Switching-time waveforms.

A9.7. Q_g, Q_{gs}, AND Q_{gd}: TOTAL GATE CHARGE, GATE–SOURCE CHARGE, AND GATE–DRAIN CHARGE

The test circuit is shown in Figure A9.9. The total gate charge has two components: the gate-source charge and the gate–drain charge (or Miller charge). Figure A9.10 shows the test waveforms.

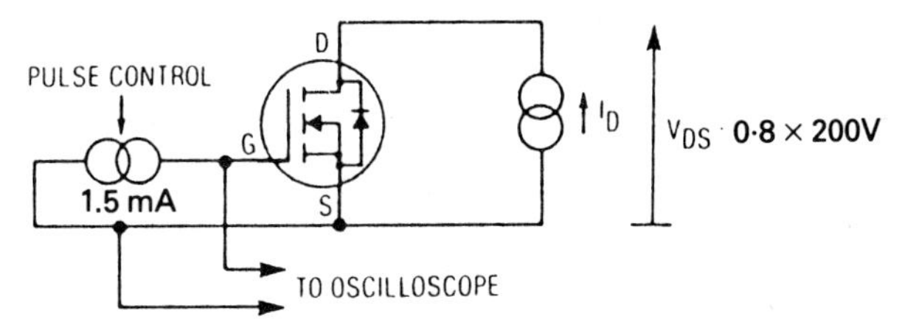

Figure A9.9. Test circuit for Q_g, Q_{gs}, and Q_{gd}.

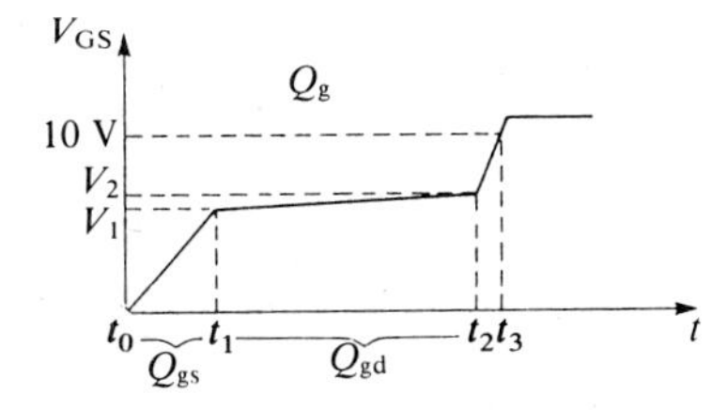

Figure A9.10. Test waveforms for gate charge.

A9.8. V_{SD}: BODY–DRAIN DIODE CONDUCTION VOLTAGE

The test circuit is shown in Figure A9.11. The current source may consist of a voltage source and a series resistor, and should be applied in short pulses (less than 300 μs) with a low duty cycle (less than 2%).

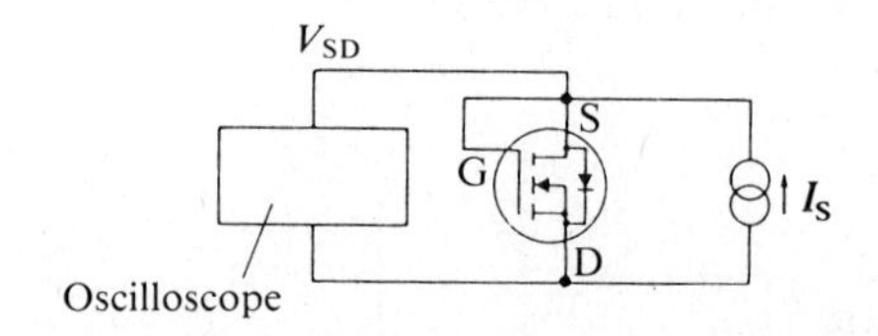

Figure A9.11. Test curve for V_{SD} and $R_{DS(on)}$.

REFERENCE

1. R. Pearce, S. Brown, and D. Grant, "Measuring HEXFET characteristics." *Int. Rectifier App. Note* **AN-957** (1986).

Index

Accelerated gate stress test, 160
Accelerated life-tests, 152, 160, 161
Accumulation layer, 41–44, 57, 82, 83, 117, 193, 336
  charge density (Q_{AL}), 435
  effect on gate-drain capacitance, 41, 43
  theory, 435–443
AC load control, 225
Activation energy, 158–162
Active area bonding, 73, 79, 365
Active clamping, 189, 218
Active state, 98, 107, 114, 115, 136, 142
Alumina, 148, 149, 356, 361, 362
Aluminum gate, 39, 40
Anti-saturation clamp, for bipolar transistors, 337, 341
Antistatic materials, 487
Arrhenius law, 158–162, 368
Assured out-going quality level (AOQL), 152
Audio amplifier, linear, 324
Automobile applications, 229, 236
Avalanche breakdown, 20, 459–465
  of body-drain diode, 86, 87, 91, 95, 103, 127, 143, 161
Avalanche capability:
  of bipolar transistors, 344
  of MOSFETS, 344
Avalanche energy:
  repetitive rating, 86, 477
  single pulse rating, 477
  test circuit, 161, 482
  use of rating, 216
Avalanche requirements in PWM inverter, 306
Average drain current rating, 476

Band diagram, 25
Bathtub curve, 152–155
Battery-driven motors, 275
Battery lifetime, with MOSFET regulator, 326
Bernard's method, 156
Beryllia, 356, 361
Bias point stabilisation, in linear amplifiers, 328
Bidirectional MOSFET (BOSFET), 226
Bipolar junction transistor (BJT), 1, 8, 17
  avalanche breakdown, 88–90
  avalanche capability, 344
  in cascade with MOSFET, 344
  in cascode with MOSFET, 350
  as charge controlled device, 333
  common emitter gain, 327
  comparison with MOSFET, 333
  drive requirements, 340
  emitter focussing, 350
  in parallel with MOSFET, 351
  parasitic, 11, 81, 85–90, 95, 149
  Safe Operating Area, 340
  saturation characteristics, 336
  switching delay times, 337
  switching waveform, 338
  transfer characteristics, 336
Blocking state, 45–51, 142
Blocking voltage, 8, 18, 19, 27, 74–76, 78, 79
Body region, 14, 26, 28, 30, 81
Body-drain diode, 11, 85, 86, 90, 91, 93, 95
  advantage in linear regulator, 323
  average current rating, 480
  controlled avalanche breakdown, 86, 87, 91, 103, 127, 143, 161
  forward voltage drop, 480
  pulsed current rating, 480
  reverse-recovery time, 480
  test for forward voltage, 496
  use in PWM drives, 284
BOSFET, 226
Braking, in dc motors, 262
Breakdown voltage, 459–465. (*See also* Blocking voltage)
  and $R_{DS(on)}$, 74–79, 94
  temperature dependence, 91, 482
  test for, 493
Breakdown, controlled avalanche, 20, 86, 87, 91, 95, 103, 127, 161
Bridge circuit, for motor drives, 265
Brushless motor drive, 267
Burn-in, 153, 160, 161

Capacitance, as a function of drain voltage, 41, 44, 110, 482
  input (C_{iss}), 84, 85, 105, 107, 479, 480
  output (C_{oss}), 84, 85, 480
  reverse transfer (C_{rss}), 84, 85, 480
Capacitor, impedance as a function of frequency, 240
Carrier mobility, 1, 5, 63 94, 97
Carrier transit time, 5, 58, 59
Cascade (Darlington) connection, of bipolar transistor and MOSFET, 344
Cascode connection, of bipolar transistor and MOSFET, 343
  of GTO and MOSFET, 352

Cell density, 70, 72
Cell geometry, 69–74, 94, 456, 457
Cell size, 69–76, 455
Cellular structure, 15, 16, 21, 69–74, 94
Channel length, 8, 13, 56–61, 65, 69, 70, 97, 446–448
  effective, 451
  modulation, 8, 38, 59
    parameter (λ), 60, 63
Channel region, 4, 5, 14, 26, 28, 51, 56–58, 445, 446
Channel resistance, 53
Channel voltage, 53, 56
Characteristics (current–voltage):
  effect of velocity saturation, 61–63, 448–453
  fuller theories, 63–65, 446–453
  in sub-threshold region, 65–68, 94
  JFET, 2
  MOSFET, 3, 6, 7
  p-channel device, 7, 92, 93
  showing operating regions, 45, 55
  simple theory, 53–60, 94
  simplified for transient analysis, 97, 98
  typical, 54, 470, 481
Charge control models, 55, 94, 333
Charge pump, 213, 230, 231
Charge:
  gate–drain, 135–138, 479
  gate–source, 135–138, 479
  reverse-recovery, of body drain diode, 86, 481
  total gate, 479
Chopper, 258
Circuit time constant, 110
Clamped inductive load, 102–105, 142
Class-D amplifier, 313, 315
  avoiding distortion, 316
  circuit topologies, 318
  efficiency, 317
  output filter, 316
CMOS, 17, 203
Combinations of MOSFET and bipolar transistor, 344
COMFET™, 20
Common source inductance, in multiple-die packages, 359
Commutation waveforms, in PWM inverter, 295, 302
Comparison, of bipolar and MOSFET application, 342
Computer-aided design (CAD), 25
Computer models of devices, 25, 377–381
Conducting state, 45–50, 142
Conduction losses, 165
  in PWM inverter, 301
Conductivity modulation, 336, 406, 408
Confidence limits, 156, 369
Contact potential, 30–32, 39, 425, 429, 433, 436, 445–449
  in degenerate semiconductor, 32
Containers, for transporting MOSFETs, 490
Cranking voltage, requirement in automobiles, 232
Crystallographic orientation, 10, 11
Cumulative failure probability density function, ($F(t)$), 154–157
Current distribution, 75, 76
Current hogging, 19
Current-mode control, 243
Current-sensing MOSFET, 93, 383
  applications, 222, 392
  in current-mode SMPS, 243
  dynamic characteristics, 388
  principle of operation, 383
  resistor sensing circuit, 390
  reverse conduction, 388
  sense ratio variation, 387
  static characteristics, 385
  structure, 383
  with virtual earth sensing circuit, 389
Curve tracer, using with MOSFETs, 365
Cut-off frequency, 97

Darlington (cascade) connection, of MOSFET and bipolar transistor, 344
Data sheet, of power MOSFET, 467
Degradation processes, 156, 159–162
Delay time:
  turn-off, 127, 128, 479
  turn-on, 105–107, 479
Depletion layer, capacitance, 67, 429, 430, 447
  charge density (Q_D), 36, 425, 427, 429, 430, 432–434, 440, 441, 445, 447, 448, 451
  voltage (V_{SCL}), 53–55
  width, 8, 425–429
Depletion mode MOSFET, 3, 4, 253
Derating, due to internal voltage drop, 360
  factor, 477
  due to skin effect, 360
Device models, 25, 56, 377–380
  equivalent circuits, 97–99, 104, 108, 113–116, 128–132, 379
Device time constant, 110
Die attachment, 148, 364
Die-bond failure, 160, 161, 372
Diffusion capacitance, 19
Diffusion equation, 139
Diffusion, carrier, 450
  impurity, 3, 146, 149
Diode pump, *see* Charge pump
Diode-recovery charge, effect of temperature, 297
Diode-recovery current, effect of temperature, 295
  in dc motor drive, 270
  limitation by inductance, 293
  methods of reducing, 263, 265
  in PWM inverters, 287

Diode-recovery dv/dt rating, 477
Diode-recovery dv/dt, 284
Diode-recovery losses, calculation of, 260, 294
DIP package, cooling of, 170
Dispersion (of lognormal distribution), 156
Dissipation in power MOSFETs, 163
Doping profile, 27, 38, 70, 76–78, 430, 459–461, 465
Double-diffusion process, 12–14, 38, 51, 69
Drain circuit (slow), 107, 143
Drain current rise time, 112
Drain drift region, 26, 74–79
Drain–source breakdown voltage rating, 478
Drain–source capacitance, bootstrapping, 324
  effect on turn-off waveform, 163
Drain–source leakage current, 159–161
Drain voltage fall time, 111, 112
Drift velocity (carrier), 56–59, 61, 64, 65, 446, 448–451
Dual voltage input circuit, 243
dv/dt, diode-recovery, 284, 340, 342, 477

Early effect, 429
Electron drift velocity, 61, 64, 65
  saturation, 61–65, 68, 69, 94, 448–452
Electron irradiation, 86, 306, 407, 411
Electron mobility, as function of longitudinal field, 64, 65, 449–452
  bulk, 459–465
  as function of normal field, 63, 450–452
  surface, 56, 58, 59, 61, 67–69, 94, 446–448
Electrostatic Discharge (ESD):
  antistatic materials, 486
  antistatic measures, 487
  discharge (ESD) of, 486
  generation of, 484
  MOSFET failure mode, 485
  nature of, 483
  precautions, 483
Electrostatic Discharge (ESD) protection, 483
  antistatic measures, 487
  conductive material, 487
  containers for MOSFETs, 490
  grounding, 491
  static dissipative material, 487
  table tops, 488
  treatment of floors, 488
  use of ionizers, 488
Energy dump circuit, 280, 283
Enhancement-mode device, 3–6
Epitaxial growth, 3, 21
Epitaxial layer, 28, 145–147, 149, 455
  resistance, 53, 54, 75–78, 145, 146, 149
Equivalent circuit, 80, 97
Equivalent parameters, of MOSFET and bipolar transistors, 366
ESD, *see* Electrostatic discharge
Etching, 3, 145, 146, 148
  anisotropic, 8, 12
  isotropic, 8
  plasma, 145, 148

Fabrication technology, 21, 145–151, 161
Failure mechanisms, 159–161
Failure probability density function ($f(t)$), 153–157
Failure rate (instantaneous) ($\lambda(t)$), 152–157
Failure units (FITS), 154, 160
Failures, due to body-drain diode conduction, 286
  induced, 373
Fall time, 100–103, 479
Fall time, 479
Fast-recovery diode MOSFETs, 305
  effect of temperature on Q_{RR}, 305
Feature size, 7, 70, 94
Fermi energy, 28, 46, 47, 51
  in degenerate semiconductor, 30, 32
Fermi function, 28
Ferrite bead, to control instability, 182, 184
Field effect, 1
Field enhancement, 12, 78, 87
Field rings, 146, 160, 161
Five-lead packages, 357
Flat-band condition, 28, 29, 33
Forced cooling, effect on thermal resistance, 170
Forward blocking condition, 46–50
Forward converter, with current sensing, 245
Forward turn-on time of body-drain diode, 481
Free-wheeling diode, 102, 105, 123, 127, 220
Frequency response, 99

Gate:
  metal, 14, 15, 39
  polycrystalline silicon, 14, 15, 28, 192
  propagation delay, 139, 149, 248
  self-aligned, 15, 148, 151
Gate bias, negative, 193, 209
  to counter radiation effects, 395
Gate capacitance, 105, 110
Gate charge, 135–138
  data, 482
  test for, 495
  test circuit, 482
Gate circuit:
  fast, 107, 117, 123, 133
  power consumption, 139
  slow, 121–123, 125, 134, 135
  slowed by source inductance, 113, 114
Gate damping resistor, 180
Gate dissipation, 165
Gate-drain charge, 135–138, 479
Gate drive, from CMOS, 203
  from TTL, 200
Gate driver:
  asymmetric, 195, 349

Gate driver (*Continued*)
bootstrap, 330
complementary emitter follower, 196
effect on stability, 188
for full bridge, 213
integrated circuit, 197
off-state voltage, 192
on-state voltage, 193
optically isolate, 308
photo-voltaic, 210
for PWM applications, 307
stored energy type, 199
transformer coupled, 210
using comparators, 197
using operational amplifiers, 197
Gate driver circuits, 191–213
Gate driver impedance, effect on stability, 182
Gate leakage current, 159, 160
Gate oxide:
breakdown voltage, 69, 477
capacitance, 18, 38, 60, 66–69, 433, 445–452
cause of damage, 186, 484
defects, 160
life data, 160, 368
lifetime calculations, 369
thickness, 7, 8, 18, 36, 38, 67, 69, 111, 401
Gate oxide damage, 191, 484
Gate resistance, 15, 99, 165
Gate-source charge, 479
Gate-source leakage current, 478
Gate-source voltage, maximum allowable, 477
Gate threshold voltage, guaranteed limits, 478
Gate-turn-off thyristor (GTO), 17–19
in cascode with MOSFET, 344, 352
Gate vias, 74, 142
Gate width (w), 21, 68–73, 455–457
Gaussian distribution, 156
GEMFET™, 20
Gradual channel approximation, 446, 450
GTO, *see* Gate turn-off thyristor
Guard-rings, 146

H-bridge driver, 213
Handling of power MOSFETs, 374
Harmonic elimination in six step drives, 279
Heatsink:
design, 163
selection, 171, 235, 371
selection, graphical method, 167
Hermetic packages, 161, 357
High temperature reverse bias (HTRB):
life data, 369
lifetime calculations, 370
test, 160
High voltage MOSFETs, 189
High-frequency:
applications, 313
limitations, 139–142
High-side switch, 212, 229
Highly accelerated temperature and humidity stress test (HAST), 161
Hole mobility, 5, 92, 95
Hole saturation drift velocity, 92, 95
Humidity, effect of, 161, 162
Hybrid devices, 361, 405

IGBT, *see* Insulated Gate Bipolar Transistor
IGT™, 20
Incandescent lamp, load characteristics, 218, 223
Inductance:
external source, 177, 265, 293, 355, 479
internal drain, 479
internal source, 479
of MOSFET packages, 355
in stored energy gate drive, 199
use of, to limit diode-recovery current, 265, 293
Induction heating, 319
series resonant circuit, 320
Inductive load, 102, 216
Infant mortalities, 153
Input capacitance, 84, 85, 105, 107, 479, 480
test for, 494
Inrush current, limiting, 218
Instability, during switching of a single MOSFET, 186
Instability, during switching of paralleled MOSFETs, 179
Insulated Gate Bipolar Transistor (IGBT), 20, 21, 93, 237
characteristics, 406
complementary devices, 408
gate drive circuit, 410
latch-up, 409
principle of operation, 405
in PWM inverters, 311
switching waveforms, 412
Interference, reducing by circuit layout, 362
Internal drain inductance, 479
Internal source inductance, 479
Inversion layer, 4, 37, 41–44, 56–59, 433–435, 437, 440–443, 445
charge density ($Q_{(IL)}$), 41, 56–59, 65, 66, 433–435, 441, 442, 445, 446, 449–451
Ion implantation, 3, 146, 149
Ionizers, 488

Junction field effect transistor (JFET), 1–3
parasitic, 69, 75, 78, 81–83, 95, 455
parasitic, effect on $R_{DS(on)}$, 81, 174
Junction temperature:
calculation of, 168
peak, 168, 174, 233, 363, 367, 478, 481

Lagrange undetermined multiplier, 463–465
Lamp dimmer, 225

Lamp life extender, 224
Leakage current:
effect of radiation on, 397
gate-source, 478
losses associated with, 164
Lifetime (mean time to failure), 154, 369
Lightly doped drain (LDD), 8, 9
Linear amplification, 45
Linear amplifier, bias point stabilisation, 328
Linear applications, 323
Linear condition, 46
Linear post-regulator, 323
Linear regulator, low dropout, 323
Load dump, requirement in automobiles, 231
Logic level MOSFETs, 219, 229. *See also* Low threshold voltage MOSFETs
Lognormal distribution function, 155, 156, 158, 161
Losses:
as a function of frequency in bipolar transistors, 343
as a function of frequency in MOSFETs, 343
Low inductance package, 363
Low threshold voltage MOSFETs, 38, 232, 401. *See also* Logic level MOSFETs

Majority carriers, 333, 340
Masking processes, 145–151
self-aligned, 15, 146, 151
Maximum allowable junction temperature, 478
MCT, see MOS-controlled thyristor
Mean time between failures (MTBF), 154
Mean time to failure (MTTF), 154, 369
Meaurements:
with curve tracer, 365
with oscilloscope, 179, 366, 388
temperature, 367
Median lifetime, 156, 158, 159, 369
Metal–oxide–semiconductor field-effect transistor (MOSFET), lateral, 8, 9, 12, 14
n-channel, 3–6
p-channel, 3, 7, 39, 92, 93
principles, 3, 45–52, 55–65, 94
selection for PWM inverters, 310
types, 3, 4, 8, 14, 21
Metallization, 3, 8, 148
Micro-stepping, 276
Miller capacitance, 107, 135–138
Minority carriers, 333, 336, 339, 406
Mobility:
carrier, 1, 63, 94, 97
electron, 5, 11, 58, 60, 446–448, 450–452, 459–465
hole, 5, 92, 95
Modelling power MOSFETs, 25, 377–381
MOS-controlled thyristor (MCT), in PWM inverters, 311
principle of operations, 412
triggering, 413

Motor drives, 257–281
Multiple-die packages, 358

Negative voltage linear regulator, 323
Noise immunity of gate circuit, 192, 207, 219, 232

On-state resistance, (*see* $R_{DS(on)}$)
Output capacitance (C_{oss}), 84, 85, 480
test for, 494
Output characteristics of MOSFET, 470, 481
Overlapping conduction (to control diode-recovery current), 288
Oxide growth, thermal, 3, 15, 146
Oxide thickness, 8, 36, 146
Oxide windows, 3, 4, 146–148

Packages, 355–358
five-lead, 357
hermetic, 357
hybrid, 361
low inductance, 363
marking of, 467
multiple-die, 358
plastic, 161, 356
surface mount, 363
Parallel combination of bipolar transistor and MOSFET, 351
Parallel-connected MOSFETs, 173
dynamic current sharing, 176
effects of threshold mismatch, 178
effects of transconductance mismatch, 179
instability during switching
for linear operation, 185
packaging, 358
parasitic oscillation, 179
reasons for, 173
stability analysis, 180
static conditions, 174
thermal model, 176
Parameters, equivalence of MOSFET and bipolar, 366
Parasitic BJT, 11, 81, 85–90, 95, 149
Parasitic capacitance:
drain–gate (Miller), 80, 84, 107
drain–source, 80, 84, 479
gate–source, 80, 84, 135–138, 149
gate-substrate, 41, 44
Parasitic components, 79–91, 99
Parasitic diode, 11, 81, 85
Parasitic inductance, drain, 84, 99, 359, 479
source, 84, 99, 479
Parasitic JFET, 69, 75, 78, 81–83, 95, 455
Parasitic oscillation, 179, 186, 291, 328, 360, 366
detection of, 184
Partially clamped inductive load, 104, 105, 123
Peak die temperature, 168, 174, 233, 363, 367, 478, 481
effect on power capability, 234

Photolithography, 3, 4, 7, 12, 145, 455
Pinch off, 3, 4, 8, 45, 59, 61, 62, 66, 68, 446, 448, 449
Pinch-base resistance ($R_{bb'}$), 87, 90, 92
Planar silicon technology, 3, 4, 8
Plastic packages, 161, 356
Polycrystalline silicon, 14, 15, 28
  resistance, 15, 139–140
Positive voltage linear regulator, 326
Potential distribution, 46–50
Power cycling, reliability data, 372
  test, 160
Power dissipation capability, 234
Power Integrated Circuits, 415
  basic, 419
  construction, 416
  example of, 418
Power supplies, 239
  losses in single-ended, 240
  quasiresonant, 250
  resonant, 19, 246–248
  switching frequency, 239
Print hammer driver, 221
Propagation delay (gate), 139–141, 149
Pulse-Width Modulation, in chopper circuits, 258
  in Class-D amplifier, 314
  in hammer drives, 221
  inverter current waveforms, 284
  inverters, 283
  in lamp dimmer, 255
  overmodulation, 309
  in power supplies, 240
  use of body-drain diode, 284
  waveform generator, 309
Pulsed drain current rating, 476
Pulsed source current rating, 480
Punch through, 8

Quality, 151, 152
  outgoing level, 152, 374, 483
Quasi-resonant power supplies, 19, 250
Quasi-saturation, in bipolar transistor, 336

Radiation hardened MOSFETs, 393, 401
Radiation, circuit designs to protect against, 399
  counter measures, 399
  effects in power MOSFETs, 394
  ionizing, 394
  non-ionizing, 398
  types, 393
Radiation-hardened devices, 93
Random failures, 153, 154, 161
$R_{DS(on)}$, 54, 55, 94, 95, 117, 148, 151, 452
  effect of slice thickness, 145
  effect of temperature, 86, 91, 92, 94, 482
  effect of throat, 69
  in p-channel devices, 92, 95
  increase as test of connection failure, 152, 159
  increase through JFET action, 81, 82, 174
  increase with blocking voltage, 18, 20, 78, 79, 94
  rating, 478
  reduced by decreasing cell size, 455
  reduced by implant, 83
  relative contributions, 74, 76
  test, 494
  variation with drain current, 482
Reach through, 459
Relay driver, 220
Reliability function ($R(t)$), 153, 154
Reliability, 78, 79, 151–161, 218, 351, 363, 364, 368, 476
  testing for, 368
Resistance, channel, 53–55
  epitaxial layer, 53–55
  on-state, *see* $R_{DS(on)}$
  pinch-base ($R_{bb'}$), 87, 90, 92
Resonant power supplies, 246–248
  power density, 246
  range of switching frequency, 247
RESURF, 418
Reverse battery connection in automobiles, 232
Reverse conduction characteristics of MOSFET, 251, 289
Reverse-recovery charge (of body-drain diode), 481
Reverse-recovery time (of body-drain diode), *vs.* voltage rating, 288
Reverse-transfer capacitance (C_{rss}), 80, 84, 107, 480
  test for, 494
Rise time, 100–103, 479
Rms value of current waveform, 166

Safe operating area (SOA), 19, 20
  of bipolar transistor, 340
  of MOSFET, 340, 481
Saturation characteristics of MOSFET, 194, 481
Saturation drift velocity, 61–65, 68, 69, 94, 95, 97
Saturation in bipolar transistor, 336
Schottky diodes, use in PWM inverters, 287
Second breakdown, 19, 20, 339, 340, 344
Self-alignment, 15, 146, 148
Series connection of MOSFETs, 188
Shoot through in inverters, 291, 320
Short channel effect, 63
Silent motor drives, 280, 412
SIT, *see* Static Induction Transistor
Slice thickness, 145
Snubber:
  for bipolar transistor, 341
  to control diode-recovery d*v*/d*t*, 287
  for MOSFET, 164, 216, 342
  polarized, 342

Snubber circuits, 107
Solid-state ballast, 313
Source inductance:
  external, 177, 265, 293, 355, 479
  parasitic, 84, 99, 479
Source resistor ballasting, 185
  graphical selection of resistors, 186
Source-drain ON-voltage ($V_{SD(on)}$), 93, 161
Special purpose MOSFETS, 383–402
SPICE, 25
  computer time requirements, 380
  limitation of MOSFET models, 377
  modelling drain-gate capacitance, 378
  modelling power MOSFETs, 377
Stability in multiple-die packages, 184
Static-dissipative materials, 487
Static Induction Thyristor (SITH), 405
Static Induction Transistor (SIT), 319, 405
Stepping motor drive, 272
Strong inversion, 34–36, 51
Sub-threshold current, 65–68
Surface charge, 1, 32–39, 432–436, 452
Surface-mount packages, 363
Surface region, 26
Surface states, 1
Switch, MOSFET used as, 215
Switched mode power supplies, common circuits, 241–243
Switching energy, 100–103
Switching frequency, 100
Switching losses, 163
  calculation in PWM inverter, 296
Switching speed, 21, 99, 142
Switching time test circuit, 482, 495
Switching waveforms, bipolar transistor, 338
  IGBT, 411
  MOSFET, 338
Symbol for power MOSFET, 467
Synchronous rectification, 250
  performance comparison, 254
Synchronous rectifier, losses in, 252
  two-terminal, 254

Temperature cycling test, 161
Temperature dependence, of bipolar transistor parameters, 91, 92, 339
  of MOSFET parameters, 94, 339
Temperature measurement:
  using body-drain diode, 367
  using $R_{DS(on)}$, 367
  using threshold voltage, 367
Temperature–humidity–bias test, 161
Test circuit, body–drain diode forward voltage, 496
  drain–source breakdown voltage, 493
  gate charge, 482, 495
  input capacitance, 494
  output capacitance, 494
  $R_{DS(on)}$, 494
  reverse transfer capacitance, 494
  switching time(s), 482, 495
  threshold voltage, 493
  unclamped inductive load, 482
Testing MOSFETs, 365–367
Thermal impedance, transient, 168, 481
Thermal resistance, 145, 148, 159
  case-to-sink, 481
  junction-to-ambient, 481
  junction-to-case, 481
  steady state, 167
Thermal runaway, 159, 167, 173
Thermally activated degredation processes, 156–161
Three-phase drive:
  brushless, 268–271
  PWM, 280, 286
  six-step, 277
Threshold voltage, 15, 34–39, 51, 58–63, 65, 67, 86, 105, 447, 448
  effect of temperature, 38, 39, 51, 91
  guaranteed limits, 478
  metal gate, 39
  shift due to radiation, 396
  test for, 493
Throat region, 81–83, 95
Thyristor, 17, 18, 21
  gate-turn-off (GTO), 17–19, 344, 352
Total gate charge, 138, 139, 478
Transconductance, compared with gain of bipolar transistor, 327
  minimum guaranteed value, 478
Transconductance (g_{fs}), 68, 69, 83, 91, 94, 151, 455
Transconductance parameter (k), 59, 64, 65
Transfer characteristics of MOSFET, 97, 481
Transient dissipation, 99–103
  at turn-off, 99
  at turn-on, 99
Transient thermal impedance, 168, 481
Transit time (carrier), 5, 58, 61, 97
Traps, 1, 86
TTL, 17, 200
Turn-off, 20, 99–103, 123–136, 142–143
  delay time, 127, 128, 142, 479
  effect of drain inductance, 127, 130–136
  effect of source inductance, 131, 132, 136
  energy, 99–103, 135, 136, 142
  ring, 132–135
Turn-on, 99–123, 136, 142, 143, 455
  controlling speed of, 290
  delay time, 105–107, 142, 479
  effect of drain inductance, 107–115, 117–125, 136
  effect of source inductance, 107, 112–115, 119–124, 136
  energy, 99–103, 117–123, 136, 142, 143

Turned-OFF state, 45–51, 142
Turned-ON state, 45–51, 142

V-groove, 8–13, 21
VDMOS, 1, 13–17, 21
  active regions, 26
  channel (body) region, 26–28
  doping profile, 27
  drain drift region, 26
  surface layer, 26
  throat, 26
Very large scale integration (VLSI), 1, 8, 18, 148
Virtual-earth current sense method, 245
Voltage downgrades, 467
Voltage rating of MOSFETS, 161, 291
  selection, 307
VVMOS, 10–13, 148

Wear out, 153, 155, 161
Wire-bond fatigue, 159, 161

Zener diode clamp for drain voltage, 127,
    132–135, 143, 216
Zero-gate-voltage drain current, 478
Zero-temperature-coefficient point, 185